T0186055

LIZARD ECOLOGY

The foraging mode of lizards has been a central theme in guiding research in lizard biology for three decades. Foraging mode has been shown to be a persuasive evolutionary force molding the diet, ecology, behavior, anatomy, biomechanics, life history, and physiology of lizards. This volume reviews the state of our knowledge on the effects of foraging mode on these and other organismal systems to show how they have evolved with foraging mode over a wide taxonomic survey of lizard groups. The reviews presented here reveal the continuous nature of foraging strategies in lizards and snakes, providing the general reader with an up-to-date review of the field, and will equip researchers with new insights and future directions for the sit-and-wait vs. wide foraging paradigm. This volume will serve as a reference book for herpetologists, evolutionary biologists, ecologists, and animal behaviorists.

STEPHEN M. REILLY is Professor in the Department of Biological Sciences and Director of the Ohio Center for Ecology and Evolutionary studies at Ohio University.

LANCE D. MCBRAYER is an Assistant Professor in the Department of Biology and Curator of Herpetology for the Savannah Science Museum Collections at Georgia Southern University.

DONALD B. MILES is Professor in the Department of Biological Sciences at Ohio University.

LIZARD ECOLOGY

The Evolutionary Consequences of Foraging Mode

Edited by

STEPHEN M. REILLY
Department of Biological Sciences, Ohio University

LANCE D. McBRAYER
Department of Biology, Georgia Southern University

DONALD B. MILES
Department of Biological Sciences, Ohio University

CAMBRIDGE UNIVERSITY PRESS
Cambridge, New York, Melbourne, Madrid, Cape Town,
Singapore, São Paulo, Delhi, Mexico City

Cambridge University Press
The Edinburgh Building, Cambridge CB2 8RU, UK

Published in the United States of America by
Cambridge University Press, New York

www.cambridge.org
Information on this title: www.cambridge.org/9781107407480

First published 2007
First paperback edition 2012

A catalogue record for this publication is available from the British Library

ISBN 978-0-521-83358-5 Hardback
ISBN 978-1-107-40748-0 Paperback

Contents

Contributors

ROGER A. ANDERSON,
Department of Biology,
Western Washington University,
Bellingham, WA 98225-9160, USA

AARON M. BAUER,
Department of Biology,
Villanova University,
Villanova, PA 19085, USA

STEVEN J. BEAUPRE,
Department of Biological Sciences,
University of Arkansas,
Fayetteville, AR 72701, USA

KEVIN E. BONINE,
Department of Ecology and
Evolutionary Biology,
University of Arizona,
PO Box 210088, Tucson,
AZ 85721, USA

TRACEY K. BROWN,
Biological Sciences Department,
California State University,
San Marcos, CA 92096-0001, USA

WILLIAM E. COOPER, JR.,
Department of Biology,
Indiana University–Purdue
University at Ft. Wayne,
Ft. Wayne,
IN 46805-1499, USA

CLAY E. CORBIN,
Department of Biological and
Allied Health Sciences,
Bloomsburg University,
Bloomsburg,
PA 17815, USA

ANTHONY HERREL,
Department of Biology,
University of Antwerp,
Universiteitsplein 1,
B-2610 Wilrijk,
Belgium

RAYMOND B. HUEY,
Department of Zoology,
University of Washington,
Seattle,
WA 98195-1800, USA

DUNCAN J. IRSCHICK,
Department of Biology,
221 Morrill South, University of
Massachusetts at Amherst,
Amherst,
MA 01003, USA

JONATHAN B. LOSOS,
Department of Biology,
Washington University, St Louis,
MO 63130-4899, USA

LANCE D. McBRAYER,
Department of Biology,
Georgia Southern University,
Box 8042, Statesboro, GA 30460,
USA

DONALD B. MILES,
Program in Ecology and
Evolutionary Biology, Department of
Biological Sciences, Ohio University,
Athens, OH 45701, USA

CHAD E. MONTGOMERY,
Department of Biological Sciences,
University of Arkansas, Fayetteville,
AR 72701, USA

KENNETH A. NAGY,
Department of Organismic Biology,
University of California,
621 S. Young Dr., Los Angeles,
CA 90095-1606, USA

GAD PERRY,
Department of Range, Wildlife,
and Fisheries Management,
Box 42125, Texas Tech University,
Lubbock, TX 79409-2125, USA

ERIC R. PIANKA,
Department of Zoology,
University of Texas, Austin,
TX 78712, USA

STEPHEN M. REILLY,
Program in Ecology and
Evolutionary Biology,
Department of Biological Sciences,
Ohio University, Athens,
OH 45701, USA

DUSTIN R. RUBENSTEIN,
Department of Neurobiology and
Behavior, Cornell University,
Ithaca, NY 14850, USA

RICHARD SHINE,
School of Biological Sciences,
University of Sydney, Sydney,
Australia, New South Wales 2006,
Australia

BIEKE VANHOOYDONCK,
Department of Biology,
University of Antwerp,
Universiteitsplein 1,
B-2610 Wilrijk, Belgium

RAOUL VAN DAMME,
Department of Biology,
University of Antwerp,
Universiteitsplein 1,
B-2610 Wilrijk, Belgium

MAREN N. VITOUSEK,
Department of Ecology and
Evolutionary Biology,
Princeton University, Princeton,
NJ 08544, USA

LAURIE J. VITT,
Sam Noble Oklahoma Museum
of Natural History, 2401
Chautaugua, Norman,
OK 73072, USA

MICHAEL WALL,
School of Biological Sciences,
University of Sydney,
Sydney, Australia,
New South Wales 2006,
Australia

MARTIN J. WHITING,
Communication & Behaviour
Research Group, School of Animal,
Plant and Environmental Sciences,
University of the Witwatersrand,
Private Bag 3, Wits 2050,
South Africa

MARTIN WIKELSKI,
Department of Ecology and
Evolutionary Biology, Princeton
University, Princeton, NJ 08544, USA

Preface

Investigations into the natural history of lizards have been a major source of discovery in many disciplines of biology, including, but not limited to, morphology, physiology, ecology, and evolution. Although lizards serve as model organisms for a variety of research topics, the group has figured prominently in ecological studies. In particular, the 1960s witnessed a proliferation of theoretical and quantitative studies in population and community ecology that were based on data collected from lizards. As a survey of papers from the past 40 years will attest, squamate reptiles continue to serve as key organisms in ecological research. The evolution and ecology of feeding behavior is one area of ecology in which conclusions emerging from studies based on lizards were most influential. Two early papers are especially relevant to the study of feeding ecology in lizards. The study of community structure and habitat use of North American desert lizards by Pianka (1966) was one of the first to suggest a classification of species into either sit-and-wait or "constantly moving" in an attempt to link resource exploitation, habitat structure and species diversity. A few years later, Schoener (1969a,b, 1971) presented models of optimal predator size based in part on "idealized" lizard predators that corresponded to "sit-and-wait" and "widely foraging" species.

Subsequently, Huey and Pianka (1981) considered the question of which ecological traits were potentially affected by differences in foraging mode. Although several papers on foraging modes came out at about the same time (Andrews, 1979; Anderson and Karasov, 1981; Regal, 1978; Toft, 1981), the Huey and Pianka paper was the first to synthesize known patterns into a coherent set of predictions. As these authors acknowledged, the ecological correlates were varied and complex. In short, the paper by Huey and Pianka (1981) crystallized the foraging mode paradigm in lizards. Their paper was important because the conditions for plasticity in foraging mode were also described; this has often been ignored by later studies. In addition, they stated that the numerous morphological and physiological adaptations associated with foraging mode may preclude shifts in behavior, or that the shifts may be asymmetric. Finally, they highlighted how foraging mode may be a fundamental trait that influences the evolution of morphology, physiology, and behavior. From the onset many researchers have questioned whether the paradigm was truly a dichotomy of two foraging states or whether it was a continuum of states with wide foraging evolving multiple times (see, for example, Pianka, 1973; Regal, 1978; Pietruszka, 1986; McLaughlin, 1989; Schwenk, 1994; Perry, 1999; Butler, 2005). We note the similarity between the sit-and-wait vs. widely foraging dichotomy in foraging ecology and the r–K dichotomy in life history theory; both were instrumental in stimulating research in their respective fields.

Advances in lizard ecology have been periodically reviewed in a series of contributory volumes published over the past 40 years: *Lizard Ecology: A Symposium* (Milstead, 1967); *Lizard Ecology: Studies of a Model Organism* (Huey *et al.*, 1983); and *Lizard Ecology: Historical and Experimental Perspectives* (Vitt and Pianka, 1994). As evidenced by these volumes, "lizards" emerged as a model system for understanding the complex trade-offs among traits in relation to foraging mode. By focusing on sit-and-wait vs. wide-foraging species, students in the fields of ecology, evolutionary biology, and animal behavior could select species with extremely diverse natural histories to test a variety of hypotheses. The breadth of investigation spurred by a dichotomous view of foraging modes is staggering, and ripe for review and evaluation. As shown by chapters in this volume, the foraging mode paradigm has guided and molded numerous studies in morphology (head shape, biomechanics, prey processing), physiology (metabolic rate, locomotor performance, muscle physiology), ecology (predator avoidance, diet choice, niche relations, habitat effects), and behavior (e.g. movement patterns, chemoreception, feeding behavior). Because so many traits appear to be correlated with foraging mode, the sit-and-wait vs. wide-foraging paradigm has emerged as a major explanatory tool for evolutionary patterns in lizard biology. After 40 years of research on how foraging mode permeates practically all aspects of a lizard's biology, it is time to step back and examine how pervasive foraging mode is in structuring trait correlations and trait trade-offs in squamate reptiles.

Accordingly, the goal of this book is to review the current understanding of the influence of foraging mode on the biology of lizards and snakes (squamates). The book as a whole is guided by three central questions.

1. What are the specific patterns in morphology, ecology and behavior that are associated with foraging mode?
2. How do these patterns relate to phylogeny? (Is it simply shared history or are things really evolving in concert?)

3. How do emergent patterns of organismal traits and behavior integrate or change with the evolution of foraging mode?

To review foraging biology of squamates, leaders across biological disciplines studying squamate biology were asked to address these questions with new and existing data. The result, we believe, is both a synthetic review of the foraging mode literature and a novel look at the evolutionary patterns seen at organismal levels across a very diverse vertebrate group.

The book is divided into two parts. Part I, *Organismal patterns of variation with foraging mode*, presents eleven chapters describing the relationships between foraging mode and many aspects of squamate biology ranging from physiology and anatomy to performance and behavior, to diet, to life history. Part II, *Environmental influences on foraging mode*, covers the influences of nocturnality (geckos), plasticity related to environmental heterogeneity, feeding in the ocean, and the myriad of relationships between food acquisition mode and habitat use.

The book concludes with a short synthesis of the general patterns identified by contributors. One of the more interesting, and controversial, findings is that wide foraging has evolved several times rather than once as a deep split in the phylogeny of lizards. Not all contributors agree with this interpretation, however. Many emergent research questions are identified throughout the book that further probe how interconnected traits have evolved, converged, and diverged with foraging mode. Like the *r–K* continuum so important in our understanding of life history evolution, the sit-and-wait vs. wide-foraging paradigm is shifting to a continuum that better represents the emerging patterns of foraging mode biology in squamates.

Producing this volume has taken longer than we had hoped but we are confident that it will help bring new students of squamate biology up to speed on the influence of foraging mode and it will help direct the questions addressed by established researchers in the future. If nothing else, all authors agree that although much progress has been made in the past 40 years, much more information is needed in most aspects of squamate foraging biology. We hope this volume will help catalyze some of this effort.

As with all such volumes it would not have been possible without the assistance of many other people. We graciously thank the contributors for their submissions and their insights; these have certainly opened our eyes about the pervasive effects of foraging mode in squamates. The production staff at Cambridge University Press has continually been wonderful in their long-term help, guidance, and patience during all stages of the preparation of this book. We offer special thanks to many of our contributing authors and the following referees for their time, patience, and expertise in providing peer reviews of chapters: Keller Autumn, Vincent Bels, Margaurite Butler, Sharon Downes, Patrick Gregory, Craig Guyer, Bill Mautz, Eric McElroy, Shannon O'Grady, Chuck Peterson, Geoff Smith, Howard Snell, Jeff Thomason, Fritz Trillmich, Raoul Van Damme, and Mike Walton. Finally, we thank our institutions for some of the intangible support necessary in order to squeeze such a project into our "day jobs".

References

Anderson, R. A. and Karasov, W. H. (1981). Contrasts in energy intake and expenditure in sit-and-wait and widely foraging lizards. *Oecologia* **49**, 67–72.

Andrews, R. M. (1979). The lizard *Corythophanes cristatus*: an extreme "sit-and-wait" predator. *Biotropica* **11**, 136–9.

Butler, M. A. (2005). Foraging mode of the chameleon, *Bradypodion pumilum*: a challenge to the sit-and-wait versus active forager paradigm? *Biol. J. Linn. Soc.* **84**, 797–808.

Huey, R. B. and Pianka, E. R. (1981). Ecological consequences of foraging mode. *Ecology* **62**, 991–9.

Huey, R. B., Pianka, E. R. and Schoener, T. W. (1983). *Lizard Ecology: Studies of a Model Organism.* Cambridge, MA: Harvard University Press.

McLaughlin, R. L. (1989). Search modes of birds and lizards – evidence for alternative movement patterns. *Am. Nat.* **133**, 654–70.

Milstead, W. W. (1967). *Lizard Ecology: A Symposium.* Columbia, MO: University of Missouri Press.

Perry, G. (1999). The evolution of search modes: ecological versus phylogenetic perspectives. *Am. Nat.* **153**, 98–109.

Pietruszka, R. D. (1986). Search tactics of desert lizards: how polarized are they? *Anim. Behav.* **34**, 1742–58.

Pianka, E. R. (1966). Convexity, desert lizards, and spatial heterogeneity. *Ecology* **47**, 1055–9.

Pianka, E. R. (1973). The structure of lizard communities. *Ann. Rev. Ecol. Syst.* **4**, 53–74.

Regal, P. J. (1978). Behavioral differences between reptiles and mammals: an analysis of activity and mental capabilities. In *Behavior and Neurology of Lizards*, ed. N. Greenberg and P. D. MacLean, pp. 183–202. Publication (ADM) 77–491. Washington, D.C.: Department of Health, Education, and Welfare.

Schoener, T. W. (1969a). Models of optimal size for solitary predators. *Am. Nat.* **103**, 277–313.

Schoener, T. W. (1969b). Optimal size and specialization in constant and fluctuating environments: an energy-time approach. *Brookhaven Symp. Biol.* **22**, 103–14.

Schoener, T. W. (1971). Theory of feeding strategies. *Ann. Rev. Ecol. Syst.* **2**, 369–404.

Schwenk, K. (1994). Why snakes have forked tongues. *Science* **263**, 1573–7.

Toft, C. A. (1981). Feeding ecology of Panamanian litter anurans: patterns in diet and foraging mode. *J. Herpetol.* **15**, 139–44.

Vitt, L. J. and Pianka, E. R. (1994). *Lizard Ecology: Historical and Experimental Perspectives.* Princeton, NJ: Princeton University Press.

Historical introduction: on widely foraging for Kalahari lizards

RAYMOND B. HUEY

Department of Zoology, University of Washington

ERIC R. PIANKA

Department of Zoology, University of Texas

This book shows that the field of foraging biology of reptiles is alive and well. We find this exciting, as we've been interested in this field for four decades. No doubt for that reason, we've been asked to describe the history of our thinking about foraging modes. How did we become involved? What were some of the salient experiences we had, and what insights of others helped channel our thinking?

When Eric began studying US desert lizards in the early 1960s, he immediately noted that the teiid *Cnemidophorus* moved much more than did all species of iguanids. This lizard world was clearly dichotomous in terms of foraging behavior. In his 1966 paper in *Ecology*, Eric coined the terms "sit-and-wait" (hereafter SW) and "widely foraging" (hereafter WF) to characterize these different behaviors.

Ray's interest in foraging behavior evolved independently about the same time. As an undergraduate at UC Berkeley in the spring of 1965, he took Natural History of the Vertebrates (taught by R. C. Stebbins and others). Students were required to do a field project: Ray studied the feeding behavior of great blue herons. In his term paper (Huey, 1965), he noted that herons "... use two distinct types of hunting whether on land or in water – stalking and still hunting." Further, he observed that herons hunting in estuaries will switch to "still hunting" when the tide is coming in, letting the moving water bring food to the birds. Thus the dichotomy of foraging behaviors he observed within a species was exactly the same as what Eric had observed between species!

In 1966 Eric headed to Australia on a postdoc, and in 1967 Ray started graduate school at the University of Texas (UT). Ray was fascinated by Eric's work (and also by Tom Schoener's) on species diversity and competition in lizards, and their studies inspired him to do his MA thesis on competitive relations of geckos from the Sechura Desert in Peru. When he read Eric's paper (Pianka,

Lizard Ecology: The Evolutionary Consequences of Foraging Mode, ed. S. M. Reilly, L. D. McBrayer and D. B. Miles. Published by Cambridge University Press. © Cambridge University Press 2007.

1966), he found that his own observations on herons fit in neatly within Eric's concept, terminology, and ideas on foraging mode.

By a remarkable and fortuitous coincidence, Eric joined the faculty at UT in fall 1968. Eric and Ray became instant friends and intellectual soul mates, given their many mutual interests (in lizards, foraging mode, and far-away deserts). The next winter Eric received an NSF grant to study species diversity of lizards in the Kalahari; and he hired Ray and Larry Coons to work as his field assistants beginning in November 1969.

Eric arrived in the Kalahari a month later, and the three of us spent two wonderful months observing and collecting lizards in the Kalahari and Namib, drinking non-hot beer, and avoiding lions, leopards, scorpions, puff adders, and cape cobras. We no longer recall who first noticed that different species of lacertids were either WF or SW – given our histories, that pattern was probably immediately obvious to us both!

In any case, Ray had two experiences later on that trip that crystallized our thinking on the ecological significance of foraging mode. First, he realized that we ourselves were widely foraging predators on lizards; and that our foraging behavior must not only increase our encounter rates with lizards, but also influence the *kinds* of lizards we'd encounter. For example, our WF movements would necessarily increase our encounters with sedentary SW lizards. Second, Ray remembers walking one day along a sand ridge, wondering why we always found a WF paired with a SW lizard in a given habitat (e.g. *Meroles suborbitalis* with *Pedioplanis namaquensis* on white interdunal street sands, and *Pedioplanis lineoocellata* with *Heliobolus lugubris* on the red sandridges), but never two WF species or two SW species paired in a habitat (see Pianka *et al.*, 1979). Although this pairing could obviously have happened by chance, Ray wondered whether foraging mode might in fact influence the kinds of prey a lizard would encounter. If so, differences in foraging mode might reduce dietary overlap and facilitate spatial overlap. As soon as he posed these issues, he predicted that WF lizards should encounter more patchily distributed and sedentary prey (e.g. termites), whereas SW lizards should encounter mainly moving prey. To our delight, this prediction was validated when we analyzed the dietary data (Huey and Pianka, 1981).

Yet another important insight emerged from our Kalahari peregrinations. All too frequently we stepped on horned adders (*Bitis caudalis*), which hunt by burying themselves (except for their eyes) in the sand, catching unsuspecting lizards that happen to wander by. These vipers responded viciously to our transgressions by "leaping" out of the sand, striking at our feet and legs. Not surprisingly, our initial response was always that of sheer terror; and we were inspired to invent the 'Kalahari two-step,' an intensely evasive dance that has curiously never caught on with the general public. In any case, once our adrenalin titers dropped, we recognized that these vipers were archetypal (and arch-villain) SW predators.

Ray later made a related, but critical, observation one day near Tsabong, Botswana, while watching a secretary bird foraging for lizards. It would walk briskly up to a bush and suddenly raise its wings, startling any nearby lizard (most likely a SW species), which it then grabbed as the lizard fled. Thus both horned adders and secretary birds preyed on lizards, but did so with polar opposite foraging modes. These differences were striking to us, and we predicted that these two predators should eat different species of lizard. Fitting neatly with our expectations, SW sand vipers preyed relatively heavily on WF lizards, whereas WF secretary birds caught disproportionate numbers of SW lizards. These field observations – and Don Broadley's (1972) remarkable dietary data – led us to propose the idea of crossovers in foraging mode between trophic levels as well as to argue that models of foraging mode needed to incorporate interactions among multiple trophic levels, not just between predator and prey (Huey and Pianka, 1981).

Ray flew back to Austin in late November 1970, and he used the long flight to begin synthesizing our ideas on the ecology of foraging mode. He scribbled these on the back of airline non-slip placemats. Over the next few weeks, we spent many fruitful hours huddled over those placemats; and we talked and talked until our ecological correlates began to gel.

Nevertheless, our foraging mode project soon went into torpor. Other Kalahari projects (and Ray's thesis work on thermoregulation) took precedence. But one important realization that became increasingly apparent to us during this period was the comparative beauty of the Kalahari system – namely, that *close relatives* differed in foraging mode. That was not at all the case in the North American deserts, where any observed difference in ecology between SW vs. WF species might reflect foraging mode, or it might merely reflect long separate phylogenetic histories (iguanids vs. teiids), or both (Huey and Pianka, 1981; Huey and Bennett, 1986, p. 85). One couldn't easily tell which (but see Anderson and Karasov, 1988).

Another reason for delaying publication was simply that we had no quantitative data on foraging movements. So on a return trip in 1975–6 generously sponsored by the National Geographic Society, we (and Carolyn Cavalier) quantified various aspects of foraging behavior of our lacertids among other studies (Huey and Pianka, 1981; Huey *et al.*, 1983). We measured time spent moving and number of moves per minute for six species of lacertid, two of which were SW and four WF foragers (see also Cooper and Whiting, 1999). With these data, we finally had all of the elements in hand for a paper.

About the same time, many others were working on related aspects of foraging mode in herps. Bennett and Gorman (1979) and Anderson and Karasov (1981) had estimated metabolic costs of different foraging modes. Bennett and Licht (1973) and Ruben (1976, 1977) explored physiological and morphological differences in reptiles with different foraging modes. Vitt and Congdon (1978) drew attention to striking differences in morphology and reproductive effort, and proposed that relative clutch mass, body shape, and foraging mode were co-evolved characters. Regal (1978) argued that WF lizards might have enhanced learning and memory, as well as larger brains. Cathy Toft (1981) noted a number of parallels in frogs, and added the fascinating observation that WF tropical anurans are typically very poisonous. Cathy's observation was especially interesting to us, because we had discovered that the juveniles of one of our WF lacertids (*Heliobolus lugubris*) mimicked a very noxious "oogpister" beetle (Huey and Pianka, 1977; Schmidt, 2004). WF animals are of course relatively conspicuous to visually hunting predators, and thus poisons and mimicry may be relatively common ways of reducing predation on such species.

Foraging mode of herps was obviously a field whose time had come by the 1970s! It was clearly and simultaneously attracting the attention of diverse herpetologists. So when we finally put our thoughts down on paper (Pianka *et al.*, 1979); Huey and Pianka, 1981) we drew on the insights of many others, not just our own. Our synthetic debt to all is clearly evident in Table 8 in our 1981 paper.

Huey returned once more to the Kalahari in 1981–2 to investigate physiological correlates of foraging mode. This time he worked with three top-flight physiological ecologists (Al Bennett, Henry John-Alder, and Ken Nagy). They found that WF vs. SW species differed in locomotor capacity (acceleration, maximal speed, and stamina) (Huey *et al.*, 1984; but see Perry, 1999), differed in field metabolic rates and feeding rates (Nagy *et al.*, 1984), and also differed in some (but not all) lower-level physiological traits (Bennett *et al.*, 1984).

Finally, Huey and Bennett (1986) were invited to contribute a chapter to a volume on predator–prey relationships in lower vertebrates (Feder and Lauder, 1986). This was right at the beginning of the "Felsenstein era" of comparative biology, marked by Felsenstein's remarkable, now classic, 1985 paper. We were already well aware of the suitability of lacertids for comparative studies of foraging mode and had noted (Huey and Pianka, 1981, p. 991) that the close relationships of these lizards provided "a substantial measure of control over phylogenetic and sensory differences." Huey and Bennett (1986) now took that a step further and speculated on the direction of evolutionary change within this lineage. They used a crude phylogeny of the lacertids to suggest that SW was derived in this group, and next explored various selective factors that might have favored this evolutionary transition. Improved phylogenies (see, for example, Harris *et al.*, 1998) and reconstructions in Perry (1999) and McBrayer (2004) seem to support the suggestion (Huey and Bennett (1986) that SW is derived.

The two of us are rapidly approaching our academic dotage (but we're not there yet!), and we take this opportunity to reflect on our own foraging-mode work as well as add a few observations on current research in this field today. On a personal level, our Kalahari studies of foraging mode were great fun to do and had a big effect on our academic development. Those studies were growth experiences for us: in the process, we taught ourselves the critical importance of integrating field studies of behavior and ecology with laboratory and field studies of physiology (energetics and locomotor capacity). Moreover, we taught ourselves the importance of conducting comparative studies within an explicit phylogenetic context. We admit to some pride, of course, in the continuing impact of our studies (Fig. A1, and this volume) and those of others involved at the same time.

When we reflect on our old work, several things strike us as worth emphasizing.

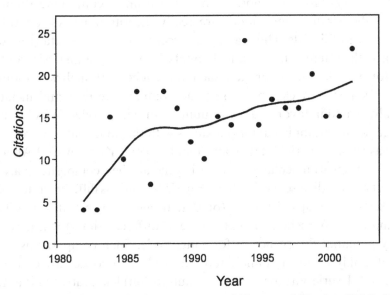

Figure A1. Number of citations per year of Huey and Pianka (1981). The curve is a non-parametric super-smoother.

First, our experiences reinforce the often-forgotten view that novel ideas can emerge from seemingly trivial natural history observations of animals in nature (Gans, 1978; Bartholomew, 1986; Greene, 1986, 2005). Who would have ever guessed that so much work would evolve in part from Eric's observations of the movement patterns of North American lizards or of Ray's of great blue herons? Or who would have thought that our stepping on horned adders would inspire the arcane idea of trophic crossovers?

Second, our work with lacertids helped make us both keenly aware of the importance of phylogeny and phylogenetic control in comparative biology. This is old news now, but it certainly was not in 1981 or even in 1986. In those "dark days of ecology" (W. L. Hodges, personal communication), comparative studies almost invariably involved distant relatives (cf. Huey, 1987), were generally restricted to only two species (Garland and Adolph, 1994), and rarely considered phylogeny (but see Greene and Burghardt, 1978!). Times have changed, and explicit phylogenetic approaches involving many species are now standard.

Third, we note that our paper is commonly viewed as advocating a strictly dichotomous view of foraging mode (Pietruszka, 1986; McLaughlin, 1989; Helfman, 1990). To be sure, we did emphasize the bimodality of foraging mode in our lacertids. But what is often overlooked is that we explicitly discussed *flexibility* in foraging mode and its consequences and constraints (see also Norberg, 1977). To us, this is an area of foraging mode that remains severely understudied (but see Dunham, 1981; Karasov and Anderson, 1984).

The flexibility of foraging mode is directly relevant to continuing attempts to quantify whether foraging modes are really dichotomous (vs. continuous) among species of lizards. Those analyses proceed by pooling data from different taxa with different habits and habitats into a single analysis, and then testing for bimodality. In our view such approaches – though well motivated and often statistically elegant – are weak comparative tests and in fact are potentially biased in favor of a continuum of foraging modes. Why? Foraging mode isn't an invariant behavior but varies with the immediate environmental conditions (Norberg, 1977; Huey and Pianka, 1981; Karasov and Anderson, 1984). If that environmental sensitivity is general, then composite analyses of movement rates of diverse species that forage in habitats as different as temperate-zone deserts and tropical forests (or that forage on substrates as different as ground or on branches; or that forage at different times) will necessarily blur any potential divergence in foraging mode. Such comparisons may be meaningful only when habit and habitat as well as phylogeny are controlled.

Moreover, Laurie Vitt (personal communication) has cautioned that foraging analyses that are based on movement rates must consider the context of movement (or lack thereof) – otherwise data can be misleading. For example, some WF species periodically punctuate their active foraging and sit for extended periods to bask or to cool (Hillman, 1969; van Berkum *et al.*, 1986). Conversely, males of some SW species start moving extensively during the breeding season (Pietruszka, 1986), but their movements are likely related to searching for mates, not food. By failing to consider the natural history context of movement, an incautious investigator quantifying movement of the above species could grossly misrepresent true foraging rates.

To us, the proper way to address this issue is to select a single habitat and taxonomic group as *the unit* for analysis, to appreciate the ecological context of observed movement patterns, and to assay whether the lizards in that habitat have dichotomous, continuous, or invariant foraging modes. Then to determine generality for lizards, one would need to iterate that procedure through additional habitats and taxa and ultimately determine the proportion of cases showing bimodality versus a continuum. We took a first step with Kalahari lacertids, but we are fully aware that seemingly limited variation in foraging behavior within most other taxa (Perry, 1999; Vitt and Pianka, 2005) hinders or even precludes parallel analyses in most environments. Unfortunately, evolution hasn't always cooperated with the analytical needs of comparative biologists (Huey, 1987)!

What's happened to our own interests since those wonderful days of our youth, foraging widely over the red dunes of the Kalahari? In his old age, Eric has switched from being a WF to a SW hunter of lizards. He notes that he can't hear or see as well as he once could, and that his bad leg can't support his new huge mass as well as it could when he weighed only 140–155 pounds. So now he pit traps lizards, almost exclusively, except for monitor lizards, large ones of which are too big to trap. But he is still avidly pursuing foraging mode. With the help of Bill Cooper and Kurt Schwenk, Pianka and Laurie Vitt recently achieved a synthesis of the history of evolutionary innovations that led to the evolution of widely foraging (Pianka and Vitt 2003; Vitt *et al.* 2003). Recently, they demonstrated that 28% of the variance in diets of 184 extant lizard species in 12 families from 4 continents can be attributed to the first split in squamate phylogeny 200 Ma when Iguania and Scleroglossa split (Vitt and Pianka 2005). Iguanians retained ancestral traits including tongue prehension and ambush foraging, whereas scleroglossans switched from tongue prehension to jaw prehension. This freed the tongue to evolve along new lines (Schwenk and Wagner, 2001), ultimately leading to much keener vomerolfaction, which in turn promoted a more active lifestyle and facilitated a shift to wide foraging.

Ray on the other hand has largely moved away from the world of lizards and foraging biology. He claims to have good reasons for doing so (see Huey, 1994), but many herpers no doubt see his departure as clear evidence of moral deterioration and premature senility – what else could explain his evolutionary transition to the worlds of fruit flies and mountain climbers? He still does field work but is no longer a WF hunter of wily lizards in remote deserts – instead he is reduced to running a trap line and collecting piddly *Drosophila* on rotting, smelly banana baits. Perhaps this book will inspire him back into the fold.

References

Anderson, R. A. and Karasov, W. H. (1981). Contrasts in energy intake and expenditure in sit-and-wait and widely foraging lizards. *Oecologia* **49**, 67–72.

Anderson, R. A. and Karasov, W. H. (1988). Energetics of the lizard *Cnemidophorus tigris* and life history consequences of food-acquisition mode. *Ecol. Monogr.* **58**, 79–110.

Bartholomew, G. A. (1986). The role of natural history in contemporary biology. *BioScience* **36**, 324–9.

Bennett, A. F. and Gorman, G. C. (1979). Population density, thermal relations, and energetics of a tropical insular lizard community. *Oecologia* **42**, 339–58.

Bennett, A. F. and Licht, P. (1973). Relative contributions of anaerobic and aerobic energy production during activity in Amphibia. *J. Comp. Physiol.* **81**, 227–88.

Bennett, A. F., Huey, R. B. and John-Alder, H. B. (1984). Physiological correlates of natural activity and locomotor capacity in two species of lacertid lizards. *J. Comp. Physiol.* **154**, 113–18.

Broadley, D. G. (1972). The horned viper *Bitis caudalis* (A. Smith) in the central Kalahari. *Botswana Notes Rec.* **4**, 263–4.

Cooper, W. E. Jr. and Whiting, M. J. (1999). Foraging modes in lacertid lizards from southern Africa. *Amph.-Rept.* **20**, 99–311.

Dunham, A. E. (1981). Populations in a fluctuating environment: the comparative population ecology of the iguanid lizards *Sceloporus merriami* and *Urosaurus ornatus*. *Misc. Publ. Mus. Zool. Univ. Michigan*, No. 158.

Feder, M. F. and Lauder, G. V. eds. (1986). *Predator-Prey Relationships. Perspectives and Approaches from the Study of Lower Vertebrates*. Chicago, IL: University of Chicago Press.

Felsenstein, J. (1985). Phylogenies and the comparative method. *Am. Nat.* **125**, 1–15.

Gans, C. (1978). All animals are interesting. *Am. Zool.* **18**, 3–9.

Garland, T. Jr. and Adolph, S. C. (1994). Why not to do two-species comparative studies: limitations on inferring adaptation. *Physiol. Zool.* **67**, 797–828.

Greene, H. W. (1986). Natural history and evolutionary biology. In *Predator-Prey Relationships: Perspectives and Approaches from the Study of Lower Vertebrates*, ed. M. F. Feder and G. V. Lauder, pp. 99–108. Chicago, IL: University of Chicago Press.

Greene, H. W. (2005). Organisms in nature as a central focus for biology. *Trends Ecol. Evol.* **20**, 23–7.

Greene, H. W. and Burghardt, G. M. (1978). Behavior and phylogeny: constriction in ancient and modern snakes. *Science* **200**, 74–7.

Harris, D. J., Arnold, E. N. and Thomas, R. H. (1998). Relationships of lacertid lizards (Reptilia: Lacertidae) estimated from mitochondrial DNA sequences and morphology. *Proc. R. Soc. Lond.* **B265**, 1939–48.

Helfman, G. S. (1990). Mode selection and mode switching in foraging animals. In *Advances in the Study of Behavior*, Vol. 19, ed. P. J. B. Slater, J. S. Rosenblatt, and C. Beer, pp. 249–98. San Diego, CA: Academic Press.

Hillman, P. E. (1969). Habitat specificity in three sympatric species of *Ameiva* (Reptilia: Teiidae). *Ecology* **50**, 476–81.

Huey, R. B. (1965). Diurnal feeding and associated behavior of Great Blue Herons in central coastal California. Unpublished term paper, University of California, Berkeley.

Huey, R. B. (1987). Phylogeny, history, and the comparative method. In *New Directions in Ecological Physiology*, ed. M. E. Feder, A. F. Bennett, W. W. Burggren and R. B. Huey, pp.76–98. Cambridge, UK: Cambridge University Press.

Huey, R. B. (1994). Introduction to Evolutionary Ecology Section. In *Lizard Ecology*, ed. L. J. Vitt and E. R. Pianka, pp. 175–182. Princeton, NJ: Princeton University Press.

Huey, R. B. and Bennett, A. F. (1986). A comparative approach to field and laboratory studies in evolutionary biology. In *Predator-Prey Relationships. Perspectives and Approaches from the Study of Lower Vertebrates*, ed. M. E. Feder and G. V. Lauder, pp. 82–98. Chicago, IL: University of Chicago Press.

Huey, R. B. and Pianka, E. R. (1977). Natural selection for juvenile lizards mimicking noxious beetles. *Science* **195**, 201–3.

Huey, R. B. and Pianka, E. R. (1981). Ecological consequences of foraging mode. *Ecology* **62**, 991–9.

Huey, R. B., Bennett, A. F., John-Alder, H. and Nagy, K. A. (1984). Locomotor capacity and foraging behavior of Kalahari lizards. *Anim. Behav.* **32**, 41–50.

Huey, R. B., Pianka, E. R. and Cavalier, C. M. (1983). Ecology of lizards in the Kalahari Desert, Africa. *Nat. Geogr. Soc. Res. Rep.* **16**, 365–70.

Karasov, W. H. and Anderson, R. A. (1984). Interhabitat differences in energy acquisition and expenditure in a lizard. *Ecology* **65**, 235–47.

McBrayer, L. D. (2004). The relationship between skull morphology, biting performance and foraging mode in Kalahari lacertid lizards. *Zool. J. Linn. Soc.* **140**, 403–16.

McLaughlin, R. L. (1989). Search modes of birds and lizards: evidence for alternative movement patterns. *Am. Nat.* **133**, 654–70.

Nagy, K. A., Huey, R. B. and Bennett, A. F. (1984). Field energetics and foraging mode of Kalahari lacertid lizards. *Ecology* **65**, 588–96.

Norberg, R. A. (1977). An ecological theory of foraging time and energetics and choice of optimal food-searching method. *J. Anim. Ecol.* **46**, 511–29.

Perry, G. (1999). The evolution of search modes: ecological versus phylogenetic perspectives. *Am. Nat.* **153**, 98–109.

Perry, G. and Pianka, E. R. (1997). Animal foraging: past, present and future. *Trends Ecol. Evol.* **12**, 360–4.

Pianka, E. R. (1966). Convexity, desert lizards, and spatial heterogeneity. *Ecology* **47**, 1055–9.

Pianka, E. R. (1971). Lizard species density in the Kalahari desert. *Ecology* **52**, 1024–9.

Pianka, E. R. and Vitt, L. J. (2003). *Lizards, Windows to the Evolution of Diversity*. Berkeley, CA: University of California Press, Berkeley.

Pianka, E. R., Huey, R. B. and Lawlor, L. R. (1979). Niche segregation in desert lizards. In *Analysis of Ecological Systems*, ed. D. J. Horn, R. Mitchell and G. R. Stairs, pp. 67–115. Columbus, OH: Ohio State University Press.

Pietruszka, R. D. (1986). Search tactics of desert lizards: how polarized are they? *Anim Behav.* **34**, 1742–58.

Regal, P. J. (1978). Behavioral differences between reptiles and mammals: an analysis of activity and mental capabilities. In *Behavior and Neurobiology of Lizards*, ed. N. Greenberg and P. D. MacLean, pp. 183–202. Washington, D.C.: Dept. of Health, Education and Welfare.

Ruben, J. A. (1976). Correlation of enzymatic activity, muscle myoglobin concentration and lung morphology with activity metabolism in snakes. *J. Exp. Zool.* **197**, 313–20.

Ruben, J. A. (1977). Some correlates of cranial and cervical morphology with predatory modes in snakes. *J. Morphol.* **152**, 89–100.

Schmidt, A. D. (2004). *Die Mimikry zwischen Eidechsen und Laufkafern.* Frankfurt am Main: Edition Chimaira.

Schwenk, K. and Wagner, G. P. (2001). Function and the evolution of phenotypic stability: connecting pattern to process. *Am. Zool.* **41**, 552–63.

Toft, C. A. (1981). Feeding ecology of Panamanian litter anurans: patterns in diet and foraging mode. *J. Herpetol.* **15**, 139–44.

van Berkum, F. H., Huey, R. B. and Adams, B. (1986). Physiological consequences of thermoregulation in a tropical lizard (*Ameiva festiva*). *Physiol. Zool.* **59**, 464–72.

Vitt, L. J. and Congdon, J. D. (1978). Body shape, reproductive effort, and relative clutch mass in lizards: resolution of a paradox. *Am. Nat.* **112**, 595–608.

Vitt, L. J. and Pianka, E. R. (2005). Deep history impacts present day ecology and biodiversity. *Proc. Nat. Acad Sci. USA* **102**, 7877–81.

Vitt, L. J., Pianka, E. R., Cooper, W. E. and Schwenk, K. (2003). History and the global ecology of squamate reptiles. *Am. Nat.* **162**, 44–60.

I

Organismal patterns of variation with foraging mode

1

Movement patterns in lizards: measurement, modality, and behavioral correlates

GAD PERRY

Department of Range, Wildlife, and Fisheries Management, Texas Tech University

> From the least to the greatest in the zoological progression, the stomach sways the world.
>
> *(Fabre, 1913)*

Introduction

To reproduce successfully, an organism must survive, attain suitable size, attract a mate, and produce viable offspring. All of these activities require that the individual obtain considerable amounts of energy. Foraging success can thus strongly impact reproductive success (Travers and Sih, 1991; Bernardo, 1994; Nilsson, 1994). Reproductive success is the fabric upon which natural selection works (Darwin, 1859). Evolutionary biologists and behavioral ecologists, starting with the pioneering work of MacArthur and Pianka (1966) and Emlen (1966), have therefore shown considerable interest in foraging behaviors and their correlates. The resulting literature is voluminous and often contentious: too much so for a single chapter, or even volume, to effectively summarize. In keeping with the theme of this book, I focus on the issue of bimodality in lizard foraging behavior, its phylogenetic background, and its putative correlates.

Before one can discuss patterns, however, methodological issues must be clarified. This chapter is therefore divided into two main sections. The first section focuses on some previously neglected methodological issues related to measurement of foraging behavior. Establishing these is crucial for ensuring data quality in the analyses that follow. The second section then concentrates on testing theoretical predictions of foraging theory, and on some conceptual consequences of what has been learned to date. Because the literature is so extensive, I frequently limit the use of references to representative examples throughout this chapter.

Lizard Ecology: The Evolutionary Consequences of Foraging Mode, ed. S. M. Reilly, L. D. McBrayer and D. B. Miles. Published by Cambridge University Press. © Cambridge University Press 2007.

Technical issues

Measuring foraging behavior: qualitative versus quantitative measures

Early discussions of foraging strategy, being theoretical in nature, did not focus on ways of quantifying such behavior. Many papers discussing foraging behaviors have subsequently provided strictly qualitative descriptions, but use of such labels can be problematical. It may mask important methodological differences between studies (Biro and Ridgway, 1995), and assumes that different observers will use the same terminology to describe similar behaviors. As an example, an extensive argument has occurred over the years regarding the number of foraging modes that should be recognized (Perry *et al.*, 1990; Mac Nally, 1994).

If authors cannot agree on the number of foraging modes, then qualitative descriptors are not likely to provide a consistent portrayal of an organism's behavior. To a reader unfamiliar with an organism, qualitative assignment of categories can, in the absence of well-defined criteria, prove quite confusing. For example, Cooper and Whiting (2000) characterized *Mabuya spilogaster* as typical of ambush foragers whereas Huey *et al.* (2001) concluded, based on subjective criteria, that it was widely foraging. Most likely, both sets of authors observed similar behaviors but, in the absence of clear numerical cut-offs, used different terms to describe them. Similarly, Huey and Pianka (1981) compared *Pedioplanis lineoocellata* with other lacertids, found it to be relatively sedentary, and called it an ambush forager. However, as Cooper (1994) points out, the same behaviors would be considered typical of an active forager when compared with those of a typical iguanid. Thus, qualitative terms such as "sit-and-wait" can be confusingly context-specific. Although they offer an easy way to identify a broad category of organisms, they are not particularly useful for communicating information about species that are not at one extreme or another. A numerical description provides a more nuanced approach, greatly reduces the chances for misunderstandings to occur, and improves statistical rigor. Careful choice of methodology can further increase statistical power by reducing variability due to improper sampling.

Pianka *et al.* (1979) were the first (and so far have been the only) authors to suggest widely used numerical parameters for describing foraging behaviors: speed; the number of moves per minute (MPM); and the percent of the time spent moving (PTM). However, subsequent work has shown that velocity, although an important parameter in other contexts, is correlated with body size for at least some taxa (see, for example, Avery *et al.*, 1987) and therefore not likely to be a good proxy for foraging behavior (Perry *et al.*, 1990). This leaves MPM and PTM, both of which have been widely utilized for describing foraging behaviors, primarily in lizards and fish. In addition, Cooper and Whiting (2000) proposed a new index, the percent of attacks on prey that occurred while the predator was moving, or PAM. This index is conceptually appealing, but there have not been enough such data collected yet to date to allow analysis here.

As Perry *et al.* (1990) and Cooper *et al.* (2001) have shown, MPM and PTM are positively correlated: species that rank high in one often also score high on the other. Perhaps because of this, many of the papers surveyed by Perry (1999) only report one or the other. Perry *et al.* (1990) and Perry (1995) none the less recommended using both, considering them to be complementary measures. For example, the calculated MPM value for an animal that made one brief move (and thus had a low value for PTM) might be very similar to that obtained from an individual that spent all of its time moving (PTM = 100%) and never paused (producing a low MPM value). In a similar vein, comparable PTM values can be produced by animals that never pause and ones that do so frequently but briefly. Thus, use of only one measure can mask biologically important differences, whereas an analysis of both provides a more complete understanding of the biology of an organism.

Measuring foraging behavior quantitatively

There is no overwhelming consensus on how foraging mode should be quantified (McLaughlin, 1989). However, Huey and Pianka (1981) described the procedure used by most workers as follows.

Foraging modes of lizards were documented by observing several individuals ($\bar{x} = 14.3$, range = 5–25) ... for at least 1 min ($\bar{x} = 5.9$ min/individual) during mid-morning ... For each lizard we recorded distance and duration of each move and duration of each stop ...

Thus, they and later workers have identified several crucial elements that must be addressed when collecting quantitative foraging data: sample size (number of individuals); data quality (length of observation); and possible confounding factors. I now summarize the current thinking on such methodological issues. When analyzing these data, I use non-parametric statistical tests whenever possible, because they require fewer assumptions. Whenever no such test is available or appropriate, I used parametric tests, first transforming the data to ensure normality (Sokal and Rohlf, 1995). Unless stated otherwise, *p* values are two-tailed.

Number of individuals

Several authors have reported data for species based on only a handful of (usually extra-long; see "length of observation" below) observations. For

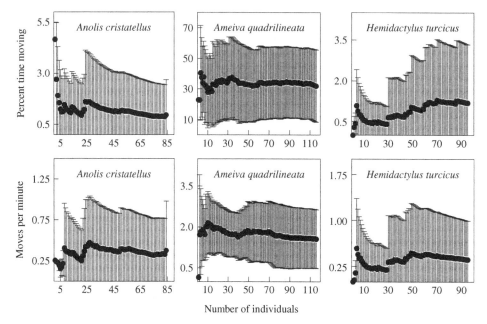

Figure 1.1. Foraging indices of three lizard species representing three major lizard clades. In each panel, the cumulative mean value for the species is displayed as a function of increasingly large numbers of individuals. Species means based on a small number of individuals tend to vary considerably from values obtained when a larger sample is assessed.

example, the data for *Mesalina guttulata* presented by Perry *et al.* (1990) are based on three individuals, but encompass 82 minutes of observation. However, the number of observations used can have a large effect on the data being reported. For example, the number of sightings is often considered a crucial element in determining home range size, and a minimum number of observations are traditionally required for data to be considered representative (see, for example, Perry and Garland, 2002). This issue has not previously been evaluated in the context of foraging behavior studies. To address this, I examined data for three species, one from each major lizard clade, for which I have a large number of observations. These preliminary findings (Fig. 1.1) suggest that a relatively small number of individuals provide a good idea of behavior in both Iguania (here represented by *Anolis cristatellus*, left panels) and Scincomorpha (*Ameiva quadrilineata*, center panels). However, data for the Gekkota (*Hemidactylus turcicus*, right panels) show a different pattern, with a much larger sample required for average values to stabilize. This is consistent with statements made by a number of authors in the past, who indicated that, unlike that of most lizards, gecko behavior is characterized by relatively long periods of sedentary behavior, alternating with short bouts of very high activity (see, for example, Perry, 1999; Werner *et al.*, 1997). These findings, albeit consistent with my subjective recollection of multiple species from each clade, are none the less based on a small number of species. Thus, they indicate that care is required to ensure that clade-specific traits be taken into account when designing studies of foraging. More information is required before broad generalizations are possible.

Length of observation

Early studies included observations as short as 1 min per individual in their analysis, although most observations were considerably longer. Concerned that short observations might not provide a representative sample of the behavioral repertoire of an organism, the authors of more recent studies (see, for example, Perry *et al.*, 1990) usually employed a 5 min minimum cut-off, with longer observations used for taxa, such as geckos, that show intermittent locomotion (see, for example, Weinstein and Full, 1999). However, this value is also arbitrary, and I am not aware of any studies to determine to what extent observation length affects the resulting indices of foraging behavior.

To explore this issue, I plotted length of observation against the indices of foraging behavior for four species for which I have large datasets and a range of observation periods, many of them too short to have been previously reported (Fig. 1.2). In all cases, shorter observations produced a much greater variability in foraging indices than did longer ones, indicating that short observations only sample part of the behavioral repertoire of the animal. In all but one case (in which the relation was marginally significant), there was a significant negative relation between observation length and the resulting foraging index (Spearman rank correlations. *Sceloporus* ($n = 74$): MPM: $\rho = 0.42$, 2-tailed $p < 0.001$ and PTM: $\rho = 0.31$, $p = 0.007$; *Anolis* ($n = 206$): MPM: $\rho = 0.22$, $p = 0.001$ and PTM: $\rho = 0.21$, $p = 0.002$; *Ameiva* ($n = 116$): MPM: $\rho = 0.41$, $p < 0.001$ and PTM: $\rho = 0.26$, $p = 0.004$; *Hemidactylus* ($n = 92$): MPM: $r = 0.22$, $p = 0.039$ and PTM: $r = 0.19$, 2-tailed $p = 0.07$). Thus, shorter observations often produced relatively high values of MPM or PTM. (It should be noted that the lines indicating trends in Fig. 1.2 were arbitrarily obtained by using a power regression, and that the most appropriate algorithm will need to be identified as more such data are available for analysis.) In addition, longer observations showed considerably less variability than did short (< 10 min) observations.

These findings are not surprising. A brief observation can document a period of relatively high or relatively low activity, producing values that are more extreme than would a longer observation sampling a more representative stretch of activity. Because such extreme values are produced, variation

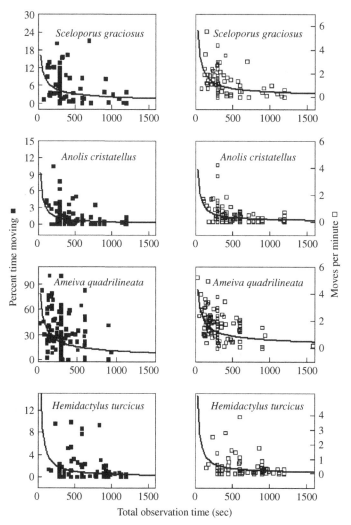

Figure 1.2. Foraging indices of four lizard species representing three major lizard clades. In each panel, the value for each individual is displayed as a function of total observation time. Short observation times result in individual values that are much more variable than those obtained from longer observations.

around the mean is greater when that mean is calculated based on relatively short observations. Thus, future studies can reduce variability and enhance statistical power by employing a minimum observation time of 10 min.

How repeatable are numerical measures of foraging behavior?

Discussions of repeatability of numerical measures of performance are common in the physiological literature (see, for example, Fuiman and Cowan,

2003; Perry *et al.*, 2004). The reason is that, whereas numerical measures such as MPM and PTM can be very precise (that is, carry many significance digits), this does not necessarily mean that they are accurate (i.e. close to the value obtained from a long-term study). The main reason for this is that different authors may use different methodologies (e.g. observation periods). Unless such observations provide a representative – and thus consistent and repeatable – description of behavior, they might be of little biological significance. However, I am not aware of any studies that have directly evaluated the repeatability of such data and the required minimum observation length. Cooper *et al.* (2001) collated published data for species that have been observed several times and concluded that MPM, and especially PTM, were generally consistent across studies. I directly addressed this question by using two relatively small datasets on *Anolis cristatellus* which I have previously collected.

In one study I observed each of 15 adult lizards for two consecutive 10 min periods (20 min continuous observation overall). Mean values were similar for first and second observations (first PTM: $\bar{x} = 0.5$, SD $= 0.74$ compared with second PTM: $\bar{x} = 0.7$, SD $= 1.70$; first MPM: $\bar{x} = 0.3$, SD $= 0.40$ compared with second MPM: $\bar{x} = 0.4$, SD $= 0.91$). Because of the small sample size, I used a non-parametric test to analyze these data. There was a significant positive correlation between foraging indices obtained during the first 10 min and the second 10 min (MPM: $r = 0.59$, $p = 0.021$; PTM: $r = 0.527$, $p = 0.043$). In the second study I also observed adult lizards ($n = 30$) repeatedly. However, this time repeated observations were separated by 2–7 days. Mean values were similar for first and second observations (first PTM: $\bar{x} = 0.6$, SD $= 1.14$ compared with second PTM: $\bar{x} = 0.7$, SD $= 1.05$; first MPM: $\bar{x} = 0.2$, SD $= 0.35$ compared with second MPM: $\bar{x} = 0.4$, SD $= 0.69$). Contrary to consecutive observations, however, foraging metrics derived from temporally disjunct observations were far from significantly correlated (MPM: Spearman's $\rho = 0.008$, $p = 0.966$; PTM: $\rho = 0.004$, $p = 0.983$).

To the extent that they suggest that relatively brief observations do provide a reliable and repeatable picture of the behavior of the animal, at least in the short term, these data are encouraging. They suggest that MPM and PTM both reliably measure individual behavior in a manner that provides biologi-cally significant information about short-term time allocation decisions. Moreover, mean values are consistent enough, when sampling bouts are compared across longer time periods, that repeatability of population indices appears to be quite satisfactory and the indices appear to convey biologically meaningful information.

The finding that observations separated by days do not produce repeatable results is more difficult to interpret. It is consistent with the observation that, even with large sample sizes and long observations, variability in MPM and PTM remains high, with the standard deviations typically being similar in magnitude to the mean values (Perry, 1999; current study, Table 1.1). This suggests that individual behavior changes over the course of days, perhaps as a function of food availability or other constraints. It is also possible that much longer observations, perhaps on the order of an hour or more, would produce values that would be repeatable on such a time scale. However, it is possible that more variation occurs in individual behavior over time than is seen among individuals when averaged indices of their long-term behaviors are compared. Much more work is required on this topic: both datasets are small and the species studied is highly sedentary. This study should be extended in two ways. First, the sample size should be expanded for *Anolis cristatellus*, so that both short-term and longer-term repeatability can be more thoroughly evaluated. Second, additional species should be similarly studied. A comparison with a more active forager could be especially informative.

Confounding factors

Whenever behavior is recorded, a variety of possible confounding factors must be taken into account. The most obvious is *observer effect* (see, for example, Sugarman and Hacker, 1980). Lizard species vary in their sensitivity to human presence. Some, such as *Anolis polylepis*, will quickly become used to a human presence, even consuming mosquitoes feeding on the observer. Others, such as *Basiliscus vittatus*, require a lengthy recovery period and observation by using binoculars from tens of meters away (Perry, 1995). When working at night with artificial lighting, care must be taken to minimize the impact on animals (for example, by using red lights that do not appear to modify animal behavior) (see, for example, Werner *et al.*, 1997). Prior knowledge of the biology of the species to be studied is therefore essential, so that sensitivity to presence of researchers can be taken into account. The same term, observer effect, is sometimes used to denote a radically different methodological impact: the differences in data reported by different researchers observing the same phenomenon. Although it is not possible to completely eliminate such differences, the goal of this section is to provide a standardized methodology that can be used to reduce interobserver variation.

Another issue in field studies is ensuring that data points are truly independent by not observing a single animal more than once. Authors have generally addressed this issue by changing locations following each observation (see, for example, Perry, 1996).

The biggest problem in quantifying foraging behavior, however, is how to ensure that only foraging behaviors are being observed. Animals do not often

Table 1.1 *Information on lizard foraging behaviors published since the compilation of Perry (1999)*

New data for species covered previously are not included. Values are rounded. Data cited as "present study" were collected on the Pacific island of Guam.

Species	Moves per minute		Percent time moving		Source
	mean	SD	mean	SD	
Agamidae					
Agama aculeata	0.2	0.26	0.0	0.00	Cooper *et al.*, 1999
Agama anchietae	0.0	—	0.0	—	Cooper *et al.*, 1999
Agama atra	0.3	0.39	1.0	0.00	Cooper *et al.*, 1999
Agama aculeata	0.6	0.68	2.0	0.00	Cooper *et al.*, 1999
Cordylidae					
Cordylosaurus subtessellatus	1.2	0.28	47.4	11.40	Cooper *et al.*, 1997
Cordylus anguina	0.4	0.21	1.9	0.93	du Toit *et al.*, 2002
Cordylus cataphractus	0.2	0.31	2.2	2.75	Mouton *et al.*, 2000
Cordylus polyzonus	0.04	0.08	0.2	0.29	Cooper *et al.*, 1997
Cordylus cordylus	0.09	0.13	0.3	0.54	Cooper *et al.*, 1997
Gerrhosaurus validus	0.7	0.24	14.9	16.48	Cooper *et al.*, 1997
Platysaurus broadleyi	1.2	—	7.9	—	Greeff and Whiting, 2000
Platysaurus capensis	1.3	1.50	6.6	6.41	Cooper *et al.*, 1997
Pseudocordylus capensis	0.6	0.49	6.8	4.90	Cooper *et al.*, 1997
Crotaphytidae					
Crotaphytus reticulatus	0.2	0.05	0.01	0.003	Husak and Ackland, 2003
Gekkonidae					
Cnemaspis kendallii	0.02	0.04	0.3	0.62	Werner and Chou, 2002
Lepidodactylus lugubris	1.1	1.05	2.0	1.99	Present study
Pachydactylus turneri	0.2	0.23	0.0	0.00	Cooper *et al.*, 1999
Phyllodactylus porphyreus	0.04	0.09	1.0	2.24	Cooper *et al.*, 1999
Rhotropus afer	0.0	0.00	0.0	0.00	Cooper *et al.*, 1999
Rhotropus barnardi	0.3	0.38	1.0	0.00	Cooper *et al.*, 1999
Rhotropus boultoni	0.3	0.34	1.0	0.00	Cooper *et al.*, 1999
Lacertidae					
Meroles ctenodactylus	3.2	1.34	29.0	13.42	Cooper and Whiting, 1999
Meroles knoxii	0.61	0.57	7.0	10.39	Cooper and Whiting, 1999

Table 1.1 *(cont.)*

Species	Moves per minute		Percent time moving		Source
	mean	SD	mean	SD	
Meroles reticulatus	0.05	0.10	0.0	0.00	Cooper and Whiting, 1999
Pedioplanis undata	1.4	0.68	50.0	20.00	Cooper and Whiting, 1999
Opluridae					
Oplurus cuvieri	0.1	—	0.3	—	Mori and Randriamahazo, 2002
Phrynosomatidae					
Sceloporus clarkii	0.2	0.29	0.8	1.34	Cooper *et al.*, 2001
Sceloporus jarrovi	0.3	0.45	0.9	6.78	Cooper *et al.*, 2001
Sceloporus magister	0.05	0.10	0.2	0.40	Cooper *et al.*, 2001
Sceloporus virgatus	0.4	0.97	0.8	1.10	Cooper *et al.*, 2001
Uta stansburiana	0.2	0.31	0.6	0.77	Cooper *et al.*, 2001
Polychrotidae					
Anolis angusticep	0.7	—	2.1	—	Irschick, 2000
Anolis evermanni	0.6	—	0.7	—	Irschick, 2000
Anolis grahami	0.8	—	1.6	—	Irschick, 2000
Anolis gundlachi	0.4	—	0.5	—	Irschick, 2000
Anolis lineatopus	0.4	—	0.5	—	Irschick, 2000
Anolis sagrei	0.3	—	0.6	—	Irschick, 2000
Anolis valencienni	0.8	—	7.2	—	Irschick, 2000
Scincidae					
Carlia fusca	2.0	1.30	17.7	17.86	Present study
Eumeces laticeps	0.6	0.30	72.4	27.00	Cooper *et al.*, 2001
Lamprolepis smaragdina	1.5	1.84	11.6	15.09	Perry and Buden, 1999
Mabuya margaritifer	1.2	0.81	8.4	7.31	Wymann and Whiting, 2002
Mabuya spilogaster	0.3	0.36	2.9	19.90	Cooper and Whiting, 2000
Mabuya striata	1.7	0.67	41.4	1.70	Cooper and Whiting, 2000
Mabuya variegata	1.2	0.88	28.8	25.60	Cooper and Whiting, 2000
Oligosoma grande	1.5	0.77	4.6	3.99	Eifler and Eifler, 1999
Teiidae					
Cnemidophorus flagellicaudus	0.7	0.13	80.7	15.13	Cooper *et al.*, 2001

Table 1.1 *(cont.)*

Species	Moves per minute		Percent time moving		Source
	mean	SD	mean	SD	
Cnemidophorus sonorae	0.8	0.35	94.7	3.60	Cooper *et al.*, 2001
Cnemidophorus uniparens	0.8	0.46	78.7	17.23	Cooper *et al.*, 2001
	—	—	36.7	—	Eifler and Eifler, 1998
Tropiduridae					
Tropidurus azureus	—	—	34.6	19.7	Ellinger *et al.*, 2001

engage in a single behavior at a time. Rather, they multitask, simultaneously combining activities such as searching for food, avoiding predators, maintaining thermal equilibrium, and patrolling a home range or territory. Various authors have attempted to minimize the impact of non-foraging behaviors in various ways. One major concern is *thermoregulation*, as ambient temperature affects all aspect of reptile behavior and physiology. Some authors (e.g. Werner *et al.*, 1997) have addressed this issue by directly measuring air temperature. Others limited observation to periods when the species is known to be active (see, for example, Cooper and Whiting, 2000) or to "peak activity hours" (Perry *et al.*, 1990). Thus, a diurnal species would not be observed at night, or during the early or late parts of the day when its activities may be highly influenced by behaviors associated with gearing up for the day or winding down for the night. Animals engaged in behaviors clearly associated with thermoregulation, such as assuming a flattened pose facing the sun, are not included in such studies for this reason. *Social behaviors* are another major concern, especially for highly territorial species such as anoles. Researchers therefore generally omit from analyses observations in which clear evidence of social interaction, such as copulation or fighting, can be found (see, for example, Perry, 1996). Despite these preventative measures, Perry (1999) cautioned that the use of the term "time allocation" might be a term preferable to "foraging behavior."

Testing theoretical predictions

Both theoretical and empirical approaches have had their successes in analyzing and predicting the behaviors of free-ranging organisms. Unfortunately, an alarmingly wide gulf has opened between proponents of these two approaches in ecology (Kareiva, 1989; Lawton, 1991; Weiner, 1995). This is also true for studies of foraging. As a consequence, many assumptions and predictions of foraging theory have not been adequately tested (see, for example, Perry and Pianka, 1997). Like all models, those at the heart of foraging theory are by definition simplifications and therefore less than perfect representations of the real world (Levins, 1966; Pease and Bull, 1992). Testing of models, preferably by comparing their predictions to empirical data, is therefore critical (MacArthur and Pianka, 1966; Bernardo, 1993; Orzack and Sober, 1994). We thus have very sophisticated foraging models at our disposal, but their assumptions are often based on the output of previous (often untested) models, rather than on empirical data. At the same time, processes suggested by field studies have not all been incorporated into the theoretical framework. The second section of this chapter begins addressing this gap between extensive theory and more limited empirical testing.

Is lizard foraging behavior interspecifically bimodal?

Theoretical work and early studies suggest that foraging behavior should be bimodal (reviewed in Perry, 1999). Starting with the seminal work of Pianka (1966), these qualitative "modes" have often been called "sit-and-wait" (SW) and "widely foraging" (WF). To date, the only one suggesting a numerical cut-off separating the two was Cooper (this volume, Chapter 8), who defined "ambush foragers to be those that spend less than 10% of the time moving during foraging hours." (However, the attribution of this criterion to Perry (1995) was incorrect (W. Cooper, personal communication).) The dichotomous view of SW–WF led McLaughlin (1989) to formulate the "syndrome hypothesis," which suggests that the prey capture strategies are part of a suite of biological differences. This "behavioral syndrome" (Sih et al., 2004) is the main focus of this book.

   The issue of the putative bimodality of foraging behaviors has become one of the most troublesome issues in the study of foraging behavior, and has been widely debated (Regal, 1978; McLaughlin, 1989; Perry et al., 1990; Pianka, 1993; Cooper, 1994; Perry, 1999). Huey and Pianka (1981) were the first to suggest that phylogeny might be a confounding factor in traditional analyses of foraging behavior, a concern later shared by McLaughlin (1989) and confirmed by Cooper (1994, 1995), Schwenk (1995), and Perry (1999). Phylogenetic analyses of qualitative data show family-level conservatism in lizard search modes (Cooper, 1994, 1995; Schwenk, 1995); a quantitative data show the same trend (Perry, 1999). Moreover, the importance of phylogeny as a predictive factor in analyses of foraging behavior is not unique to lizards, but rather appears to be a general trend (Perry and Pianka, 1997). The most comprehensive analysis of the importance of phylogeny in determining lizard foraging behavior was conducted by Perry (1999), who concluded that iguanians were broadly characterized by a sedentary lifestyle, whereas autarchoglossans exhibited a wide range of foraging behaviors, including some highly active phenotypes, and that no bimodality can be shown.

Quite a few additional data have been collected since the study of Perry (1999), many of them by Cooper and his colleagues (Table 1.1). Most studies focused on a single species, and thus have little to say about the issue of bimodality. Two recent studies concluded that lizard foraging behaviors were at least locally bimodal, although the species included might none the less represent opposite ends of a continuum of foraging behaviors (Cooper and Whiting, 2000; Cooper *et al.*, 2001). However, when additional data on the same genus are added from other locations (Perry, 1995, 1999; Wymann and Whiting, 2002), no indication of bimodality, or of a phylogenetic trend within the genus *Mabuya*, is apparent (Fig. 1.3). As Cooper and Whiting (2000) state, the wide range of foraging behaviors displayed by *Mabuya* affords an excellent model system in which to test hypotheses about putative correlates.

To reassess the conclusion of Perry (1999), I added all new data (Table 1.1) to those tabulated by Perry (1999). The results (Figs. 1.4, 1.5) do not indicate the existence of bimodality in either MPM or PTM (see Perry, 1999 for a discussion of the limitations of available statistical tests of bimodality). Within each major clade, the notion of bimodality in lizard foraging behaviors remains unsupported. In the same way that ecologists still discuss "*r*" and "*K*" strategists while acknowledging they represent the ends of a continuum, the terms "sit-and-wait" and "widely foraging" remain conceptually useful. At times, as when analyzing dichotomous variables (e.g. presence or absence of a tongue flicking in response to chemical cues), (Cooper, this volume, Chapter 8), these terms are also analytically convenient. However, they are far too ambiguous to be descriptively advantageous, especially for taxa showing intermediate behaviors. The challenge we are facing is thus not to seek a dichotomous set of correlates of foraging "modes." Rather, the challenge is a more complex and nuanced attempt to disentangle the effects of phylogeny and time allocation, and to address them in a quantitative manner (see, for example, Chapters 2, 4, 9, 10, and 15).

Population differences in foraging behavior

Several studies (see, for example, Cooper, 1994; Perry, 1999) have documented the phylogenetically conservative nature of foraging behavior, suggesting a

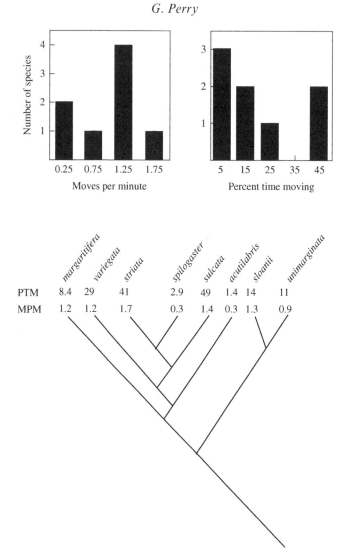

Figure 1.3. Phylogeny and foraging behavior of skinks in the genus *Mabuya*. Values on the x axis of both bar graphs are medians for each category. Neither the number of moves per minute nor the percent of time spent moving show a clear bimodal pattern (upper panels), nor is a clear phylogenetic pattern apparent (lower panel). The phylogeny is based on Mausfeld *et al.* (2000) and Honda *et al.* (2003), and the data for the eight species are taken from Perry (1999), Cooper and Whiting (2000) and Wymann and Whiting (2002).

large genetic component in determining the overall way in which organisms search for food. In contrast, the classic view espoused in theoretical, optimality-based investigations is that such behaviors are highly labile and context-specific (see, for example, Krebs and Kacelnik, 1991; Stephens and Krebs, 1986). For example, Pereira *et al.* (2003) suggest that competition among

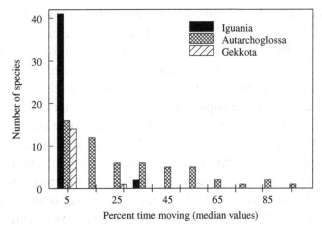

Figure 1.4. The distribution of percent time moving values in 112 lizard species. Data are divided by major clade, and do not indicate the existence of bimodality in any of them. See Table 1.1 for sources of data.

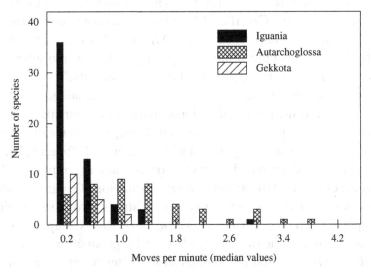

Figure 1.5. The distribution of moves per minute values in 116 lizard species. Data are divided by major clade, and do not indicate the existence of bimodality in any of them. See Table 1.1 for sources of data.

foragers for limited food resources should lead to intrapopulational variability in home range size as a function of prey patchiness and unequal fighting abilities. To some extent, the two views refer to different organizational levels: phylogenetic tests most often contrast taxa, whereas optimality analyses typically focus on within-species comparisons. None the less, they offer an apparent contradiction for two reasons. First, phylogenetic analyses of ecologically relevant traits can be used intraspecifically (see, for example, Wiens *et al.*,

1999). Second, populations and species are not dichotomously different, but rather part of a continuum of increasing reproductive isolation and genetic differentiation. It therefore appears unlikely (but not impossible) that the high degree of flexibility implied by a highly labile intraspecific trait will show strong conservatism when viewed at the interspecific level.

One way to approach the apparent dichotomy between views of foraging as highly labile on the one hand and highly conservative on the other is to compare the foraging behaviors of multiple populations within a single species, taking into account differences in ambient conditions (and, ideally, genetics) among them. Parameters such as food availability and population density are likely to have important impacts on time allocation. Two studies (Perry and Buden, 1999; Greeff and Whiting, 2000), have attempted to do so. Perry and Buden (1999) found no differences among three populations of the Pacific skink *Lamprolepis smaragdina*, one natural and two probably recently introduced, but acknowledged that the time since the populations were genetically separated "was too short for behavioral differences to evolve due to genetic isolation alone." Greeff and Whiting (2000) compared three populations of *Platysaurus broadleyi* in South Africa and found that "males and females feeding on figs moved more frequently and spent more time moving than insect-feeding lizards, regardless of the insect density."

Both studies described above were carried out on a small geographical and temporal scale, and neither involves large numbers of animals. A better test would involve populations separated for much longer, and ideally comprising individuals exposed to a variety of ambient conditions. If foraging behavior is strongly genetically determined, such environmental and temporal differences could drive the evolution of variable foraging behaviors. Geographically widespread species experience a wide range of conditions throughout their range. For example, the teiid lizard *Cnemidophorus tigris* shows adaptive differences in thermoregulation, time of activity, diet, and reproduction (Pianka, 1970), and *Anolis cristatellus* from dry islands show better water retention abilities than do those from relatively wet islands (Dmi'el *et al.*, 1997; Perry *et al.*, 1999, 2000). I have collected two datasets on widespread taxa experiencing a variety of ambient conditions.

To test the hypothesis that foraging behavior will be intraspecifically labile, I studied the foraging behavior of multiple populations of *Anolis cristatellus* in the British Virgin Islands from 1993 to 1996. The resulting database encompasses 630 individuals from 16 islands. To assess the overall importance of year and island identity on MPM and PTM values, I used a two-way ANOVA. Although some differences are immediately apparent among islands in average values of MPM and PTM, variability is high at all of them (Fig. 1.6).

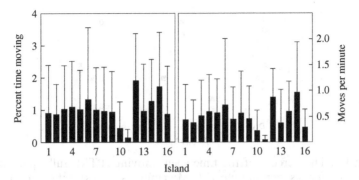

Figure 1.6. Foraging indices of 630 *Anolis cristatellus* from 16 island populations in the British Virgin Islands. Data shown are the mean and standard deviation for each, based on data collected between 1993 and 1996. Sample sizes are given in parentheses in the key, below. There were no significant differences among islands or study years. 1, Guana (230); 2, Anegada (20); 3, Tortola bridge (49); 4, Sage Mountain (54); 5, Virgin Gorda (28); 6, Necker (52); 7, Beef (34); 8, Peter (26); 9, Norman (44); 10, Little Thatch (17); 11, Great Camanoe (6); 12, Little Jost (10); 13, Big Jost (14); 14, Fallen Jerusalem (19); 15, George Dog (13); 16, Great Dog (14).

Consequently, neither the effects of year nor those of island were statistically significant for either foraging index ($p > 0.05$ in all four cases). Thus, I found no support for the expected pattern of behavioral flexibility among islands.

The test above is arguably weak, because it does not explicitly take into account island characteristics that might affect foraging behavior. However, data for 543 of the 630 individuals (86%) originated were collected on ten islands for which aridity data were available from related work (Dmi'el *et al.*, 1997; Perry *et al.*, 1999, 2000). I could therefore also test the hypothesis that populations from wetter islands would show behaviors different from those on dry islands, because of differences in food availability. Figure 1.7 presents the observations of 206 individuals observed in 1993, a pattern similar to that seen in other years and overall. Correlation analyses of the full dataset did not reveal any significant relation between island aridity and either MPM or PTM in this much extended database (MPM: $\rho = 0.03$, $p = 0.54$; PTM: $\rho = 0.004$, $p = 0.92$). Thus, I found no support for the expected pattern of behavioral flexibility (whether short-term or evolutionary) mirroring the physiological plasticity shown previously at both levels (Dmi'el *et al.*, 1997; Perry *et al.*, 1999, 2000). An even stronger test would involve an explicitly phylogenetic comparison of these populations. Unfortunately, no intraspecific phylogeny is available for this species.

This work shows that *Anolis cristatellus* shows little sign of behavioral flexibility in foraging mode. A more limited study of *A. carolinensis* similarly

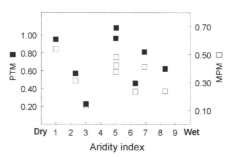

Figure 1.7. The percent of the time spent moving (PTM, full squares) and number of moves per minute (MPM, empty squares) exhibited by nine populations of *Anolis cristatellus* from the British Virgin Islands. The aridity index represents a range from relatively dry to relatively wet locations. Although the amount of food available is expected to be greater in wet locales, lizards do not show any consistent patterns in either index of foraging behavior. See Dmi'el *et al.* (1997) for details on the aridity index, the localities studied, and the lizard populations at each.

shows no differences among four populations introduced to the tropical environments of Micronesia and Hawaii and those native to Texas (G. Perry, unpublished). The inflexible nature of foraging behavior in *Anolis* is especially surprising, given that the same populations of *A. cristatellus* studied here show both annual and island differences in water loss rates, and that both phenotypic plasticity and genetic differentiation appear to be responsible (Dmi'el *et al.*, 1997; Perry *et al.*, 1999, 2000). Such physiological plasticity appears limited, with some non-plastic genetic components differing among populations (Perry *et al.*, 2000). None the less, it is curious that a physiological trait shows more plasticity than does foraging behavior (this study) or reproductive mode (Fitch, 1970), both of which are directly related to reproductive fitness yet show little or no variation as conditions change. However, these findings are consistent with the view of lizard behavior as highly genetically constrained, presented by Cooper (1994) and Perry (1999).

In a second test of the same hypothesis, I studied multiple populations from two species of *Sceloporus*, *S. graciosus* (155 individuals, 7 populations) and *S. occidentalis* (213 individuals, 9 populations), at altitudes ranging from 100 to 2500 m in California and Nevada (Fig. 1.8). Altitudinal gradients offer a natural experiment by providing divergent conditions at relatively close geographical proximity. At high elevations, activity seasons are relatively short, a phenomenon especially relevant to ectotherms such as lizards. Selection should favor the appearance of local behavioral phenotypes that help compensate for this. I therefore hypothesized that high-altitude lizards would

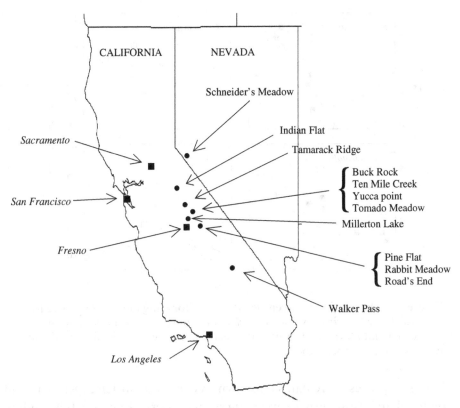

Figure 1.8. A map of California and Nevada, showing the study sites (circles) used for documenting the effect of elevation on the foraging behaviors of *Sceloporus graciosus* and *S. occidentalis*. Several major cities (squares) are added to aid in placing these locations.

forage more actively, as this would allow them to increase food intake rate and partly offset the effects of altitude. Populations from higher altitudes were generally more active (Fig. 1.9). I used a parametric ANCOVA to test the relation between foraging behaviors and altitude within each species and to compare the two species' response. I used each individual's foraging behavior as the response variable, species as the factor, and altitude as the covariate. The analysis showed a significant altitudinal effect for both MPM ($F_{1,365} = 10.84$, $p = 0.001$) and PTM ($F_{1,365} = 9.10$, $p = 0.003$) that is independent of taxon identity. The two species differed significantly in altitude-corrected MPM ($F_{1,365} = 7.50$, $p = 0.006$), but there was no species effect in PTM ($F_{1,365} = 1.10$, $p = 0.294$).

Thus, both species of *Sceloporus* show interpopulation differences in their foraging behaviors. Whether these changes occur in the short term, as individuals modify their behavior, or are a long-term result of population-level

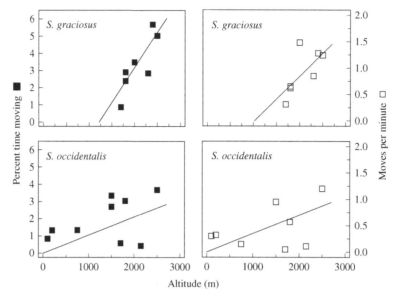

Figure 1.9. The relation between elevation and foraging indices is positive for both *Sceloporus graciosus* and *S. occidentalis* observed in the mountains of California and Nevada. However, the relation is only statistically significant in *S. graciosus* (upper panels).

evolutionary processes, remains unknown. A common-garden experiment will be required to distinguish between these processes. However, this finding stands in contrast to those described above for *Anolis cristatellus* and the less well-studied *A. carolinensis*. It is therefore clear that some species have the ability to modify their foraging behaviors in accordance with environmental conditions, as predicted by foraging theory (see, for example, Krebs and Kacelnik, 1991; Stephens and Krebs, 1986). The foraging behaviors of others, however, appear to be greatly constrained, as suggested by the phylogenetic conservatism identified by Cooper (1994) and Perry (1999).

Are predator and prey activity patterns negatively correlated?

What is the most profitable way of searching for a resource? This question has a critical importance in many biological contexts, such as finding mates or food (Bell, 1991). Despite this, it was first mathematically studied in the context of military operations (Koopman, 1956, in Gerritsen and Strickler, 1977). In biology, this issue was not quantitatively addressed until two decades later. Studying zooplankton, Gerritsen and Strickler (1977) asked what tactics would best allow predators to obtain nourishment. Two optimal strategies emerged from their theoretical analysis: a stationary predator feeding on roving prey, and a mobile predator that specialized on slow-moving prey. These tactics are equivalent to the SW and WF strategies of Pianka (1966). I have argued (Perry 1995, 1999, this chapter) that the distinction between WF and SW is artificial and often counterproductive. However, the terms remain useful relative descriptors, and their use here should be interpreted as denoting relative position along the continuum ranging from completely sedentary to always moving.

The idea that SW predators should mainly feed on widely foraging organisms and wide foragers should prey on stationary prey is here termed the predator–prey locomotor crossover hypothesis, or crossover hypothesis for short. It was popularized by Huey and Pianka (1981). Although it has been widely accepted and cited (see, for example, Perry *et al.*, 1990; Pianka, 1993; Cooper, 1994), this intuitive model has never been thoroughly tested. The accepted view suggests that the two strategies may have evolved to allow better handling of prey with bimodal movement patterns (Fitzpatrick, 1981; Huey & Pianka, 1981; Pianka, 1986; McLaughlin, 1989). Why prey should show a bimodal locomotor pattern is unclear, nor has it ever been demonstrated. In this section I test this hypothesis by using data on both the behavior of lizards, foraging for food, and that of their prey. To allow quantitative analysis I use quantitative data on lizard foraging behaviors and develop a novel technique for quantifying prey behavior.

I obtained data on lizard foraging from published studies, unpublished material provided by several authors, and original fieldwork (Table 1.2). Studies have shown that lizard diet can vary among sites and years (Perry and Brandeis, 1992). Consequently, I primarily used data originating in simultaneous studies of foraging behavior and diet. To increase sample size, I also incorporated separate datasets if they originated in the same geographic area and were collected during the same season(s). Published reports based on very small samples ($n < 10$) were omitted.

Although quantitative data on lizard foraging behavior have been available in the literature for over twenty years, no similar indices are available to describe the locomotor behavior of their prey. In the absence of even qualitative descriptions for most taxa, I devised a quantitative description to fill this gap. Rather than concentrate on individual prey species, the metric concentrates on the diet of the predator and describes the overall tendency of its prey to move.

The framework, whereby each prey movement pattern ranged from zero (completely stationary) to 6 (very mobile), was established with the help of P.J. DeVries (Director, Center for Biodiversity Studies, Milwaukee Public Museum). Dr DeVries and five additional entomologists (Drs L.E. Gilbert, R.W. Patrock, M.C. Singer, and E.L. Vargo, University of Texas at Austin; Dr. C.R. Nelson, Brigham Young University) were asked, based on their

Table 1.2 *Sources of data on lizard diet and foraging behavior*

Species	Source for	
	Diet	Foraging
Corytophanidae		
Corytophanes cristatus	Andrews, 1979a	Perry, 1999
Iguanidae		
Cyclura pinguis	Goodyear, unpubl.	Perry, 1999
Iguana iguana	Rand *et al.*, 1990	Perry, 1999
Lacertidae		
Acanthodactylus schreiberi	Avital, 1981	Perry *et al.*, 1990
Acanthodactylus scutellatus	Avital, 1981	Perry *et al.*, 1990
Heliobolus lugubris	Pianka *et al.*, 1979	Pianka *et al.*, 1979
Ichnotrophis squamulosa	Pianka *et al.*, 1979	Pianka *et al.*, 1979
Meroles suborbitalis	Pianka *et al.*, 1979	Pianka *et al.*, 1979
Mesalina guttulata	Orr *et al*, 1979	Perry *et al.*, 1990; Werner, unpubl.
Nucras tessellata	Pianka *et al.*, 1979	Pianka *et al.*, 1979
Pedioplanis lineoocellata	Pianka *et al.*, 1979	Pianka *et al.*, 1979
Pedioplanis namaquensis	Pianka *et al.*, 1979	Pianka *et al.*, 1979
Phrynosomatidae		
Phrynosoma mcallii	Turner and Medica, 1982	Muth, 1992
Phrynosoma modestum	Shaffer and Whitford, 1981	Shaffer and Whitford, 1981
Sceloporus graciosus	Perry, unpubl.	Perry, 1999
Uma inornata	Durtsche, 1992	Durtsche, unpubl.
Polychrotidae		
Anolis auratus	Magnusson and da Silva, 1993	Magnusson *et al.*, 1985
Anolis cristatellus	Perry, unpubl.	Perry, 1995
Anolis limifrons	Andrews, 1979b	Andrews, 1979b
Anolis oculatus	Andrews, 1979b	Andrews, 1979b
Anolis polylepis	Perry, 1996	Perry, 1996
Anolis punctatus	Gasnier *et al.*, 1994	Gasnier *et al.*, 1994
Anolis stratulus	Lister, 1981	Reagan, 1986
Scincidae		
Ctenotus taeniolatus	Taylor, 1986	Taylor, 1986
Mabuya acutilabris	Castanzo and Bauer, 1993	Castanzo and Bauer, 1993
Tiliqua rugosa	Dubas and Bull, 1991	Dubas, 1987
Teiidae		
Ameiva ameiva	Magnusson and da Silva, 1993	Magnusson *et al.*, 1985
Ameiva exsul	Lewis, 1989	Lewis and Saliva, 1987
Ameiva quadrilineata	Hirth, 1963	Perry, 1999

Table 1.2 *(cont.)*

Species	Source for	
	Diet	Foraging
Cnemidophorus deppii	Vitt *et al.*, 1993	Vitt *et al.*, 1993
Cnemidophorus lemniscatus	Magnusson and da Silva, 1993	Magnusson *et al.*, 1985
Cnemidophorus sexlineatus	Paulissen, 1987	Paulissen, 1987
Cnemidophorus tigris	Anderson, 1993	Anderson, 1993
Kentropyx calcarata	Gasnier *et al.*, 1994	Gasnier *et al.*, 1994
Kentropyx striatus	Magnusson and da Silva, 1993	Magnusson *et al.*, 1985
Tropiduridae		
Plica umbra	Gasnier *et al.*, 1994	Gasnier *et al.*, 1994
Uranoscodon superciliosa	Gasnier *et al.*, 1994	Gasnier *et al.*, 1994

familiarity with the prey taxa, to independently assign each prey type a rank based on this system. Ranks were based on each prey species' locomotor behavior during the time it is normally eaten by the predators studied. Thus scorpions, highly mobile prey when active but motionless during the day, were assigned a low movement rank if eaten by a diurnal predator. An additional factor taken into account was the amount of searching necessary to locate prey. *Messor* ants, which are relatively mobile, were nevertheless assigned a low rank. This was done because they forage in large groups on regular paths; once an ant trail is located, a predator need not move to obtain many additional prey items. The mobility values assigned by the six entomologists were highly similar and strongly correlated with those initially assigned by P. J. DeVries and myself (range for individual entomologists: $r^2 = 0.63$–0.88). Because this metric is somewhat subjective, it adds "statistical noise" to the analyses that follow. However, there are two important reasons to use it. First, no better index exists, and thus no other method is available for even beginning to address the current question. Second, the high degree of similarity between ranks independently assigned by individual entomologists, and the wide geographical experience represented by these experts, suggests that these average values provide a biologically meaningful, if preliminary and imprecise, description of prey movement. So although these assessments are subjective and imprecise, they offer a first approximation for quantitative tests of the crossover hypothesis. Since many taxa are encompassed and entomologists appear to broadly agree on their relative rankings, occasional deviations should not be large enough to overwhelm major effects.

Table 1.3 *Mobility ranking of prey found in diets of lizard species listed in Table 1.2.*

See text for methods used to produce these values. Invertebrate nomenclature follows Borror *et al.* (1989).

Rank	Taxa			
0	Gastropoda	plants	pellets	
	shed skins	eggs	cocoons	
1	Diplopoda	Phasmida	Thysanura	all larvae
2	Apterygota	Chilopoda	Collembola	Dermaptera
	Embioptera	Isopoda	Isoptera	Myriapoda
	Scorpiones	Thysanoptera	Zoraptera	Annelida
	slow Araneae (fam. Clubionidae, Gnaphosidae)			
	other Arachnida (not Araneae or Scorpiones)			
3	Neuroptera			
	slow Formicidae (*Messor*)			
	slow Coleoptera (fam. Chrysomelidae, Cuculionidae, Pselaphidae)			
	slow Hemiptera (fam. Aphidae, Miridae, Lygiidae)			
3.5	Solifugae unidentified Araneae, Hemiptera, Homoptera			
4	Orthoptera			
	medium Formicidae (*Camponotus, Leptothorax, Paratichina, Phidole, Plagiolaepus*)			
	fast Hemiptera (fam. Cicadelidae, Membracidae, Reduviidae)			
	unidentified Formicidae and Coleoptera			
5	Amphipoda Blattidae Decapoda Mantodea			
	fast Formicidae (*Cataglyphis, Chromatogaster*)			
	fast Coleoptera (fam. Carabidae, Scarabaeidae, Staphilinidae, Tenebrionidae)			
	fast Araneae (fam. Lycosidae, Oxyopidae, Salticidae)			
6	flying organisms: Diptera, Hymenoptera (except non-winged Formicidae), Lepidoptera, Odonata, winged Isoptera			

Dietary data were obtained from the literature (Table 1.2). Prey composition is sometimes reported by volume and sometimes by numbers. I used either in this analysis, but preferred the more commonly provided percentage of total volume when both were reported. Unfortunately, using both measures is likely to add some noise to the analysis. I used the average of all movement values assigned a particular taxon and termed it the movement index, M_i (Table 1.3).

I then used prey composition and movement index for each prey type to calculate a single value (termed H_j) representing the weighted average of the locomotor tendencies of all the items in the diet of each lizard species. This value can be mathematically expressed as

$$H_j = \Sigma M_i F_i \qquad \text{(Eq. 1.1)}$$

Table 1.4 *Food movement index* (H$_j$) *and average foraging parameters of 37 lizard species*

MPM, moves per minute; PTM, percent time moving. See Table 1.2 for data sources and Table 1.3 for movement indices of individual prey items.

Species	MPM	PTM	H$_j$
Corytophanidae			
Corytophanes cristatus	0.01	0.5	240
Iguanidae			
Cyclura pinguis	0.20	5.1	3
Iguana iguana	0.40	2.8	0
Lacertidae			
Acanthodactylus schreiberi	1.54	30.5	340
Acanthodactylus scutellatus	1.01	7.7	327
Heliobolus lugubris	2.97	57.4	255
Ichnotrophis squamulosa	3.10	54.6	267
Meroles suborbitalis	1.83	13.5	293
Mesalina guttulata	0.46	35.1	307
Nucras tessellata	2.90	50.2	270
Pedioplanis lineoocellata	1.54	14.3	304
Pedioplanis namaquensis	2.78	53.5	267
Phrynosomatidae			
Phrynosoma mcallii	—	32.0	359
Phrynosoma modestum	0.12	—	359
Sceloporus graciosus	1.31	5.8	354
Uma inornata (male)	—	2.4	169
(female)	—	1.9	201
Polychrotidae			
Anolis auratus	0.01	—	340
Anolis cristatellus (male)	0.43	1.5	382
(female)	0.45	0.9	353
(juvenile)	0.21	0.5	339
Anolis limifrons	0.09	—	352
Anolis oculatus	0.20	—	344
Anolis polylepis (male)	0.24	1.0	320
(female)	0.42	1.1	362
(juvenile)	0.43	1.5	356
Anolis punctatus	0.14	—	370
Anolis stratulus (wet season)	0.72	—	318
(dry season)	1.16	—	328
Scincidae			
Ctenotus taeniolatus	—	79.0	300
Mabuya acutilabris	0.29	—	323
Tiliqua rugosa	—	10.4	0

Table 1.4 (cont.)

Species	MPM	PTM	H_j
Teiidae			
Ameiva ameiva	0.64	26.7	292
Ameiva exsul	—	41.2	205
Ameiva quadrilineata	1.26	18.0	359
Cnemidophorus deppii	—	62.8	342
Cnemidophorus lemniscatus	0.51	—	315
Cnemidophorus sexlineatus (adult)	—	47.9	348
(juvenile)	—	67.7	346
Cnemidophorus tigris	1.62	87.0	239
Kentropyx calcarata	0.86	—	383
Kentropyx striatus	0.24	—	296
Tropiduridae			
Plica umbra	0.19	—	372
Uranoscodon superciliosa	0.02	—	329

where M_i is the movement index and F_i is the percentage of each prey type in the diet of predator species j. The summation is carried out over all prey types in the diet of the predator. Calculated values of H_j ranged from zero in several strictly herbivorous species to 372 in *Plica umbra* (Table 1.4). Except in the herbivorous species, however, H_j rarely exceeded 400 or fell below 200. Thus, no predaceous species, active or sedentary, specialized on truly sedentary or highly mobile prey.

I used version 5.0 of the PDTREE program (Garland *et al.*, 1993) to analyze these data, employing the composite phylogeny presented by Perry and Garland (2002), with additional information from Reeder *et al.* (2002) on *Cnemidophorus* and Reeder (2003) on skinks (Fig. 1.10). PDTREE calculates independent contrasts (Felsenstein, 1985) in each trait, thus removing phylogenetic pseudoreplication (Perry and Pianka, 1997). Because branch lengths were not known, I used Grafen's (1989) arbitrary branch lengths after verifying they were appropriate using the diagnostics provided by Garland *et al.* (1992). Following Perry (1995), herbivorous species were included in the dataset. Because some authors excluded herbivorous species from past analyses, calculations were repeated both with and without them.

In Fig. 1.11 I plot the contrasts in predator and prey locomotor behaviors against one another, including all taxa. Data were analyzed by using Pearson parametric regressions calculated through the origin. No significant relation was found when considering MPM ($n = 34$ contrasts, $r = 0.034$, $p = 0.29$). The relation between contrasts in H_j and PTM was significant ($n = 30$, $r = 0.192$,

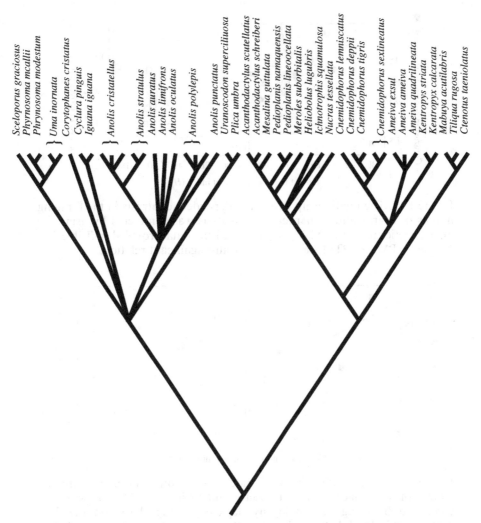

Figure 1.10. A composite phylogeny for lizard species included in the study of the relationship between predator and prey foraging behaviors. Slightly different subsets of this phylogeny were used for analyzing moves per minute and percent time moving, as some species only had data for one index or the other. See text for phylogenetic sources.

$p = 0.014$), but removal of a single, extreme value (Fig. 1.11) changed the outcome markedly ($n = 29$, $r = 0.039$, $p = 0.296$). Similar results were obtained when the analyses were repeated without herbivorous species. I also repeated the analysis while taking into account the possibility that one of the two major clades represented might show a difference whereas the other did not. However, major clade was not a significant factor in either MPM or PTM (ANCOVA, $p > 0.05$ in both cases). Thus, a broad analysis across Lacertilia shows little

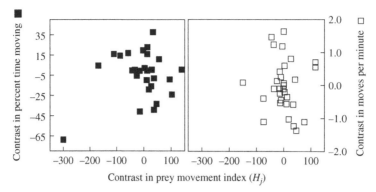

Figure 1.11. The relationship between independent contrasts in the foraging indices of lizards and their prey. The analysis was conducted by using the phylogenetic tree depicted in Fig. 1.10 and the data provided in Table 1.4. Neither MPM nor PTM shows a statistically significant relation.

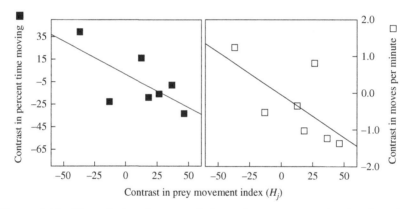

Figure 1.12. The relation between independent contrasts in the foraging indices of seven lacertid lizard species and their prey. The analysis was conducted by using the phylogeny depicted in Fig. 1.10 and the data provided in Table 1.4. As predicted by the crossover hypothesis, both MPM and PTM were negatively correlated with prey activity measured by H_j.

support for the crossover hypothesis: sedentary predators are not more likely to feed on active prey than are active predators. This is, perhaps, not surprising: ambushers will presumably feed on all available prey, including some that are relatively inactive, and active predators will likewise also sample both active and less active prey found as they forage.

Although I found no overall support for the predator–prey locomotor crossover hypothesis, I repeated the analysis for the family Lacertidae (Fig. 1.12). The crossover hypothesis was originally proposed for this group (Huey and Pianka, 1981), which was also the only family with a relatively large number of taxa ($n = 7$ species) and a relatively well resolved phylogenetic tree.

Because of the small sample size and the directional prediction, I used 1-tailed tests (Pearson parametric correlations). As required for phylogenetic analyses, all regressions were computed through the origin.

The relationship within Lacertidae was, as predicted by Huey and Pianka (1981), significantly negative (MPM: $r = -0.664$, $p = 0.05$; PTM: $r = -0.680$, $p = 0.05$). Thus, the situation within Lacertidae was markedly different from that for lizards as a whole: as they become more active, lacertids are indeed increasingly more likely to feed on inactive prey. This disparity suggests that different constraints and mechanisms affect foraging behaviors within different families, and that the overall variance is great enough to swamp the pattern showed by the Lacertidae. This is consistent with the findings of Reilly and McBrayer (Chapter 10), which show a range of morphological adaptations for prey capture within this clade. Unfortunately, either a paucity of data or a poorly resolved tree prevents a similar analysis from being carried out in other clades. Such analyses, which will help clarify whether other lizards show similar flexibility in their foraging behaviors, must await collection of further ecological and phylogenetic data.

Conclusions

This chapter has two aims. In the first section I address a number of methodological issues related to the measurement of lizard foraging behavior / time allocation. Analyses of several large datasets, albeit in a small number of species, indicate that:

(1) For non-gekkonid lizards, observations of a relatively small number of individuals ($n \geq 15$) may provide a good approximation of mean values obtained from larger samples (Fig. 1.1). However, gecko behavior is likely to be more variable in the short term, requiring a much larger sample size ($n \geq 50$). This generalization needs to be tested based on a larger sample size from each clade. In addition, the relation should be evaluated on a species-by-species basis in order to determine appropriate sample sizes for future studies.

(2) In all taxa for which data are available, observations lasting five minutes or less provide estimators of time allocation that have higher variance than those from longer observations (Fig. 1.2). I recommend that future studies use observation periods of ten minutes or more per individual whenever possible.

(3) Quantitative metrics are superior to qualitative ones for describing lizard foraging behaviors. The number of moves per minute and the percent of time spent moving are both useful quantitative metrics that have often been reported. Additional metrics, such as PAM, have been proposed, but the available data are too few to evaluate their biological usefulness.

The second section of this chapter details three tests of hypotheses proposed over the years in the context of lizard foraging behaviors, their distributions, and their putative correlates. Analyses carried out in this section show that:

(1) Lizard foraging behavior shows no obvious sign of bimodality, once the dichotomy between Iguania and Autarchoglossa is accounted for (Figs. 1.3–1.5).
(2) Whereas some taxa appear incapable of modifying their foraging behaviors in response to immediate conditions (*Anolis*, Figs. 1.6, 1.7), others show more behavioral plasticity (*Sceloporus*, Figs. 1.8, 1.9).
(3) Phylogenetic analyses indicate that, although there is no support for the predator–prey crossover hypothesis when all data are analyzed, the family Lacertidae does indeed demonstrate the negative relation predicted by foraging theory (Figs. 1.10–1.12).

These findings suggest a complex situation in which phylogeny is a crucial element. Some genera and even families show little variation in their foraging behaviors. For example, *Anolis* species show little variation in their foraging behaviors at both the interspecific (see, for example, Perry, 1999; Irschick, 2000, present study) and the intraspecific (present study) level. *Anolis* are also highly conservative in their reproductive strategy (all species laying a single egg per clutch: Fitch, 1970) and locomotor performance (Irschick, 2000). Other taxa, such as members of the genus *Sceloporus*, show considerably more variation in both foraging behavior and its correlates. Interestingly, phrynosomatid lizards are also rather variable in reproductive output (Fitch, 1970) and locomotor physiology (Bonine and Garland, 1999). Data are too sketchy, at this point, to evaluate the generality of this observation.

Despite the great increase in recent research on this topic, the number of species studied thus far remains a small fraction of extant taxa, and many groups remain totally unstudied. Data are still completely lacking for anguid lizards and snakes, for example. This, makes drawing broad conclusions difficult, but I believe that a pattern may be emerging. Some clades, especially within the Iguania and exemplified by *Anolis*, are highly conservative in multiple traits, including foraging behaviors. It is tempting to think of such groups as "primitive," retaining a broad array of ancestral characters and having relatively little flexibility. (The genus *Anolis* has none the less shown recurring, if somewhat repetitive and limited in scope, adaptive radiations in morphology [Losos *et al.*, 1998]. Despite its conservative ecology and behavior, *Anolis* has been greatly evolutionarily successful.) Other groups within the Iguania, such as phrynosomatids, show greater variability in morphology, physiology, ecology, and behavior. Nevertheless, they remain constrained in some ways, as evidenced by the similarity in one index of *Sceloporus* foraging behavior but not the other. In more recent (= derived) clades, and especially within the

Autarchoglossa in general and Lacertidae in particular, fewer and fewer traits appear to be thus constrained. This breaking up of the behavioral syndrome (sensu Sih *et al.*, 2004) allows for increasing diversity in form and function, as exemplified by the wide range of foraging behaviors and the support for the crossover hypothesis shown by this group (Perry, 1999, and this chapter). Thus, the process of evolution in lizard foraging behaviors and their correlates was probably not sudden and singular, but rather encompassed a number of stages during the evolutionary history of the lizard clade.

Acknowledgements

Data reported in this chapter have been collected over about 15 years with the involvement of many colleagues. I thank Y. L. Werner and E. R. Pianka for many discussions of lizard foraging behaviors. K. R. LeVering assisted with data collection and provided invaluable comments on this manuscript, as did W. E. Cooper. T. Garland, Jr., provided important advice on phylogenetic analyses. Work was conducted with permission from landowners and regulatory authorities, including Henry and Gloria Jarecki, the Schneider family, R. Branson, the Costa Rican Direccion General de Vida Silvestre and Servicio de Parques Nacionales, the US Forest Service, the US National Park Service, the US Fish and Wildlife Service, and the Texas Department of Parks and Wildlife. P. J. DeVries provided crucial assistance in devising the movement index for prey, H_j. Support for this work was provided by the Conservation Agency through multiple grants from the Falconwood Foundation, by the Lorraine I. Stengl Endowment to the Department of Zoology, University of Texas at Austin, by a Hartman Award, by a grant from the Institute of Latin American Studies at the University of Texas, by funds provided by the Department of Zoology, University of Texas at Austin, by a Michael Guyer postdoctoral fellowship at the University of Wisconsin, Madison, by NSF grant DEB-9307804 to M. C. Singer, and by NSF grants IBN-9723758 and DEB-9981967 to T. Garland, Jr. This is manuscript T-9-1008 of the College of Agricultural Sciences and Natural Resources, Texas Tech University.

References

Anderson, R. A. (1993). An analysis of foraging in the lizard, *Cnemidophorus tigris*. In *Biology of Whiptail Lizards (Genus Cnemidophorus)*, ed. J. W. Wright and L. J. Vitt, pp. 83–116. Norman, OK: Oklahoma Museum of Natural History.

Andrews, R. M. (1979a). The lizard *Corytophanes cristatus*: an extreme "sit-and-wait" predator. *Biotropica* **11**, 136–9.

Andrews, R. M. (1979b). Evolution of life histories: a comparison of *Anolis* lizards from matched island and mainland habitats. *Breviora* **454**, 1–51.

Avery, R. A., Mueller, C. F., Jones, S. M., Smith, J. A. and Bond, D. L. (1987). The movement patterns of lacertid lizards: a comparative study. *J. Herpetol.* **21**, 324–9.

Avital, E. (1981). Resource partitioning between two lizard species of the genus *Acanthodactylus* living in the same area of sands. M.Sc. thesis, Hebrew University of Jerusalem. (In Hebrew.)

Bell, W. J. (1991). *Searching Behaviour*. London: Chapman and Hall.

Bernardo, J. (1993). Determinants of maturation in animals. *Trends Ecol. Evol.* **8**, 166–73.

Bernardo, J. (1994). Experimental analysis of allocation in two divergent, natural salamander populations. *Am. Nat.* **143**, 14–38.

Biro, P. A. and Ridgway, M. S. (1995). Individual variation in foraging movement in a lake population of young-of-the-year brook charr (*Salvelinus fontinalis*). *Behaviour* **132**, 57–74.

Bonine, K. E. and Garland, T., Jr. (1999). Sprint performance of phrynosomatid lizards, measured on a high-speed treadmill, correlates with hindlimb length. *J. Zool. Lond.* **248**, 255–65.

Borror, D. J., Triplehorn, C. A. and Johnson, N. F. (1989). *An Introduction to the Study of Insects*. Philadelphia, PA: Saunders.

Castanzo, R. A. and Bauer, A. M. (1993). Diet and activity of *Mabuya acutilabris* (Reptilia: Scincidae) in Namibia. *Herpetol. J.* **3**, 130–5.

Cooper, W. E. Jr. (1994). Prey chemical discrimination, foraging mode, and phylogeny. In *Lizard Ecology: Historical and Experimental Perspectives*, ed. L. J. Vitt and E. R. Pianka, pp. 95–116. Princeton, NJ: Princeton University Press.

Cooper, W. E. Jr. (1995). Foraging mode, prey chemical discrimination, and phylogeny in lizards. *Anim. Behav.* **50**, 973–85.

Cooper, W. E. Jr. and Whiting, M. J. (1999). Foraging modes in lacertid lizards from southern Africa. *Amph.-Rept.* **20**, 299–311.

Cooper, W. E. Jr. and Whiting, M. J. (2000). Ambush and active foraging modes both occur in the scincid genus *Mabuya*. *Copeia* **2000**, 112–18.

Cooper, W. E. Jr., Vitt, L. J., Caldwell, J. P. and Fox, S. F. (2001). Foraging modes of some American lizards: relationships among measurement variables and discreteness of modes. *Herpetologica* **57**, 65–76.

Cooper, W. E. Jr., Whiting, M. J. and Van Wyk, J. H. (1997). Foraging modes of cordyliform lizards. *S. Afr. J. Zool.* **32**, 9–13.

Cooper, W. E. Jr., Whiting, M. J., Van Wyk, J. H. and Mouton, P. F. N. (1999). Movement- and attack-based indices of foraging mode and ambush foraging in some gekkonid and agamine lizards from southern Africa. *Amph.-Rept.* **20**, 391–9.

Darwin, C. (1859). *On the Origin of Species*. London: John Murray.

Dmi'el, R., Perry, G. and Lazell, J. (1997). Evaporative water loss in nine insular populations of the *Anolis cristatellus* group in the British Virgin Islands. *Biotropica* **29**, 111–16.

Dubas, G. (1987). Biotic determinants of home range size of the scincid lizard *Trachydosaurus rugosus* (Gray). Ph.D. thesis, Flinders University.

Dubas, G. and Bull, C. M. (1991). Diet choice and food availability in the omnivorous lizard, *Trachydosaurus rugosus*. *Wild. Res.* **18**, 147–55.

Durtsche, R. D. (1992). Feeding time strategies of the fringe-toed lizard, *Uma inornata*, during breeding and non-breeding seasons. *Oecologia* **89**, 85–9.

du Toit, A., Mouton, P. L. N., Geertsema, H. and Flemming, A. F. (2002). Foraging mode of serpentiform, grass-living cordylid lizards: a case study of *Cordylus anguina*. *Afr. Zool.* **37**, 141–9.

Eifler, D. A. and Eifler, M. A. (1998). Foraging behavior and spacing patterns of the lizard *Cnemidophorus uniparens*. *J. Herpetol.* **32**, 24–33.

Eifler, D. A. and Eifler, M. A. (1999). Foraging behavior and spacing patterns of the lizard *Oligosoma grande*. *J. Herpetol.* **33**, 632–9.

Ellinger, N., Schlatte, G., Jerome, N. and Hödl, W. (2001). Habitat use and activity patterns of the neotropical arboreal lizard *Tropidurus* (= *Uracentron*) *azureus werneri* (Tropiduridae). *J. Herpetol.* **35**, 395–402.

Emlen, J. M. (1966). The role of time and energy in food preference. *Am. Nat.* **100**, 611–17.

Fabre, J. H. (1913). *The Life of the Fly*. London: Hodder and Stoughton.

Felsenstein, J. (1985). Phylogenies and the comparative method. *Am. Nat.* **125**, 1–15.

Fitch, H. S. (1970). Reproductive cycles in lizards and snakes. *Univ. Kansas Mus. Nat. Hist. Misc. Pub.* **52**, 1–247.

Fitzpatrick, J. W. (1981). Search strategies of tyrant flycatchers. *Anim. Behav.* **29**, 810–21.

Fuiman, L. A. and Cowan, J. H. (2003). Behavior and recruitment success in fish larvae: repeatability and covariation of survival skills. *Ecology* **84**, 53–67.

Garland, T. Jr., Dickerman, A. W., Janis, C. M. and Jones, J. A. (1993). Phylogenetic analysis of covariance by computer simulation. *Syst. Biol.* **42**, 265–92.

Garland, T. Jr., Harvey, P. H. and Ives, A. R. (1992). Procedures for the analysis of comparative data using phylogenetically independent contrasts. *Syst. Biol.* **41**, 18–32.

Gasnier, T. R., Magnusson, W. E. and Lima, A. P. (1994). Foraging activity and diet of four sympatric lizard species in a tropical rainforest. *J. Herpetol.* **28**, 187–92.

Gerritsen, J. and Strickler, J. R. (1977). Encounter probabilities and community structure in zooplankton: a mathematical model. *J. Fish. Res. Board Can.* **34**, 73–82.

Grafen, A. (1989). The phylogenetic regression. *Phil. Trans. R. Soc. Lond.* B**326**, 19–157.

Greeff, J. M. and Whiting, M. J. (2000). Foraging-mode plasticity in the Augrabies flat lizard *Platysaurus broadleyi*. *Herpetologica* **56**, 402–7.

Hirth, H. F. (1963). The ecology of two lizards on a tropical beach. *Ecol. Monogr.* **33**, 83–112.

Honda, M., Ota, H., Köhler, G. *et al.* (2003). Phylogeny of the lizard subfamily Lygosominae (Reptilia: Scincidae), with special reference to the origin of the New World taxa. *Gen. Genet. Syst.* **78**, 71–80.

Huey, R. B. and Pianka, E. R. (1981). Ecological consequences of foraging mode. *Ecology* **62**, 991–9.

Huey, R. B., Pianka, E. R. and Vitt, L. J. (2001). How often do lizards run on empty? *Ecology* **82**, 1–7.

Husak, J. F. and Ackland, E. N. (2003). Foraging mode of the reticulate collared lizard, *Crotaphytus reticulatus*. *Southw. Nat.* **48**, 282–6.

Irschick, D. J. (2000). Comparative and behavioral analyses of preferred speed: *Anolis* lizards as a model system. *Physiol. Biochem. Zool.* **73**, 428–37.

Kareiva, P. (1989). Renewing the dialogue between theory and experiments in population ecology. In *Perspectives in Ecological Theory*, ed. J. Roughgarden, R. M. May and S. A. Levin, pp. 68–88. Princeton, N.J.: Princeton University Press.

Krebs, J. R. and Kacelnik, A. (1991). Decision-making. In *Behavioural Ecology*, 3rd edn, ed. J. R. Krebs and N. B. Davies, pp. 105–36. London: Blackwell Scientific.

Lawton, J. (1991). Ecology as she is done, and could be done. *Oikos* **61**, 289–90.

Levins, R. (1966). The strategy of model building in population biology. *Am. Sci.* **54**, 421–31.

Lewis, A. R. (1989). Diet selection and depression of prey abundance by an intensively foraging lizard. *J. Herpetol.* **23**, 164–70.

Lewis, A. R. and Saliva, J. (1987). Effects of sex and size on home range, dominance, and activity budgets in the Puerto Rican teiid *Ameiva exsul* (Lacertilia: Teiidae). *Herpetologica* **43**, 374–83.

Lister, B. C. (1981). Seasonal niche relationships of rain forest anoles. *Ecology* **62**, 1548–60.

Losos, J. B., Jackman, T. R., Larson, A., de Queiroz, K. and Rodríguez-Schettino, L. (1998). Contingency and determinism in replicated adaptive radiations of island lizards. *Science* **279**, 2115–18.

MacArthur, R. H. and Pianka, E. R. (1966). On optimal use of a patchy environment. *Am. Nat.* **100**, 603–9.

Mac Nally, R. C. (1994). On characterizing foraging versatility, illustrated by using birds. *Oikos* **69**, 95–106.

Magnusson, W. E. and da Silva, E. V. (1993). Relative effects of size, season and species on the diets of some Amazonian savanna lizards. *J. Herpetol.* **27**, 380–5.

Magnusson, W. E., de Paiva, L. J., da Rocha, R. M. *et al.* (1985). The correlates of foraging mode in a community of Brazilian lizards. *Herpetologica* **41**, 324–32.

Mausfeld, P., Vences, M., Schmitz, A. and Veith, M. (2000). First data on molecular phylogeography of scincid lizards of the genus *Mabuya*. *Molec. Phylog. Evol.* **17**, 11–14.

McLaughlin, R. L. (1989). Search modes of birds and lizards: evidence for alternative movement patterns. *Am. Nat.* **133**, 654–70.

Mori, A. and Randriamahazo, H. J. A. R. (2002). Foraging mode of a Madagascan iguanian lizard, *Oplurus cuvieri cuvieri*. *Afr. J. Ecol.* **40**, 61–4.

Mouton, P. F. N., Geertsema, H. and Visagie, L. (2000). Foraging mode of a group-living lizard, *Cordylus cataphractus* (Cordylidae). *Afr. Zool.* **35**, 1–7.

Muth, A. (1992). Development of baseline data and procedures for monitoring populations of the flat-tailed horned lizard, *Phrynosoma mcallii*. Final report for California Dept. of Fish and Game contract 86/87 C-2056 and 87/88 C-2056.

Nilsson, J.-A. (1994). Energetic bottle-necks during breeding and the reproductive cost of being too early. *J. Anim. Ecol.* **63**, 200–8.

Orr, Y., Shachak, M. and Steinberger, Y. (1979). Ecology of the small spotted lizard (*Eremias guttulata guttulata*) in the Negev desert (Israel). *J. Arid Env.* **2**, 151–61.

Orzack, S. H. and Sober, E. (1994). How (not) to test an optimality model. *Trends Ecol. Evol.* **9**, 265–7.

Paulissen, M. A. (1987). Optimal foraging and intraspecific diet differences in the lizard *Cnemidophorus sexlineatus*. *Oecologia* **71**, 439–46.

Pease, C. M. and Bull, J. J. (1992). Is science logical? *BioScience* **42**, 293–8.

Pereira, H. M., Bergman, A. and Roughgarden, J. (2003). Socially stable territories: the negotiation of space by interacting foragers. *Am. Nat.* **161**, 143–52.

Perry, G. (1995). The evolutionary ecology of lizard foraging: a comparative study. Ph.D. dissertation, University of Texas at Austin.

Perry, G. (1996). The evolution of sexual dimorphism in the lizard *Anolis polylepis* (Iguania): evidence from intraspecific variation in foraging behavior and diet. *Can. J. Zool.* **74**, 1238–45.

Perry, G. (1999). The evolution of search modes: ecological versus phylogenetic perspectives. *Am. Nat.* **153**, 98–109.

Perry, G. and Brandeis, M. (1992). Variation in stomach contents of the gecko *Ptyodactylus hasselquistii guttatus* in relation to sex, age, season and locality. *Amph.-Rept.* **13**, 275–82.

Perry, G. and Buden, D. W. (1999). Notes on the ecology, behavior and color variation of the green tree skink, *Lamprolepis smaragdina* (Lacertilia: Scincidae), in Micronesia. *Micronesica* **31**, 263–73.

Perry, G. and Garland, T. Jr. (2002). Lizard home ranges revisited: effects of sex, body size, diet, habitat, and phylogeny. *Ecology* **83**, 1870–85.

Perry, G. and Pianka, E. R. (1997). Foraging behaviour: past, present and future. *Trends. Ecol. Evol.* **12**, 360–4.

Perry, G., Dmi'el, R. and Lazell, J. (1999). Evaporative water loss in insular populations of the *Anolis cristatellus* group (Reptilia: Sauria) in the British Virgin Islands II: the effects of drought. *Biotropica* **31**, 337–43.

Perry, G., Dmi'el, R. and Lazell, J. (2000). Evaporative water loss in insular populations of *Anolis cristatellus* (Reptilia: Sauria) in the British Virgin Islands III: a common garden experiment. *Biotropica* **32**, 722–8.

Perry, G., Lampl, I., Lerner, A. *et al.* (1990). Foraging mode in lacertid lizards: variation and correlates. *Amph.-Rept.* **11**, 373–84.

Perry, G., LeVering, K., Girard, I. and Garland, T. Jr. (2004). Locomotor performance and social dominance in male *Anolis cristatellus*. *Anim. Behav.* **67**, 37–47.

Pianka, E. R. (1966). Convexity, desert lizards, and spatial heterogeneity. *Ecology* **47**, 1055–9.

Pianka, E. R. (1970). Comparative autecology of the lizard *Cnemidophorus tigris* in different parts of its geographic range. *Ecology* **51**, 703–20.

Pianka, E. R. (1986). *Ecology and Natural History of Desert Lizards*. Princeton, NJ: Princeton University Press.

Pianka, E. R. (1993). The many dimensions of a lizard's ecological niche. In *Lacertids of the Mediterranean Region*, ed. E. D. Valakos, W. Böhme, V. Pérez-Mellado and P. Maragou, pp. 121–54. Athens: Hellenic Zoological Society.

Pianka, E. R., Huey, R. B. and Lawlor, L. R. (1979). Niche segregation in desert lizards. In *Analysis of Ecological Systems*, ed. D. J. Horn, R. Stairs and R. D. Mitchell, pp. 67–115. Columbus, OH: Ohio State University.

Rand, A. S., Dugan, B. A., Monteza, H. and Vianda, D. (1990). The diet of a generalized folivore: *Iguana iguana* in Panama. *J. Herpetol.* **24**, 211–14.

Reagan, D. P. (1986). Foraging behavior of *Anolis stratulus* in a Puerto Rican rain forest. *Biotropica* **18**, 157–60.

Reeder, T. W. (2003). A phylogeny of the Australian *Sphenomorphus* group (Scincidae: Squamata) and the phylogenetic placement of the crocodile skinks (*Tribolonotus*): A Bayesian approaches to assessing congruence and obtaining confidence in maximum likelihood inferred relationships. *Molec. Phylog. Evol.* **27**, 384–97.

Reeder, T. W., Cole, C. J. and Dessauer, H. C. (2002). Phylogenetic relationships of whiptail lizards of the genus *Cnemidophorus* (Squamata: Teiidae): A test of monophyly, reevaluation of karyotypic evolution, and review of hybrid origins. *Am. Mus. Nov.* **3365**, 1–61.

Regal, P. J. (1978). Behavioral differences between reptiles and mammals: an analysis of activity and mental capabilities. In *Behavior and Neurology of Lizards*, ed. N. Greenberg and P. D. Maclean, pp. 183–202. Rockville, MD: NIMH.

Schwenk, K. (1995). Of tongues and noses: chemoreception in lizard and snakes. *Trends. Ecol. Evol.* **10**, 7–12.

Shaffer, D. T. Jr. and Whitford, W. G. (1981) Behavioral responses of a predator, the round-tailed horned lizard, *Phrynosoma modestum* and its prey, honey pot ants, *Myrmecocystus* spp. *Am. Midl. Nat.* **105**, 209–16.

Sih, A., Bell, A. M., Johnson, J. C. and Ziemba, R. E. (2004). Behavioral syndromes: an integrative overview. *Quart. Rev. Biol.* **79**, 241–77.

Sokal, R. R. and Rohlf, F. J. (1995). *Biometry*, 3rd edn. New York: Freeman.

Stephens, D. W. and Krebs, J. R. (1986). *Foraging Theory*. Princeton, NJ: Princeton University Press.

Sugarman, R. A. and Hacker, R. A. (1980). Observer effects on collared lizards. *J. Herpetol.* **14**, 188–90.

Taylor, J. A. (1986). Food and foraging behavior of the lizard, *Ctenotus taeniolatus. Australian J. Ecol.* **11**, 49–54.

Travers, S. E. and Sih, A. (1991). The influence of starvation and predators on the mating behavior of a semiaquatic insect. *Ecology* **72**, 2123–36.

Turner, F. B. and Medica, P. A. (1982). The distribution and abundance of the flat-tailed horned lizard (*Phrynosoma mcallii*). *Copeia* **1982**, 815–23.

Vitt, L. J., Zani, P. A., Caldwell, J. C. and Durtsche, R. D. (1993). Ecology of the whiptail lizard *Cnemidophorus deppii* on a tropical beach. *Can. J. Zool.* **71**, 2391–400.

Weiner, J. (1995). On the practice of ecology. *J. Ecol.* **83**, 153–8.

Weinstein, R. B. and Full, R. J. (1999). Intermittent locomotion increases endurance in a gecko. *Physiol. Biochem. Zool.* **72**, 732–9.

Werner, Y. L. and Chou, L. M. (2002). Observations on the ecology of the arrhythmic equatorial gecko *Cnemaspis kendallii* in Singapore (Sauria: Gekkoninae). *Raffles Bull. Zool.* **50**, 185–96.

Werner, Y. L., Okada, S., Ota, H., Perry, G. and Tokunaga, S. (1997). Varied and fluctuating foraging modes in nocturnal lizards of the family Gekkonidae. *Asia. Herpetol. Res.* **7**, 153–65.

Wiens, J. J., Reeder, T. W. and De Oca, A. N. M. (1999). Molecular phylogenetics and evolution of sexual dichromatism among populations of the Yarrow's spiny lizard (*Sceloporus jarrovii*). *Evolution* **53**, 1884–97.

Wymann, M. N. and Whiting, M. J. (2002). Foraging ecology of rainbow skinks (*Mabuya margaritifer*) in southern Africa. *Copeia* **2002**, 943–57.

2

Morphology, performance, and foraging mode

DONALD B. MILES

Department of Biological Sciences, Ohio University

JONATHAN B. LOSOS

Department of Biology, Washington University

DUNCAN J. IRSCHICK

Department of Ecology and Evolutionary Biology, Tulane University

Introduction

The feeding behavior of an animal is a fundamental attribute that has major fitness implications. Individual variation in the success with which prey is acquired ultimately affects growth rate, survivorship and reproductive success. Foraging success is linked to search behavior, which profoundly influences the variety and number of prey encountered (Pianka, 1966, 1973; Pietruszka, 1986). The ability to find prey therefore affects food intake and ultimately an individual's energy budget. The importance of foraging success is manifest in the myriad of behaviors animals display for the searching, pursuit, and capture of prey. Different modes of search behavior also entail costs. The duration of time spent foraging and the type of habitat an animal searches for food affects the risk of predation and ability to avoid predators (Huey and Pianka, 1981).

Not surprisingly, the analysis of foraging behavior has been an important topic in ecology and evolutionary biology (Schoener, 1971; Gerritsen and Strickler, 1977; Stephens and Krebs, 1986; Perry and Pianka, 1997). Past investigations proposed a dichotomy in foraging patterns based on observations of consistent and prominent differences in behaviors among species in how prey are pursued and captured (see McLaughlin, 1989; Vitt and Pianka, this volume, Chapter 5; Perry, this volume, Chapter 1). Most species have been classified as either "ambush" ("sit-and-wait") predators or widely (active) foraging predators (Pianka, 1966; Regal, 1978; Huey and Pianka, 1981) based on qualitative examination of activity patterns in the field. This division has been reinforced, in part, by empirical data on distances moved, percentage time moving, and moves per minute (Cooper *et al.*, 1999, 2001; McLaughlin, 1989; Butler, 2005). However, the existence of bimodal foraging modes is controversial and is contingent on the method of analysis. Many authors agree that the dichotomy may be artificial and that species search modes are extremes on a

Lizard Ecology: The Evolutionary Consequences of Foraging Mode, ed. S. M. Reilly, L. D. McBrayer and D. B. Miles. Published by Cambridge University Press. © Cambridge University Press 2007.

continuum. For example, recent comparative analyses based on expansive data-sets have not supported a bimodal pattern of foraging mode (Perry, 1999, this volume, Chapter 1). Despite the debate surrounding the threshold values used to categorize species into either group and whether indeed a bimodal pattern to foraging modes exists, there is general agreement about the general character-istics and correlates of each foraging mode.

Foraging mode as a paradigm

Foraging mode has become established as a major paradigm in ecology. One reason for the importance is that several morphological, ecological, and behavioral traits covary with foraging mode in a predictable way (Huey and Pianka, 1981; Pietruszka, 1986; Perry and Pianka, 1997; Perry, 1999). Numerous studies have described the general characteristics associated with foraging mode. Differences in search behavior have favored the evolution of traits related to detection, capture, and consumption of prey (Cooper, 1997; Vitt *et al.*, 2003). Ambush foragers (SW) are relatively sedentary and seek mainly mobile prey by using visual cues; prey processing involves lingual prehension. In contrast, active foragers (AF) move over great distances seek-ing prey, use both vision and olfaction to detect prey (Cooper, this volume, Chapter 8) and rely primarily on jaw prehension to subdue prey. The tongue of AF species has become specialized for detecting chemical cues (Cooper, 1995, 1997, this volume, Chapter 8). Because the type of prey encountered by sit-and-wait and active foraging species may vary in physical and textural proper-ties, e.g. hardness and size, other differences associated with foraging mode include robustness of the skull and bite performance. Consequently, recent investigations have illuminated how foraging mode imposes constraints on tongue structure, skull morphology, bite performance, and prey processing mechanics (Reilly and McBrayer, this volume, Chapter 10; McBrayer and Corbin, this volume, Chapter 9; McBrayer, 2004; McBrayer and Reilly, 2002; Schwenk, 2000).

   In addition, species that vary in foraging mode exhibit consistent differences in duration of activity and energy requirements (Huey and Pianka, 1981; Pianka, 1986; Vitt and Pianka, this volume, Chapter 5; Brown and Nagy, this volume, Chapter 4). Active foraging species move across large areas, remain active for longer periods of time and have higher energy budgets (Anderson and Karasov, 1981). In contrast, sit-and-wait predators have relat-ively smaller home ranges and lower daily energy expenditures (Nagy *et al.*, 1984; Brown and Nagy, this volume, Chapter 4). The differences in activity and energy budgets as well as the variation in predator milieu also constrain life history traits, such as reproductive effort (in lizards, Vitt and Congdon, 1978; Vitt and Price, 1982; Dunham *et al.*, 1988; in snakes, Webb *et al.*, 2003). For example, high rates of movement in AF species result in greater energy expenditure, which is balanced by the consumption of locally abundant sedentary prey. However, a high capacity for mobility imposes constraints on reproductive patterns, such that AF species tend to have smaller clutches and lower reproductive investment per clutch than SW species. The longer duration of activity by active foraging species presumably increases their risk of mortality due to predation, which further enhances the constraints on reproductive output.

Morphology, performance, and foraging mode

Body shape and locomotor performance are two key traits that are assumed to correlate with foraging mode (Huey and Pianka, 1981; Perry and Pianka, 1997. A large body of evidence supports the hypothesis that morphology affects locomotor performance (see, for example, Losos, 1990a; Garland and Losos, 1994; Miles, 1994; Melville and Swain, 2000b); thus it is expected that foraging mode should constrain the correlated evolution between morphology and locomotor performance.

Apart from studies that have examined variation in tail length and autotomy in relation to foraging mode (Vitt, 1983; Zani, 1996), only one study has compared morphological adaptations in body form and limb elements between AF and SW species (Thompson and Withers, 1997). However, this study was restricted to species of lizards in the family Varanidae in Australia. Thus, there is a paucity of data and analyses that have tested predictions about variation in body form in relation to foraging mode. Furthermore, the predictions listed in Huey and Pianka (1981) are relatively broad and lack details about the expected modifications of the elements of the fore- and hindlimbs. The absence of information on the covariation of morphology and foraging mode requires the generation of specific predictions based on general studies of morphology and locomotor performance (Garland and Losos, 1994; Miles, 1994; Aerts *et al.*, 2000; Herrel *et al.*, 2002). Because sit-and-wait species are predicted to use crypsis rather than flight to avoid predators, the reduced dependency on speed may favor the evolution of more varied body plans and limb dimensions. Thus, body shape may reflect a trade-off between movement on different substrates and speed. For example, many SW species inhabit rocky or arboreal habitats; hence, the limb morphology may reflect adaptations for climbing and clinging. Therefore, in climbing species, the limbs may represent a compromise between the requirement for rapid acceleration and movement on substrates with different widths and orientations. Terrestrial species often wait for prey within a burrow or beneath shrubs in fairly open habitats; hence selection may favor morphological adaptations that enhance rapid acceleration (Losos, 1990a; Zaaf and Van Damme, 2001; Herrel *et al.*, 2002). One prediction is that these lizards should have long hindlimbs and short forelimbs. In addition, the distal limb elements should be longer than the proximal elements, which would favor attaining high velocities. In contrast, active foraging species should exhibit modifications that reduce the cost of locomotion and enhance endurance and maneuverability. These two characteristics would be necessary for moving over great distances without expending a high amount of energy and facilitating the ability of avoiding predators. Therefore, we predict AF species to have long limbs with elongated distal segments. Furthermore, we expect no difference in the relative lengths of the fore- and hindlimb, which would facilitate maneuverability. A slender form favors the positioning of the legs closer to the midline of the body, which increases stride length and reduces the cost of transport (Garland and Losos, 1994; Van Damme *et al.*, 2003). Slender bodies would also favor the ability to make rapid turns, and a long tail could reduce the capture success of a predator (Vanhooydonck and Van Damme, 1999). Because most active foraging species are terrestrial, a slender body and long tail may facilitate movement of species through shrub and grassy habitats (Wiens and Slingluff, 2001). Interestingly, elongated tails have been hypothesized to reduce the success of a predator capturing an actively foraging species and not to enhance locomotor performance (Huey and Pianka, 1981; Vitt, 1983).

Predictions regarding the locomotor performance consequences of foraging mode are well established in the literature. Huey and Pianka (1981) first predicted that sit-and-wait foragers should capture prey by rapidly darting from a perch. This behavior would select for greater sprint capacity. Active foragers are predicted to have enhanced endurance as an adaptation for moving greater distances or longer activity periods (Huey and Pianka, 1981; Huey and Bennett, 1986; Garland, 1993; Garland and Losos, 1994). Furthermore, the two locomotor traits are expected to be negatively correlated (Garland and Losos, 1994; Vanhooydonck *et al.*, 2001). Despite the widely cited patterns of variation in morphology and locomotion between species that differ in foraging mode, there are few studies that have examined these issues.

Relatively more information has been published on the association between locomotor performance and foraging mode. The foraging mode paradigm predicts that sit-and-wait foragers will have higher sprint capacities and active foragers greater endurance (Huey *et al.*, 1984). There is equivocal support for the prediction of enhanced sprint speed in SW species and greater endurance in

AF species in the literature. One of the first studies documenting locomotor differences between foraging modes involved a comparison between two lacertid species from the Kalahari Desert (Huey *et al.*, 1984). The sit-and-wait foraging species *Eremias lineoocellata* exhibited faster sprint speeds, whereas the active foraging species *Eremias lugubris* had enhanced endurance (Huey *et al.*, 1984). Additional support for a relationship emerged from comparative analyses of interspecific variation in endurance. Indirect evidence for an association between foraging mode and endurance emerged from a study of locomotor performance in relation to natural activity in lizards (Hertz *et al.*, 1988). Actively foraging lizards were found to move longer distances than sit-and-wait species, and average daily movement distances were correlated with endurance in an analysis of nine species of lizard from four different families. Additional evidence supporting high endurance capacities of actively foraging lizards was provided by Garland (1993) in an analysis of performance in *Cnemidophorus tigris*. The endurance capacity of *C. tigris* was greater than other species including other active foraging species and sit-and-wait species. Garland (1993) concluded that the high endurance capacity of *C. tigris* was an adaptation for foraging. This conclusion was based on the data for endurance capacity presented in Garland (1993) as well as data from other taxa including the Gila Monster (*Heloderma suspectum*) and varanid lizards. However, it is possible that the pattern found by Garland (1993) was confounded with phylogeny. A subsequent analysis of endurance, based on an expanded dataset, found considerable interspecific variation and suggested that larger species had greater endurance, after controlling for the phylogenetic relatedness of the taxa included in the study (Garland, 1994). Although it was not included in the analysis, inspection of Figure 11.5 of Garland (1994) showed active foraging species having enhanced endurance capacities. Garland (1999) provided additional evidence supporting the association between performance and foraging mode, when he documented a significant, positive correlation between endurance capacity and moves per minute (MPM) and percent time moving (PTM). Thus, species characterized by frequent movements and a greater proportion of time involved in locomotion had greater endurance capacity (Garland, 1999).

Other studies have either disputed the evidence supporting the observation that locomotor performance differs between foraging modes or failed to find a significant difference in performance. A re-examination of the African lacertid data by Perry (1999) found no difference in sprint speed when comparing the species of *Eremias*. Based on the available phylogenetic hypothesis for the Lacertidae, Perry (1999) argued that the appropriate comparison with *E. lineoocellata* was the species *E. namaquensis*, which results in similar values for initial and maximum velocities in the active foraging and sit-and-wait

species. The most recent extensive analysis of interspecific variation in locomotor performance showed no difference in sprint speed between SW and
AF foragers (Vanhooydonck and Van Damme, 2001). Their analysis included
94 species from a diversity of lizard families including the Iguanidae,
Agamidae, Chameleonidae, Gekkonidae, Scincidae, Teiidae, and Lacertidae.
The sit-and-wait and active foraging species were shown to have nearly
identical values for sprint speed whether or not phylogeny was included in
the analysis. Curiously, the correlation between sprint speed and the behavioral traits associated with foraging mode (moves per minute, MPM, and
PTM) has yet to be investigated.

In this chapter we describe how morphology and locomotor performance
covaries with foraging mode. We are specifically interested in describing the
patterns of variation in foraging mode among lizards based on published
values for two variables (PTM and MPM) and then determining whether
there are morphological and locomotor performance differences between foraging categories. As many studies have pointed out, interspecific variation in
foraging mode tends to be confounded with phylogeny (Dunham *et al.*, 1988;
Perry, 1999). Specifically, foraging mode is, in many instances, fixed or invariant within a family. For example, species in the family Iguanidae and
Agamidae are ambush foragers, whereas species in the families Teiidae and
Helodermatidae are active foragers. Consequently, any difference in performance with respect to foraging mode may reflect phylogenetic inertia rather than
unique adaptations for prey localization. However, at least three families have
species that vary in foraging mode: the Gekkonidae, Scincidae, and Lacertidae
(Cooper and Whiting, 2000). Evidence also suggests that foraging mode can
vary seasonally within species (Whiting, this volume, Chapter 13). Hence a key
question is how many evolutionary transitions have occurred in foraging
mode. In addition, if morphology and locomotor performance are part of a
suite of traits related to foraging mode, as implied by the paradigm, then there
should be a concomitant change in these traits with the shift in foraging
behavior. Our first goal was to reconstruct the ancestral (nodal) character
states by using parsimony analysis in order to infer the evolutionary transitions in foraging mode. Our second goal was to ascertain whether locomotor
performance differed between sit-and-wait and active foragers. We initially
focus our analyses on a dichotomous characterization of foraging mode in
order to compare our results with previous studies. Specifically, based on
previous studies, we predicted that (a) SW foragers should have greater sprint
speeds than AF species; and (b) AF species should have higher values of
endurance relative to SW species. We make these comparisons using species
values and then control for the effects of phylogeny. Because numerous studies

emphasize that SW and AF are endpoints on a continuum, we also performed analyses on the continuous characters that define foraging modes. Specifically, we examined the association between locomotor performance and the behavioral variables that have been used to classify foraging mode: moves per minute (MPM) and percent time moving (PTM). In particular we tested whether locomotor performance is co-adapted with foraging behavior and which locomotor traits are associated with the key variables used to objectively define foraging modes. We then compare the evolutionary transitions in foraging mode with those for locomotor performance. If the shift in locomotor performance is to be considered an adaptation for feeding behavior, then we predict that the change in locomotor performance should accompany an evolutionary transition in foraging behavior. In addition, we test the hypothesis that the different foraging modes have diverged in morphology. Finally, we use multivariate methods to describe whether the relation between morphology and performance is affected by foraging mode. If foraging mode did not influence the association between morphology and performance, then one would expect a common slope for both sit-and-wait and active foraging species. However, if foraging mode has influenced the association between morphology and performance, then we expect different relations among the variables for each category.

We next move from a macroevolutionary perspective to examine the relationship between foraging behavior and locomotor performance within *Anolis*, a well-studied group of lizards, for which a substantial database is available on behavioral and locomotor activity in the field.

Materials and methods

Foraging mode

Our analysis of foraging mode was based on published estimates of foraging behavior. Specifically, we relied on data on PTM and MPM presented by Perry (1999, this volume, Chapter 1) as well as recently published values. Species with values of PTM < 10% and MPM < 1.0 were categorized as sit-and-wait foragers. We relied on the foraging mode data presented in Dunham *et al.* (1988) and Zani (2000) for taxa that lacked published estimates of PTM or MPM. It is important to note that activity patterns during periods of "undisturbed" activity are a composite of many different influences, most of which are difficult to discern (Jayne and Irschick, 2000). For example, lizards may be moving in response to an unseen predator, in response to a rival male that may not be seen by the observer, and so forth. In this sense, measures of movement during undisturbed periods, such as moves per minute (MPM), reflect not only "foraging" patterns, but also many other non-foraging activities. Nevertheless, an assumption of our approach is that the vast majority of undisturbed movements reflect an underlying behavioral strategy, which is reflected in the typical paradigm of "sit-and-wait" versus "actively foraging".

Performance

We surveyed the literature for papers presenting data on sprint speed or endurance or both. Following Van Damme and Vanhooydonck (2001), we excluded estimates of sprint speed made on high-speed treadmills or racetracks with distances between photocells exceeding 0.50 m. Published estimates of sprint speed and endurance were supplemented by using our own unpublished work (list of references available from D.B.M.).

Statistical analyses of locomotor performance and foraging mode

We transformed the data prior to analysis in order to reduce skew and generate an acceptable fit to a normal distribution. A $\log_{10}$ transformation for snout–vent length (SVL), mass, maximum velocity and endurance resulted in a normal distribution. Moves per minute required a square root transformation, whereas percentage time moving required a negative inverse square root transformation $\dfrac{-1}{(\sqrt{PTM + 1})}$.

We first tested the null hypothesis that sit-and-wait foraging species did not differ from active foraging species in velocity and stamina by using an ANCOVA. Snout–vent length (SVL) was used as our measure of body size. Our rationale was that body size often affects locomotor performance (see Miles, 1994; Garland, 1994), hence any difference in performance as a consequence of foraging mode could be an artifact of body size. We tested for the homogeneity of slopes prior to statistically comparing the means after adjusting for the effects of body size.

Morphology

We selected taxa from each of the major lineages of lizards with the purpose of sampling the range of ecological (e.g. foraging mode) and morphological diversity within the order. We measured 250 specimens (species) from 20 lizard families. One male specimen from each species was included. Using digital calipers, we measured 13 external measurements on each specimen: snout–vent length (SVL), tail length, body width (maximum width of the body), depth of the body at the pectoral girdle, width of the body at the pectoral girdle, humerus length, antebrachium length, manus length, length of the longest digit (IV), shank length, crus length, foot length, length of the longest hind toe (digit IV), and tail length (see Losos and Miles, 2002 for more detailed descriptions of measurements).

Body size variation among the species in our sample was considerable (32–400 mm). Because morphological differences in foraging mode could be confounded with body size, we elected to focus on variation in morphological shape. We removed the effects of size by using Mosimann's (1970) geometric-mean method. First we calculated a size variable based on the geometric mean of the 13 traits. Each variable was $\log_{10}$-transformed, which allowed us to define the geometric mean and the simple arithmetic mean of all 13 variables. A size-free estimate of each variable was obtained by subtracting our index of size from each trait.

We conducted two analyses based on the morphological shape variables. First, we evaluated the patterns of morphological covariation by using a principal components (PC) analysis. This allowed us to compare the different lineages in a morphological space. We next determined whether active foraging species differed from sit-and-wait species in morphology by a canonical discriminant analysis. We entered all 12 shape variables into a canonical discriminant function analysis and extracted classification functions from the pooled within-group covariance matrix. A test of the hypothesis that the group centroids were not significantly different was then evaluated. We used a cross-validation method to determine the goodness of fit of the estimated classification function. Finally we examined the covariation between morphology, locomotor performance and foraging mode by using the method of partial least squares (PLS), an extension of multiple regression that allows multiple response variables. The method attempts to find a linear combination of the predictor (morphological) variables that maximizes the variation in the response (locomotor performance) variables. Partial least squares is the preferred method of analysis when the number of observations is low relative to the number of variables. The output comprises a series of factors that best explains the patterns of covariation among the variables. The number of factors to retain is determined by a cross-validation approach. The optimal solution would be the number of factors that minimizes the predicted error sum-of-squares. Our source of morphological variables was the family means from the three PC axes. We used the family means for sprint speed, endurance, MPM and PTM as the response variables. All analyses were conducted by using PC SAS 9.1.

Phylogeny

It is well known that interspecific analyses require statistical procedures that are phylogenetically informed (Felsenstein, 1985, 1988; Garland *et al.*, 1993). We developed a composite phylogenetic hypothesis for the species of lizard included in our analyses by using results from several studies (see phylogeny in Fig. 2.1). Relationships among the major lineages of the lizards (e.g. Iguania, Gekkota,

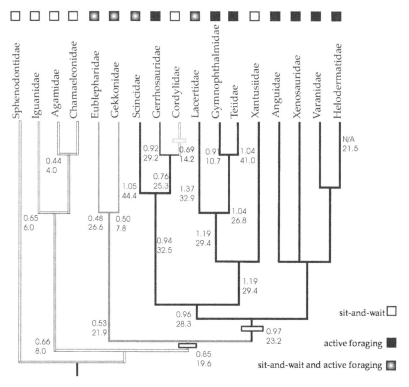

Figure 2.1. Best compromise phylogenetic hypothesis showing relationships of extant lizard families. Our hypothesis is a composite based on Estes *et al.* (1988), Evans (2003), Frost and Etheridge (1989), Hedges *et al.* (1991), Reeder and Wiens (1996), Reeder *et al.* (2002), Rest *et al.* (2003), and Wiens and Reeder (1997). The boxes adjacent to the family names represent the observed foraging mode for each group (open boxes, sit-and-wait foragers; black boxes, active foragers; gray boxes, families that have both foraging modes present). Numbers adjacent to the branches are the inferred values for moves per minute (MPM; upper) and percent time moving, (PTM; lower) at each node. The open bar at the branch leading to the autarchoglossans represents an early transition from sit-and-wait foraging to active foraging based on the maximum parsimony character mapping. The gray bar at the Cordylidae represents a reversal from AF to SW foraging. The shaded bar at the branch leading to the scleroglossans is the hypothesized transition to AF from SW based on the CART analysis (see text).

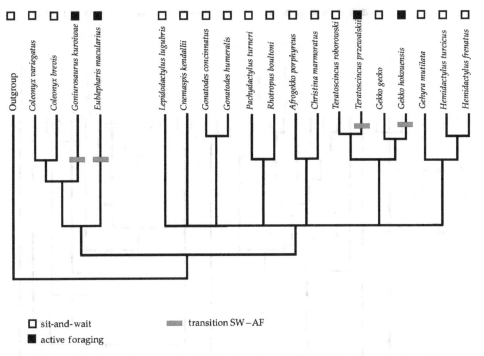

Figure 2.2. Best compromise phylogenetic tree displaying the relationships among species in the Gekkota. Species relationships were based on Kluge (1987), Han *et al.* (2004), and Zaaf and Van Damme (2001).

Autarchoglossa) were determined from Estes *et al.* (1988), Evans (2003) and Rest *et al.* (2003). An alternative phylogenetic hypothesis for the relationships of lizards was presented by Townsend *et al.* (2004). However, the results and conclusions of the comparative analyses were not materially altered when using the alternative topology. More detailed information about evolutionary relationships of species in each of the clades that have sit-and-wait and active foraging species (Gekkota, Scincoidea, Varanoidea) were estimated from a variety of sources (references given in Figs. 2.2–2.4). Thus, we were able to examine evolutionary transitions in morphology, performance, and foraging mode.

Our composite phylogeny was based on multiple studies and we therefore lacked divergence times for certain key clades. Consequently, we set all branch lengths equal to one. Previous investigations have shown that analyses based on actual divergence times do not substantially alter the results of the comparative phylogenetic analyses (Diaz-Uriarte and Garland, 1998).

Phylogenetic comparative analyses

We used two methods for assessing the phylogenetic signal in morphology, performance, and foraging mode. First, we employed character mapping to

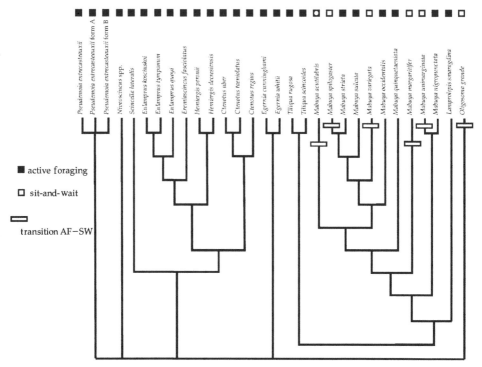

Figure 2.3. Phylogeny for Scincidae (based on Honda *et al.*, 1999, 2003; Mausfeld *et al.* 2000; Melville and Swain, 2000a, b; Reeder, 2003; Whiting *et al.*, 2003).

reconstruct the character states at ancestral nodes. We used these values to infer evolutionary transitions in foraging mode (including MPM and PTM) and locomotor performance. In one reconstruction, we scored foraging mode as a binary character and inferred ancestral values by using linear parsimony and likelihood model of character states. We rooted the tree with *Sphenodon* as the presumed outgroup for squamates (Rest *et al.*, 2003). Based on previous analyses, we coded *Sphenodon* as a sit-and-wait forager (Vitt *et al.*, 2003). Foraging mode was treated as an ordered character. Ancestral values for SVL, sprint speed, endurance, foraging mode, MPM, and PTM were inferred by using squared-change parsimony with all branches assumed to be equal. The reconstruction of ancestral traits based on squared-change parsimony is equivalent to maximum likelihood estimates when branch lengths are assumed to be equal (Schluter *et al.*, 1997). All parsimony analyses were performed by using Mesquite (Maddison and Maddison, 2004) and the PDAP module for Mesquite (Midford *et al.*, 2003).

Standardized independent contrasts were calculated by using the PDAP module in Mesquite (Midford *et al.*, 2003; Maddison and Maddison, 2004) for

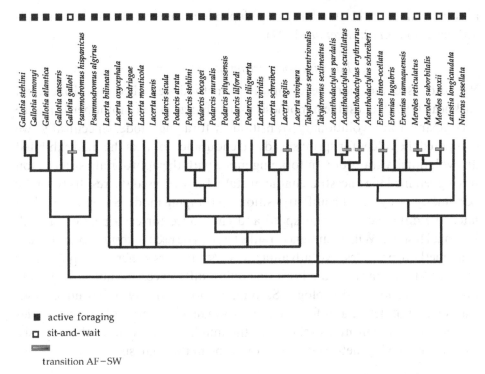

■ active foraging
□ sit-and-wait
▬
transition AF–SW

Figure 2.4. Phylogenetic hypothesis for the Lacertidae. The evolutionary relationships among some lineages are still unresolved and our tree represents the most recent result. The phylogeny for Lacertidae is based on Arnold (1989), Fu (2000), Harris and Arnold (1999, 2000), Harris *et al.* (1998), Lamb and Bauer (2003), and Ota *et al.* (2002).

SVL mass, sprint speed, endurance, MPM, and PTM. We compared sprint speed and endurance differences between SW and AF species by using a phylogenetic analysis of variance. We based our analyses of locomotor performance on size-adjusted values for sprint speed and endurance, i.e. the residuals from a regression analysis of performance as the response variable and body size (SVL) as the predictor variable. We generated simulated data to determine the statistical significance of the difference in locomotor performance between the two foraging modes. Our approach relied on the use of phylogenetic permutations to create 5000 random datasets that incorporate the phylogenetic relatedness of the species (see Lapointe and Garland, 2001). These random datasets were analyzed by using PDANOVA to obtain a null distribution of the *F*-statistic. If the observed F (F_{obs}) exceeded the upper ninety-fifth percentile of the null distribution, then the differences would be considered significant.

Diagnostic tests were performed by using the PDAP module of Mesquite. Standardization of the independent contrasts was verified by plotting the absolute values of the standardized independent contrasts versus their standard deviation (Garland *et al.*, 1992).

Pairwise comparison of sister taxon

We also used the pairwise comparison method described by Moller and Birkhead (1992) to test the hypothesis that a change in locomotor performance accompanied an evolutionary transition in foraging mode. Specifically, we compared the values for sprint speed and endurance for taxa that had exhibited an evolutionary transition in foraging mode with the appropriate sister taxon having retained the ancestral character state. First we used the results from the parsimony analysis to identify transitions in foraging mode. Second, we calculated the difference in sprint speed and endurance for each pair of taxa. A binomial test or a Wilcoxon signed-ranks test can be used to test for significance. This method provides two advantages. First, factors other than phylogeny may affect comparative analyses, including differences in morphology, life history, behavior, and ecology. Second, phylogenetically independent contrasts may not detect a difference in performance among foraging categories if there are few transitions relative to the number of tip taxa in the analysis. That is, the phylogenetic signal may overwhelm the analysis.

Classification and regression tree analysis (CART)

One focus of the foraging mode paradigm has been the inference of the timing of divergence in feeding behavior and the consequent evolutionary transitions in functional morphology, life history, and ecology (Vitt *et al.*, 2003). This question is critical because of the implicit hypothesis that the shift in foraging mode and the evolution associated traits was a major innovation responsible, in part, for the adaptive diversification of scleroglossan lizards. Therefore, rather than using parsimony methods to reconstruct the ancestral character states based on whether the species were SW or AF, we used an alternative method to infer the diversification of foraging mode. We began by deriving a classification function for assigning species to either SW or AF. Classification trees are a non-parametric method for categorizing samples into *a priori* categories. Specifically, we determined the values of sprint speed and endurance that predicted whether a species was either SW or AF. The algorithm involves a successive divisive partitioning of the data so as to maximize predictive accuracy. The partitioning is compared with a χ^2 or Gini index goodness-of-fit criterion to assess the significance of the classification (Breiman *et al.*, 1983). CART can use multiple, continuous predictor variables and determine the subset that best explains the partitioning of samples into the outcome variable. The major advantages of CART methods include the ease in interpretation of the results and the lack of restrictive assumptions in other parametric multivariate methods, such as discriminant functions. The result of a CART analysis is a hierarchical decision tree that one may use to classify subsequent samples. One potential problem with CART is the potential to overfit the data (Clark and Pregibon, 1992). Therefore, we estimated the optimal tree size by using repeated cross-validation (Venables and Ripley, 1994). We then estimated the nodal values of MPM and PTM by using squared-change parsimony and then used the classification rule to assign each ancestor to either SW or AF. We used the CART algorithm in JMP version 5.1 to conduct this analysis.

Intraspecific analysis of morphology, performance, and foraging mode

The above treatment focused on macroevolutionary comparisons among species that are distantly related to one another. However, although such comparisons offer the advantage of comparing taxa that are highly divergent in foraging mode and morphology (see, for example, Perry, 1999), these same comparisons do not provide the finer resolution for understanding how traits co-evolve. For this latter goal, comparing characteristics within a well-defined clade (e.g. within a single genus) offers a better opportunity for evaluating the detailed evolution of traits.

Fortunately, previous authors have compiled a substantial amount of data on the morphology, performance, and movement patterns of Caribbean *Anolis* lizards, thus providing an excellent opportunity for examining interspecific correlations among these variables (Losos, 1990a, b; Irschick, 2000). Caribbean *Anolis* lizards have radiated into a variety of distinct "ecomorphs," with each ecomorph type exhibiting unique morphological, ecological, and behavioral characteristics (Williams, 1983; Losos, 1990a, 1994; Irschick *et al.*, 1997). For example, trunk–ground anoles tend to have long hindlimbs and a long tail and occupy broad perches (e.g. tree trunks) close to the ground. At the other extreme, twig anoles tend to have short hindlimbs and a short tail and occupy narrow perches higher in the canopy. These divergent ecomorphs have arisen repeatedly within the Caribbean, with the majority of the adaptive radiation occurring on the larger Greater Antillean islands of Cuba and Hispaniola (Losos *et al.*, 1998; Jackman *et al.*, 1999, 2002).

Previous studies have documented variation among anole species in movement variables such as percentage time moving (PTM) and moves per minute (MPM); at least some of this variation can be attributed to differences among ecomorphs (Losos, 1990a, b; Irschick, 2000). For example, trunk–ground anoles tend to spend relatively little time moving (i.e. low PTM), whereas twig anoles tend to spend more time moving (i.e. relatively high PTM). In this regard, anoles offer an excellent opportunity to understand the evolution of foraging mode in a more detailed context; previous broad-scale analyses have uniformly treated *Anolis* lizards (indeed, practically all lizards within the subfamily Polychrotinae) as sit-and-wait predators (see, for example, Perry, 1999). Simple inspection of values of PTM for different ecomorphs reveals that this is not the case; for example, the twig anole *A. valencienni* spends about 7.2% of its time moving compared with only about 0.64% for the trunk–ground anole *A. sagrei*, equaling about an 11-fold difference in PTM (Irschick, 2000).

Based upon a phylogenetic tree of the eight anole species (see phylogeny in Fig. 2.5), we calculated independent contrasts for both movement variables (MPM and PTM), both measures of performance capacity (maximum speed and sprint sensitivity), and relative hindlimb length, resulting in seven contrasts for each of these variables. Independent contrast analyses used branch lengths (see Irschick and Losos, 1999), and all regression lines were forced through the origin. As above, we conducted the appropriate diagnostic tests to ensure the standardization was appropriate. We then conducted bivariate correlations between these variables in the following manner: (1) Maximum

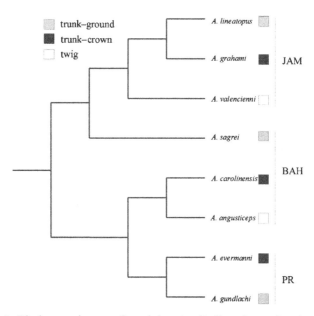

Figure 2.5. Phylogenetic tree for eight *Anolis* lizard species (see text for details). Locality headings next to species names represent the Caribbean islands on which these species occur, although *A. carolinensis* was studied in Louisiana. JAM, Jamaica; PR, Puerto Rico; BAH, Bahamas.

speed and sprint sensitivity versus PTM and MPM (four correlations); (2) relative hindlimb length versus PTM and MPM (two correlations). This analysis thus uses a total of six correlations, which we regard as small enough that we did not employ any corrections for multiple comparisons. Because the number of comparisons is already very low (6 df for each test), the correlation tests are already conservative.

However, no study has examined how movement patterns, morphology, and performance capacity are related for different species of *Anolis* lizards. Losos (1990b) examined simple relations among size-adjusted morphology, movement characteristics (percentage of moves that were walks, runs, or jumps), and maximum sprint speed in 14 Caribbean anole species. Irschick (2000) examined evolutionary correlations for eight anole species (some of which were the same as in Losos, 1990b) among average preferred speed during undisturbed movements versus several simple habitat measures (e.g. perch diameter use). However, neither Irschick (2000) nor Losos (1990b) specifically examined relations among more standard movement measures of foraging activity (e.g. MPM, PTM), size-adjusted morphology, and performance capacity.

Here, we focus on the same eight anole species that Irschick (2000) examined, but specifically address correlations between MPM, PTM, morphology, and performance capacity. We focused on the dataset of Irschick (2000), rather than that of Losos (1990b) or Perry (1999) because the former dataset was gathered from videotaped focal observations, whereas the latter dataset was gathered from transcribed focals based on visual observation. Thus, the videotaped analysis provides a more detailed opportunity to examine correlates of MPM and PTM. The eight anole species examined by Irschick (2000) are especially suitable for evolutionary comparisons because they form three ecomorphs (three trunk–ground, three trunk–crown, and two twig anoles), and each of these species is only distantly related to the other members of the same ecomorph category (Fig. 2.5). In this regard, these eight species provide a high degree of statistical power because of the repeated evolution of phenotypes. In addition, two measures of ecologically relevant performance capacity (maximum sprint speed on a flat surface, and ability to sprint well on surfaces of varying diameter [sprint sensitivity]) are also available for all eight species (taken from Losos and Sinervo, 1989; Irschick and Losos, 1999). Sprint sensitivity is the slope of the line between surface diameter (log) and mean maximum speed (log) on each diameter. Thus, species with high sprint sensitivities tend to decline in maximum speed dramatically among different diameters; species with low sprint sensitivities tend to decline only slightly in speed among different diameters (Irschick and Losos, 1999). Thus, these two performance measures define potentially independent aspects of locomotor performance: the ability to run rapidly on only a single surface, and the ability to run effectively (e.g. with a small decline in speed) on a variety of surfaces. To define movement patterns, we used moves per minute (MPM) and percentage time moving (PTM) based on videotaped focal observations for each of these eight species (taken from Irschick, 2000). We focused on a single morphological variable (size-adjusted hindlimb length = the residual value of hindlimb length for each species based on log–log regressions of hindlimb length (y-axis) vs. snout–vent length (x-axis)) for the eight species (taken from Losos, 1990a). We focused on relative hindlimb dimensions because this variable (i) has been shown to be biomechanically linked to both sprint sensitivity and maximum speed (Losos and Sinervo, 1989; Irschick and Losos, 1999), and (ii) varies dramatically among anole ecomorphs (Losos, 1990a).

Results

Variation in foraging mode among lizard families

We obtained 117 values for MPM and 110 values for PTM from our survey of the literature (Table 2.1). The majority of species (67%) had values for MPM less than 1.0. Over 70% of the observations for PTM were 10% or below. Most previous studies categorized a species as an active forager if the value of PTM was 10% or greater. Based on this criterion, all of the Iguania families would be considered sit-and-wait foragers (Fig. 2.1). Species in the family Cordylidae would also be classified as sit-and-wait foragers (Table 2.1, Fig. 2.1). There are scant data available for the Xantusiids, hence their classification is tentative. There are four families, Eublepharidae, Gekkonidae, Lacertidae and Scincidae (Figs. 2.2–2.4), with the broadest range of values for MPM and PTM. Consequently, these families have representative species in each of the foraging categories (Table 2.1). Species in the families Gerrhosauridae, Gymnophthalmidae, Teiidae, Varanidae, and Helodermatidae would be strict active foragers (Fig. 2.1, Table 2.1).

Evolutionary transitions in foraging mode

The results for the ancestor reconstruction analysis are presented in Fig. 2.1. Sit-and-wait foraging is supported as the ancestral behavior when *Sphenodon* is used as the outgroup. Both maximum parsimony and likelihood reconstruction methods revealed similar patterns for the reconstruction of foraging mode. The results from maximum parsimony show that a major shift in

Table 2.1 *Means and standard errors (in parentheses) of body size, locomotor performance, and movement behavior among lizard families*

Foraging Mode: SW, sit-and-wait; AF, Active Foraging. *n*, Number of species included in the analysis; SVL, Snout–vent length, a measure of body size; MPM, moves per minute; PTM, percentage time moving.

Family	Foraging mode	*n*	SVL (mm)	Velocity (m/s)	Stamina (s)	MPM	PTM
Iguania							
Agamidae	SW	8	111.08 (6.84)	2.47 (0.06)	—	0.35 (0.08)	1.5 (0.93)
Corytophanidae	SW	2	185.00 (27.00)	2.15 (1.02)	—	0.14 (0.13)	6.0 (0.12)
Crotaphytidae	SW	3	105.75 (5.75)	—	809.00 (283.00)	0.85 (0.66)	0.25 (0.15)
Hoplocercidae	SW	1	153.00	1.99	276.00		—
Opluridae	SW	1	153.0	—		0.97	2.0
Phrynosomatidae	SW	20	67.56 (3.32)	2.17 (0.15)	181.78 (22.79)	0.66 (0.20)	4.14 (2.04)
Polychrotidae	SW	19	55.28 (5.07)	1.68 (0.11)	119	0.33 (0.07)	1.4 (0.58)
Tropiduridae	SW	6	88.98 (6.05)	2.38 (0.27)	224.00	0.07 (0.06)	3.4
Scleroglossa							
Anguidae	AF	1	110.40	1.11	391.00	—	—
Cordylidae	SW	7	95.0 (12.57)	1.94 (0.27)	—	0.48	4.3 (1.26)
Eublepharidae	SW	2	64.3 (2.3)	1.51 (0.02)	72.0 (6.0)	0.57	—
	AF	2	128.1	0.66	—	0.39	23.0
Gekkonidae	SW	1	63.94 (6.23)	1.48 (0.17)	69.67 (8.57)	0.33 (0.08)	3.0 (0.74)
	AF	18	66.96 (7.32)	—	504.00	0.35	25.5
Gerrhosauridae	AF	2		—	—	0.95	31.2 (16.25)
Gymnophthalmidae	AF	1	—	—	—	0.91	10.7
Helodermatidae	AF	2	342.00 (41.0)	1.63 (0.07)	940.00 (286.00)	0.95	22.0
Lacertidae	SW	4	57.84 (5.91)	2.39 (0.25)	208.74 (65.26)	0.95 (0.29)	11.54 (2.36)
	AF	34	73.16 (6.06)	2.07 (0.12)	315.95 (103.72)	1.88 (0.25)	41.2 (4.65)
Scincidae	SW	2	79.4 (17.50)	2.37		0.85 (0.29)	4.32 (1.51)
	AF	32	86.15 (11.87)	1.02 (0.09)	1155.8 (282.11)	1.26 (0.16)	26.0 (6.09)
Teiidae	AF	14	89.39 (7.65)	3.27 (0.40)	1587.00 (402.62)	0.95 (0.18)	61.6 (9.08)
Varanidae	AF	2	314.00 (96.00)	3.95 (0.07)	900	—	—
Xantusidae	SW	1	95.3	1.57	87.00	—	—

foraging mode from sit-and-wait to active foraging occurred in the autarcho-glossans and not at the ancestor for scleroglossans as suggested by other studies. The ancestral condition for gekkotans was ambiguous; data for more species are necessary to resolve the evolutionary transition in foraging mode. The character state for the ancestral node for the Iguania was sit-and-wait. A reversal from AF to SW was found to occur at the common ancestor for the Cordylidae. Maximum likelihood (ML) ancestor states were similar. Sit-and-wait foraging is ancestral according to the ML estimates. Support for a SW to AF transition in the Gekkota is weak (nearly 1:1 in favor of sit-and-wait). However, the transition to AF at the ancestral node for scleroglossans is strongly supported (9.9:1 in favor of active foraging). The single reversal from AF to SW in the Cordylidae is also strongly supported (9.7:1).

Whereas the evolutionary transitions in foraging mode among families occurred early in the history of lizards, the ancestral reconstruction analysis found multiple transitions within families. Reconstructions based on maximum parsimony indicate that a shift to active foraging from a sit-and-wait ancestor occurred once in the Eublepharidae (*Goniurosaurus* and *Eublepharis*) (Fig. 2.2) and twice in the Gekkonidae (*Teratoscincus przewalskii* and *Gekko hokuensis*) (Fig. 2.2). Maximum likelihood also strongly supported this pattern. Because of the limited amount of data on movement behavior during foraging for the gecko, the ancestral reconstruction should be considered tentative. Conversely, parsimony and maximum likelihood reconstructions identified twelve transitions from AF to SW within two scleroglossan families, Scincidae and Lacertidae. Interestingly, foraging mode appears to be fairly plastic within these two families. The majority of species in the Scincidae are active foragers. However, five transitions from AF to SW were revealed in skinks (Fig. 2.3). With the exception of *Oligosoma grande*, the remaining transitions were in the genus *Mabuya*. All were strongly supported by maximum likelihood estimates. An additional seven transitions were uncovered in the Lacertidae (Fig. 2.4). One transition occurred in the genus *Gallotia* (*G. galloti*), one in the genus *Lacerta* (*L. agilis*), two in the genus *Acanthodactylus* (*A. scutellatus* and *A. erythrurus*), one in the genus *Eremias* (*E. lineo-ocellata*), and two in the genus *Meroles* (*M. reticulatus* and *M. knoxii*). Because of the proximity of the transitions toward the tips of the tree in both the Scincidae and Lacertidae, we infer that all of the transitions represent relatively recent shifts in foraging mode. It is possible that the ancestral reconstructions for the Scincidae and Lacertidae may have a high degree of uncertainty because of the numerous transitions in foraging mode. However, the likelihood ratios for all of the recent ancestors exceed 7.4, which is the recommended criterion for the significance of a transition (Schluter *et al.*, 1997).

Evolutionary changes in moves per minute (MPM) and percentage time moving (PTM)

Data on the hypothesized ancestral values for moves per minute and percent time moving were presented in Fig. 2.1. The common ancestors to the Iguanidae and Agamidae had low values for both MPM and PTM (Fig. 2.1). The values for MPM and PTM were also low at the node that defines the Iguania (= Agamidae, Chameleonidae, and Iguanidae). The first node with values for PTM that exceeds 10% occurs at the common ancestor of the Scleroglossa (= Gekkota + Autarchoglossa), although the value for MPM is still below 1.0. The Gekkota present an interesting case, in that the Eublepharidae have high values for PTM, but the magnitude of MPM is similar to that of the Iguania. Conversely, the Gekkonidae have low values for both MPM and PTM, which are similar to those for the Iguania. With the exception of the Cordylidae, all of the nodes within the autarchoglossans had high values for MPM and PTM.

The relationship between locomotor performance and foraging mode

Non-phylogenetic analysis

Velocity We found that sit-and-wait foraging species were significantly faster than active foraging species (Fig. 2.6A) ($F_{1,142} = 6.08$; $p < 0.01$). Because sprint speed could be influenced by body size, we investigated whether the observed difference between foraging categories was an artifact of size. Maximum velocity and body size (SVL) were significantly correlated ($r = 0.36$, $F_{1,141} = 21.9$; $p < 0.0001$). We tested for the homogeneity of slopes and noted

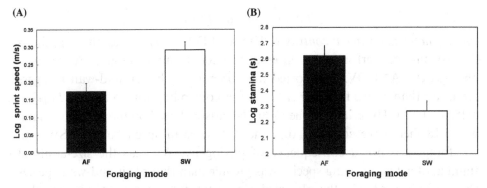

Figure 2.6. Differences in locomotor performance between foraging modes. Foraging mode is presented on the *x*-axis (SW, sit-and-wait; AF, active foraging). (A) Sprint speed; (B) endurance. Error bars are ± 1 s.e.

no difference between foraging categories (slope for SW = 0.39 ± 0.08, AF = 0.45 ± 0.12; $F_{1,141} = 0.14$; $p = 0.70$). However, the difference in velocity was not a consequence of variation in body size (ANCOVA for foraging mode $F_{1,141} = 11.64$, $p < 0.001$).

Stamina A preliminary regression analysis revealed that stamina and body size (SVL) were significantly correlated ($r = 0.54$; $F_{1,56} = 23.6$, $p < 0.001$). Therefore, we tested this hypothesis by an ANCOVA. A homogeneity of slopes test confirmed that the two foraging categories shared similar slopes ($F_{1,54} = 1.39$; $p = 0.24$). Active foraging species had greater values for stamina than sit-and-wait foragers, even after adjusting for the covariate (Fig. 2.6B) ($F_{1,55} = 12.61$; $p < 0.001$).

Trade-off between sprint speed and stamina There was no evidence for a trade-off between sprint speed and stamina ($r = -0.13$, df = 40, $p = 0.48$) based on untransformed values. However, the correlation between sprint speed and stamina after adjusting for body size was nearly significant ($r = -0.27$; $p = 0.07$). We did not detect a trade-off between velocity and stamina whether the analysis was based on all the species or grouped by foraging mode or, where there were sufficient observations, by family. Although the correlations were not significant, sit-and-wait foraging species exhibited a positive correlation between speed and stamina ($r = 0.41$, $p = 0.12$) and active foraging species had the expected negative correlation ($r = -0.31$, $p = 0.15$). Among the family comparisons only the Lacertidae showed a negative correlation between sprint speed and stamina (Spearman's $r = -0.49$, $n = 11$, $p = 0.03$). Interestingly, the Teiidae exhibited a positive correlation between speed and stamina (Spearman's $r = 0.90$, $n = 14$, $p = 0.03$).

Phylogenetic analyses

Phylogenetic independent contrasts We used PDANOVA to test the hypothesis that locomotor performance differed between foraging modes. A one-way phylogenetic ANCOVA supported the observation that sit-and-wait foragers are faster than active foraging species after controlling for body size ($F_{1,112} = 8.25$, $p < 0.01$). The critical value of the F-statistic, based on the permuted data, was 5.18, hence we conclude that the difference in speed between SW and AF foragers is not a consequence of phylogeny. Whereas the size-adjusted stamina of active foraging species was greater than that of sit-and-wait species ($F_{1,55} = 3.62$, $p < 0.05$), the observed F-statistic did not exceed the critical F ($F_{crit} = 6.85$), which suggests the difference in stamina is a consequence of phylogeny.

Table 2.2 *Pairwise comparisons derived from the evolutionary transitions in foraging mode*

The focal taxon is the species or family that exhibited a shift in foraging mode as identified by the ancestor reconstruction analysis. "Transition" describes the type of evolutionary change in foraging mode, either sit-and-wait to active foraging or the reverse. The sister taxon is the nearest relative that did not change in foraging mode. The direction of change is the difference between sprint speed between the focal taxon and its sister taxon; ↑, there was an increase in speed, ↓, there was a decrease in speed.

Focal taxon	Transition	Sprint speed (m/s)	Sister taxon	Sprint speed (m/s)	Direction of change
Eublepharis macularus	SW→AF	0.66	*Coleonyx brevis*	1.53	↓
Gallotia galloti	AF→SW	1.93	*G. stehleni*	3.3	↓
Lacerta agilis	AF→SW	1.68	*L. vivipara*	0.89	↑
Acanthodactylus scutellatus	AF→SW	2.795	*A. pardalis*	2.62	↑
A. erythrurus	AF→SW	3.3	*Psammodromus algirus*	2.53	↑
Eremias lineo-ocellata	AF→SW	2.46	*L. lugubris*	1.72	↑
Mabuya spilogaster	AF→SW	2.37	*M. striata*	2.1	↑
Cordylidae	AF→SW	1.94	Gerrhosauridae	1.52	↑

Sister taxon comparisons The phylogenetic independent contrasts analysis revealed significant differences in sprint speed, but not stamina, between the foraging mode categories. One weakness of independent contrasts is that the analysis emphasizes changes spread throughout the phylogeny. Consequently, factors other than foraging mode may vary among taxa when conducting a large cross-species analysis. Furthermore, most of the transitions in foraging mode represented recent evolutionary events. Hence, there is the opportunity to determine whether a transition in foraging mode involves a concomitant transition in locomotor performance. We therefore compared the performance traits of sister taxa that differed in foraging mode. The previous analysis of ancestor reconstruction identified 16 evolutionary transitions in foraging mode. We had sprint speed data for eight of the 16 transitions. Seven of the eight comparisons involved changes in sprint speed in the predicted direction, which is statistically significant (Table 2.2, binomial test, $p = 0.035$). For example, two gekkonid species showed a decrease in speed that accompanied the shift from sit-and-wait to active foraging. Most of the species that showed a transition from active foraging to sit-and-wait showed an increase in sprint speed. For example, *Mabuya spilogaster* increased in sprint speed (2.37 m/s) compared with its sister taxon *M. occidentalis* (1.36 m/s). The sole exception to the pattern was *Gallotia galloti* (1.93 m/s), whose value for sprint speed was lower than that of its sister taxon, *G. simonyi* (3.3 m/s). Only three pairwise comparisons were available for stamina. One of the comparisons involved an SW–AF transition (i.e. *Teratoscincus przewalskii*), which entailed an increase in stamina compared with the closest species with data on stamina (*Hemidactylus turcicus*). Two other comparisons involved AF–SW transitions. In each case stamina was lower in the sit-and-wait species; for example, the lacertid species *Eremias lineoocellata* had a lower value for stamina (276 s) compared with its congener *E. lugubris* (1500 s).

Does locomotor performance predict moves per minute or percentage time moving?

Non-phylogenetic analyses

There was no significant relation between sprint speed and MPM (Table 2.3). However, when SVL is included as a covariate in the analysis the result was a significant and positive association ($r = 0.39$, df $= 41$, $p = 0.008$), (Fig. 2.7A). Faster species tended to move more often on average compared with slower species. Interestingly, SVL and MPM were significantly and negatively correlated ($r = -0.47$, df $= 64$, $p < 0.01$). Thus, larger species were less likely to move frequently. Species that moved frequently tended to have higher levels of stamina. An *a priori* prediction is that sprint speed and PTM should be negatively related. Sprint speed was not significantly associated with PTM,

Table 2.3 *Correlations among body size, locomotor performance, and foraging behavior*

Correlations above the diagonal are based on the values for terminal taxa. Correlations below the diagonal were calculated from independent contrasts. All correlations were based on appropriately transformed data. Numbers in parentheses indicate the sample size for each correlation. Symbols: $^{+}0.05 < p < 0.10$; $^{*}p < 0.05$; $^{**}p < 0.01$; $^{***}p < 0.001$.

Trait	SVL	Sprint speed	Endurance	MPM	PTM
SVL	—	0.36^{***} (143)	0.58^{***} (59)	-0.16 (78)	0.02 (78)
Sprint speed	0.34^{*} (116)	—	0.26^{+} (47)	0.38^{*} (43)	0.04 (41)
Endurance	0.16 (57)	-0.32^{+} (45)	—	0.44^{+} (26)	0.71^{***} (26)
MPM	-0.17 (64)	-0.19 (32)	0.41^{*} (26)	—	0.58^{***} (102)
PTM	-0.10 (67)	-0.21 (31)	0.70^{***} (23)	0.44^{***} (89)	—

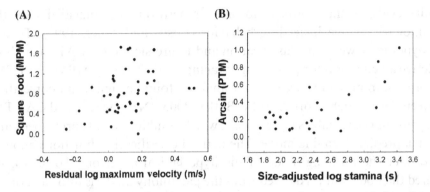

Figure 2.7. The association between locomotor performance and foraging behavior. (A) Moves per minute (square root transformed) versus log (sprint speed). (B) Percentage time moving (arcsin transformed) versus log (endurance). Regressions were forced through the origin.

even with SVL included as a covariate. However, for those families with multiple observations (Gekkonidae, Lacertidae, Phrynosomatinae, Polychrotinae, and Scincidae), the trend between sprint speed and PTM was negative (see below for the results of the phylogenetic analysis).

We found no relation between MPM and stamina (Table 2.3). However, removing the effects of body size yielded a significant correlation ($r = 0.55$, df $= 23$, $p < 0.05$). As predicted by the foraging mode paradigm, endurance and PTM were significantly related ($r = 0.82$, $F_{1,23} = 22.67$, $p < 0.001$) (Fig. 2.7B). The relation was significant after controlling for the effects of body size ($r = 0.64$, df $= 25$, $p < 0.001$) and MPM and PTM were significantly correlated ($r = 0.58$, $p < 0.01$). This strong correlation suggests that both MPM and PTM should be used in defining the foraging mode of lizards and other taxa (see Butler, 2005).

Phylogenetic analyses

The contrasts for sprint speed were not significantly correlated with the contrasts for MPM or PTM (Table 2.3). However, the contrasts for endurance were positively and significantly related to MPM (Table 2.3). As predicted by the foraging mode paradigm, endurance and PTM were significantly related (Table 2.3). The relation was significant after controlling for the effects of body size. Independent contrasts for MPM and PTM were also significantly correlated (Table 2.3).

Evolutionary transition in locomotor performance in relation to foraging behavior

The analyses above suggest that physiological performance and foraging behavior are co-adapted. However, phylogenetic definitions of adaptation require evidence that shifts in locomotor performance coincided with the evolutionary change in feeding behavior (Losos and Miles, 1994). To test this hypothesis, we used Classification and Regression Tree (CART) analysis to derive an equation for predicting foraging mode by using MPM and PTM. We used both measures because our analyses found a significant correlation between the two behaviors (see also Butler, 2005). Next, we applied CART to the performance variables to determine which combination of sprint speed and stamina predicts foraging mode. The derived classification functions allowed us to infer the foraging mode of the hypothetical ancestors. To do so, we entered the nodal values derived from the parsimony analysis into each of the classification functions.

Only one variable, PTM, was important in classifying species. Sit-and-wait species had values of PTM $< 10.4\%$ ($G^2 = 20.5$, $p < 0.05$, $r^2 = 0.84$); all other species were categorized as active foragers. The classification tree based on locomotor performance required three steps to categorize the species. The first step placed species into the active foraging category if sprint speed was below $1.02\,\mathrm{m/s}$ ($G^2 = 5.07$, $p < 0.01$, $r^2 = 0.72$); faster species could be either SW or AF. The second step involved a placing species into sit-and-wait foragers if sprint speed exceeded $1.02\,\mathrm{m/s}$ and endurance was below $276\,\mathrm{s}$ ($G^2 = 29.9$, $p < 0.05$, $r^2 = 0.70$).

Applying the classification function to the nodal values revealed that the ancestor to the Scleroglossa, with a PTM of 19%, would be an active forager. The estimated ancestral values for sprint speed ($1.69\,\mathrm{m/s}$) and stamina ($546\,\mathrm{s}$) are also consistent with an active forager. Hence the evolutionary transition in foraging mode coincides with a shift in locomotor performance. The gekkonids and cordylids have performance characteristics corresponding to that of SW species. In addition, the Xantusiidae have performance values consistent with a SW forager. However, this is based on a single species, more data are required to substantiate this inference. In summary, each of the major transitions in foraging mode has been accompanied by a shift in locomotor performance.

Morphological differences associated with foraging mode

Three axes were required to explain 80% of the variation (Table 2.4). The first PC axis (43% of total variation) described a contrast between the hindfoot, longest toe and tail length against the variables that described "stockiness," i.e. body width and pectoral width. Thus, at one end of this axis we see mainly stocky species with short tails and elongated distal limb elements, e.g. species in the Iguania (Fig. 2.8). Species with narrower bodies, e.g. Lacertidae and

Table 2.4 *Results of a principal components analysis extracted from a covariance matrix of 12 size-adjusted morphological variables*

	PC axis		
Variable	1	2	3
Body width	−0.74	−0.34	−0.50
Pectoral depth	−0.53	−0.39	0.19
Pectoral width	−0.74	−0.28	−0.52
Humerus	−0.53	0.29	0.62
Radius	−0.56	0.43	0.49
Hand	0.24	0.74	−0.14
Longest digit	0.36	0.73	−0.07
Femur	−0.39	0.17	0.50
Shank	−0.19	0.38	0.35
Foot	0.76	0.37	−0.24
Fourth toe	0.81	0.37	−0.30
Tail length	0.72	−0.66	0.17
Eigenvalue	0.047	0.028	0.014
Variance	43%	26%	12%

Teiidae, are positioned at the opposite end of the axis. The traits that affected the second PC axis (26% of total variation) were mainly related to the forelimb (e.g. length of the distal part of the forelimb and the length of the hand and longest digit of the forelimb) and tail length. This axis contrasted species with relatively short forelimbs and toes (e.g. Anguidae and Scincidae) against species that have longer forelimbs, e.g. Tropidurinae, Hoplocercinae, and Polychrotinae at the positive end (Fig. 2.8). The third axis, 12% of total variation, contrasted elements of the fore- and hindlimbs against those of body and pectoral width. Thus, narrow-bodied species with long limbs were placed at the positive side of the axis (e.g. Corytophaninae and Crotaphytinae) but species in the Gerrhosauridae and Oplurinae were placed at the negative end. We used the scores from each PC axis in a one-way ANOVA with foraging mode as the main effect. The means for sit-and-wait and active foraging differed on all three axes (PC 1 $F_{1,\,245} = 11.2$; PC 2 $F_{1,245} = 85.0$, and PC 3 $F_{1,245} = 13.02$; all $p < 0.001$). Thus, active foraging species were significantly different from sit-and-wait species along all three axes of variation.

Canonical discriminant analysis significantly separated the AF from SW species (Wilk's $\lambda = 0.49$, $F_{12,242} = 19.6$; $p < 0.001$). Active foraging species were characterized by relatively long tails, but relatively short forelimbs and hind-limbs (Table 2.5). We tested whether the pattern of separation was due to

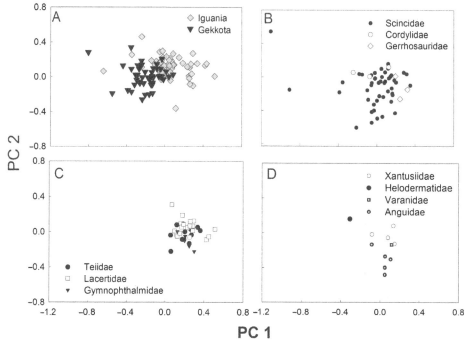

Figure 2.8. Plot of the first two principal components axes. Individual panels present the position of species from the different lizard families included in the study. (A) Families and subfamilies of the Iguania and Gekkota; (B) Gerrhosauridae, Cordylidae, and Scincidae, (C) Lacertidae, Gymnophthalmidae, and Teiidae; (D) Xantusiidae, Anguidae, Varanidae, and Helodermatidae.

phylogenetic affinity by conducting an additional canonical analysis using appropriate clade-specific contrasts (Table 2.5). This approach is similar to the canonical phylogenetic ordination described by Giannini (2003), in which a tree matrix is coded as a binary indicator variable for all the monophyletic groups. There was a major divergence in morphology that corresponds with the Iguania–Scleroglossa contrast. If the difference in morphology between AF and SW species was unrelated to phylogeny, then the loadings for each canonical analysis (foraging mode and Iguania–Scleroglossa contrast) should differ once phylogeny is explicitly incorporated into the analysis. Whereas the previous test compared two clades, it did not explicitly incorporate phylogenetic information. We tested for the effects of phylogeny by performing a canonical variates analysis in Mesquite (Rhetenor Module). The analysis found a significant difference between sit-and-wait and active foragers ($F_{1,244} = 130.00$, $p < 0.001$). The loadings for the analysis were qualitatively similar to the non-phylogenetic canonical variates.

Table 2.5 *Results from a canonical variates analysis*

The columns present the results of a canonical analysis for the following contrasts: foraging mode (AF vs. SW), Iguania versus scleroglossans (Ig–Sclero), gekkotans versus autarchoglossans (Gek–Auto), Gekkota versus Scincoidea (Gek–Scin), Scincoidea versus Lacertoidea (Scin–Lacer), and Lacertidae versus Teiidae (Lacer–Teiidae).

	Contrast					
Trait	AF vs. SW	Ig–Sclero	Gek–Auto	Gek–Scin	Scin–Lacer	Lacer–Teiidae
Body width	−0.13	−0.29	0.13	0.06	0.46	−0.34
Pectoral depth	−0.19	−0.12	0.01	−0.04	0.36	−0.32
Pectoral width	−0.05	−0.42	0.33	0.31	0.29	−0.12
Humerus	0.42	0.15	0.49	0.47	0.35	0.05
Radius	0.43	0.04	0.55	0.58	−0.14	0.07
Hand	0.15	0.37	0.01	0.03	−0.29	0.29
Longest digit	0.29	0.26	0.05	0.11	−0.35	0.53
Femur	0.53	0.19	0.48	0.43	0.40	−0.19
Shank	0.60	0.42	0.36	0.39	−0.13	0.11
Foot	0.05	0.45	−0.38	−0.37	−0.49	0.09
Fourth toe	−0.03	0.24	−0.29	−0.31	−0.29	0.27
Tail length	−0.39	−0.13	−0.56	−0.54	−0.27	−0.09
Wilks' λ	0.49	0.46	0.29	0.28	0.66	0.379
F-ratio	19.6	24.24	30.2	23.5	3.94	3.41
p	0.001	0.001	0.001	0.001	0.001	0.01

A second facet of the morphological analysis entails determining whether the observed variation in morphology is related to performance and foraging mode. We computed the mean values of three PC axes, sprint speed, endurance, MPM, and PTM for each family. These mean values were used in a partial least squares analysis to determine the correlation between each variable. Retaining a single factor resulted in the lowest value for the predicted error sum-of-squares. There was no significant improvement in the goodness-of-fit when adding another 1 or 2 factors. Sprint speed and endurance were significantly associated with PC 1 ($r = 0.56^{**}$ and 0.57^{**}, respectively). Endurance and MPM had high loadings with PC 2 ($r = -0.51^{*}$). Only MPM and PTM were correlated with PC 3 ($r = -0.49^{*}$ and -0.72^{*}, respectively). Thus, narrow-bodied species with long tails were characterized by high sprint capacities and endurance. High values for MPM and PTM were linked with families whose species had narrow bodies, long tails and relatively short distal elements of the forelimb (e.g. Anguidae, Scincidae, Teiidae), whereas low values were linked with families that had stocky bodies, long limb elements and short tails (several families and subfamilies in the Iguania).

Table 2.6 *Pearson correlation values between two movement variables (MPM, PTM), two measures of performance capacity (maximum sprint speed and sprint sensitivity), and one morphological measure (residual hindlimb length) for the eight anole species shown in Fig. 2.5.*
*, $p < 0.05$ (all tests have 6 df).

	MPM	PTM
Maximum sprint speed	−0.75*	−0.61
Sprint sensitivity	−0.72*	−0.90*
Relative hindlimb length	−0.69	−0.87*

Co-evolution of movement, morphology, and performance within an Anolis lizard clade

The contrasts for body size (mean snout–vent length) were not significantly correlated with the contrasts for either movement variable (MPM, Pearson $r = 0.48$, PTM, $r = 0.41$, $p > 0.20$ for both tests, 6 df), or either performance measure (see Irschick and Losos, 1998). This result makes intuitive sense, as these eight species were chosen because they are approximately similar in overall size. Thus, we used non-size-adjusted values in all bivariate correlations. Percentage time moving and moves per minute are significantly and positively correlated among species ($r = 0.85$, $p < 0.05$, 6 df). Thus, species that possess a high value of MPM also tend to have high values of PTM. Indeed, patterns were similar for MPM and PTM with regard to their relations with morphology and performance (Table 2.6), although correlations were generally stronger for PTM. The PTM was significantly and negatively related with sprint sensitivity, meaning that species that spent a large fraction of their time moving (e.g. the twig anole *A. valencienni*) also tended to sprint effectively across a range of surface diameters (Fig. 2.9A). Similarly, species with high values of PTM tend to have relatively long hindlimbs (Fig. 2.9B). The PTM was not significantly correlated with maximum speed, however (Table 2.6), although it showed a strong negative relation with this variable (MPM showed a significant negative relation with maximum speed) (Table 2.6). The differences between MPM and PTM partly reflect the fact that PTM varies more markedly among species, especially in comparison with MPM (Fig. 2.10), thus leading to somewhat weaker correlations between MPM and ecological and morphological variables. Nevertheless, an overall conclusion from these analyses is that anole species that tend to move for relatively long periods of time

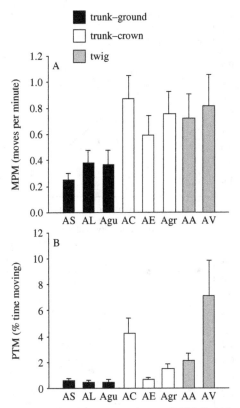

Figure 2.9. Mean values of moves per minute (MPM) (A) and percent time moving (PTM) (B) for eight anole species (see Irschick [2000] for details). Error bars are + 1 s.e.

(or make relatively large numbers of movements per unit time) tend also to have relatively short hindlimbs, move at low maximum speeds, and decline in speed relatively little among different surface diameters.

Discussion

Variation in foraging mode

Our analysis of the distribution of foraging mode revealed a more complex scenario than a SW–AF dichotomy. Results from the ancestor reconstruction based on the binary scoring of foraging mode showed a shift in behavior that occurred early in the evolution of squamate reptiles, i.e. at the autarchoglossans. Mapping the two traits MPM and PTM on the phylogeny suggests that the divergence in foraging mode occurred even earlier, i.e. at the node for scleroglossans; this result is consistent with other studies (Vitt and Pianka,

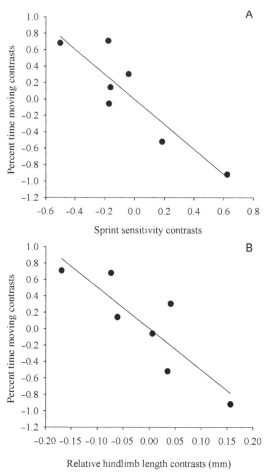

Figure 2.10. Scatter plots of independent contrasts of PTM (y-axis) versus (A) sprint sensitivity (x-axis) and (B) residual hindlimb length for the eight anole species shown in Fig. 2.5. Regressions were forced through the origin.

2005; Vitt *et al.*, 2003). In addition, reconstruction of ancestral values revealed 16 transitions in foraging mode. Four transitions involved a shift from SW to AF in the Gekkota. Multiple shifts from AF to SW were also evident, particularly in the families Lacertidae and Scincidae. We note that the direction of evolutionary change in foraging mode depends on the method of ancestor reconstruction. Both parsimony and likelihood analysis found a substantial amount of uncertainty regarding the ancestral value of foraging mode in the early diversification of the Gekkota. Interestingly, all of the transitions in foraging mode occurred relatively recently in the history of each lineage. Despite the tendency for foraging mode to have a strong phylogenetic signal, i.e. to be relatively stable within each family, the evidence for plasticity in foraging mode provides an opportunity to investigate the ultimate selective and proximate ecological factors that may favor the transition in foraging mode. This is one aspect of the foraging mode paradigm that has been neglected. The reconstruction of the continuous traits MPM and PTM demonstrated that the ancestor to the families Scincidae, Gerrhosauridae, Lacertidae, and Teiidae moved frequently. The low amount of movement during foraging is evident in the Iguania and the Cordylidae. Thus, the behaviors associated with foraging mode evolved early in the history of lizards.

We stress that these results are based only on a limited sample of lizard species. Few data are available on MPM or PTM for most species. Indeed, the majority of estimates are based on a limited sampling regime (less than 5 min) on a restricted number of individuals (Butler, 2005). Significantly, we are missing data for MPM or PTM for some critical taxa, such as species in the Gekkonidae, Eublepharidae, Xantusiidae, and Varanidae. Despite the assertion that SW and AF are endpoints on a continuum, the use of a dichotomous variable will not yield additional insights into foraging behavior. Instead, analyses should focus on more quantitative variables to unambiguously categorize the foraging behavior of lizards (and other taxa). Additional traits should be measured (e.g. velocity during foraging, mean distance moved), to develop a more realistic characterization of differences in foraging behavior related to selection of ambush sites versus prey searching and pursuit (see Cooper *et al.*, 2005). We highlight these weaknesses not to diminish the conclusions emerging from our analysis, but rather to suggest that additional data are likely to reveal more plasticity in foraging mode and increase the power of future comparative analyses.

Locomotor performance and foraging mode

The evolution of specialization in locomotor performance is a key aspect of the foraging mode paradigm. Each mode of acquiring prey entails either high sprint capacity or the ability to maintain high amounts of aerobic activity, as estimated by stamina. Sit-and-wait foragers had significantly faster sprint speeds than active foragers, after adjusting for differences in size in a non-phylogenetic analysis. The stamina of active foragers was greater than that of the sit-and-wait species. Both of these results were obtained without controlling for phylogeny. After controlling for phylogenetic relationships, only the differences in sprint speed persisted. The phylogenetic ANOVA had results for stamina that were in the predicted direction, but the difference was only marginally significant. These results are consistent with those of Huey *et al.* (1984), who found that a sit-and-wait forager (*Eremias [Pedioplanis] lineo-ocellata*) had greater sprint speeds than its sister taxon, *E. [Heliobolus] lugubris*. This difference was challenged by Perry and Pianka (1997) and Perry (1999), who suggested that the comparison used by Huey *et al.* (1984) was inappropriate. This debate highlights the importance of avoiding two-species comparisons (Garland and Adolph, 1994). We suggest that two-species comparisons are often critical for initial hypothesis generation (and indeed the work of Huey *et al.* was inspiration for all that followed), but such studies should be followed by broad-scale comparisons.

Interestingly, our results differed from those obtained by Van Damme and Vanhooydonck (2001), who found no difference in sprint speed in relation to foraging mode. This difference is intriguing, because of the nearly 80% overlap in the species included in each study. There are two potential explanations for the discrepancy. First, we included taxa from lineages not included in their study. For example, our analysis included species from the Cordylidae, Anguidae, Helodermatidae, Xantusiidae, and Gymnophthalmidae. Second, we noted that several taxa in Van Damme and Vanhooydonck (2001) coded as AF are actually sit-and-wait species (e.g. *Mabuya spilogaster*). We note that two different approaches to controlling for the effects of phylogeny yielded similar results. However, it is clear that the dataset for locomotor performance has numerous gaps. For example, data on stamina for the Gerrhosauridae are lacking. Locomotor performance data for this group are critical, because the sister taxon (Cordylidae) has evolved sit-and-wait foraging from an active foraging ancestor (see Whiting, this volume, Chapter 13). Consequently, key evolutionary transitions in the sister-taxon comparison were excluded.

Despite the absence of performance data for key taxa, the patterns show that (a) ambush foragers have higher sprint speeds than widely foraging species and (b) the evolutionary transition from SW to AF involves a decrease in speed and an increase in endurance. The paucity of data on endurance makes it difficult to definitively state whether the recent transitions from SW to AF involved the evolution of greater endurance. In at least one example, *Teratoscincus przewalskii*, an AF species, had a higher value for endurance than other closely related SW gekkonids.

Foraging mode is repeatedly discussed as a continuous variable with SW and AF representing extreme values. However, most analyses focus on the dichotomous classification, which prevents examination of the nuances in foraging behavior. For example, Butler (2005) provided evidence for at least four different foraging strategies. We used the variables underlying the classification of species into each foraging mode to assess the link with locomotor performance. This approach provides the ability to link locomotor performance with behaviors related to foraging. We found a positive association between sprint speed and MPM. One prediction emerging from the foraging mode paradigm is that species with low values of PTM should have high sprint capacities. Therefore, we expected a negative correlation between speed and PTM, but instead there was no association. In contrast, endurance was positively correlated with PTM, but uncorrelated with MPM. In general, the correlations were similar when controlling for the effects of phylogeny. The positive correlation between sprint speed and MPM was not significant once the effects of phylogeny were removed (Table 2.3). Our results are concordant with Garland (1999), who also demonstrated a positive correlation between endurance and PTM. Such a correlation is evidence for co-adaptation between physiology and behavior. In a previous paper, Garland (1993) asked whether high endurance capacity was an adaptation for wide foraging. The answer to the question would require showing that high endurance is a general character of all active foragers, not only a few select taxa, e.g. species in the family Teiidae. The significant correlation between stamina and an index of activity in the field across a wide range of taxa provides evidence in support of an adaptive explanation. In addition, activities other than movement may be associated with high stamina. For example, many studies have argued that stamina may provide an overall index into the aerobic capacity of an individual (see Garland and Losos, 1994; Garland, 1999; Robson and Miles, 2000; Sinervo *et al.*, 2000; Miles *et al.*, 2001). In contrast, the costs of travel over the course of a day may be less than the estimated capacity derived from measurements made on a treadmill. One neglected aspect is that multiple tasks may require enhanced aerobic capacity. The pursuit of prey may only entail a fraction of the organism's time and energy budget. Hence, an individual that has a high aerobic capacity may pursue activities that require sustained performance, e.g. ability to dig burrows, excavate hidden prey, or process large prey, that may also be tied to high endurance capacity. Nevertheless, we have shown that lizards that move extensively while searching for prey have also evolved the physiological capacity to maintain high activity. We agree with Garland (1999) in hoping that our analysis will stimulate additional fieldwork on collecting additional data on field behavioral and locomotor performance.

Enhanced stamina capacity would be considered an adaptation for active foraging if the change in the trait occurred simultaneously with a shift in feeding behavior. However, if stamina showed an evolutionary change before or after the transition in behavior, then the inference would be that the change in performance was an exaptation and indirectly affected foraging behavior. The ancestral reconstruction of foraging mode placed the shift to active foraging as early as the appearance of ancestral autarchoglossans. We found that a shift in performance, specifically stamina, was coincident with the change in foraging behavior. Based on this pattern we conclude that evolving enhanced locomotor performance, specifically endurance capacity, was an important selective factor in the evolution of foraging behavior.

A trade-off between sprint speed and endurance is a key prediction of the foraging mode syndrome. An earlier analysis of sprint speed and endurance based on 12 lacertid species revealed support for a trade-off between speed and endurance (Vanhooydonck *et al.*, 2001). Our comparative analysis did not find evidence for such a trade-off. The absence of a trade-off suggests that speed and endurance may be decoupled traits (Robson and Miles, 2000). However, examination of correlations within families found significant negative correlations for the Polychrotidae and Lacertidae. Interestingly, teiid lizards had a positive correlation between speed and endurance. This Olympian behavior may be explained by the types of habitats teiid lizards often exploit when foraging and their risk of predation. Macroteiid species such as *Aspidoscelis tigris* often feed in open habitats, which increases their exposure to predators. On the one hand, searching for prey over great distances, as seen in active foraging species (e.g. *A. tigris*) is facilitated by high endurance. Furthermore, feeding in open environments enhances the probability of encountering a predator. Therefore, high sprint capacities may be required for avoiding predators. Active foraging species that exploit habitats with complex micro-environments and numerous refugia from predators should relax selection favoring high sprint performance. One would predict a strong correlation between survivorship and sprint speed in teiids.

Finally, a myriad of factors may influence individual variation in locomotor performance. That is, multiple ecological tasks may require enhanced speed or high endurance. As Garland (1993, 1999) and others (Robson and Miles, 2000) have highlighted, there are many behaviors that require enhanced endurance, including the patrolling of a home range, territorial defense, searching for and acquiring mates, and locating food. Sprint speed may also be used for multiple reasons, including predator escape and subduing prey. Definitive evidence regarding the link between locomotor performance and foraging mode should include a path analysis that decomposes variation in locomotor performance into separate factors. This would provide an opportunity to determine the relative importance of locomotor performance with each task.

Morphological variation

We first investigated the patterns of covariation among 12 morphological traits by using principal components analysis. The placement of species in the morphological space shows a clear phylogenetic component. Examination of Fig. 2.8 shows a clustering of the major lineages that is consistent with the major lineages of lizards. Interestingly, the Iguania have species that range over most of the morphological space. In addition, there was a partitioning of the Scincoidea (= Gerrhosauridae + Cordylidae + Scincidae) from the Lacertoidea (= Lacertidae + Gymnophthalmidae + Teiidae) along the second PC axis. Another pattern emerging from the PC analysis is that the morphological variation reflects adaptation to different habitats. Finally, the patterns of covariation appear to be consistent with the predictions for adaptations to different modes of foraging. The loadings for the first two axes describe suites of morphological traits that covary as predicted by theoretical studies of foraging behavior. The means for SW and AF species were significantly different. The morphological differences included active foraging species having a longer tail and narrower body. Sit-and-wait species had shorter tails, longer hindlimbs and elongated distal limb elements, which would facilitate rapid acceleration and faster speed. In total, the characteristics of the active foraging species are consistent with the traits favoring elevated stamina.

Discriminant functions analysis is a more statistically appropriate method for comparing morphological differences between groups. We followed Vitt and Pianka (2005) in comparing morphological variation between several contrasts identified from the phylogenetic tree (Fig. 2.1). Interestingly, body form was not the major trait that separated SW and AF species. Rather, tail length and limb proportions were the traits that had the highest loadings. Active foraging species had longer tails, but relatively shorter limbs, than sit-and-wait species. We note that the discriminant analysis using phylogenetic contrasts as the classification variable identified different morphological traits that separated these groups. Specifically, the split between Iguania and scleroglossans entails differences in body form (body width and pectoral width) and distal elements of the fore- and hindlimb. Interestingly, two groups that have been considered to represent convergence in form (Lacertidae and Teiidae) were significantly different in morphology. Finally, a phylogenetically based canonical variates analysis supported the morphological differences obtained in the non-phylogenetic multivariate tests.

A variety of studies have demonstrated correlations between morphology and locomotor performance. However, few have tested whether the ecomorphological association is concordant between different ecological groups. It is not sufficient to show that SW and AF species differ in morphology. One must ascertain that the variation in limb length or body form results in a difference in performance. Demonstrating significant covariation between morphology and the locomotor performance traits associated with feeding behavior would strengthen the adaptive significance of the differences between foraging modes. We showed that the morphological and performance traits found to differ between foraging strategies were also correlated. Interestingly, both sprint speed and stamina were associated with the first PC axis. However, high endurance was characteristic of species with narrow bodies and long tails, whereas faster speeds were characteristic of species with shorter tails, relatively short forelimbs relative to the hindlimb, and longer distal elements of the hindlimb. Endurance and MPM were associated with the second PC axis. Again, tail length was associated with performance as inferred from the PC loadings (Table 2.4). Species with high values of MPM also had longer limbs, but not differences between the fore- and hindlimb. Clearly, the covariation between morphology, performance, and behavior demonstrates that the morphological differences between SW and AF species have functional consequences.

Intraspecific variation

Our analyses show that movement patterns are indeed correlated with both performance capacity and relative limb dimensions in a well-defined clade of *Anolis* lizards. To a large extent, the documented relations among movement patterns, hindlimb dimensions, and sprinting performance reflect an ecological and behavioral transition among anole ecomorphs. At one end of the extreme, trunk–ground anoles are characterized by relatively long hindlimbs, high sprint sensitivities, and low values of PTM and MPM. In many ways, trunk–ground anoles fit the classic view of a highly polygynous "sit-and-wait" predator. Inspection of Fig. 2.9 shows that there are exceptions to this trend, however. For example, the trunk–crown anole *A. carolinensis* is noteworthy in having unusually high values of both MPM and PTM. Trunk–ground males are highly conspicuous, intensely territorial, and have relatively small home ranges that typically encompass the home ranges of various females (Schoener, 1968; Losos, 1990a). Because trunk–ground anole males occupy broad surfaces (e.g. large tree trunks), their lifestyle also emphasizes high-speed running, both to capture prey and to elude predators (Irschick and Losos, 1998). Thus, these species tend to be more sedentary and rely on actively moving prey, i.e. a sit-and-wait foraging strategy. On the other extreme, twig anoles are characterized by their relatively short hindlimbs, low sprint sensitivities, and high values of PTM and MPM. Behavioral studies of the twig anole *A. valencienni* show that this species has extremely large home ranges, is relatively non-territorial, and is also highly cryptic (Hicks and Trivers, 1983). Indeed, many aspects of the morphology of twig anoles are designed for crypsis, such as their pale coloration (which blends in nicely with the twig surface) and their short tails, which are aligned with the branch surface (Irschick and Losos, 1996). This cryptic lifestyle is enhanced by their use of narrow twigs, which requires short legs to increase their sure-footedness. Consequently, their morphology prevents their relying on speed to avoid predators or capture prey. These interrelated variables have favored a wide foraging behavior to search extensively for hidden prey (Losos *et al.*, 2003). In conclusion, our analyses show that studying a well-defined clade can shed considerable light on the evolution of movement patterns. Indeed, this is the power of the phylogenetic approach. Factoring out the variation that characterizes large clades presents the opportunity to investigate associations between characters as they evolve. A key lesson from our analysis is that a uniform treatment of one group as "sit-and-wait" or "actively foraging" may overshadow many interesting biological differences among species. We suggest that, parallel with important macro-evolutionary studies of movement patterns (Perry, 1999), researchers also investigate the evolution of movement patterns in detail within well-defined and closely related animal groups.

Acknowledgements

This chapter benefited from the comments of two anonymous reviewers. Our research has been generously supported by grants from the National Science Foundation.

References

Aerts, P. R., Van Damme, R., Vanhooydonck, B., Zaaf, A. and Herrel, A. (2000). Lizard locomotion: how morphology meets ecology. *Netherl. J. Zool.* **50**, 261–77.

Arnold, E. N. (1989). Towards a phylogeny and biogeography of the Lacertidae: relationships within an old-world family of lizards derived from morphology. *Bull. Brit. Mus. (Nat. Hist.) Zool.* **55**, 209–57.

Anderson, R. A. and Karasov, W. H. (1981). Contrast in energy intake and expenditure in sit and wait and widely foraging lizards. *Oecologia* **49**, 67–72.

Breiman, J. H., Fridman, J. H., Olshen, R. A. and Stone, C. J. (1983). *Classification and Regression Trees*. Belmont, CA: Wadsworth.

Butler, M. A. (2005). Foraging mode of the chameleon, *Bradypodion pumilum*: a challenge to the sit-and-wait versus active forager paradigm? *Biol. J. Linn. Soc.* **84**, 797–808.

Clark, L. and Pregibon, D. (1992). Tree-based models. In *Statistical Models*, ed. J. M. Chambers and T. J. Hastie, pp. 377–419. Pacific Grove, CA: Wadsworth.

Cooper, W. E. Jr. (1995). Foraging mode, prey chemical discrimination, and phylogeny in lizards. *Anim. Behav.* **50**, 973–85.

Cooper, W. E. Jr. (1997). Correlated evolution of prey chemical discrimination with foraging, lingual morphology, and vomeronasal chemoreceptor abundance in lizards. *Behav. Ecol. Sociobiol.* **41**, 257–65.

Cooper, W. E. Jr. and Whiting, M. J. (2000). Ambush and active foraging modes both occur in the scincid genus *Mabuya*. *Copeia* **2000**, 112–18.

Cooper, W. E. Jr., Whiting, M. J., Van Wyk, J. H. and Mouton, P. le F. N. (1999). Movement- and attack-based indices of foraging mode and ambush foraging in some gekkonid and agamine lizards from southern Africa. *Amph.-Rept.* **20**, 391–9.

Cooper, W. E. Jr., Vitt, L. J., Caldwell, J. P. and Fox, S. F. (2001). Foraging modes of some American lizards: relationships among measurement variables and discreteness of modes. *Herpetologica* **57**, 65–76.

Cooper, W. E. Jr., Vitt, L. J., Caldwell, J. P. and Fox, S. F. (2005). Relationships among foraging variables, phylogeny, and foraging modes, with new data for nine North American lizard species. *Herpetologica* **61**, 250–9.

Diaz-Uriarte, R. and Garland, T. Jr. (1998). Effects of branch length errors on the performance of phylogenetic independent contrasts. *Syst. Biol.* **47**, 654–72.

Dunham, A. E., Miles, D. B. and Reznick, D. N. (1988). Life history patterns in squamate reptiles. In *Biology of the Reptilia*, vol. 16, ed. C. Gans and R. B. Huey, pp. 331–86. New York: A. R. Liss.

Estes, R., de Queiroz, K. and Gauthier, J. (1988). Phylogenetic relationships within Squamata. In *Phylogenetic Relationships of the Lizard Families: Essays Commemorating Charles L. Camp*, ed. R. Estes and G. Pregill, pp. 119–281. Stanford, CA: Stanford University Press.

Evans, S. E. (2003). At the feet of dinosaurs: the early history and radiation of lizards. *Biological Reviews* **78**, 513–51.

Felsenstein, J. (1985). Phylogenies and the comparative method. *Amer. Nat.* **125**, 1–15.

Felsenstein, J. (1988). Phylogenies and quantitative characters. *Ann. Rev. Ecol. Syst.* **19**, 445–71.

Frost, D. R. and Etheridge, R. (1989). A phylogenetic analysis and taxonomy of iguanian lizards. *Misc. Publ. Mus. Nat. Hist. Univ. Kansas* **81**, 1–65.

Fu, J. (2000). Towards the phylogeny of the family Lacertidae: why 4708 base pairs of mtDNA sequences cannot draw the picture. *Biol. J. Linn. Soc.* **71**, 203–17.

Garland, T. Jr. (1993). Locomotor performance and activity metabolism of *Cnemidophorus tigris* in relation to natural behaviors. In *Biology of Whiptail Lizards (Genus Cnemidophorus)*, ed. J. W. Wright and L. J. Vitt, pp. 163–210. Norman, OK: Oklahoma Museum of Natural History.

Garland, T. Jr. (1994). Phylogenetic analyses of lizard endurance capacity in relation to body size and temperature. In *Lizard Ecology: Historical and Evolutionary Perspectives*, ed. L. J. Vitt and E. R. Pianka, pp. 207–36. Princeton, NJ: Princeton University Press.

Garland, T. Jr. (1999). Laboratory endurance predicts variation in field locomotor behaviour among lizard species. *Anim. Behav.* **57**, 77–83.

Garland, T. Jr. and Adolph, S. C. (1994). Why not to do two-species comparative studies: limitations on inferring adaptation. *Physiol. Zool.* **67**, 797–828.

Garland, T. Jr. and Losos, J. B. (1994). Ecological morphology of locomotor performance in squamate reptiles. In *Ecological Morphology: Integrative Organismal Biology*, ed. P. C. Wainwright and S. M. Reilly, pp. 240–302. Chicago, IL: University of Chicago Press.

Garland, T. Jr., Dickerman, A. W., Janis, C. M. and Jones, J. A. (1993). Phylogenetic analysis of covariance by computer simulation. *Syst. Biol.* **42**, 265–92.

Garland, T. Jr., Harvey, P. H. and Ives, A. R. (1992). Procedures for the analysis of comparative data using phylogenetically independent contrasts. *Syst. Biol.* **41**, 18–32.

Gerritsen, J. and Strickler, J. R. (1977). Encounter probabilities and community structure in zooplankton: a mathematical model. *J. Fish. Res. Board Can.* **34**, 73–82.

Giannini, N. P. (2003). Canonical phylogenetic ordination. *Syst. Biol.* **52**, 684–95.

Han, D., Zhou, K. and Bauer, A. M. (2004). Phylogenetic relationships among gekkotan lizards inferred from C-*mos* nuclear DNA sequences and a new classification of the Gekkota. *Biol. J. Linn. Soc.* **83**, 353–68.

Harris, D. J. and Arnold, E. N. (1999). Relationships of wall lizards, *Podarcis* (Reptilia: Lacertidae) based on mitochondrial DNA sequences. *Copeia* **1999**, 749–54.

Harris, D. J. and Arnold, E. N. (2000). Elucidation of the relationships of the spiny-footed lizards, *Acanthothdactylus* spp. (Reptilia: Lacertidae) using mitochondrial DNA sequence, with comments on their biogeography and evolution. *J. Zool.* **252**, 351–62.

Harris, D. J., Arnold, E. N. and Thomas, R. H. (1998). Relationships of lacertid lizards (Reptilia: Lacertidae) estimated from mitochondrial DNA sequences and morphology. *Proc. R. Soc. Lond.* **B265**, 1939–48.

Hedges, S. B., Bezy, R. L. and Maxson, L. R. (1991). Phylogenetic relationships and biogeography of Xantusiid lizards, inferred from mitochondrial DNA sequences. *Molec. Biol. Evol.* **8**, 767–80.

Herrel, A., Meyers, J. J. and Vanhooydonck, B. (2002). Relations between microhabitat use and limb shape in phrynosomatid lizards. *Biol. J. Linn. Soc.* **77**, 149–63.

Hertz, P. E., Huey, R. B. and Garland, T. Jr. (1988). Time budgets, thermoregulation, and maximal locomotor performance: are ectotherms Olympians or boy scouts? *Amer. Zool.* **28**, 927–38.

Hicks, R. A. and Trivers, R. L. (1983). The social behavior of *Anolis valencienni*. In *Advances in Herpetology and Evolutionary Biology: Essays in Honor of Ernest E. Williams*, ed. A. G. J. Rhodin and K. I. Miyata, pp. 570–95. Cambridge, MA: Museum of Comparative Zoology, Harvard University.

Honda, M., Ota, H., Kobayashi, M. *et al.* (1999). Evolution of Asian and African lygosomine skinks of the *Mabuya* group (Reptilia: Scincidae): A molecular perspective. *Zool. Sci.* **16**, 979–84.

Honda, M., Ota, H., Köhler, G. *et al.* (2003). Phylogeny of the lizard subfamily Lygosominae (Reptilia: Scincidae), with special reference to the origin of the New World taxa. *Genes Gen. Syst.* **78**, 71–80.

Huey, R. B. and Pianka, E. R. (1981). Ecological consequences of foraging mode. *Ecology* **62**, 991–9.

Huey, R. B. and Bennett, A. F. (1986). A comparative approach to field and laboratory studies in evolutionary biology. In. *Predator-Prey Relationships: Perspectives and Approaches From the Study of Lower Vertebrates*, ed. M. E. Feder and G. V. Lauder, pp. 82–98. Chicago, IL: University of Chicago Press.

Huey, R. B., Bennett, A. F., John-Alder, H. B. and Nagy K. A. (1984). Locomotor capacity and foraging behaviour of Kalahari lacertid lizards. *Anim. Behav.* **32**, 41–50.

Irschick, D. J. (2000). Comparative and behavioral analyses of preferred speed: *Anolis* lizards as a model system. *Physiol. Biochem. Zool.* **73**, 428–37.

Irschick, D. J. and Losos, J. B. (1996). Morphology, ecology, and behavior of the twig anole *Anolis angusticeps*. In *Contributions to West Indian Herpetology: a Tribute to Albert Schwartz*, ed. R. Powell and R. W. Henderson, pp. 291–301.

Contributions in Herpetology, vol. 12. Ithaca, NY: Society for the study of Amphibians and Reptiles (SSAR).

Irschick, D. J. and Losos, J. B. (1998). A comparative analysis of the ecological significance of maximal locomotor performance in Caribbean *Anolis* lizards. *Evolution* **52**, 219–26.

Irschick, D. J., and Losos, J. B. (1999). Do lizards avoid habitats in which performance is submaximal? The relationship between sprinting capabilities and structural habitat use in Caribbean anoles. *Amer. Nat.* **154**, 293–305.

Irschick, D. J., Vitt, L. J., Zani, P. A. and Losos, J. B. (1997). A comparison of evolutionary radiations in mainland and Caribbean *Anolis* lizards. *Ecology* **78**, 2191–203.

Jackman, T. R., Irschick, D. J., de Queiroz, K., Losos, J. B. and Larson, A. (2002). Molecular phylogenetic perspective on evolution of lizards of the *Anolis grahami* series. *J. Exp. Zool.* **294**, 1–16.

Jackman, T. R., Larson, A., de Queiroz, K. and Losos, J. B. (1999). Phylogenetic relationships and tempo of early diversification in *Anolis* lizards. *Syst. Biol.* **48**, 254–85.

Jayne, B. C. and Irschick, D. J. (2000). A field study of incline use and preferred speeds for the locomotion of lizards. *Ecology* **81**, 2969–83.

Kluge, A. G. (1987). Cladistic relationships in the Gekkonoidea (Squamata, Sauria). *Misc. Publ. Mus. Zool. Univ. Mich.* **173**, 1–54.

Lamb, T. and Bauer, A. M. (2003). *Meroles* revisted: complementary systematic inference from additional mitochondrial genes and a complete taxon sampling of southern Africa's desert lizards. *Molec. Phylogenet. Evol.* **29**, 360–4.

Lapointe, F.-J. and Garland, T. Jr. (2001). A generalized permutation test for the analysis of cross-species data. *J. Classif.* **18**, 109–27.

Losos, J. B. (1990a). The evolution of form and function: morphology and locomotor performance in West Indian *Anolis* lizards. *Evolution* **44**, 1189–203.

Losos, J. B. (1990b). Concordant evolution of locomotor behavior, display rate, and morphology in *Anolis* lizards. *Anim. Behav.* **39**, 879–90.

Losos, J. B. (1994). Integrative approaches to evolutionary ecology: *Anolis* lizards as model systems. *Ann. Rev. Ecol. Syst.* **25**, 467–93.

Losos, J. B. and Miles, D. B. (1994). Adaptation, constraint and the comparative method: phylogenetic issues and methods. In *Ecological Morphology: Integrative Organismal Biology*, ed. P. C. Wainwright and S. M. Reilly, pp. 60–98. Chicago, IL: The University of Chicago Press.

Losos, J. B. and Miles, D. B. (2002). Testing the hypothesis that a clade has adaptively radiated: iguanid lizard clades as a case study. *Am. Nat.* **160**, 147–57.

Losos, J. B. and Sinervo, B. (1989). The effects of morphology and perch diameter on sprint performance of *Anolis* lizards. *J. Exp. Biol.* **145**, 23–30.

Losos, J. B., Butler, M. and Schoener, T. W. (2003). Sexual dimorphism in body size and shape in relation to habitat use among species of Caribbean *Anolis* lizards. *In Lizard Social Behavior*, ed. S. F. Fox, J. K. McCoy and T. A. Baird, pp. 356–80. Baltimore, MD: Johns Hopkins University Press.

Losos, J. B. Jackman, T. R., Larson, A., de Queiroz, K. and Rodriquez-Schettino, L. (1998). Contingency and determinism in replicated adaptive radiations of island lizards. *Science* **279**, 2115–18.

McBrayer, L. D. (2004). The relationship between skull morphology, biting performance and foraging mode in Kalahari lacertid lizards. *Zool. J. Linn. Soc.* **140**, 403–16.

McBrayer, L. D. and Reilly, S. M. (2002). Prey processing in lizards: Behavioral variation in sit-and-wait and widely foraging taxa. *Can. J. Zool.* **80**, 882–92.

McLaughlin, R. L. (1989). Search modes of birds and lizards: evidence for alternative movement patterns. *Amer. Nat.* **133**, 654–70.

Maddison, W. P. and Maddison, D. R. (2004). Mesquite: A modular system for evolutionary analysis. Ver 1.06. http://mesquiteproject.org.

Mausfeld, P., Vences, M., Schmitz, A. and Veith M. (2000). First data on the molecular phylogeography of Scincid lizards of the genus *Mabuya*. *Molec. Phylogenet. Evol.* **17**, 11–14.

Melville, J. and Swain, R. (2000a). Mitochondrial DNA-sequence based phylogeny and biogeography of the snow skinks (Squamata: Scincidae: *Niveoscincus*) of Tasmania. *Herpetologica* **56**, 196–208.

Melville, J. and Swain, R. (2000b). Evolutionary relationships between morphology, performance, and habitat openness in the lizard genus *Niveoscincus* (Scincidae: Lygosominae). *Biol. J. Linn. Soc.* **70**, 667–83.

Midford, P. E., Garland, T. Jr. and Maddison, W. P. (2003). PDAP Package for Mesquite Ver. 1.05. http://mesquiteproject.org.

Miles, D. B. (1994). Covariation between morphology and locomotory performance in Sceloporine lizards. In *Lizard Ecology: Historical and Evolutionary Perspectives*, ed. L. J. Vitt and E. R. Pianka, pp. 207–36. Princeton, NJ: Princeton University Press.

Miles, D. B., Snell, H. L. and Snell, H. M. (2001). Interpopulation variation in endurance of Galapagos lava lizards *Microlophus albemarlensis*: evidence for an interaction between natural and sexual selection. *Evol. Ecol. Res.* **3**, 795–804.

Moller, A. P. and Birkhead, T. R. (1992). A pairwise comparative method as illustrated by copulation frequency in birds. *Amer. Nat.* **139**, 644–56.

Mosimann, J. E. (1970). Size allometry: size and shape variables with characterizations of the lognormal and generalized gamma distributions. *J. Amer. Stat. Assoc.* **65**, 930–45.

Nagy, K. A., Huey, R. B. and Bennett, A. F. (1984). Field energetics and foraging mode of Kalahari lacertid lizards. *Ecology* **65**, 588–96.

Ota, H., Honda, M., Chen, S.-L. *et al.* (2002). Phylogenetic relationships, taxonomy, character evolution and biogeography of the lacertid lizards of the genus *Takydromus* (Reptilia: Squamata): a molecular perspective. *Biol. J. Linn. Soc.* **76**, 493–509.

Perry, G. (1999). The evolution of search modes: ecological versus phylogenetic perspectives. *Amer. Nat.* **153**, 98–109.

Perry, G. and Pianka, E. R. (1997). Animal foraging: past, present and future. *Trends Ecol. Evol.* **12**, 360–4.

Pianka, E. R. (1966). Convexity, desert lizards and spatial heterogeneity. *Ecology* **47**, 1055–9.

Pianka, E. R. (1973). The structure of lizard communities. *Ann. Rev. Ecol. Syst.* **4**, 53–74.

Pianka, E. R. (1986). *Ecology and Natural History of Desert Lizards*. Princeton, NJ: Princeton University Press.

Pietruszka, R. D. (1986). Search tactics of desert lizards: how polarized are they? *Anim. Behav.* **34**, 1742–58.

Reeder, T. W. (2003). A phylogeny of the Australian *Sphenomorphus* group (Scincidae: Squamata) and the phylogenetic placement of the crocodile skinks (*Tribolonotus*): Bayesian approaches to assessing congruence and obtaining confidence in maximum likelihood inferred relationships. *Molec. Phylog. Evol.* **27**, 384–97.

Reeder, T. W., Cole, C. J. and Dessauer, H. C. (2002). Phylogenetic relationships of Whiptail lizards of the genus *Cnemidophorus* (Squamata: Teiidae): a test of monophyly, reevaluation of karyotypic evolution, and review of hybrid origins. *Amer. Mus. Nov.* **3365**, 1–61.

Reeder, T. W. and Wiens, J. J. (1996). Evolution of the lizard family Phrynosomatidae as inferred from diverse types of data. *Herpetol. Monogr.* **10**, 43–84.

Regal, P. J. (1978). Behavioral differences between reptiles and mammals: an analysis of activity and mental capacities. In *Behavior and Neurology of Lizards* ed. N. Greenberg and P. D. Maclean, pp. 183–202. Washington, D.C.: Department of Health, Education, and Welfare.

Rest, J. S., Ast, J. C., Austin, C. C. *et al.* (2003). Molecular systematics of primary reptilian lineages and the tuatara mitochondrial genome. *Molec. Phylog. Evol.* **29**, 289–307.

Robson, M. A. and Miles, D. B. (2000). Locomotor performance and dominance in male tree lizards, *Urosaurus ornatus. Funct. Ecol.* **14**, 338–44.

Schluter, D., Price, T., Mooers, A. Ø. and Ludwig, D. (1997). Likelihood of ancestor character states in adaptive radiation. *Evolution* **51**, 1699–711.

Schoener, T. W. (1968). The *Anolis* lizards of Bimini: resource partitioning in a complex fauna. *Ecology* **49**, 704–26.

Schoener, T. W. (1971). Theory of feeding strategies. *Ann. Rev. Ecol. Syst.* **2**, 369–404.

Schwenk, K. (2000). An introduction to tetrapod feeding. In *Feeding: Form, Function and Evolution in Tetrapod Vertebrates*, ed. Kurt Schwenk, pp. 21–61. San Diego, CA: Academic Press.

Sinervo, B., Miles, D. B., DeNardo, D., Frankino, T. and Klukowski, M. (2000). Testosterone, endurance, and Darwinian fitness: natural and sexual selection on the physiological bases of alternative male behaviors in side-blotched lizards. *Horm. Behav.* **38**, 222–33.

Stephens, D. W. and Krebs, J. R. (1986). *Foraging Theory.* Princeton, NJ: Princeton University Press.

Thompson, G. G. and Withers, P. C. (1997). Comparative morphology of western Australia monitor lizards (Squamata: Varanidae). *J. Morphol.* **233**, 127–52.

Townsend, T., Larson, A., Louis, E. and Macey, R. J. (2004). Molecular phylogenetics of Squamata: the position of snakes, Amphisbaenians, and Dibamids and the root of the Squamate tree. *Syst. Biol.* **53**, 735–57.

Van Damme, R. and Vanhooydonck, B. (2001). Origins of interspecific variation in lizard sprint capacity. *Funct. Ecol.* **15**, 186–202.

Van Damme, R. B., Vanhooydonck, B., Aerts, P. and De Vree, F. (2003). Evolution of lizard locomotion: context and constraint. In *Vertebrate Biomechanics and Evolution*, ed. V. L. Bels, J.-P. Gasc and A. Casinos, pp. 267–83. Oxford: BIOS Scientific Publishers.

Vanhooydonck, B. and Van Damme, R. (1999). Evolutionary relationships between body shape and habitat use in lacertid lizards. *Evol. Ecol. Res.* **1**, 785–805.

Vanhooydonck, B. and Van Damme, R. (2001). Evolutionary trade-offs in locomotor capacities in lacertid lizards: are splendid sprinters clumsy climbers? *J. Evol. Biol.* **14**, 46–54.

Vanhooydonck, B., Van Damme, R. and Aerts, P. (2001). Speed and stamina trade-off in Lacertid lizards. *Evolution* **55**, 1040–8.

Venables, W. N. and Ripley, B. D. (1994). *Modern Applied Statistics in S-Plus*. New York: Springer-Verlag.

Vitt, L. J. (1983). Tail loss in lizards: the significance of foraging and predator escape modes. *Herpetologica* **39**, 151–62.

Vitt, L. J. and Congdon, J. D. (1978). Body shape, reproductive effort, and relative clutch mass in lizards: resolution of a paradox. *Amer. Nat.* **112**, 595–608.

Vitt, L. J. and Pianka, E. R. (2005). Deep history impacts present-day ecology and biodiversity. *Proc. Nat. Acad. Sci. USA* **102**, 7877–81.

Vitt, L. J., Pianka, E. R., Cooper, W. E. and Schwenk, K. (2003). History and the global ecology of squamate reptiles. *Amer. Nat.* **162**, 44–60.

Vitt, L. J. and Price, H. J. (1982). Ecological and evolutionary determinants of relative clutch mass in lizards. *Herpetologica* **38**, 237–55.

Webb, J. K., Brook, B. W. and Shine, R. (2003). Does foraging mode influence life history traits? A comparative study of growth, maturation, and survival of two species of sympatric snakes from south-eastern Australia. *Aust. Ecol.* **28**, 601–10.

Whiting, A. S., Bauer, A. M. and Sites, J. W. Jr. (2003). Phylogenetic relationships and limb loss in sub-Saharan African scincine lizards (Squamata: Scincidae). *Molec. Phylogen. Evol.* **29**, 582–98.

Wiens, J. J. and Reeder, T. W. (1997). Phylogeny of the spiny lizards (*Sceloporus*) based on molecular and morphological evidence. *Herpetol. Monogr.* **11**, 1–101.

Wiens, J. J. and Slingluff, J. L. (2001). How lizards turn into snakes: A phylogenetic analysis of body form evolution in Anguid lizards. *Evolution* **55**, 2303–18.

Williams, E. E. (1983). Ecomorphs, faunas, island size, and diverse end points in island radiations of *Anolis*. In *Lizard Ecology: Studies of a Model Organism*, ed. R. B. Huey, E. R. Pianka and T. W. Schoener, pp. 326–70. Cambridge, MA: Harvard University Press.

Zaaf, A. and Van Damme, R. (2001). Limb proportions in climbing and ground-dwelling geckos (Lepidosauria, Gekkonidae): a phylogenetically informed analysis. *Zoomorphology* **121**, 45–53.

Zani, P. A. (1996). Patterns of caudal autotomy evolution in lizards. *J. Zool. Lond.* **240**, 201–20.

Zani, P. A. (2000). The comparative evolution of lizard claw and toe morphology and clinging performance. *J. Evol. Biol.* **13**, 316–25.

3

Physiological correlates of lizard foraging mode

KEVIN E. BONINE

Department of Ecology and Evolutionary Biology, University of Arizona

Introduction

Everything that an animal does, from feeding to escaping predation, is influenced by underlying physiological traits. In this chapter, I focus on the insights gained by examining physiology in the context of lizard foraging mode. I will discuss apparent relationships between physiological traits and foraging mode and identify areas where we might expect to uncover explanatory correlates in the future. Understanding these relationships allows us to learn more about the evolution of lizard foraging and the related evolutionary processes and selective factors that act on underlying physiological components.

Many biologists have addressed the importance of evolutionary and comparative physiology (see, for example, Prosser, 1950; Diamond, 1993; Garland and Carter, 1994; Natochin and Chernigovskaya, 1997; Feder *et al.*, 2000). Physiology can be rather broadly defined to include many integrated and hierarchical suborganismal traits that manifest in organismal function. An incomplete list of these traits and processes includes enzyme activity, membrane selectivity, establishment of ion gradients, ATP production, cellular respiration, lung ventilation, aerobic capacity, Q_{10} effects, lactate buffering, pH balance, sprint speed, digestive efficiency, and many other processes involved in homeostasis that ultimately contribute to the survival and reproduction (Darwinian fitness) of organisms (see Table 3.1 for a list of physiology-related traits likely to inform the study of foraging modes). Importantly, these myriad suborganismal traits do not work in isolation, nor are they typically involved in only one aspect of organismal function (see, for example, Bennett, 1989; Garland and Losos, 1994; Rose and Lauder, 1996b). Hence, it is important to consider suites of characters and their potential co-adaptation (see, for example, Price and Langen, 1992; Bauwens *et al.*, 1995) as well as the potential constraints (see, for example, Zera and Harshman, 2001; Van Damme *et al.*,

Lizard Ecology: The Evolutionary Consequences of Foraging Mode, ed. S. M. Reilly, L. D. McBrayer and D. B. Miles. Published by Cambridge University Press. © Cambridge University Press 2007.

Table 3.1 *Physiological traits (broadly defined) that may prove fruitful for understanding correlates of foraging mode in lizards*

Performance traits	Suborganismal traits
Locomotion	Circulation / oxygen transport
Endurance	Lung capacity
Sprint speed	Heart mass
Aerobic capacity	Hematocrit / hemoglobin
Anaerobic scope	Muscle physiology
Activity	Muscle mass
Body temperature / temperature breadth	Muscle fiber types
Standard / basal metabolic rate	Myoglobin
Water loss rates (cutaneous / respiratory)	Enzyme activity
Supercooling / freeze tolerance	Blood / muscle buffering
Neural Processing	Other
Nervous system integration	Kidney function
Orientation / homing	Liver mass
Chemoreception	Olfaction
	Visual acuity

2003) placed on different components of an organism. For example, as discussed by Cooper in Chapter 8, the evolution of lizard feeding followed a couple of paths, one of which involves increased use of the tongue for chemoreception, and concomitant changes in tongue length and forkedness, vomeronasal activity, and increased discriminatory ability among chemicals in potential foods. Potential constraints for such a lizard now include reduced abilities of the tongue for actually manipulating prey.

This chapter will highlight known and potential correlations among physiological and performance characters primarily related to the locomotor apparatus required for foraging. I will also address shifts in traits or their function and potential subsequent reduction in utility of those traits for other activities. Examination of physiological characters should not be made in isolation but should be considered in the context of evolutionary history and much of the information presented in this volume, especially in conjunction with the morphological and performance correlates of foraging mode (Miles *et al.*, this volume, Chapter 2), which thereby presents a fuller picture of forms and functions involved in lizard foraging abilities.

A useful framework for examining physiology and foraging mode, the Morphology → Performance → Fitness paradigm, was first presented by Arnold (1983) and then expanded to incorporate Behavior (between performance and fitness; Garland *et al.*, 1990a; Garland and Losos, 1994); the performance abilities of an animal may not be expressed if the behavior of the animal intercedes. For example, a lizard may have the morphological and physiological machinery (e.g. long legs and a high proportion of fast-twitch glycolytic muscle) for high-speed locomotion, but if the lizard chooses to sit still then selection will be acting on the behavior and not the performance abilities. This M→P→B→F paradigm is useful in that it allows for the separation of the suborganismal characters that constrain and determine whole-animal performance from the evolutionary processes acting on the whole animal. Once this operational dichotomy has been created we can begin to amass data (and many data have already been collected) to understand what traits are important for performance abilities and then how evolutionary history has acted on those performance abilities. Much of this chapter will involve the M → P portion of this paradigm.

Some areas of lizard physiology likely to be important for foraging abilities have been the focus of much attention involving many species, whereas other areas are especially ripe for further in-depth analysis. The physiological underpinnings of any aspect of lizard biology could easily warrant a volume or two. In this chapter I will touch on the highlights, assess what we know, and suggest areas worthy of increased effort. Because most relevant data exist for limbed lizards, I will not address the many groups of limbless lizards, including the most prominent limbless group, the snakes (see, for example, Savitzky, 1980; Greene, 1997; Wiens and Slingluff, 2001).

Foraging mode and physiology

Previously, several workers have addressed foraging mode and physiological traits, often with just a few species. One of the early studies, presented next, will be useful as a reference point throughout this chapter.

Studying two species of lacertid from the Kalahari, Bennett, Huey, and John-Alder (1984) found that the widely foraging species *Eremias* (now *Heliobolus*) *lugubris*, had higher maximal oxygen consumption and greater relative heart mass and hematocrit, traits expected to be more developed in a species adapted to spending a greater portion of its activity time moving around. Conversely, the sit-and-wait species *E.* (now *Pedioplanis*) *lineoocellata*, had higher anaerobic scope, measured as the level of lactate accumulating in the body after burst exercise. Interestingly, several hindlimb muscle traits did not differ between these two species: hindlimb muscle mass as a proportion of body mass, myoglobin concentration, citrate synthase or myofibrillar ATPase activity, and isometric contractile properties of the iliofibularis muscle. This two-species comparison highlights what several other researchers have found subsequently: some expected characters vary with foraging mode, yet many do not. We are thus reminded that traits rarely evolve in isolation or for one sole purpose. Furthermore, measuring the contractile properties of one muscle, although an important first step, does not allow us to understand what other muscles in the hindlimb (or the forelimbs, tail, and trunk) are contributing to locomotor abilities. Huey *et al.* (1984) found that *E. lugubris* and *E. lineoocellata* do indeed differ in burst sprinting abilities and endurance capacity in the directions predicted by their foraging modes. However, another widely foraging species, *E.* (now *Pedioplanis*) *namaquensis*, did not have similarly high endurance capacity, leading the authors to suggest that behavior may be a more plastic trait than suborganismal components of locomotor performance abilities. Indeed, several researchers have suggested that behaviors may evolve before other traits (Blomberg *et al.*, 2003).

The comparison of two species above also highlights an important issue regarding statistical inference. If the expected null hypothesis is that of no difference in trait values between two species, then observed differences between species will be in the predicted, alternative direction 50% of the time even if the proposed adaptive explanation is wrong. The result is dramatic inflation of the Type I error rate (Garland and Adolph, 1994). As we shall see, the subject to which this chapter is devoted is ready for a broad-based phylogenetic analyses of foraging mode and underlying physiological and performance characteristics. The examination of only a few species or traits will not allow us to understand the correlates of foraging mode across the diversity of extant lizards.

Whole-animal performance measures

Energy balance

In a large-scale review of data for stomach contents of 18 223 individuals of 127 lizard species from four continents, Huey *et al.* (2001; see also Vitt and Pianka, this volume, Chapter 5) found that widely foraging species have a greater frequency of empty stomachs, suggesting negative energy balance, as compared with sit-and-wait foragers. Top predators had a higher frequency of empty stomachs compared with species that feed lower on the trophic cascade, and nocturnal species had more empty stomachs than did diurnal species. Lizards specializing on ants did not often have empty stomachs, similar to diurnal termite specialists. Neither diet breadth nor body size was correlated with frequency of empty stomachs. However, certain families had species with more empty stomachs than other families, indicating a need for phylogenetically based analyses of these data, as noted by Huey *et al.* (2001), because the observed relationships may be driven more by phylogenetic relationships than other ecological processes. The initial findings about foraging mode, nocturnality, and trophic level are exciting and warrant further detailed investigation. One relevant example that should further spur research is the evolution of nocturnality in geckos that has been accompanied by a decreased cost of locomotion, and increased maximal aerobic speed and maximum oxygen consumption, at reduced temperatures when compared with diurnal relatives (Autumn *et al.*, 1999). Findings like these should be integrated with an explicit phylogenetic analysis of underlying suborganismal traits to see whether foraging mode explains, in part, the evolution of nocturnality in geckos (Cooper, 1995) and in other lizard groups.

Other earlier examinations of foraging mode in relation to energy flux used doubly labeled water (Anderson and Karasov, 1981; Nagy *et al.*, 1984) for more accurate measurement of foraging efficiency and activity costs. For the *Eremias* pair discussed earlier, the actively foraging *E. lugubris* had four times higher energy expenditure while foraging, but also had twice the overall food gain (Huey and Pianka, 1981). Across six species of lacertid, the wide foragers had 1.3–1.5 times greater activity expenditure compared with sit-and-wait foragers, but they also averaged 1.3–2.1 times greater food intake (Huey and Pianka, 1981). Similar patterns of energy expenditure and energy gain were observed for *Cnemidophorus* (now *Aspidoscelis*) *tigris* (widely foraging) and *Callisaurus draconoides* (sit-and-wait) (Anderson and Karasov, 1981), as well as for *Cnemidophorus* (now *Aspidoscelis*) *exsanguis* (widely foraging) and *Sceloporus jarrovi* (sit-and-wait) (Andrews, 1984; see also Nagy, 1983). Subsequent to these early studies other researchers have accurately measured energy budgets in a number of individual species, but until this volume (Brown and Nagy, Chapter 4) no synthetic compilation of these data was applied to questions of energy balance and foraging mode across a broader taxonomic range (see Nagy *et al.*, 1999; Nagy and Shemanski, 2003).

Sprint speed

Sprint speed has been shown to correlate with a host of traits (see Miles *et al.*, this volume, Chapter 2) including body mass and body temperature (Van Damme and Vanhooydonck, 2001), hindlimb span (Bonine and Garland, 1999 and references therein), caudifemoralis longus muscle length (Zani, 1996), and muscle fiber area (Gleeson and Harrison, 1988). More important for our focus in this chapter, across 111 species for which sprint speed has been measured on a racetrack, burst sprint speed was unrelated to foraging mode (Van Damme and Vanhooydonck, 2001). This result contrasts with earlier findings between the pair of *Eremias* species (Huey *et al.*, 1984). A phylogeneti-cally informed analysis, as well as incorporation of metrics such as moves per minute or percent time moving, rather than the dichotomous sit-and-wait or actively foraging categories (see discussion in Van Damme and Vanhooydonck, 2001), is needed to more definitively assess the relation between sprinting abilities and foraging mode (see, for example, Miles *et al.*, Chapter 2).

A further consideration is the method by which sprint speeds are measured. Racetrack estimates of sprint speed are often lower than those provided by high-speed treadmills, especially for faster species (Bonine and Garland, 1999; Irschick and Garland, 2001; K. Bonine and T. Garland, in preparation). When attempting to correlate burst sprint performance with other biological char-acters, it is important to accurately estimate *maximal* performance abilities in order to elevate the signal to noise ratio and thereby identify where animals and physiological systems are truly constrained (see, for example, Garland and Losos, 1994). Losos *et al.* (2002) highlight other important considerations when measuring and interpreting sprint performance data.

Endurance

The early analyses of *Eremias* indicated that foraging mode was indeed corre-lated with endurance ability for some species, but not for others (Huey *et al.*, 1984). Across 57 species of lizard, endurance was positively correlated with body mass and body temperature (Garland, 1994), using either phylo-genetically correct (Garland *et al.*, 1993) or phylogenetically uninformed analyses. Upon publication of a broad dataset quantifying lizard foraging behavior and indicating that phylogenetic relationships were the greatest predictor of foraging mode (Perry, 1999, this volume, Chapter 1), these endur-ance data were re-analyzed in relation to foraging mode. Across the subset of species for which datasets overlapped, laboratory endurance time measured on a treadmill at 1 km/h was correlated with both percent time moving ($n = 15$ species) and daily movement distance ($n = 11$ species) (Garland, 1999) even when phylogeny was taken into account. Thus, for this relatively small sam-pling of species (there are more than 4800 species of lizard, excluding snakes [Pough *et al.*, 2004]) endurance does indeed seem to be correlated with com-ponents of foraging mode. A logical next step is asking whether the predictors of endurance also predict variation in foraging mode.

The ways in which lizards move around in their natural environment has not always been synonymous with the ways in which biologists have attempted to measure whole-animal performance (Irschick, 2000; Irschick and Garland, 2001). To measure endurance in the laboratory, researchers most often attempt to keep an animal moving constantly at a fixed speed in order to make meaningful comparisons across species (Garland, 1994; Miles *et al.*, this volume, Chapter 2; but see Pinch and Claussen, 2003). However, intermittent locomotion seems to be a more commonly observed phenomenon in natural settings and can increase aerobic endurance and activity capacity (Weinstein and Full, 1999; Weinstein, 2001; Gleeson and Hancock, 2002). Applying knowledge of intermittent behavior (see Kramer and McLaughlin, 2001) and variations in propensity to pause across species, in addition to variation in laboratory endurance capacity discussed above, could prove fruitful for understanding important correlates of foraging mode and activity-related behaviors.

Temperature

The influence of temperature on biological function, especially in ectotherms, is well documented (see, for example, Dawson, 1967; Bennett, 1980). However, the relationship between field-active body temperature and foraging mode needs to be thoroughly examined. An early synthetic look at thermoregulation and body temperature placed lizards into broad categories such as "species in which activity temperatures are relatively high and constant" (Avery, 1982, p. 100). Many other publications, older ones cited by Avery (1982), report field active body temperatures for a few species at a time. A relatively untapped mine of temperature data has been amassed by at least one researcher over the past decade or more (Robert Espinoza, unpublished) and could be extremely useful for answering questions about foraging mode and other aspects of lizard biology.

The influence of temperature seems to differ for aerobic vs. anaerobic capacities in lizards. Aerobic capacities studied in several species are strongly influenced by temperature, with Q_{10} values in the neighborhood of 2–2.5 as would be expected from temperature-dependent chemical reactions (Bennett, 1982; Autumn *et al.*, 1999). However, body size influences the effect of temperature on aerobic capacity in some species: larger species are temperature-insensitive across relatively high temperatures (Bennett, 1982). Interestingly, temperature does not seem to be as important for anaerobic capacity at the whole animal level (Q_{10} not different from 1) (Bennett, 1982). At the muscle level (skeletal muscle in *Dipsosaurus dorsalis*), rates of force development, contraction, and relaxation are temperature-dependent, with Q_{10} values around 2 (Bennett, 1985). However, the maximal force produced is not temperature-dependent (Bennett, 1985). Gaining a better understanding of the thermal dependence of whole-animal vs. suborganismal physiological traits, across a broad range of lizard taxa, will inform our discussion of temperature and foraging habits.

Physiological limits imposed by temperature constrain activity periods and geographic distribution (see, for example, Miles, 1994; Burke *et al.*, 2002; Porter *et al.*, 2002). In addition to, and consistent with, maximal performance variation, temperature also influences behavioral choices in many species, such as when to flee (see, for example, Bulova, 1994; Cooper, 2000, 2003; Amo *et al.*, 2003) or when to forage (see, for example, Angilletta, 2001). Integrating temperature, physiology, and behavior with foraging mode will be a fruitful line of research when enough data are compiled to perform broad-scale tests of foraging correlates.

Water loss

Water balance follows readily from a discussion of temperature. As body temperature increases so does evaporative water loss, especially as metabolism and respiration also increase. Water is gained from free water (including condensation and rain harvesting (see, for example, Peterson, 1998)), pre-formed water in food items, or as a byproduct of food metabolism. Water is lost via urine and fecal excretion (and salt excretion in herbivorous species such as *Dipsosaurus dorsalis*, *Amblyrhynchus cristatus*, and others) (Peaker and Linzell, 1975), evaporation from wet surfaces of the mouth and eyes, evaporation across the skin, and from reproductive products such as eggs. Studies of water loss across lizard skin have found dramatic differences among species (see summary in Buttemer, 1990). Ecological correlations have been hypothesized for several groups, including *Coleonyx* (Dial and Grismer, 1992), *Cnemidophorus* (now *Aspidoscelis*) (Cullum, 1997), and *Agama* (Dmi'el, 2001), as well as for many studies of individual species. Similar to temperature, constraints mediated through water balance influence the evolution of physiological traits and adaptive behaviors, as well as determining activity time (daily, seasonally, and in response to climate change) and location (microhabitat selection and geographic distribution) across multiple scales. Again, a synthetic analysis of water loss rates in the context of broader phylogenetic relationships and foraging mode would be a helpful addition to the growing body of knowledge.

Aerobic capacity

As discussed above, maximal oxygen consumption was greater in the actively foraging *H. lugubris* than in the sit-and-wait *P. lineoocellata*. Suborganismal physiological traits, relative heart mass, and hematocrit in the *Eremias* (now *Pedioplanis*) study, expected to facilitate differences in maximal oxygen consumption, also differed between species (Bennett *et al.*, 1984). Thompson and Withers (1997) presented data for nine species of *Varanus* suggesting that higher maximal oxygen consumption was positively correlated with degree of arboreality and foraging distance. Other variables hypothesized to be important for aerobic capacity, and therefore for foraging mode, include hemoglobin, myoglobin, maximum aerobic speed, and citrate synthase activity (see outlined list in Garland, 1993). Datasets of these traits for many groups of species have been published; notably, Garland (1993) presents data on heart mass for 16 species, hematocrit for 35 species, hemoglobin for 39 species, maximum aerobic speed for 15 species, and maximal oxygen consumption for 15 species. Addition of other species and an explicit test of these traits across lizard taxa remain to be done; whether the correlations, and lack thereof, observed for *Eremias* will hold is an interesting empirical question.

The aerobic capacities of *Heloderma* provide an interesting cautionary tale. Helodermatid lizards have low resting metabolic rates (Beck and Lowe, 1994), but higher than expected endurance at 1.0 km/h, high aerobic scope, and lower than expected maximal sprint speed for their body size (John-Alder *et al.*, 1983; Beck *et al.*, 1995). Beck *et al.* (1995) note that the aerobic capacities, which are surprisingly high for a relatively sedentary lizard, may be important during prolonged male–male wrestling to gain access to females. High aerobic capacity in *Heloderma* may have little to do with foraging behavior (although they are reported to move 21.5% of the time when active (cited in Perry [1999]). The ecology of individual species or populations should inform our assessment of correlates of foraging mode.

Anaerobic scope

The aerobic capacity of lizards is approximately an order of magnitude lower than that of similarly sized mammals (Bennett, 1978). However, abilities for short-term burst speeds are comparable (see, for example, Bennett, 1991), indicating an increased utilization of anaerobic metabolic pathways (see, for example, Bennett, 1978). Indeed, in the species that have been evaluated, the ability to elevate blood lactate concentrations and then to replenish energy stores from accumulated lactate is dramatically greater in lizards (and other reptiles) than in mammals (Bennett, 1978; Gleeson and Dalessio, 1989; Gleeson, 1996; Donovan and Gleeson, 2001). Reliance on short-term burst activity, and the fact that burst speed seems relatively temperature-insensitive (see above), is one successful strategy adopted by lizards. The details of anaerobic capacity and variation among species and populations will require expanded datasets.

Suborganismal physiology

Thus far I have primarily reviewed traits measured at the level of the whole animal. Many researchers have also focused on underlying physiological traits that are predicted to explain variation in whole-animal performance abilities. Often these traits are specific to a certain part of the body (hindlimb, lungs, blood, muscle, etc.) and many of these traits are measured on tissue or blood samples or on sacrificed animals. In the next section I will explore an interesting subset of areas that has informed our knowledge of lizard physiology and physiology in the context of foraging behaviors. It is important to remember that selection and evolution are acting on whole animals and not directly on underlying traits, thus observed variations in suborganismal components may or may not be manifest in whole-animal performance or behavioral traits (see, for example, Irschick and Garland, 2001) and our analyses, therefore, require a multi-tiered approach.

Respiration

Providing oxygen to working tissues is an important component of aerobic capacity and endurance as discussed above. Respiratory capacity, especially during exercise, varies across species. Lung ventilation and oxygen uptake increased to match increased levels of running exercise (submaximal but above maximum aerobic speed) in five individual *Varanus exanthematicus*, but not in five individual *Iguana iguana*. After laboratory exercise bouts, the *Varanus* quickly returned to pre-exercise respiratory rates, but the *Iguana* increased ventilation rate and oxygen uptake (Wang *et al.*, 1997; see also "excess post-exercise oxygen consumption" discussion in Gleeson and Hancock, 2002). It seems that the *Iguana*, in contrast to the *Varanus*, were unable to match increased oxygen demand with adequate ventilation at moderate and higher speeds (Wang *et al.*, 1997). Several researchers have hypothesized that a mechanical constraint exists in some lizard species because trunk and hypaxial musculature are employed in lateral bending during locomotion and therefore may not be able to simultaneously expand the thoracic cavity to ventilate the lungs (Carrier, 1989; see review in Boggs, 2002). Several species, including *Varanus*, may utilize buccal pumping and positive pressure ventilation during locomotion to overcome this functional restriction (Al-Ghamdi *et al.*, 2001; Frappell *et al.*, 2002; Boggs, 2002). The role of buccal pumping across species remains to be determined, and the speeds at which some lizards employ positive pressure ventilation may be restricted to speeds below maximal aerobic (Boggs, 2002).

Other sources of variation important for adequate oxygen delivery include lung volume and surface area (Perry, 1998) and the role of blood shunting (typically right-to-left) at the ventricle (intracardiac) and within the lung (intrapulmonary) (Wang *et al.*, 1998). Broad-scale phylogenetic analyses of these traits, and their effects on foraging mode, will undoubtedly reveal many interesting patterns, but the data for most species remain to be collected.

Muscle physiology

Skeletal muscle has received much attention over the past several decades as a likely source of variation pertaining to locomotor performance differences. Characteristics such as contraction time, twitch and tetanic tension, relaxation time, and fatigue resistance all measure performance properties of individual muscles *in vitro*. Electromyography (EMG) allows for *in vivo* examination of muscle participation in various activities. Other measured variables include muscle size, enzyme (e.g. citrate synthase or myosin ATPase) activities, and lactate accumulation (see, for example, Gleeson and Harrison, 1988). One area of focus has been muscle fiber-type composition (see, for example, Putnam *et al.*, 1980; Bonine *et al.*, 2001) because different fiber types are better suited to certain types of work. Slow oxidative (SO) fibers are typically fatigue-resistant but do not contract as quickly or produce as much force as fast glycolytic (FG) fibers (which are not fatigue-resistant). Intermediate fibers, fast oxidative–glycolytic (FOG), have intermediate properties (see, for example, Gleeson *et al.*, 1980b; Gleeson and Johnston, 1987). These three fiber types also have different densities of mitochondria, enzymes, myoglobin, and capillaries (Guthe, 1981; Gleeson and Harrison, 1986). Given that there are constraints on total muscle mass, different muscles have different proportions of fibers (see Peter *et al.*, 1972 for description of three fiber types in mammalian skeletal muscle) likely correlated with their functional roles.

Within a given muscle, different portions may be recruited for different tasks. In *Varanus exanthematicus* at 37 °C, the red (oxidative) portion of the iliofibularis muscle is regularly recruited at slow speeds and reaches a recruitment plateau at about 1.5 km/h. The white (fast-glycolytic) portion of the same muscle is more often recruited at higher speeds: above 1.3 km/h (Jayne *et al.*, 1990). Not coincidentally, the maximum aerobic speed for *V. exanthematicus* is about 1.2 km/h (Gleeson *et al.*, 1980a). Therefore, understanding the relative contributions of different muscles, and different fibers within muscles, should provide insight into correlated locomotor performance variation across species.

To date, most studies on muscle properties have focused on a few muscles and a few species. The first comparative analysis of lizard muscle examined temperature-dependent isometric twitch kinetics in the gastrocnemius and the iliofibularis of four species (Putnam and Bennett, 1982). More recently, the measured variation in iliofibularis muscle fiber-type composition among eleven species of phrynosomatid lizard (Bonine *et al.*, 2001) is suggestive of a locomotor performance relationship, but direct tests have not been published. The only multispecies test for correlations between physiological traits of muscle and locomotor performance abilities was for the two *Eremias* (now *Heliobolus* and *Pedioplanis*) species discussed at the beginning of this chapter (Bennett *et al.*, 1984). Other studies have focused on individual species (see, references in Table 1 of Bonine *et al.*, 2001). The time is ripe for a synthetic examination of muscle properties and locomotor performance abilities in a broad phylogenetic context. Bonine and colleagues are collecting iliofibularis muscle fiber-type data on additional species (mostly native to the southwestern USA) beyond their 11 phrynosomatids in pursuit of a multivariate analysis of correlates of locomotor performance abilities and other behavioral manifestations such as foraging mode. Raoul Van Damme (University of Antwerp, Belgium) and colleagues are working specifically on correlates, including caudifemoralis muscle properties, of foraging behavior across many species of Lacertidae. The analyses and results of these two efforts, and others, should be enlightening.

A brief overview of the known muscle-fiber variation in phrynosomatids (Bonine *et al.*, 2001) provides an example of why suborganismal physiological traits may prove useful in broad-scale analyses. Among 11 species, representing all three subclades of Phrynosomatidae (Fig. 3.1), the iliofibularis muscle, important for hindlimb retraction during locomotion (Jayne *et al.*, 1990), varies in muscle fiber-type composition (Fig. 3.2) using mean values for species from measures of four individuals per species. Importantly, the variation in fiber-type composition is greater between the two most closely related subclades, the horned lizards and the sand lizards, than between the *Sceloporus* group and either the horned or sand lizards (Fig. 3.3). The greatest variation results from a trade-off between proportion of FG and proportion of FOG because the proportion of SO is less variable (<1% to 17%) (see Fig. 3.2). Thus, overall, it seems likely that muscle fiber-type composition is a fairly labile trait, especially considering that the likely ancestral trait value was a mean value of the variation observed in extant animals weighted by phylogeny (approximately 44% FG, 44% FOG, 12% SO) (Bonine, 2001). For this sample of lizards, evolution has acted quickly (<25 million years [Montanucci, 1987 and references therein] compared with *c.* 200 million years for all lizards; see

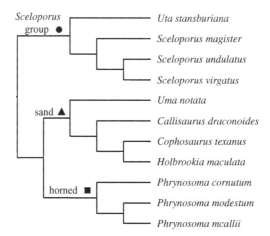

Figure 3.1. Phylogenetic relationships among 11 species of Phrynosomatidae. Branch lengths represent relative, not actual, divergence times. Modified from Bonine *et al.* (2001).

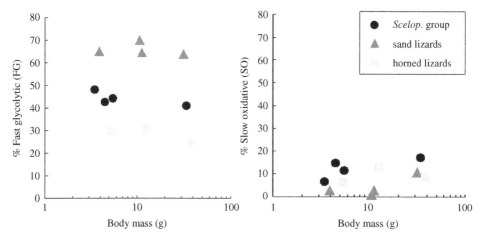

Figure 3.2. Iliofibularis muscle fiber-type composition based on species mean values of 11 phrynosomatid lizard species comprising three subclades. Modified from Bonine *et al.* (2001).

references in Pough *et al.* [2004]) to alter the relative proportion of muscle fiber types in an important locomotor muscle, suggesting that this is a very informative area for further research. Furthermore, preliminary analyses indicate that variation in muscle fiber-type composition is much greater than variation in hematocrit, heart mass, or liver mass for these same species (K. Bonine, T. Garland, and T. Gleeson, unpublished). Expanding the muscle fiber dataset to include other species, along with measures of muscle function and explicit tests of correlations with locomotor performance and foraging mode, will

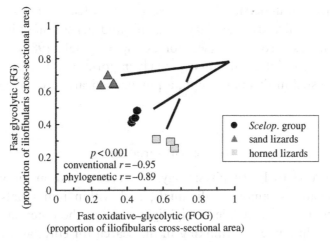

Figure 3.3. Observed trade-off in proportion of fiber types in the iliofibularis muscle of 11 phrynosomatid lizard species. Relative phylogenetic relationships are denoted by the connected black lines forming a "tree" in the upper right-hand portion of the figure. Modified from Bonine *et al.* (2001).

greatly enrich our understanding of muscle fiber variation and its broader functional and evolutionary relationships.

Other physiological traits

Many other suborganismal physiological traits may prove fruitful for analyses in a broader context; several of them have been referred to already in this chapter. Garland (1993) compiled a useful set of data (and suggested other fruitful research avenues) for many suborganismal traits that could be employed in future synthetic analyses. Further examination of enzyme activity levels, especially those pertaining to aerobic or anaerobic locomotor abilities, might provide insight across a broad range of lizard taxa. The capacity of muscle and blood to buffer changes in pH or other metabolite concentrations resulting from exercise (see, for example, Mitchell and Gleeson, 1985; Snyder *et al.*, 1995) may influence locomotor abilities and thus be a candidate for modification by natural selection. Metabolic pathways involved in glucose and glycogen metabolism may be important for foraging-related differences among species (Wickler and Gleeson, 1993). Similarly, the ability to meta- bolize different food items (e.g. effects of herbivory; see Herrel, this volume, Chapter 7) or tolerate prey defenses (see, for example, Schmidt *et al.*, 1989) may affect or constrain the function of blood and other tissues in other capacities. Kidney and liver function vary among species; a concerted effort should be made to understand how these organs affect energetics and metabolism across all lizards. Capillary diameter and density may be related to aerobic capacities across taxa (see, for example, Pough, 1980). Many other suborganismal physiological traits are likely important and await, as do most of the traits listed in this paragraph, a phylogenetically diverse collection of relevant data.

Sources of variation

For traits related to lizard physiology, observed variation among species, among populations, among individuals, and within individuals can have many origins. It is worthwhile to highlight some of the potential sources of variation to help guide our thinking in the search for correlates of foraging mode (see Table 3.2). Some of these sources of variation are the explicit focus of attempts to understand adaptation to different foraging modes. Other sources of variation may complicate or hinder the search for adaptive traits. The importance of phylogenetic relationships in foraging behavior has been well documented (Perry, 1999, this volume, Chapter 1). Other sources of variation may not receive the attention they deserve.

The adaptationist approach (see, for example, Rose and Lauder, 1996a; and as discussed by Gould and Lewontin, 1979) generally seeks species-level differences that would explain differences in resource use, density, geographic

Table 3.2 *Sources of variation in physiological traits related to foraging mode*

Attempts to correlate physiology with feeding mode should consider, or control for, many of these factors.

Variation across individuals
Species- and higher-level taxonomic divisions
Population
Weather and climate

Within-individual variation
Body size
Ontogeny
Sex
Reproductive state
Plasticity / training
Season
Body condition: nutrition, hydration, parasites, tail regeneration, etc.
Daily circadian rhythms

distribution, etc. However, these differences, even if real, can be difficult to measure because of other sources of variation. As Perry highlights in Chapter 1, population-level differences in foraging behavior can be substantial. Other behaviors and more traditional physiological variables also exhibit population-level differences (see, for example, Sinervo and Losos, 1991; Bulova, 1994; Van Damme *et al.*, 1998; Angilletta, 2001; Perry, this volume, Chapter 1). It becomes a somewhat philosophical debate as to whether or not biologists should seek to restrict examinations of a species to one population in order to reduce the influence of among-population level variation, or to embrace the variation within a species as measured across populations and arrive at some mean value, with greater variance, as the trait value for that species. Alternatively, researchers could sample two or more populations from each species and keep them as separate "tips" for analysis. This has been suggested as a way to look at "micro" and "macro" evolution simultaneously (Garland *et al.*, 1992). The variation among populations is the result of interacting genetic and environmental factors; understanding the relative contribution of each, at the level of individual populations within species, will require a heroic input of energy and resources.

Within individuals, we can identify several potential sources of variation that many biologists seeking answers at the level of the two halves (Iguania and Scleroglossa) of the entire lizard phylogeny would hope to control. For example, decades of allometric analyses has shown us that body size contributes substantially to variation at many levels of biological function (see, for example, Calder, 1984; Garland, 1984; Schmidt-Nielsen, 1984; Garland and Else, 1987; Irschick and Jayne, 2000). Typically, once we understand the influence of body size, we explore other physiological relationships by using residuals from regressions on body mass. We know that muscle-fiber diameter scales with body size in phrynosomatid lizards (Bonine *et al.*, 2001). Is there variation in muscle fiber size, once body mass has been accounted for, that explains variation in a behavior such as foraging mode? A broad-scale analysis awaits further data; current knowledge of iliofibularis muscle fibers and foraging metrics overlap for only nine species (Perry, 1999; this volume, Chapter 1; Bonine *et al.*, 2001), making analyses premature.

The sex of an animal is known to influence a host of traits and behaviors (see, for example, Marler and Moore, 1989; M'Closkey *et al.*, 1990; Cullum, 1998; Lailvaux *et al.*, 2003). Within each sex, the reproductive status of an individual, influenced by changes in hormone activity, will alter behavior (basking, territorial defense and aggression, mate searching, etc.) (Carpenter, 1967; Marler and Moore, 1988) and performance abilities (see, for example, Sinervo *et al.*, 1991; Olsson *et al.*, 2000; Shine, 2003). Irrespective of reproductive status, seasonal variation can alter physiological and behavioral traits (Garland and Else, 1987). Similarly, the age of individuals has a direct effect on most traits that have been measured with an eye toward uncovering age-related differences (see, for example, Van Berkum *et al.*, 1989; Miles *et al.*, 1995; Irschick *et al.*, 2000). Understanding these sex-, seasonal-, and reproductive-based differences can allow researchers to attempt to control for variation in the trait of interest (e.g., foraging mode) by only studying adults of one sex in one season. Alternatively, this rich variation can be a source of biologically meaningful information that is often species-, or population-, specific but which can greatly enhance our understanding of the interplay between sex, age, and reproductive state and a trait such as burst sprint speed (see, for example, Irschick, 2000; Olsson *et al.*, 2000). Further complicating the issue are considerations of social status (see, for example, Garland *et al.*, 1990b; Perry *et al.*, 2004) and recent social interactions of individuals (see, for example, Trigosso-Venario *et al.*, 2002; Yang & Wilczynski, 2002; Korzan and Summers, 2004).

Developmental plasticity and training effects are other potential sources of variation. The few attempts to understand endurance training in lizards have met with mixed results (Gleeson, 1979; Garland *et al.*, 1987; Conley *et al.*, 1995; see also Holloszy and Coyle, 1984; Miller and Camilliere, 1981). Understanding plasticity will require common-garden type experimental protocols. The condition of individuals can obviously influence measurable traits and behaviors (see, for example, Downes and Shine, 1999). But, how effective individuals, populations, or species are at coping with stresses such as disease (see, for example, Ribas *et al.*, 1998), malnutrition, or dehydration (see, for example, Cullum, 1997, 1998) could very well be adaptive (although perhaps difficult for biologists to measure) and explain variation in foraging mode or related behavioral, morphological, and physiological traits. Work on pathogens and their effects has revealed information on the role of organisms such as malaria in shaping proximate behaviors and potential selection pressures in *Sceloporus* species (see, for example, Schall *et al.*, 1982; Schall, 1983; J. Foufopoulos and K. Bonine, unpublished). Tail status is another potential indicator of individual health and population-level predation pressures. Loss of a tail may mean loss of stored nutrients and water, may delay or impede growth and reproduction during tail regeneration, and may influence mate choice. Effects, including increased predation risk (Downes and Shine, 2001), highlight the need to consider the influence of other, perhaps subtler, physiological variations across individuals, especially those more difficult for researchers to measure, such as intestinal parasites or specific nutrient deficiencies.

Finally, daily variation in behavior, metabolism, circulating hormone levels, etc. all influence physiological and behavioral traits. A well-designed study will measure variables at the appropriate time of day, consistently, for all study subjects. Again, understanding variation across the course of a day can bring to light important changes that lizards either need to accommodate, or have developed to persist in their environment. Understanding causality is another and more difficult challenge, as is the case for much of the correlational and observational data discussed in this chapter.

Conclusions

Lizard foraging behaviors are likely the complex result of microhabitat, natural selection, and phylogenetic relatedness (Felsenstein, 1985; Garland *et al.*, 1992; 1999; see discussion of phylogenetic inertia and phylogenetic signal in Blomberg *et al.*, 2003). Teasing apart all of the relevant variables will require a substantial research investment. Fortunately, the knowledge to be gained en route to a broad-scale synthetic analysis of lizard foraging correlates will be important and meritorious on its own. Findings from studies of a few species (e.g. the Kalahari lacertids discussed earlier) (Huey and Pianka, 1981; Bennett *et al.*, 1984; Huey *et al.*, 1984; Nagy *et al.*, 1984) lead us to believe that many of the physiological traits we predict to relate to locomotor performance or other traits or behaviors related to foraging will be robust predictors of variation across species. I have highlighted some of those traits for broader samples in this chapter. However, other likely physiological characters do not seem to correlate with either performance or foraging mode. The complex interactions of traits which serve many functions and operate across many scales requires a synthetic, multivariate, and broad-based approach to reveal important and robust relationships within lizards in the context of foraging mode.

The success of books such as the *Lizard Ecology* series (Milstead, 1967; Huey *et al.*, 1983; Vitt and Pianka, 1994) and the recently published *Lizards: Windows to the Evolution of Diversity* (Pianka and Vitt, 2003) reveal the general appeal of lizard biology to scientists as well as to a broader audience. Exciting innovations in scientific tools such as microarrays and genetic engineering (see review in Feder *et al.*, 2000) should allow future in-depth analysis of the mechanistic drivers behind variation in physiology and behavior. These new findings will have to be integrated with, and not replace, an increasing understanding of the role of phylogenetic relatedness, physiological function, and whole-animal performance (Table 3.1) upon which evolution and natural selection act. Research in the near future should reveal important patterns and relationships that generally inform the different and related components of the Morphology → Performance → Behavior → Fitness paradigm, and the study of foraging mode specifically.

Acknowledgments

Ted Garland, Angela Urbon, and an anonymous reviewer provided helpful comments and suggestions on earlier drafts of this chapter. The Department of Ecology and Evolutionary Biology and the Wildlife and Fisheries Resources Program in the School of Natural Resources at the University of Arizona provided valuable in-kind support.

References

Al-Ghamdi, M. S., Jones, J. F. and Taylor, E. W. (2001). Evidence of a functional role in lung inflation for the buccal pump in the agamid lizard *Uromastyx aegyptius microlepis*. *J. Exp. Biol.* **204**, 521–31.

Amo, L., Lopez, P. and Martin, J. (2003). Risk level and thermal costs affect the choice of escape strategy and refuge use in the Wall Lizard, *Podarcis muralis*. *Copeia* **2003**, 899–905.

Anderson, R. A. and Karasov, W. H. (1981). Contrasts in energy intake and expenditure in sit-and-wait and widely foraging lizards. *Oecologia* **49**, 67–72.

Andrews, R. M. (1984). Energetics of sit-and-wait and widely-searching lizard predators. In *Vertebrate Ecology and Systematics: A Tribute to Henry S. Fitch*, ed. R. A. Seigel, L. E. Hunt, J. L. Knight, L. Malaret, and N. L. Zuschlag, pp. 137–45. Lawrence, Kansas: Univ. Kans. Mus. Nat. Hist.

Angilletta, M. J. Jr. (2001). Variation in metabolic rate between populations of a geographically widespread lizard. *Physiol. Biochem. Zool.* **74**, 11–21.

Arnold, S. J. (1983). Morphology, performance and fitness. *Am. Zool.* **23**, 347–61.

Autumn, K., Jindrich, D., DeNardo, D. and Mueller, R. (1999). Locomotor performance at low temperature and the evolution of nocturnality in geckos. *Evolution* **53**, 580–99.

Avery, R. A. (1982). Field studies of body temperatures and thermoregulation. In *Biology of the Reptilia*, vol. 12, ed. C. Gans and F. H. Pough, pp. 93–166. New York: Academic Press.

Bauwens, D., Garland, T. Jr., Castilla, A. M. and Van Damme, R. (1995). Evolution of sprint speed in lacertid lizards: morphological, physiological, and behavioral covariation. *Evolution* **49**, 848–63.

Beck, D. D. and Lowe, C. H. (1994). Resting metabolism of helodermatid lizards: allometric and ecological relationships. *J. Comp. Physiol.* **B164**, 124–9.

Beck, D. D., Dohm, M. R., Garland, T. Jr., Ramirez-Bautista, A. and Lowe, C. H. (1995). Locomotor performance and activity energetics of helodermatid lizards. *Copeia* **1995**, 586–607.

Bennett, A. F. (1978). Activity metabolism of the lower vertebrates. *Ann. Rev. Physiol.* **40**, 447–69.

Bennett, A. F. (1980). The thermal dependence of lizard behaviour. *Anim. Behav.* **28**, 752–62.

Bennett, A. F. (1982). The energetics of reptilian activity. In *Biology of the Reptilia*, vol. 13, ed. C. Gans and F. H. Pough, pp. 155–99. London: Academic Press.

Bennett, A. F. (1985). Temperature and muscle. *J. Exp. Biol.* **115**, 333–44.

Bennett, A. F. (1989). Integrated studies of locomotor performance. In *Complex Organismal Functions: Integration and Evolution in Vertebrates*, ed. D. B. Wake and G. Roth, pp. 191–202. Chichester: John Wiley and Sons.

Bennett, A. F. (1991). The evolution of activity capacity. *J. Exp. Biol.* **160**, 1–23.

Bennett, A. F., Huey, R. B. and John-Alder, H. B. (1984). Physiological correlates of natural activity and locomotor capacity in two species of lacertid lizards. *J. Comp. Physiol.* **B154**, 113–18.

Blomberg, S. P., Garland, T. Jr. and Ives, A. R. (2003). Testing for phylogenetic signal in comparative data: behavioral traits are more labile. *Evolution* **57**, 717–45.

Boggs, D. (2002). Interactions between locomotion and ventilation in tetrapods. *Comp. Biochem. Physiol.* **133A**, 269–88.

Bonine, K. E. (2001). Morphological and physiological predictors of lizard locomotor performance: a phylogenetic analysis of trade-offs. Ph.D. dissertation, University of Wisconsin, Madison.

Bonine, K. E. and Garland, T. Jr. (1999). Sprint performance of phrynosomatid lizards, measured on a high-speed treadmill, correlates with hindlimb length. *J. Zool. Lond.* **248**, 255–65.

Bonine K. E., Gleeson, T. T. and Garland, T. Jr. (2001). Comparative analysis of fiber-type composition in the iliofibularis muscle of phrynosomatid lizards (Squamata). *J. Morphol.* **250**, 265–80.

Bulova, S. J. (1994). Ecological correlates of population and individual variation in antipredator behavior of two species of desert lizards. *Copeia* **1994**, 980–92.

Burke, R. L., Hussain, A. A., Storey, J. M. and Storey, K. B. (2002). Freeze tolerance and supercooling ability in the Italian wall lizard, *Podarcis sicula*, introduced to Long Island, New York. *Copeia* **2002**, 836–42.

Buttemer, W. A. (1990). Effect of temperature on evaporative water loss of the Australian tree frogs *Litoria caerulea* and *Litoria chloris*. *Physiol. Zool.* **63**, 1043–57.

Calder, W. A. (1984). *Size, Function, and Life History*. Cambridge, MA: Harvard University Press.

Carpenter, C. C. (1967). Aggression and social structure in iguanid lizards. In *Lizard Ecology: A Symposium*, ed. W. W. Milstead, pp. 87–105. Columbia: University of Missouri Press.

Carrier, D. R. (1989). Ventilatory action of the hypaxial muscles of the lizard *Iguana iguana*: a function of slow muscle. *J. Exp. Biol.* **143**, 435–57.

Conley, K. E., Christian, K. A., Hoppeler, H. and Weibel, E. R. (1995). Heart mitochondrial properties and aerobic capacity are similarly related in a mammal and a reptile. *J. Exp. Biol.* **198**, 739–46.

Cooper, W. E. Jr. (1995). Prey chemical discrimination and foraging mode in gekkonoid lizards. *Herp. Monogr.* **9**, 120–9.

Cooper, W. E. Jr. (2000). Effect of temperature on escape behaviour by an ectothermic vertebrate, the keeled earless lizard (*Holbrookia propinqua*). *Behaviour* **137**, 1299–315.

Cooper, W. E. Jr. (2003). Risk factors affecting escape behavior by the desert iguana, *Dipsosaurus dorsalis*: speed and directness of predator approach, degree of cover, direction of turning by a predator, and temperature. *Can. J. Zool.* **81**, 979–84.

Cullum, A. (1997). Comparisons of physiological performance in sexual and asexual whiptail lizards (genus *Cnemidophorus*): implications for the role of heterozygosity. *Am. Nat.* **150**, 24–47.

Cullum, A. (1998). Sexual dimorphism in physiological performance of whiptail lizards (genus *Cnemidophorus*). *Physiol. Zool.* **71**, 541–52.

Dawson, W. R. (1967). Interspecific variation in physiological responses of lizards to temperature. In *Lizard Ecology: A Symposium*, ed. W. W. Milstead, pp. 230–257. Columbia: University of Missouri Press.

Dial, B. E. and Grismer, L. L. (1992). A phylogenetic analysis of physiological-ecological character evolution in the lizard genus *Coleonyx* and its implications for historical biogeographic reconstruction. *Syst. Biol.* **41**, 178–95.

Diamond, J. M. (1993). Evolutionary physiology. In *The Logic of Life: The Challenge of Integrative Physiology*, ed. C. A. R. Boyd and D. Noble, pp. 89–111. Oxford: Oxford University Press.

Dmi'el, R. (2001). Skin resistance to evaporative water loss in reptiles: a physiological adaptive mechanism to environmental stress or a phyletically dictated trait? *Israel J. Zool.* **47**, 55–67.

Donovan, E. R. and Gleeson, T. T. (2001). Evidence for facilitated lactate uptake in lizard skeletal muscle. *J. Exp. Biol.* **204**, 4099–106.

Downes, S. and Shine, R. (1999). Do incubation-induced changes in a lizard's phenotype influence its vulnerability to predators? *Oecologia* **120**, 9–18.

Downes, S. and Shine, R. (2001). Why does tail loss increase a lizard's later vulnerability to snake predators? *Ecology* **82**, 1293–303.

Feder, M. E., Bennett, A. F. and Huey, R. B. (2000). Evolutionary physiology. *Ann. Rev. Ecol. Syst.* **31**, 315–41.

Felsenstein, J. (1985). Phylogenies and the comparative method. *Am. Nat.* **125**, 1–15.

Frappell, P. B., Schultz, T. J. and Christian, K. A. (2002). The respiratory system in varanid lizards: determinants of O_2 transfer. *Comp. Biochem. Physiol.* **133A**, 239–58.

Garland, T. Jr. (1984). Physiological correlates of locomotory performance in a lizard: an allometric approach. *Am. J. Physiol.* **247**, R806–15.

Garland, T. Jr. (1993). Locomotor performance and activity metabolism of *Cnemidophorus tigris* in relation to natural behaviors. In *Biology of Whiptail Lizards (Genus Cnemidophorus)*, ed. J. W. Wright and L. J. Vitt, pp. 163–210. Norman, OK: Oklahoma Museum of Natural History.

Garland, T. Jr. (1994). Phylogenetic analyses of lizard endurance capacity in relation to body size and body temperature. In *Lizard Ecology: Historical and Experimental Perspectives*, ed. L. J. Vitt and E. R. Pianka, pp. 237–59. Princeton, NJ: Princeton University Press.

Garland, T. Jr. (1999). Laboratory endurance capacity predicts variation in field locomotor behaviour among lizard species. *Anim. Behav.* **57**, 77–83.

Garland, T. Jr. and Adolph, S. C. (1994). Why not to do two-species comparative studies: limitations on inferring adaptation. *Physiol. Zool.* **67**, 797–828.

Garland, T. Jr. and Carter, P. A. (1994). Evolutionary physiology. *Ann. Rev. Physiol.* **56**, 579–621.

Garland, T. Jr. and Else, P. L. (1987). Seasonal, sexual, and individual variation in endurance and activity metabolism in lizards. *Am. J. Physiol.* **252**, R439–49.

Garland, T. Jr. and Losos, J. B. (1994). Ecological morphology of locomotor performance in squamate reptiles. In *Ecological Morphology: Integrative Organismal Biology*, ed. P. C. Wainwright and S. M. Reilly, pp. 240–302. Chicago, IL: University of Chicago Press.

Garland, T. Jr., Bennett, A. F. and Daniels, C. B. (1990a). Heritability of locomotor performance and its correlates in a natural population. *Experientia* **46**, 530–3.

Garland, T. Jr., Dickerman, A. W., Janis, C. M. and Jones, J. A. (1993). Phylogenetic analysis of covariance by computer simulation. *Syst. Biol.* **42**, 265–92.

Garland, T. Jr., Else, P. L., Hulbert, A. J. and Tap, P. (1987). Effects of endurance training and captivity on activity metabolism of lizards. *Am. J. Physiol.* **252**, R450–6.

Garland, T. Jr., Hankins, E. and Huey, R. B. (1990b). Locomotor capacity and social dominance in male lizards. *Funct. Ecol.* **4**, 243–50.

Garland, T. Jr., Harvey, P. H. and Ives, A. R. (1992). Procedures for the analysis of comparative data using phylogenetically independent contrasts. *Syst. Biol.* **41**, 18–32.

Garland, T. Jr., Midford, P. E. and Ives, A. R. (1999). An introduction to phylogenetically based statistical methods, with a new method for confidence intervals on ancestral values. *Am. Zool.* **39**, 374–88.

Gleeson, T. T. (1979). The effects of training and captivity on the metabolic capacity of the lizard *Sceloporus occidentalis. J. Comp. Physiol.* **129**, 123–8.

Gleeson, T. T. (1996). Post-exercise lactate metabolism: a comparative review of sites, pathways, and regulation. *Ann. Rev. Physiol.* **58**, 565–81.

Gleeson, T. T. and Dalessio, P. M. (1989). Lactate: a substrate for reptilian muscle gluconeogenesis following exhaustive exercise. *J. Comp. Physiol.* **160**, 331–8.

Gleeson, T. T. and Hancock, T. V. (2002). Metabolic implications of a 'run now, pay later' strategy in lizards: an analysis of post-exercise oxygen consumption. *Comp. Biochem. Physiol.* **133A**(2), 259–67.

Gleeson, T. T. and Harrison, J. M. (1986). Reptilian skeletal muscle: fiber-type composition and enzymatic profile in the lizard, *Iguana iguana. Copeia* **1986**, 324–32.

Gleeson, T. T. and Harrison, J. M. (1988). Muscle composition and its relationship to sprint running in the lizard *Dipsosaurus dorsalis. Am. J. Physiol.* **255**, R470–7.

Gleeson, T. T. and Johnston, I. A. (1987). Reptilian skeletal muscle: contractile properties of identified, single fast-twitch and slow fibers from the lizard *Dipsosaurus dorsalis. J. Exp. Zool.* **242**, 283–90.

Gleeson, T. T., Mitchell, G. S. and Bennett, A. F. (1980a). Cardiovascular responses to graded activity in the lizards *Varanus* and *Iguana. Am. J. Physiol.* **239**, R174–9.

Gleeson, T. T., Putnam, R. W. and Bennett, A. F. (1980b). Histochemical, enzymatic, and contractile properties of skeletal muscle fibers in the lizard *Dipsosaurus dorsalis. J. Exp. Zool.* **214**, 293–302.

Gould, S. J. and Lewontin, R. C. (1979). The spandrels of San Marco and the Panglossian paradigm: a critique of the adaptationist programme. *Proc. R. Soc. Lond.* **B205**, 581–98.

Greene, H. W. (1997). *Snakes: the Evolution of Mystery in Nature.* Berkeley, CA: University of California Press.

Guthe, K. F. (1981). Reptilian muscle: fine structure and physiological parameters. In *Biology of the Reptilia*, vol. 11, ed. C. Gans and T. S. Parsons, pp. 265–354. New York: Academic Press.

Holloszy, J. O. and Coyle, E. F. (1984). Adaptations of skeletal muscle to endurance exercise and their metabolic consequences. *Exercise Physiol.* **56**, 831–8.

Huey, R. B. and Pianka, E. R. (1981). Ecological consequences of foraging mode. *Ecology* **62**, 991–9.

Huey, R. B., Bennett, A. F., John-Alder, H. B. and Nagy, K. A. (1984). Locomotor capacity and foraging behaviour of Kalahari lacertid lizards. *Anim. Behav.* **32**, 41–50.

Huey, R. B., Pianka, E. R. and Schoener, T. W. ed. (1983). *Lizard Ecology: Studies of a Model Organism.* Cambridge, MA: Harvard University Press.

Huey, R. B., Pianka, E. R. and Vitt, L. J. (2001). How often do lizards "run on empty"? *Ecology* **82**, 1–7.

Irschick, D. J. (2000). Effects of behaviour and ontogeny on the locomotor performance of a West Indian lizard, *Anolis lineatopus*. *Funct. Ecol.* **14**, 438–44.

Irschick, D. J. and Garland, T. Jr. (2001). Integrating function and ecology in studies of adaptation: investigations of locomotor capacity as a model system. *Ann. Rev. Ecol. Syst.* **32**, 367–96.

Irschick, D. J. and Jayne, B. C. (2000). Size matters: ontogenetic differences in the three-dimensional kinematics of steady-speed locomotion in the lizard *Dipsosaurus dorsalis*. *Exp. Biol.* **203**, 2133–48.

Irschick, D. J., Macrini, T. E., Koruba, S. and Forman, J. (2000). Ontogenetic differences in morphology, habitat use, behavior and sprinting capacity in two West Indian *Anolis* lizard species. *J. Herp.* **34**, 444–51.

Jayne, B. C., Bennett, A. F. and Lauder, G. V. (1990). Muscle recruitment during terrestrial locomotion: how speed and temperature affect fibre type used in a lizard. *J. Exp. Biol.* **152**, 101–28.

John-Alder, H. B., Lowe, C. H. and Bennett, A. F. (1983). Thermal dependence of locomotory energetics and aerobic capacity of the gila monster (*Heloderma suspectum*). *J. Comp. Physiol.* B**151**, 119–26.

Korzan, W. J. and Summers, C. H. (2004). Serotonergic response to social stress and artificial social sign stimuli during paired interactions between male *Anolis carolinensis*. *Neuroscience* **123**, 835–45.

Kramer, D. L. and McLaughlin, R. L. (2001). The behavioral ecology of intermittent locomotion. *Amer. Zool.* **41**, 137–53.

Lailvaux, S. P., Alexander, G. J. and Whiting, M. J. (2003). Sex-based differences and similarities in locomotor performance, thermal preferences, and escape behaviour in the lizard *Platysaurus intermedius wilhelmi*. *Physiol. Biochem. Zool.* **76**, 511–21.

Losos, J. B., Creer, D. A. and Schulte, J. A., II. (2002). Cautionary comments on the measurement of maximum locomotor capabilities. *J. Zool. Lond.* **258**, 57–61.

Marler, C. A. and Moore, M. C. (1988). Evolutionary costs of aggression revealed by testosterone manipulations in free-living male lizards. *Behav. Ecol. Sociobiol.* **23**, 21–6.

Marler, C. A. and Moore, M. C. (1989). Time and energy costs of aggression in testosterone-implanted free-living male Mountain Spiny Lizards (*Sceloporus jarrovi*). *Physiol. Zool.* **62**, 1334–50.

M'Closkey, R. T., Deslippe, R. J., Szpak, C. P. and Baia, K. A. (1990). Ecological correlates of the variable mating system of an iguanid lizard. *Oikos* **59**, 63–9.

Miles, D. B. (1994). Population differentiation in locomotor performance and the potential response of a terrestrial organism to global environmental change. *Am. Zool.* **34**, 422–6.

Miles, D. B., Fitzgerald, L. A. and Snell, H. L. (1995). Morphological correlates of locomotor performance in hatchling *Amblyrhynchus cristatus*. *Oecologia* **103**, 261–4.

Miller, K. and Camilliere, J. J. (1981). Physical training improves swimming performance of the African clawed frog *Xenopus laevis*. *Herpetologica* **37**, 1–10.

Milstead, W. W., ed. (1967). *Lizard Ecology: A Symposium*. Columbia, MO: University of Missouri Press.

Mitchell, G. S. and Gleeson, T. T. (1985). Acid-base balance during lactic-acid infusion in the lizard *Varanus salvator*. *Resp. Physiol.* **60**, 253–66.

Montanucci, R. R. (1987). *A Phylogenetic Study of the Horned Lizards, Genus* Phrynosoma, *Based on Skeletal and External Morphology.* Contributions in Science 390. Los Angeles, CA: Natural History Museum.

Nagy, K. A. (1983). Ecological energetics. In *Lizard Ecology: Studies of a Model Organism,* ed. R. B. Huey, E. R. Pianka and T. W. Schoener, pp. 24–54. Cambridge, MA: Harvard University Press.

Nagy, K. A. and Shemanski, D. R. (2003). Locomotor activity costs in free-living animals. *Comp. Biochem. Physiol.* **134**A(Suppl. 1), S36.

Nagy, K. A., Huey, R. B. and Bennett, A. F. (1984). Field energetics and foraging mode of Kalahari lacertid lizards. *Ecology* **65**, 588–96.

Nagy, K. A., Girard, I. A. and Brown, T. K. (1999). Energetics of free-ranging mammals, reptiles and birds. *Ann. Rev. Nutr.* **19**, 247–77.

Natochin, Y. V. and Chernigovskaya, T. V. (1997). Evolutionary physiology: history, principles. *Comp. Biochem. Physiol.* **118**A, 63–79.

Olsson, M., Shine, R. and Bak-Olsson, E. (2000). Locomotor impairment of gravid lizards: Is the burden physical or physiological? *J. Evol. Biol.* **13**, 263–8.

Peaker, M. and Linzell, J. L. (1975). *Salt Glands in Birds and Reptiles.* Monographs of the Physiological Society 32. Cambridge: Cambridge University Press.

Perry, S. F. (1998). Lungs: comparative anatomy, functional morphology, and evolution. In *Biology of the Reptilia,* vol. 19, ed. C. Gans and A. S. Gaunt, pp. 1–80. Ithaca, NY: Society for the Study of Amphibians and Reptilians.

Perry, G. (1999). The evolution of search modes: ecological versus phylogenetic perspectives. *Am. Nat.* **153**, 98–109.

Perry, G., LeVering, K., Girard, I. and Garland, T. Jr. (2004). Locomotor performance and social dominance in male *Anolis cristatellus. Anim. Behav.* **67**, 37–47.

Peter, J. B., Barnard, R. J., Edgerton, V. R., Gillespie, C. A. and Stemple, K. E. (1972). Metabolic profiles of three fiber types of skeletal muscle in guinea pigs and rabbits. *Biochemistry* **11**, 2627–33.

Peterson, C. C. (1998). Rain-harvesting behavior by a free-ranging desert horned lizard (*Phrynosoma platyrhinos*). *Southwest. Nat.* **43**, 391–4.

Pianka, E. R. and Vitt, L. J. (2003). *Lizards: Windows to the Evolution of Diversity.* Berkeley, CA: University of California Press.

Pinch, F. C. and Claussen, D. L. (2003). Effects of temperature and slope on the sprint speed and stamina of the Eastern Fence Lizard, *Sceloporus undulatus. J. Herp.* **37**, 671–9.

Porter, W. P., Sabo, J. L., Tracy, C. R., Reichman, O. J. and Ramankutty, N. (2002). Physiology on a landscape scale: plant-animal interactions. *Integ. Comp. Biol.* **42**, 431–53.

Pough, F. H. (1980). The advantages of ectothermy for tetrapods. *Am. Nat.* **115**, 92–112.

Pough, F. H., Andrews, R. M., Cadle, J. E. *et al.* (2004). *Herpetology,* 3rd edn. Upper Saddle River, NJ: Prentice Hall.

Price, T. and Langen, T. (1992). Evolution of correlated characters. *Trends Ecol. Evol.* **7**, 307–10.

Prosser, C. L., ed. (1950). *Comparative Animal Physiology.* Philadelphia, PA: W. B. Saunders Co.

Putnam, R. W. and Bennett, A. F. (1982). Thermal dependence of isometric contractile properties of lizard muscle. *J. Comp. Physiol.* **147**, 11–20.

Putnam, R. W., Gleeson, T. T. and Bennett, A. F. (1980). Histochemical determination of the fiber composition of locomotory muscles in a lizard, *Dipsosaurus dorsalis. J. Exp. Zool.* **214**, 303–9.

Ribas, S. C., Rocha, C. F. D., Teixeira-Filho, P. F. and Vicente, J. J. (1998). Nematode infection in two sympatric lizards (*Tropidurus torquatus* and *Ameiva ameiva*) with different foraging tactics. *Amph.-Rept.* **19**, 323–30.

Rose, M. R. and Lauder, G. V., ed. (1996a). *Adaptation*. San Diego, CA: Academic Press.

Rose, M. R. and Lauder, G. V. (1996b). Post-spandrel adaptationism. In *Adaptation*, ed. M. R. Rose and G. V. Lauder, pp. 1–8. San Diego, CA: Academic Press.

Savitzky, A. H. (1980). Role of venom delivery strategies in snake evolution. *Evolution* **34**, 1194–204.

Schall, J. J. (1983). Lizard malaria: parasite-host ecology. In *Lizard Ecology: Studies of a Model Organism*, ed. R. B. Huey, E. R. Pianka and T. W. Schoener, pp. 84–100, 437–9, 488. Cambridge, MA: Harvard University Press.

Schall, J. J., Bennett, A. F. and Putnam, R. W. (1982). Lizards (*Sceloporus occidentalis*) infected with malaria: physiological and behavioral consequences. *Science* **217**, 1057–9.

Schmidt, P. J., Sherbrooke, W. C. and Schmidt, J. O. (1989). The detoxification of ant (*Pogonomyrmex*) venom by a blood factor in horned lizards (*Phrynosoma*). *Copeia* **1989**, 603–7.

Schmidt-Nielsen, K. (1984). *Scaling: Why Is Animal Size So Important?* Cambridge: Cambridge University Press.

Shine, R. (2003). Effects of pregnancy on locomotor performance: an experimental study on lizards. *Oecologia* **136**, 450–6.

Sinervo, B. and Losos, J. B. (1991). Walking the tight rope: arboreal sprint performance among *Sceloporus occidentalis* lizard populations. *Ecology* **72**, 1225–33.

Sinervo, B., Hedges, R. and Adolph, S. C. (1991). Decreased sprint speed as a cost of reproduction in the lizard *Sceloporus occidentalis*: variation among populations. *J. Exp. Biol.* **155**, 323–36.

Snyder, G. K., Nestler, J. R., Shapiro, J. I. and Huntley, J. (1995). Intracellular pH in lizards after hypercapnia. *Am. J. Physiol.* **268**, R889–95.

Thompson, G. G. and Withers, P. C. (1997). Standard and maximal metabolic rates of goannas (Squamata: Varanidae). *Physiol. Zool.* **70**, 307–23.

Trigosso-Venario, R., Labra, A. and Niemeyer, H. (2002). Interactions between males of the lizard *Liolaemus tenuis*: roles of familiarity and memory. *Ethology* **108**, 1057–64.

van Berkum, F. H., Huey, R. B., Tsuji, J. S. and Garland, T. Jr. (1989). Repeatability of individual differences in locomotor performance and body size during early ontogeny of the lizard *Sceloporus occidentalis* (Baird & Girard). *Funct. Ecol.* **3**, 97–105.

Van Damme, R. and Vanhooydonck, B. (2001). Origins of interspecific variation in lizard sprint capacity. *Funct. Ecol.* **15**, 186–202.

Van Damme, R., Aerts, P. and Vanhooydonck, B. (1998). Variation in morphology, gait characteristics and speed of locomotion in two populations of lizards. *Biol. J. Linn. Soc.* **63**, 409–27.

Van Damme, R., Vanhooydonck, B., Aerts, P. and De Vree, F. (2003). Evolution of lizard locomotion: context and constraint. In *Vertebrate Biomechanics and Evolution*, ed. V. L. Bels, J. P. Gasc and A. Casinos, pp. 267–82. Oxford: BIOS Scientific Publishers.

Vitt, L. J. and E. R. Pianka, ed. (1994). *Lizard Ecology: Historical and Experimental Perspectives*. Princeton, NJ: Princeton University Press.

Wang, T., Carrier, D. R. and Hicks, J. W. (1997). Ventilation and gas exchange in lizards during treadmill exercise. *J. Exp. Biol.* **200**, 2629–39.

Wang, T., Smits, A. W. and Burggren, W. W. (1998). Pulmonary function in reptiles. In *Biology of the Reptilia*, vol. 19, ed. C. Gans and A. S. Gaunt, pp. 297–374. Ithaca, NY: Society for the Study of Amphibians and Reptiles.

Weinstein, R. B. (2001). Terrestrial intermittent exercise: common issues for human athletics and comparative animal locomotion. *Am. Zool.* **41**, 219–28.

Weinstein, R. B. and Full, R. J. (1999). Intermittent locomotion increases endurance in a gecko. *Physiol. Biochem. Zool.* **72**, 732–9.

Wickler, S. J. and Gleeson, T. T. (1993). Lactate and glucose metabolism in mouse (*Mus musculus*) and reptile (*Anolis carolinensis*) skeletal muscle. *Am. J. Physiol.* **264**, R487–91.

Wiens, J. J. and Slingluff, J. L. (2001). How lizards turn into snakes: a phylogenetic analysis of body-form evolution in anguid lizards. *Evolution* **55**, 2303–18.

Yang, E. J. and Wilczynski, W. (2002). Relationships between hormones and aggressive behavior in green anole lizards: an analysis using structural equation modeling. *Horm. Behav.* **42**, 192–205.

Zani, P. A. (1996). Patterns of caudal-autotomy evolution in lizards. *J. Zool. Lond.* **240**, 201–20.

Zera, A. J. and Harshman, L. G. (2001). The physiology of life history trade-offs in animals. *Ann. Rev. Ecol. Syst.* **32**, 95–126.

4

Lizard energetics and the sit-and-wait vs. wide-foraging paradigm

TRACEY K. BROWN

Biological Sciences Department, California State University

KENNETH A. NAGY

Department of Organismic Biology, University of California

Introduction

The acquisition and allocation of energy can be two of the most critical processes in an organism's life. The amount of energy an individual obtains from the environment can determine not only its survival, but also whether it can engage in reproduction, a very evolutionarily important activity. Among vertebrates, there exists a fairly definitive dichotomy between the high-energy requirements of endotherms and the more energetically "conservative" ectotherms. Within the ectothermic reptiles, the possibility of a similarly clear-cut dichotomy, that between ambushing and actively foraging modes, has been repeatedly discussed, and is the focus of this volume.

Over twenty years ago, Huey and Pianka (1981) asked the question "what are the extra energetic costs of foraging widely?" The observation that wide-foragers travel longer distances through their environment creates the expectation of higher overall daily energy needs (Huey and Pianka, 1981). At the time, Huey and Pianka (1981) estimated that daily maintenance costs could be up to 1.5 times higher in an actively foraging species, a value validated subsequently by Nagy *et al.* (1984).

The validity and demonstrability of the sit-and-wait (SW) and widely foraging (WF) dichotomy has drawn much scientific attention. A quick perusal of the table of contents of this book can attest to the body of knowledge resulting from such research efforts. While many foraging studies focus on the physiological, morphological or behavioral correlates of foraging mode, here we examine a metric that encompasses all of these aspects at the whole-organism level: field metabolic rate. The purposes of this chapter are twofold: an exploration of the expectation of higher energy needs in WF lizards, and an instructive summary of relevant field studies of energy requirements of lizards.

Lizard Ecology: The Evolutionary Consequences of Foraging Mode, ed. S. M. Reilly, L. D. McBrayer and D. B. Miles. Published by Cambridge University Press. © Cambridge University Press 2007.

The measurement of field metabolic rate (FMR, in units of kJ expended per day) of free-ranging reptiles using the doubly labeled water (DLW) method (Lifson and McClintock, 1966) has become increasingly popular over the past 25 years, largely owing to decreased isotope costs and analytical advancements (Nagy, 1975, 1983a). The first review of such studies (Nagy, 1983b) included data from nine species of iguanian lizard and yielded a statistically significant allometric regression. The equation for that relation was subsequently used in many studies to predict daily energy expenditures of lizards during their activity season. Since then, more than 40 other lizard species, representing several families and habitats, have been studied by using DLW (summarized by Nagy *et al.*, 1999). With these results, along with even newer data on more species, it is now possible to compare statistically the daily energy expenditures of free-living SW and WF lizards. If substantial energetic differences do exist between the two foraging modes, then the allometric equations for each foraging mode can be calculated separately, and be used for assessing newly studied species as well as for predicting, developing, and testing bioenergetic hypotheses.

We have taken a historical approach in this chapter. First we examine the early studies comparing two or more distantly related lizard species, and then a later study where the phylogenetic distance between a SW and a WF species was minimized. Next, we review all relevant DLW studies to try to detect general trends in, and differences between, SW and WF species of lizard. Finally, the recent determination of the phylogenetic relationships among many lacertilians allows us to explore the effects of phylogeny on energy usage.

Two-species comparisons among lizards

The first study comparing field metabolic rates of lizards having different foraging modes was done on two desert species, the wide-foraging whiptail lizard (*Cnemidophorus tigris*) and the sit-and-wait zebra-tailed lizard (*Callisaurus draconoides*) (Anderson and Karasov, 1981). Both species are diurnal insectivores, and they were sympatric at the field site used in the study. This ecological overlap largely controlled for possible variation in FMR associated with habitat and climate. The whiptails had much higher daily energy expenditures (relative to body mass), primarily because of their higher level of activity while abroad during the day. Both species had similar daily maintenance costs, but activity costs were much higher for the whiptails. Even though the whiptails spent only half as much time above ground as did the zebra-tailed lizards, the intensity of the whiptails' activity resulted in much higher 24 h energy expenditure (150%) than that of the SW zebra-tailed lizards. Moreover, the whiptails consumed much more food and achieved a higher foraging efficiency (metabolizable energy gained per unit of energy spent while foraging) than did the zebra-tailed lizards. Unfortunately, these two species are rather distantly related, being not only in different families (Teiidae and Phrynosomatidae, respectively) but also in different "major" clades (Scleroglossa and Iguania, respectively). This taxonomic distance makes it difficult to evaluate whether the metabolic differences between the species are associated primarily with foraging mode or phylogeny (Garland and Adolph, 1994).

A solution to this dilemma was obtained in a study of two species of diurnal, insectivorous lizard in the Kalahari sand dunes of southern Africa by Nagy *et al.* (1984). These similar-sized, sympatric species, one a WF hunter and the other a SW ambusher, are not only in the same family (Lacertidae) but were (at the time of the study) in the same genus (*Eremias*). Although the WF bushveld lizard (now *Heliobolus lugubris*), and the SW spotted sand lizard (now *Pedioplanis lineoocellata*) are now in different genera, the two species are still closely related through their familial tie. As with the study on whiptails and zebra-tails, the WF bushveld lizard had a substantially higher field metabolic rate than did the SW spotted sand lizard. Although the WF species was abroad only about one fourth of the time (2.75 h each day) that the SW species was (10.25 h), the WF species was working hard while above ground; its activity period metabolic rate was 12 times its resting metabolic rate at that body temperature (vs. only 2.8 times resting in the SW species). The extra effort expended by the WF species paid off with a greater daily intake of food. In fact, the WF species not only obtained enough extra food energy while foraging to pay its higher daily energy expenditures for metabolism, it obtained more surplus food energy (greater profit) than did the SW species, and it grew nearly twice as fast as a result.

More detailed laboratory studies of these two species have revealed differences in behavioral capabilities (Huey *et al.*, 1984) and physiological properties (Bennett *et al.*, 1984). Evaluation of their locomotor abilities by using small racetracks and treadmills revealed that the WF hunter has better endurance at low speeds, and although the SW ambusher can sprint faster, it exhausts more quickly. At the physiological level, the WF species had a larger heart and a higher amount of hemoglobin in its blood (hematocrit), allowing for higher rates of oxygen transport. Although both species had the same resting metabolic rates, muscle masses, isolated muscle performance properties, and basic muscle biochemistry, the WF hunter had a higher maximum aerobic metabolic rate, yet the SW ambusher could tolerate a higher lactate concentration in its body (higher anaerobic scope). Thus, the WF hunter is physiologically better at sustaining a high level of aerobic activity, whereas the SW ambusher is better at short but quick bouts of mostly anaerobically supported activity. This is somewhat similar to the different physiological capabilities of trained human marathon runners and human sprinters. Unfortunately, no other species pairs among the reptiles have been studied in this kind of breadth and detail so far.

Field metabolic rate and foraging mode

Overall methodology

To explore the relationship between FMR and foraging mode, we searched the literature for all accounts of energy expenditure in free-ranging reptiles. Many estimates of daily energy expenditures have been based on field time budgets and laboratory measurements of metabolic rates during various states of activity (Tinkle and Hadley, 1975; Huey and Pianka, 1981). Such estimates can have large errors (Nagy, 1989). Herein, we used only FMR results from DLW studies, which should have errors of 10% or less (Nagy, 1980, 1989). Nagy *et al.* (1999) reported a summary of such studies of lizard field metabolic rates that were published up until August 1998. To these species we added more recently published studies (up to July 2004) in compiling our database.

Field metabolic rates have been measured by using DLW in over 55 species of lizard (Nagy *et al.*, 1999) (Fig. 4.1). However, for the purposes of the present comparisons we established several criteria before including a study (species) in the final database (Fig. 4.1). The SW and WF foraging mode categories typically refer to insectivorous (carnivorous) species, as it would be difficult to fit a herbivorous lizard into this paradigm. Thus, species considered to be largely or absolutely herbivorous were excluded and are treated elsewhere (see Herrel, Chapter 7; Vitousek *et al.*, Chapter 16; Nagy *et al.*, 1999). The FMR data were included only for adult lizards that were studied during the species' normal activity season (usually spring and/or summer). If more than one value for FMR and body mass was available for a species (e.g. separate values for males and females within a study, or several studies on the same species), a single mean for that species was calculated by averaging all of the reported mean values. When necessary, the FMR data were converted from CO_2 to joules by using the factor 25.7 J/ml CO_2 for insectivores and carnivores (Nagy, 1983a).

For the analyses reported here, we categorized species as either sit-and-wait/ambushers (SW) or active/wide-foragers (WF) based on information given in the particular published FMR study, field guides, personal communication and general knowledge and experience. Occasionally, certain species would be reported alternately as a SW and a WF (e.g. *Elgaria multicarinatus*). Although

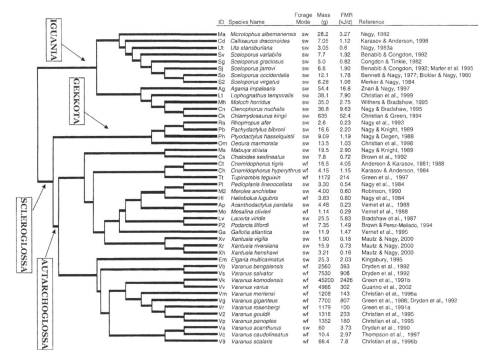

Figure 4.1. Body masses, field metabolic rates (FMR), and a composite phylogeny for lizard species included in analyses. Foraging mode classification is indicated as either ambushing (SW) or wide foraging (WF). Each species has an assigned two-letter code (ID) for its identification in Fig. 4.2. See text for phylogenetic sources and description of foraging mode classification.

we typically used the foraging mode that was reported by the study's author, we explored the effect of such alternate categorization during the statistical analyses. It would be ideal to replace the categorical values assigned here (SW, WF) with a continuous variable more representative of each species' foraging efforts. Although metrics such as moves per minute and percent time moving are becoming available for many lizard species (see Perry, Chapter 2), there were insufficient movement data for the species that have been studied with DLW.

Our overall approach to data analysis was twofold; the initial analyses were done by using conventional statistical methods. However, as Fig. 4.1 demonstrates, foraging mode is highly evolutionarily conservative, leading to the possibility that field metabolic rates might also show such a phylogenetic signal (Blomberg *et al.*, 2003). In addition, conventional statistical methods assume that each data point is independent. Because lizards (and most organisms) have evolved in a hierarchical fashion, mean values for species cannot be considered independent (Garland *et al.*, 1993). Thus, the data were secondarily analyzed with independent contrast analysis (ICA) techniques (Felsenstein, 1985) to incorporate the phylogenetic relatedness of the species. We discuss the results of each approach below, along with a description of the ICA methodology. SPSS 11.0 for Windows (SPSS, Inc., copyright © 2001 was used for regression analyses, *t*-tests and analyses of covariance (ANCOVA). SigmaPlot 8.0 (SPSS, Inc., copyright © 2002) was used to generate the figures. In all situations, significance was accepted at the $\alpha = 0.05$ level.

Conventional statistical analyses

Body size (expressed as mass) has been shown to be the most important variable influencing FMR in terrestrial vertebrates (Nagy *et al.*, 1999). As expected, the overall regression of log FMR (kJ/day) on log body mass (g) for all species was highly significant (Fig. 4.2) and was described by the following equations:

$$\log \text{FMR} = -0.810 + 0.952 \, (\log \text{ mass}) \qquad \text{(Eq. 4.1)}$$

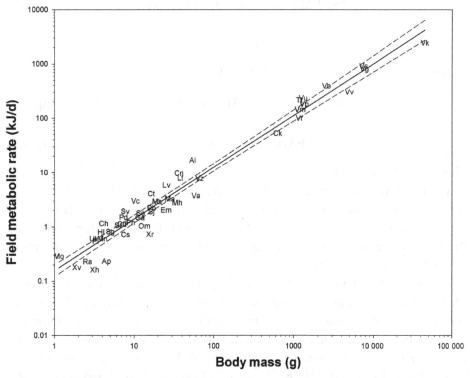

Figure 4.2. Least squares regression of $\log_{10}$ field metabolic rate (kJ/d) on $\log_{10}$ body mass (g). Two-letter codes identify individual species (see Fig. 4.1 for key). Dashed line is 95% confidence interval.

or in exponential form,

$$FMR = 0.155(mass)^{0.952} \qquad \text{(Eq. 4.2)}$$

with $r^2 = 0.963$, $F_{(1, 44)} = 1132$, $p < 0.0001$ and $n = 46$.

A visual examination of the overall regression (Fig. 4.2) suggested that certain groups (e.g. large lizards over 500 g, xantusiids) might have substantial effects on the slope of the regression. Upon excluding these groups individually from the overall regression, the values of the slope and the coefficient of determination (r^2) changed slightly, but not significantly, from the original regression (Eq. 4.1). Even if we excluded xantusiids and gekkotans, two groups whose foraging modes and energy demands are often ambiguous, the regression did not change significantly.

To explore the effects of foraging mode on FMR we calculated independent regressions of log FMR (kJ/day) on log body mass (g) for each group (Fig. 4.3A). Both relations were highly significant (p values for each were <0.000 1 via an F-test). For SW lizards:

$$\log FMR = -0.932 + 1.014(mass), \qquad \text{(Eq. 4.3)}$$

$$\text{or } FMR = 0.117(mass)^{1.014} \qquad \text{(Eq. 4.4)}$$

with $r^2 = 0.855$ and $n = 29$. For WF lizards:

$$\log FMR = -0.531 + 0.875(mass), \qquad \text{(Eq. 4.5)}$$

$$\text{or } FMR = 0.294(mass)^{0.875} \qquad \text{(Eq. 4.6)}$$

with $r^2 = 0.990$ and $n = 17$.

Are these relations significantly different? A statistical comparison via analysis of covariance (ANCOVA) indicated that the slopes of the two regression lines were not different ($p = 0.081$), but after accounting for body mass effects on the basis of the common slope, the two groups do differ in elevation ($F = 8.0$, df $= 1,43$, $p = 0.007$).

To confirm the ANCOVA results and to estimate the average difference in FMR between WF and SW lizards, we adjusted the whole-animal FMR values in Fig. 4.1 for body mass effects by dividing the FMR values by body mass in g to the exponent 0.952 (common allometric slope from Eq. 4.1). The mean mass-adjusted FMR of WF lizards was $0.204 \pm 0.069 \, kJ \, g^{-0.952} \, day^{-1}$, and for SW lizards it was $0.155 \pm 0.079 \, kJ \, g^{-0.952} \, day^{-1}$. These means differed significantly (t-test, $t = -2.108$, df $= 44$, 2-tailed $p = 0.041$, 1-tailed $p = 0.035$). This comparison indicates that the average wide foraging lizard spends 32% more energy

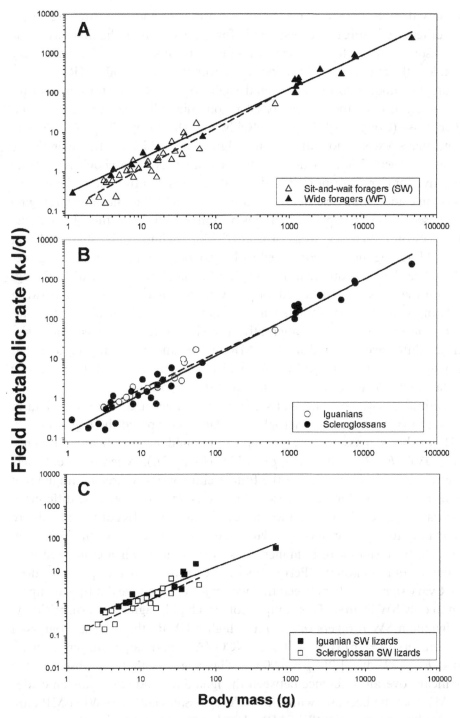

Figure 4.3. Least squares regressions of $\log_{10}$ field metabolic rate (kJ/d) on $\log_{10}$ body mass (g). Symbols as indicated in each graph. (A) Comparison of lizards grouped by foraging mode; (B) lizards grouped by major taxonomic clade; (C) ambushing (SW) lizards only, grouped by major clade. See text for statistics.

each day than does the average lizard that uses a sit-and-wait foraging mode. This difference is substantial, especially for species living in habitats with low food resources. Prior to any discussion of these results, however, it is necessary to explore the importance of phylogeny on foraging mode and FMR.

Foraging mode in lizards has been thought to be evolutionarily conservative and largely tied to the ancient basal taxonomic split between Iguania and Scleroglossa (Cooper, 1994, but see Reilly and McBrayer, Chapter 9). Nearly all iguanians, except those of the family Iguanidae (which are herbivores), are ambush foragers. Many of the scleroglossan families (autarchoglossan clade) are active foragers (except Cordylidae and perhaps Xantusiidae), whereas the gekkotan clade within Scleroglossa is considered to be intermediate in foraging mode (see Bauer, Chapter 12 in this volume). What are the phylogenetic influences on the energy needs of lizards? McLaughlin (1989) suggested that because foraging mode is correlated with taxonomy, comparisons among predators (lizards) using different foraging modes may be confounded by phylogeny. Furthermore, he suggested that any physiological differences between predators with different foraging modes may be a result of different evolutionary histories rather than alternative foraging strategies. Given that a significant difference is found in the FMR of SW and WF foragers, and that foraging mode is taxonomically conservative, perhaps the FMR differences lie at a taxonomic level between iguanians and scleroglossans.

Continuing with conventional statistics, we categorized the lizard data in Fig. 4.1 into iguanian and scleroglossan clades, compared the regressions of FMR on body mass for each group, and found no significant difference (ANCOVA, $F = 2.31$, df $= 1,43$, $p = 0.135$) (Fig. 4.3B). When these data sets were re-analyzed after deleting the four gecko species to create an iguanian versus autarchoglossan comparison, there was still no detectable difference. This was unexpected given that we suspect the higher FMRs of WF lizards are due mainly to higher costs of activity, and that autarchoglossans have been shown to have significantly higher rates of movement when compared with iguanians and gekkotans (Perry, 1999). We then asked the question: does FMR vary significantly between the two major clades if we limit the comparison to only SW lizards? This comparison was highly significant; surprisingly, the iguanian SW foragers ($n = 13$) had higher FMRs than the scleroglossan SW foragers ($n = 16$) (Fig. 4.3C); (ANCOVA, equal slopes, different intercepts; $F = 9.29$, df $= 1,26$, $p = 0.0052$). This suggests that the absence of a significant overall difference between the iguanians and scleroglossan clades in FMR may be because, within the scleroglossan clade, the WF FMRs are tempered by the very low SW FMRs. Do these findings remain after accounting more thoroughly for phylogeny?

Independent contrast analyses

The practice of incorporating phylogenetic information into comparative statistical analyses has become increasingly commonplace, especially thanks to the continual "evolution" of computer programs to aid in such calculations. The ideology and methodology behind using independent contrasts (IC) (Felsenstein, 1985) in phylogenetic statistical comparisons has been well documented elsewhere (Garland *et al.*, 1993, 1999) and is discussed only briefly here.

All IC calculations were done on $\log_{10}$ transformed data, owing to the curvilinear effect of body mass on FMR. The overall phylogeny for the 46 lizard species (Fig. 4.1) was constructed from several sources (Kluge, 1987; Estes and Pregill, 1988; Arnold, 1989; Wiens and Reeder, 1997; Clobert *et al.*, 1998; Honda *et al.*, 2000; Macey *et al.*, 2000; Ast, 2001, and personal communications from W. Hodges, A. Larson, E. Pianka, and A. Whiting). Absolute divergence times were not available for our tree, so we used the PDTREE program from the PDAP suite (Version 6.0, Garland *et al.*, 1993) to determine the most appropriate branch length manipulations and to calculate IC values for use in our statistical comparisons. Using diagnostic plots of the absolute value of the standardized independent contrast vs. its standard deviation, we selected Nee's transformation (S. Nee, personal communication in Purvis, 1995) as the most suitable branch length manipulator for our dataset. The subsequent IC values were imported into SPSS, to duplicate many of the comparisons done previously with conventional statistical analyses following the methods discussed in Garland *et al.* (1993). All regressions on contrast data were necessarily forced through the origin (Garland *et al.*, 1992). To incorporate the foraging mode and clade categories, standardized contrasts for dummy variables (e.g. SW = 0, WF = 1 or Iguania = 0, Scleroglossa = 1) were created in PDTREE and used in a multiple regression along with body mass on FMR as described in Garland *et al.* (1993).

Using the PHYSIG.M program (Blomberg *et al.*, 2003), we detected a significant overall degree of phylogenetic signal among both body mass ($K = 1.328$, $p < 0.001$) and mass-adjusted FMR (mass scaled to 0.952 exponent) data ($K = 0.574$, $p = 0.008$). As suggested by Blomberg *et al.* (2003), the existence of such a strong indicator of phylogenetic bias in our database further justified the consideration of phylogeny in our results and interpretations.

As with the conventional statistical comparisons (Eq. 4.1), the overall regression for the IC data ($n = 45$) of FMR on body mass for all species was highly significant ($F = 363.6$, df $= 1,44$, $p < 0.000\,1$) (Fig. 4.4). When these data were coded for foraging mode, we found that FMR not only remained significantly different between SW and WF lizards ($F = 17.2$, df $= 1,43$, $p = 0.000\,15$), but this difference was even more significant than that found by the conventional

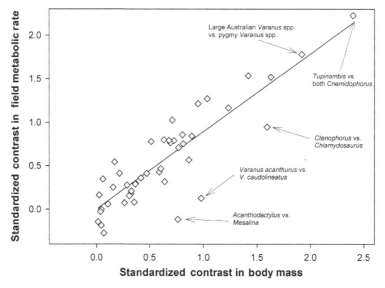

Figure 4.4. Bivariate scatterplot of relation between standardized independent contrasts in $\log_{10}$ field metabolic rate (FMR) and $\log_{10}$ body mass. Contrasts represent differences in FMR and body mass between two species, between a single species and a particular clade, or between two clades. Increasing distance away from the origin along the *x*-axis indicates the magnitude of the difference in body mass for each contrast. Deviation from the regression line (either above or below) indicates the magnitude of difference in FMR for each contrast. See text for discussion of indicated contrasts.

ANCOVA (recall $p = 0.007$). Similar to the conventional results, the IC approach also failed to detect a difference ($p = 0.671$) in FMR based on data categorized by major clade (iguanians vs. scleroglossans). Conversely, the conventionally significant difference between iguanian SW lizards and scleroglossan SW lizards was not supported in the IC analyses of the same data ($p = 0.326$).

Effects of foraging mode on field metabolic rate

It appears the answer to Huey and Pianka's (1981) original posit is that the costs of being a widely foraging lizard are substantial. The data used in the present comparisons include many possible sources of variation: wide variety of body mass values, within-population FMR variation, habitat effects, DLW methodological errors, differences in field and laboratory techniques, etc. Despite all of this possible inherent variation, there exists an overriding, highly significant effect of foraging mode on FMR, especially when phylogenetic relationships are explicitly accounted for.

How exactly do differences in foraging behavior translate into FMR differences? Could it be due to a generally greater level of metabolic activity at the cellular level in WF lizards, such as occurs in endothermic vertebrates in comparison with ectotherms? This question could be addressed by comparing the standard or resting metabolic rates (SMR, RMR) of the species listed in Fig. 4.1, to look for higher SMRs in WF species. However, such a compilation of literature values for species with measured FMR would presently be incomplete and is beyond the scope of this chapter. Bennett and Dawson (1976) reviewed the early literature on reptilian metabolism and found no differences in SMRs of reptiles that correlated with foraging mode. In a later review of reptilian SMR, Andrews and Pough (1985) found higher SMRs in squamates that were day-active predators and generally lower SMRs in reclusive and nocturnal predators, but the categories of SW and WF were not specifically examined. Beck and Lowe (1994) also found significantly low SMRs in the reclusive helodermatids. These lizards forage widely when active, but are surface active only about 70–200 hours per year. No differences in SMR were found between WF and SW species of lizards in the two detailed studies cited above (see "Two-species comparisons among lizards"). Thus, it appears that the higher FMR in WF lizards is probably not explicitly due to a generally higher level of maintenance metabolism, although it is perhaps debatable how much of a high-speed, high-endurance lizard's metabolic machinery is active under standardized conditions.

The costs are more likely due to a greater expenditure of energy on activity (mainly locomotion) while abroad and foraging. This finding is in accordance with expectations based on theory (Huey and Pianka, 1981) and it is consistent with results of the two detailed studies described above (Anderson and Karasov, 1981; Nagy *et al.*, 1984). Moreover, in the few single-species studies where the ratio of FMR to 24 h integrated SMR is reported, most SW lizards have ratios around 2, whereas the WF species generally have ratios of 3–4. Although there is much overlap in FMR:SMR ratio between groups, this trend toward polarization suggests that the WF species are working relatively harder to make a living than are the SW species. On this basis, we suspect that many WF lizard species have higher daily activity costs, despite spending less time foraging each day, by having much more intense locomotory activity during periods above ground than do many SW lizard species (see Chapters 2 and 3 for detailed consideration of foraging time and evaluation of locomotion costs).

Differences in locomotory costs likely are a major source of the energetic differences between the two foraging groups (Anderson and Karasov, 1981). The ecological cost of transport (ECT) is the energetic cost of locomotion expressed as a percentage of total daily energy expenditure (Garland, 1983;

Christian *et al.*, 1997). The ECT, although of minimal importance in mammals, is often a substantial proportion of a reptile's daily energy costs (Christian *et al.*, 1997). In a summary of several studies, Christian *et al.* (1997) found that movement costs in lizards generally represent 19.6% (range 3%–36%) of the total daily energy expenditure, but can occasionally be as high as 61% (*Varanus pantopes*) (Christian *et al.*, 1995). Because the ECT can be a substantial proportion of the overall energy budgets of lizards, the greater amount of time typically spent moving by wide foraging species likely results in substantially higher overall rates of daily energy expenditure when compared with sit-and wait species.

Phylogenetic patterns among lizard FMR

The compilation of such a large database of species presents a unique opportunity to look for patterns at various taxonomic resolutions. Neither the conventional nor the phylogenetically informed analyses detected a significant effect of a major clade (Iguania vs. Scleroglossa) on FMR. Other phylogenetic analyses also found no significant difference between major clades in various lizard demographic traits once body size (snout–vent length) was accounted for (Clobert *et al.*, 1998). Although the scleroglossans are about a 50:50 mix of SW to WF species, all of the iguanines (other half of the tree) are SW species. For ICA analyses, the distribution of traits across a tree can substantially affect a test's power (Vanhooydonck and Van Damme, 1999). The power of a test will be much greater in a tree that has the trait (here foraging mode) of interest equally spread among its tips, than in a tree that has the trait highly clustered on opposite sides. Indeed, the failure of the ICA to support the conventional analyses' detection of significantly higher FMR in iguanine sit-and-wait species than in scleroglossan sit-and-wait species may be due to such power issues. More likely, however, is that the patterns lie at higher taxonomic levels (up the tree vs. down).

Many morphological and behavioral traits, especially with respect to chemosensory abilities, characterize the foraging type of a lizard. Through an extensive evaluation of feeding behaviors and chemosensory abilities, Reilly and McBrayer (this volume, Chapter 10) identify five major groups that include foraging mode "convergences" (Iguania, Gekkota, Scincoidea, Lacertoidea and Anguimorpha; see Fig. 10.5, this volume). To look for similar clustering patterns among the FMR data, we plotted the standardized residuals from the overall regression of log FMR on log body mass grouped by family (Fig. 4.5) and calculated the mean mass-adjusted FMR for each family (Table 4.1).

Table 4.1 *Mean field metabolic rates (FMR, mass adjusted as* $kJ\,g^{-0.952}\,day^{-1}$ *)* *of lizard families for which the activity season FMR of two or more species have been determined*

SC, Scleroglossa; IG, Iguania.

Family	Major clade	n	Mean	SD
Xantusiidae	SC	3	0.067	0.025
Gekkonidae	SC	4	0.119	0.034
Scincidae	SC	2	0.136	0.049
Varanidae	SC	12	0.167	0.072
Phrynosomatidae	IG	7	0.188	0.045
Lacertidae	SC	8	0.186	0.069
Agamidae	IG	5	0.227	0.122
Teiidae	SC	3	0.277	0.020

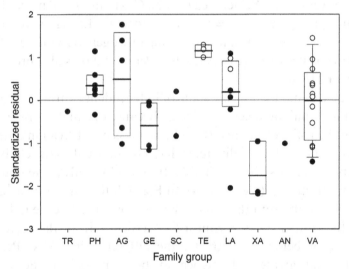

Figure 4.5. Box plot of standardized residuals of the overall $\log_{10}$ FMR on $\log_{10}$ body mass linear regression. Solid circles indicate ambushing species (SW) and open circles indicate wide-foragers (WF). Values have been grouped by family: TR, Tropiduridae; PH, Phrynosomatidae; AG, Agamidae; GE, Gekkonidae; SC, Scincidae; TE, Teiidae; LA, Lacertidae; XA, Xantusiidae; AN, Anguidae; VA, Varanidae.

Although there is some evidence for general clustering similarities perhaps within the Iguania, the level of family seems to be more important and the species-level variation is obvious (Fig. 4.5). For example, there is a fourfold difference within the Lacertoidea (Teiidae, Lacertidae, Xantusiidae) between mean Teiidae and Xantusiidae FMR (Table 4.1). However, the Teiidae and most of the Lacertidae species, which share many derived characteristics apart from Xantusiidae (Fig. 9.5, this volume), seem to cluster together above the mean residual line (Fig. 4.5). The mean residual value of Lacertidae, a family that contains both SW and WF species, would undoubtedly be higher if not for *Acanthodactylus pardalis*. In fact, when new regression and standardized residual values are calculated for the family Lacertidae only ($n = 8$), the residual value of *A. pardalis* is significantly lower than that of the other lacertid species (Dixon's test for outliers, $p < 0.05$) (Sokal and Rohlf, 1981). The *Acanthodactylus* genus is one of the few that has both SW and WF members, but the SW-foraging *A. pardalis* does not have unusually low SMR values (Duvdevani and Borut, 1974). It is suggested that the low FMR values of *A. pardalis* may result from its relatively short daily activity period (4.5 h/d) (Vernet *et al.*, 1988).

Small sample sizes make it difficult to compare the species within the other foraging "convergences" Scincoidea ($n = 2$), Gekkota ($n = 4$, no eublepharids) or Anguimorpha (Anguidae $n = 1$ vs. Varanidae $n = 12$). However, it appears that family-level conservatism found among phylogenetic analyses of foraging mode (behavior) in lizards (Cooper, 1994; Perry, 1999), is also present among many of the FMR data.

A particularly enlightening outcome of independent contrast analyses is the identification of unique species or clades. Because the analysis itself works by subtracting the traits of interest (here body mass and FMR) in a descending hierarchical fashion, those individuals that have unusual traits relative to their phylogenetic neighbors are identifiable. In our ICA analysis we noted several comparisons of interest (see Fig. 4.4). In Fig. 4.4, the further away a contrast value is horizontally from the origin, the greater the difference in body mass. Similarly, the further away a contrast is vertically from the regression line (either above or below), the greater the difference in FMR. Perhaps not surprisingly, the greatest difference in body mass was detected when *Tupinambis teguixin* (1172 g) was compared with the *Cnemidophorus* genus (mean mass 10.3 g) (Fig. 4.4). The large difference in body mass between the pygmy and non-pygmy Australian varanids was also detected. As discussed above in the residual analyses, the low FMR of *Acanthodactylus pardalis* is apparent even when compared only with its "sister" *Mesalina olivieri*.

Another species we had predicted *a priori* to have a low FMR for its family was *Varanus acanthurus*. This extremely sedentary species is the lone ambush forager among the varanids in this dataset (Dryden *et al.*, 1990). The reported FMR of this species was nearly 50% lower than that expected for any SW lizard (Eq. 4.4) of its size. We ran an ICA analysis on just the varanid family coded by foraging mode (Garland *et al.*, 1993) and found *V. acanthurus* to be significantly different ($F = 23.5$, df $= 1,9$, $p = 0.0009$). Part of the large contrast difference (Fig. 4.4) may stem from the fact that the sister species (at present) to *V. acanthurus* is *V. caudolineatus*, a species that has the highest measured maximal metabolic rate (V_{O_2}) of any squamate (Thompson and Withers, 1997). The measured FMR of *V. caudolineatus* was 30% greater than predicted for a WF lizard of the same body size (Eq. 4.6) and it had the highest standardized residual FMR among the varanids (Fig. 4.5, highest Varanidae residual). Although *V. acanthurus* belongs to a genus well known for its aerobic capacities, possesses a highly evolved vomerolfactory system to aid in prey detection (Reilly and McBrayer, Chapter 10), and descends from an evolutionary lineage that is effectively synonymous with wide foraging, it apparently also has an unusually low SMR (Thompson and Withers, 1994, 1997). To what degree the low FMR results from a low SMR is difficult to determine without detailed temperature and activity profiles. Nevertheless, it appears that behavior, including foraging activity, can be a more important determinant of energy expenditure than a species' evolutionary history.

This conclusion points to the interesting variation in metabolic rates that may occur in those families containing species that use both foraging modes. In the study of two confamilial lizard species (Nagy *et al.*, 1984), SMRs were identical, but the WF species had a much higher FMR than the SW species. Several lizard families, such as Scincidae, Polychrotidae and Gekkonidae, are poorly represented in these comparisons. Further energetic comparisons within families that encompass both ambush and active foraging species (Lacertidae, Scincidae, Pygopodidae, Gekkonidae), or even within a genus or species that possesses or switches between the two modes, such as *Acanthodactylus*, *Mabuya*, *Meroles*, *Pedioplanis* (Cooper and Whiting, 2000), and perhaps *Gambelia* (Tollestrup, 1982; Pietruszka, 1986) will allow greater understanding of the relationship between foraging mode and FMR, while minimizing possible taxonomic effects.

Conclusions

Present evidence supports the prediction that wide foraging lizards have higher field metabolic rates than do sit-and-wait lizards. In all three field studies on sympatric WF and SW squamate species of similar sizes, the WF species had shorter daily activity periods, were more intensely active while abroad, and had much higher daily energy expenditures. Most of this difference was attributable to higher locomotion costs in the WF species, although some of the difference was due to generally higher body temperatures and/or to somewhat higher standard or resting metabolic rates in one or two of the three species. The review and analysis of field metabolic rates of 46 species of lizard indicates that WF species have significantly higher daily energy expenditures, typically about 32% greater than SW species. Although there is little evidence supporting a phylogenetic influence on foraging mode energetics at the major clade level (e.g. iguanian versus scleroglossan or autarchoglossan lizards), there are strong associations with taxonomy at the family level. Major gaps in the foraging energetics data include several important, but as yet unstudied, lizard families, and those genera that include both WF and SW species. Similar FMR patterns between SW and WF snakes seem likely (Secor and Nagy, 1994) and are discussed elsewhere (Beaupre and Montgomery, Chapter 11).

Currently, the FMR data do appear to support the existence of a dichotomy between SW and WF lizards. However, the robustness and endurance of this viewpoint is uncertain. As more foraging behavior data (moves per minute, or percent time spent moving) become available, it may be possible to replace the foraging mode category with a continuous variable. If the intensity of foraging is truly the major determinant of FMR in lizards, then there should be a strong correlation between foraging behavior and FMR.

References

Anderson, R. A. and Karasov, W. H. (1981). Contrasts in energy intake and expenditure in sit-and-wait and widely foraging lizards. *Oecologia* **49**, 67–72.

Anderson, R. A. and Karasov, W. H. (1988). Energetics of the lizard *Cnemidophorus tigris* and life history consequences of food-acquisition mode. *Ecol. Monogr.* **58**, 79–110.

Andrews, R. M. and Pough, F. H. (1985). Metabolism of squamate reptiles: allometric and ecological relationships. *Physiol. Zool.* **58**, 214–31.

Arnold, E. N. (1989). Towards a phylogeny and biogeography of the Lacertidae: relationships within an Old-World family of lizards derived from morphology. *Bull. Br. Mus. Nat. Hist. (Zool.)* **55**, 209–57.

Ast, J. C. (2001). Mitochondrial DNA evidence and evolution in Varanoidea (Squamata). *Cladistics* **17**, 211–26.

Beck, D. D. and Lowe, C H. (1994). Resting metabolism of helodermatid lizards: allometric and ecological relationships. *J. Comp. Physiol.* B**164**, 124–9.

Benabib, M. and Congdon, J. D. (1992). Metabolic and water-flux rates of free-ranging tropical lizards. *Sceloporus variabilis. Physiol. Zool.* **65**, 788–802.

Bennett, A. F. and Dawson, W. R. (1976). Metabolism. In *Biology of the Reptilia*, vol. 5, ed. C. Gans and W. R. Dawson, pp. 127–223. London: Academic Press.

Bennett, A. F. and Nagy, K. A. (1977). Energy expenditure in free-ranging lizards. *Ecology* **58**, 697–700.

Bennett, A. F., Huey, R. B. and John-Alder, H. (1984). Physiological correlates of natural activity and locomotor capacity in two species of lacertid lizards. *J. Comp. Physiol.* B**154**, 113–18.

Bickler, P. E. and Nagy, K. A. (1980). Effect of parietalectomy on energy expenditure in free-ranging lizards. *Copeia* **1980**, 923–5.

Blomberg, S. P., Garland, T. Jr. and Ives, A. R. (2003). Testing for phylogenetic signal in comparative data: behavioral traits are more liable. *Evolution* **57**, 717–45.

Bradshaw, S. D., Saint Girons, H., Naulleau, G. and Nagy, K. A. (1987). Material and energy balance of some captive and free-ranging reptiles in western France. *Amph.-Rept.* **8**, 129–42.

Brown, R. P. and Perez-Mellado, V. (1994). Ecological energetics and food acquisition in dense Menorcan islet populations of the lizard *Podarcis lilfordi*. *Funct. Ecol.* **8**, 427–34.

Brown, R. P., Thorpe, R. S. and Speakman, J. R. (1992). Comparisons of body size, field energetics, and water flux among populations of the skink *Chalcides sexlineatus*. *Can. J. Zool.* **70**, 1001–6.

Christian, K. and Green, B. (1994). Seasonal energetics and water turnover of the frillneck lizard, *Chlamydosaurus kingii*, in the wet-dry tropics of Australia. *Herpetologica* **50**, 274–81.

Christian, K. A., Baudinette, R. V. and Pamula, Y. (1997). Energetic costs of activity by lizards in the field. *Funct. Ecol.* **11**, 392–7.

Christian K., Bedford, G., Green, B. *et al.* (1999). Physiological ecology of a tropical dragon, *Lophognathus temporalis*. *Aust. J. Ecol.* **24**, 171–81.

Christian, K. A., Bedford, G., Green, B., Schultz, T. and Newgrain, K. (1998). Energetics and water flux of the marbled velvet gecko (*Oedura marmorata*) in tropical and temperate habitats. *Oecologia* **116**, 336–42.

Christian, K. A., Corbett, L. K., Green, B. and Weavers, B. W. (1995). Seasonal activity and energetics of two species of varanid lizards in tropical Australia. *Oecologia* **103**, 349–57.

Christian, K., Green, B., Bedford, G. and Newgrain, K. (1996a). Seasonal metabolism of a small, arboreal monitor lizard, *Varanus scalaris*, in tropical Australia. *Aust. J. Zool.* **240**, 383–96.

Christian, K. A., Weavers, B. W., Green, B. and Bedford, G. S. (1996b). Energetics and water flux in a semiaquatic lizard, *Varanus mertensi*. *Copeia* **1996**, 354–62.

Clobert, J., Garland, T. Jr. and Barbault, R. (1998). The evolution of demographic tactics in lizards: a test of some hypotheses concerning life history evolution. *J. Evol. Biol.* **11**, 329–64.

Congdon, J. D. and Tinkle, D. W. (1982). Energy expenditure in free-ranging sagebrush lizards (*Sceloporus graciosus*). *Can. J. Zool.* **60**, 1412–16.

Cooper, W. E. Jr. (1994). Prey chemical discrimination, foraging mode, and phylogeny. In *Lizard Ecology: Historical and Experimental Perspectives*, ed. L. J. Vitt and E. R. Piank, pp. 95–116. Princeton, NJ: Princeton University Press.

Cooper, W. E. Jr. and Whiting, M. J. (2000). Ambush and active foraging modes both occur in the Scincid genus *Mabuya*. *Copeia* **2000**, 112–18.

Dryden, G., Green, B., King, D. and Losos, J. (1990). Water and energy turnover in a small monitor lizard, *Varanus acanthurus*. *Aust. Wild. Res.* **17**, 641–6.

Dryden, G. L., Green, B., Wikramanayake, E. D. and Drysen, K. G. (1992). Energy and water turnover in two tropical varanid lizards, *Varanus bengalensis* and *V. salvator*. *Copeia* **1992**, 102–7.

Duvdevani, I. and Borut, A. (1974). Oxygen consumption and evaporative water loss in four species of *Acanthodactylus* (Lacertidae). *Copeia* **1974**, 155–64.

Estes, R. and Pregill, G. (1988). *Phylogenetic Relationships of the Lizard Families: Essays Commemorating Charles L. Camp*. Stanford, CA: Stanford University Press.

Felsenstein, J. (1985). Phylogenies and quantitative characters. *Am. Nat.* **125**, 1–15.

Garland, T. Jr. (1983). Scaling the ecological cost of transport to body mass in terrestrial mammals. *Am. Nat.* **121**, 571–87.

Garland, T. Jr. and Adolph, S. C. (1994). Why not to do two-species comparative studies: limitations on inferring adaptation. *Physiol. Biochem. Zool.* **67**, 797–828.

Garland, T. Jr., Dickerman, A. W., Janis, C. M. and Jones, J. A. (1993). Phylogenetic analysis of covariance by computer simulation. *Syst. Biol.* **42**, 265–92.

Garland, T. Jr., Harvey, P. H. and Ives, A. R. (1992). Procedures for the analysis of comparative data using phylogenetically independent contrasts. *Syst. Biol.* **41**, 18–32.

Garland, T. Jr., Midford, P. E. and Ives, A. R. (1999). An introduction to phylogenetically based statistical methods, with a new method for confidence intervals on ancestral values. *Amer. Zool.* **39**, 374–88.

Green, B., Dryden, G. and Dryden, K. (1991a). Field energetics of a large carnivorous lizard, *Varanus rosenbergi. Oecologia* **88**, 547–51.

Green, B., Herrera, E., King, D. and Mooney, N. (1997). Water and energy use in a free-living tropical, carnivorous lizard, *Tupinambis teguixin. Copeia* **1997**, 200–3.

Green, B., King, D. and Butler, H. (1986). Water, sodium and energy turnover in free-living perenties, *Varanus giganteus. Aust. Wild. Res.* **13**, 589–96.

Green, B., King, D., Braysher, M. and Saim, A. (1991b). Thermoregulation, water turnover and energetics of free-living komodo dragons, *Varanus komodoensis. Comp. Biochem. Physiol.* **99A**, 97–102.

Guarino, F., Georges, A. and Green, B. (2002). Variation in energy metabolism and water flux of free-ranging male lace monitors, *Varanus varius* (Squamata: Varanidae). *Physiol. Biochem. Zool.* **75**, 294–304.

Honda, M., Ota, H., Kobayashi, M. *et al.* (2000). Phylogenetic relationships of the family Agamidae (Reptilia: Iguania) inferred from mitochondrial DNA sequences. *Zool. Sci.* **17**, 527–37.

Huey, R. B. and Pianka, E. R. (1981). Ecological consequences of foraging mode. *Ecology* **62**, 991–9.

Huey, R. B., Bennett, A. F., John-Alder, H. and Nagy, K. A. (1984). Locomotor capacity and foraging behavior of Kalahari lacertid lizards. *Anim. Behav.* **32**, 41–50.

Karasov, W. H. and Anderson, R. A. (1984). Interhabitat differences in energy acquisition and expenditure in a lizard. *Ecology* **65**, 235–47.

Karasov, W. H. and Anderson, R. A. (1998) Correlates of average daily metabolism of field-active zebra-tailed lizards (*Callisaurus draconoides*). *Physiol. Zool.* **71**, 93–105.

Kingsbury, B. A. (1995). Field metabolic rates of a eurythermic lizard. *Herpetologica* **51**, 155–9.

Kluge, A. G. (1987). Cladistic relationships in the Gekkonoidea (Squamata, Sauria). *Misc. Publ., Mus. Zool. Univ. Mich., Ann Arbor* No. **173**, 54 pp.

Lifson, N. and McClintock, R. (1966). Theory of use of the turnover rates of body water for measuring energy and material balance. *J. Theor. Biol.* **12**, 46–74.

Macey, J. R., Schulte, J. A. II, Larson, A. *et al.* (2000). Evaluating trans-tethys migration: an example using acrodont lizard phylogenies. *Syst. Biol.* **49**, 233–56.

Marler, C. A., Walsberg, G., White, M. L. and Moore, M. (1995). Increased energy expenditure due to increased territorial defense in male lizards after phenotypic manipulation. *Behav. Ecol. Socio biol.* **37**, 225–31.

Mautz, W. J. and Nagy, K. A. (2000). Xantusiid lizards have low energy, water, and food requirements. *Physiol. Biochem. Zool.* **73**, 480–7.

McLaughlin, R. L. (1989). Search modes of birds and lizards: Evidence for alternative movement patterns. *Am. Nat.* **133**, 654–70.

Merker, G. P. and Nagy, K. A. (1984). Energy utilization by free-ranging *Sceloporus virgatus* lizards. *Ecology* **65**, 575–81.

Nagy, K. A. (1975). Water and energy budgets of free-living animals: measurement using isotopically labeled water. In *Environmental Physiology of Desert Organisms*, ed. N. F. Hadley, pp. 227–45. Stroudsburg, PA: Dowden, Hutchinson and Ross.

Nagy, K. A. (1980). CO_2 production in animals: analysis of potential errors in the doubly labeled water method. *Am. J. Physiol.* **238**, R466–73.

Nagy, K. A. (1982). Energy requirements of free-living iguanid lizards. In *Iguanas of the World: Their Behavior, Ecology and Conservation*, ed. G. M. Burghardt and A. S. Rand, pp. 49–59. Park Ridge, NJ: Noyes Publishers.

Nagy, K. A. (1983a). *The Doubly Labeled Water ($^3HH^{18}O$) Method: a Guide to Its Use*. Los Angeles, CA: University of California, Publ. No. 12-1417. Available online from: kennagy@biology.ucla.edu.

Nagy, K. A. (1983b). Ecological energetics. In *Lizard Ecology: Studies of a Model Organism*, 2nd edn, ed. R. B. Huey, E. R. Pianka and T. W. Schoener, pp. 24–54. Cambridge, MA: Harvard University Press.

Nagy, K. A. (1989). Field bioenergetics: accuracy of models and methods. *Physiol. Zool.* **62**, 237–52.

Nagy, K. A. and Bradshaw, S. D. (1995). Energetics, osmoregulation, and food consumption by free-living desert lizards, *Ctenophorus* (= *Amphibolurus*) *nuchalis*. *Amph.-Rept.* **16**, 25–35.

Nagy, K. A. and Degen, A. A. (1988). Do desert geckos conserve energy and water by being nocturnal? *Physiol. Zool.* **61**, 495–9.

Nagy, K. A. and Knight, M. H. (1989). Comparative field energetics of a Kalahari skink (*Mabuya striata*) and gecko (*Pachydactylus bibroni*). *Copeia* **1989**, 13–17.

Nagy, K. A., Girard, I. and Brown, T. K. (1999). Energetics of free-ranging mammals, reptiles, and birds. *Ann. Rev. Nutr.* **19**, 247–77.

Nagy, K. A., Huey, R. B. and Bennett, A. F. (1984). Field energetics and foraging mode of Kalahari lacertid lizards. *Ecology* **65**, 588–96.

Nagy, K. A., Seely, M. K. and Buffenstein, R. (1993). Surprisingly low field metabolic rate of a diurnal desert gecko *Rhoptropus afer*. *Copeia* **1993**, 216–19.

Perry, G. (1999). The evolution of search modes: ecological versus phylogenetic perspectives. *Am. Nat.* **153**, 98–109.

Pietruszka, R. D. (1986). Search tactics of desert lizards: how polarized are they? *Anim. Behav.* **34**, 1742–58.

Purvis, A. (1995). A composite estimate of primate phylogeny. *Phil. Trans. R. Soc. Lond.* **B348**, 405–21.

Robinson, M. D. (1990). Summer field energetics of the Namib desert dune lizard *Aporosaura anchietae* (Lacertidae), and its relation to reproduction. *J. Arid Environ.* **18**, 207–16.

Secor, S. M. and Nagy, K. A. (1994). Bioenergetic correlates of foraging mode for the snakes *Crotalus cerastes* and *Masticophis flagellum*. *Ecology* **75**, 1600–14.

Sokal, R. R. and Rohlf, F. J. (1981). *Biometry*, 2nd edn. New York: W. H. Freeman and Co.

Thompson, G. G. and Withers, P. C. (1994). Standard metabolic rates of two small Australian varanid lizards (*Varanus caudolineatus* and *V. acanthurus*). *Herpetologica* **50**, 494–502.

Thompson, G. G. and Withers, P. C. (1997). Standard and maximal metabolic rates of goannas (Squamata: Varanidae). *Physiol. Zool.* **70**, 307–23.

Thompson, G. G., Bradshaw, S. D., and Withers, P. C. (1997). Energy and water turnover rates of a free-living and captive goanna, *Varanus caudolineatus* (Lacertilia: Varanidae). *Comp. Biochem. Physiol.* **116A**, 105–11.

Tinkle, D. W. and Hadley, N. F. (1975). Lizard reproductive effort, caloric estimates and comments on its evolution. *Ecology* **56**, 427–34.

Tollestrup, K. (1982). Growth and reproduction in two closely related species of leopard lizards, *Gambelia silus* and *Gambelia wislizenii*. *Amer. Mid. Nat.* **108**, 1–20.

Vanhooydonck, B. and Van Damme, R. (1999). Evolutionary relationships between body shape and habitat use in lacertid lizards. *Evol. Ecol. Res.* **1**, 785–805.

Vernet, R., Castanet, J. and Baez, M. (1995). Comparative water flux and daily energy expenditure of lizards of the genus *Gallotia* (Lacertidae) from the Canary Islands. *Amph.-Rept.* **16**, 55–66.

Vernet, R., Grenot, C. and Nouira, S. (1988). Water flux and energy metabolism in a population of Lacertidae from the Kerkenna islands (Tunisia). *Can. J. Zool.* **66**, 555–61.

Wiens, J. J. and Reeder, T. W. (1997). Phylogeny of the spiny lizards (Sceloporus) based on molecular and morphological evidence. *Herp. Monogr.* **11**, 1–101.

Withers, P. C. and Bradshaw, S. D. (1995). Water and energy balance of the thorny devil *Moloch horridus*: is the devil a sloth? *Amph.-Rept.* **16**, 47–54.

Znari, M. and Nagy, K. A. (1997). Field metabolic rate and water flux in free-living Bibron's agama (*Agama impalearis*, Boettger, 1874) in Morocco. *Herpetologica* **53**, 81–8.

5

Feeding ecology in the natural world

LAURIE J. VITT

Sam Noble Oklahoma Museum of Natural History and Zoology Department,
University of Oklahoma

ERIC R. PIANKA

Section of Integrative Biology, School of Biological Sciences,
University of Texas at Austin

Introduction

Foraging mode, originally defined on the basis of clear differences in behaviors used to find and capture prey (MacArthur and Pianka, 1966; Pianka, 1966; Schoener, 1971) has become a central paradigm in lizard ecology (see, for example, Huey and Pianka, 1981; Vitt and Congdon, 1978; Cooper, 1994a,b, 1995a,b; Perry, 1999; Perry and Pianka, 1997; Perry et al., 1990). Sit-and-wait (often referred to as "ambush") foragers pursue prey detected visually from short distances, often returning to the same perch after capturing a prey item. Wide (often referred to as "active") foragers move through the environment in search of prey that are often hidden, using a combination of visual and chemical cues to locate and discriminate prey. Trade-offs between energy invested in capture versus search for these two foraging modes are key elements of optimal foraging theory (MacArthur and Pianka, 1966; Charnov, 1976; Kamil, 1983). Identification of this foraging dichotomy has stimulated lizard research in many areas, including ecology, behavior, life histories, and physiology, to mention a few.

The foraging mode paradigm is much more complex than previously envisioned, as evidenced by research presented in other chapters in this book. For example, what appeared to be a sharp historical separation of foraging modes (see, for example, Pianka and Vitt, 2003; Vitt et al., 2003) is replete with exceptions embedded in major clades, suggesting either loss of or multiple origins of traits often linked to foraging mode (see, for example, Cooper, 1997; Cooper et al., 1997). Empirical data on components of foraging mode (e.g. percent time moving) also reveal a much more complex pattern (Perry, 1999; this volume, Chapter 1). Prey detection systems (visual versus olfactory) (Cooper et al., 1997; Schwenk, 1995, 2000a,b), jaw function (Schwenk, 2000b; McBrayer and Reilly, 2002), behavior (Anderson and Karasov, 1981; Anderson and Vitt, 1990), and even thermal physiology are associated loosely with

Lizard Ecology: The Evolutionary Consequences of Foraging Mode, ed. S. M. Reilly, L. D. McBrayer and D. B. Miles. Published by Cambridge University Press. © Cambridge University Press 2007.

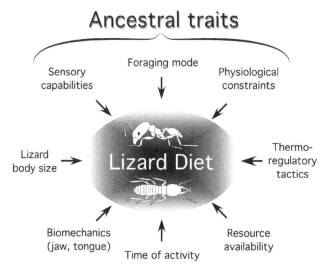

Figure 5.1. Possible sources of variation in lizard diets.

foraging mode but not necessarily predictable based exclusively on variation in foraging behavior.

We comment on major evolutionary and non-evolutionary factors that affect the kinds of prey eaten by lizards. We use data from our combined studies on lizards conducted in African, Australian, and North American deserts by ERP and the New World Tropics by LJV. Specifically, we comment on eight factors that appear *a priori* most likely to influence lizard diets: lizard body size, biomechanics of feeding structures, thermoregulatory tactics, time of activity, sensory capabilities, physiological constraints, foraging mode, and resource availability, all of which are constrained by phylogeny either directly or indirectly (Fig. 5.1). Clearly, these are not independent and the relative impact of each on the others remains largely unstudied. We comment on this as well. Finally, and most importantly, we emphasize earlier analyses (Vitt *et al.*, 2003; Vitt and Pianka, 2005) that show that a large portion of differences in diets among lizard clades is historical, and likely tied to several key events in the evolutionary history of sensory systems and differential involvement of the tongue and jaws in prey prehension.

Prior to addressing each of these, a comment on the nature of dietary data is necessary. Most lizards eat a diversity of invertebrates, along with some plant material (see, for example, Cooper and Vitt, 2002). Some species feed primarily on leafy vegetation (e.g. Iguanidae) (Iverson, 1979; Durtsche, 2000; Espinoza *et al.*, 2004), some are specialists on specific invertebrates (e.g. some tropidurids, phrynosomatids, and agamids specialize on ants) (Vitt and Zani, 1996a; Vitt *et al.*, 1997; Pianka and Parker, 1975; Pianka and Pianka 1970), while still others feed primarily on other vertebrates (e.g. *Lialis*, *Heloderma*, and most varanids) (Auffenberg, 1978; Shine, 1986; Beck, 1990; Beck and Lowe, 1991; Pianka, 1994; Pianka and King, 2004). Many other lizard species have very broad diets. Unlike many other kinds of data, dietary data are extremely complex. Size of prey eaten by an individual lizard over a period of time or by a population of lizards at one time consists of a distribution that is usually log-normal and a set of taxonomic categories varying in both frequency and size. Thus no single number can easily be assigned as representative of a particular species' diet. Moreover, for some lizard species, diets vary considerably as a function of availability of prey as evidenced by seasonal and spatial variation in diets (see, for example, Sexton *et al.*, 1972; Dunham, 1980; Vitt, 1991; Miranda and Andrade, 2003; Rocha, 1996; Van Wyk, 2000; Pianka, 1970; Vitt and Colli, 1994), whereas for other species, diets vary little from time to time or geographically (e.g. *Phrynosoma*: Pianka and Parker, 1975; Sherbrooke, 1981 or *Plica umbra*: Vitt *et al.*, 1997). Complicating dietary data even further is the observation that diets sometimes change ontogenetically (see, for example, Mautz and Nagy, 1987; Durtsche *et al.*, 1997; Durtsche 2000).

Finally, a comment is necessary on evolutionary patterns of prey prehension in squamates, because two hypotheses exist. One (McBrayer and Reilly, 2002; Reilly and McBrayer, this volume, Chapter 10) maintains that squamate ancestors possessed both lingual and jaw prehension and thus both are ancestral. Reilly and McBrayer (this volume, Chapter 10) use that argument as a basis for concluding that "lingual prehension" in some skinks is primitive retention rather than a functional reversal, even though the few skinks in which this has been observed are deeply nested within clades using jaw prehension. Wagner and Schwenk (2000) and Schwenk (2000b) consider obligate lingual prehension ancestral and obligate jaw prehension derived in scleroglossans with loss of lingual prehension associated with use of the tongue for other functions (e.g. cleaning in gekkotans and chemical sampling in autarchoglossans). Even though some iguanians use their jaws to pick up unusually large prey (e.g. *Crotaphytus* and *Gambelia*), this is not the obligate jaw prehension observed in scleroglossans, the tongue is involved, and insectivorous iguanians rarely eat large prey (see Vitt *et al.*, 2003). To us, the evidence summarized by Schwenk's (2000b) arguments is much more compelling and provides a more parsimonious view of the evolution of prey prehension in squamates. Use of tongues in prey capture in highly derived skinks, for example, more likely represents reversals, and the underlying lingual mechanism is most certainly different from that found in iguanians. For this chapter, we follow Schwenk (2000b) and leave the continuing debate to functional morphologists. However, when we return to historical (phylogenetic) factors affecting lizard diets, we re-examine some of our earlier conclusions in the context of a very different and provocative new view of squamate evolution (Townsend *et al.*, 2004).

Methods

Methods for collection of lizards and identification and measurements of prey appear elsewhere (see, for example, Pianka, 1973, 1986; Vitt and Zani, 1996b). We consider both prey sizes and types in our analyses, noting that they likely are not entirely independent. Because prey sizes were log-normally distributed, data were $\log_{10}$ transformed prior to statistical tests. We then compared prey size among target groups (foraging modes, clades, etc.) with an ANOVA. These analyses fail to account for differences in sample sizes among species or potential effects of lizard body size on prey size. To adjust for differences in prey size associated with lizard body size, we calculated mean body size (snout–vent length, SVL) and mean prey size for all lizard species, $\log_{10}$-transformed the variables, and conducted analyses of covariance (ANCOVA) with SVL as the covariate.

We briefly comment on some changes we made to our original datasets. Initial prey categories for desert and neotropical lizards were nearly identical, which allows us to re-analyze our data at various taxonomic levels. The original neotropical lizard data set included 30 broad prey types, whereas the original desert lizard data set included 20 broad prey types. Some more detailed dietary datasets for desert lizards are also available. Relatively few prey categories accounted for most of the diets of all lizards included. We specifically selected data on target species to make points in our discussion of factors affecting lizard diets. Thus, our examples, by design, in some cases represent extremes. Earlier, we combined data to come up with generalized diets for specific clades (Pianka and Vitt, 2003; Vitt *et al.*, 2003). In doing so, we restricted our analysis to mean percent utilization by volume of the seven most important prey categories for all lizards: ants (A), beetles (B), grasshoppers and crickets (G), non-ant hymenopterans (H), insect larvae, pupae, and eggs (L), spiders (S), and termites (T). Ants were treated as a unique category (rather than including them with other hymenopterans) because they exhibit their own morphotype, are highly diverse and abundant, and because some lizards specialize on them (i.e. lizards discriminate ants from other hymenopterans).

Finally, to reconstruct the history of dietary change in lizards, we combined our datasets and constructed a large data matrix consisting of diets of 184 species using 27 prey categories. These data constituted the dependent variable.

The independent variable was the clade representation of each species. Because lizard size affects diet and covaries with clade, average lizard species snout–vent length was entered as a covariate (see Vitt and Pianka, 2005). We discovered an error in dietary data for one species (*Anolis n. scypheus*) that we had overlooked in a previous analysis (Vitt and Pianka, 2005) and corrected it. We applied a canonical phylogenetic ordination based on Canonical Correspondence Analysis (CCA) (Giannini, 2003), a multivariate ordination procedure that directly associates variation in one matrix (lizard diets in this case) to variation in another (lizard phylogeny in this case). Thus, in this analysis, we ask whether an association exists between dietary composition and identified divergence points in the evolutionary history of lizards (see Vitt and Pianka, 2005 for detailed methods). The CCA was performed with CANOCO 4.5 (Ter Braak and Smilauer, 2002). We used symmetric scaling and unimodal methods and downweighted rare prey categories. In a stepwise procedure, each variable was then tested by using 9999 Monte Carlo permutations to obtain F and p values. After each significant variable was included in the model, the subsequent variable that most reduced variance was tested and included if statistically significant ($p < 0.05$). This procedure was followed until subsequent variables were no longer significant.

Lizard body size

In general, larger lizards eat larger prey (Fig. 5.2). However, much remains hidden in such regressions. On a purely statistical basis, the relation between

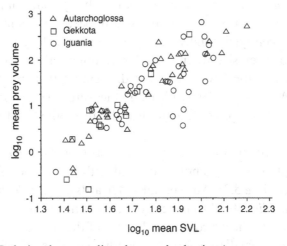

Figure 5.2. Relation between lizard mean body size (snout–vent length) and mean prey size for neotropical lizards. Each point represents a lizard species.

Table 5.1 *ANCOVAs with log$_{10}$ mean SVL as the covariate, clade level as the class variable, and log$_{10}$ mean prey volume as the dependent variable*

We retained full effects in analyses in which interaction terms were insignificant ($p > 0.05$) because interaction terms were marginally significant. Xantusiidae and Anguidae were removed from the third (family-level) analysis because each was represented by a single species. Clades are in order of mean prey size (largest to smallest).

	Slope test		Intercept	
Clades used	*F* value	*p* value	*F* value	*p* value
Iguania, Scleroglossa (1, 1, 82)	3.89	0.0521	4.49	0.0654
Iguania, Gekkota, Autarchoglossa (1, 2, 80)	3.95	0.0231	162.74	< 0.0001
Teiid, Scinc, Iguan, Gekk, Gymno (4, 4, 74)	2.39	0.0585	113.39	< 0.0001

prey size and lizard size varies among higher- and lower-level clades (Table 5.1). Reasons underlying this relationship are complex. We point to a few here and return to more later on in this chapter.

Lizards that are "generalists" (e.g. *Ameiva ameiva*) tend to eat a wide variety of prey sizes, and prey size is often associated with lizard size (Fig. 5.3) such that their diet usually consists of many small prey items and some large ones. Most likely, head size (as it affects gape) and biomechanics of the jaw and food processing structures limits the absolute size of prey that can be eaten. However, exactly what determines the maximum prey size that a lizard can take remains unclear. Lizards can eat some unusually large prey volumetrically if they are long and narrow. For example, a *Sceloporus clarkii* might eat a centipede (*Scolopendra heros*) longer than its body that completely fills its stomach, but it may not be able to eat a scarab beetle with a mass of only one third that of the centipede. Individual prey volume or length may not be the best measures of prey size when considering potential constraints on maximum prey size. For the few lizards that dismember their prey (e.g. *Varanus komodoensis*) (Auffenberg, 1978, 1981), no measures of prey size are relevant; rather, the size of the piece that the lizard can swallow determines size of portions.

Prey size of lizards that are dietary specialists (e.g. *Plica umbra*) appears to vary considerably less and is not associated with lizard body size, or at least to a lesser degree (Fig. 5.3). Ant-specialized lizards (e.g. *Moloch*, *Phrynosoma*, *Phrynocephalus*, *Plica*) tend to have broad, short heads and eat prey that are very small relative to the lizard's head. Length of the epipterygoid bone and to a lesser degree, vertical diameter of the mandible in the skull of horned lizards are negatively associated with percent of ants in the diet, suggesting that

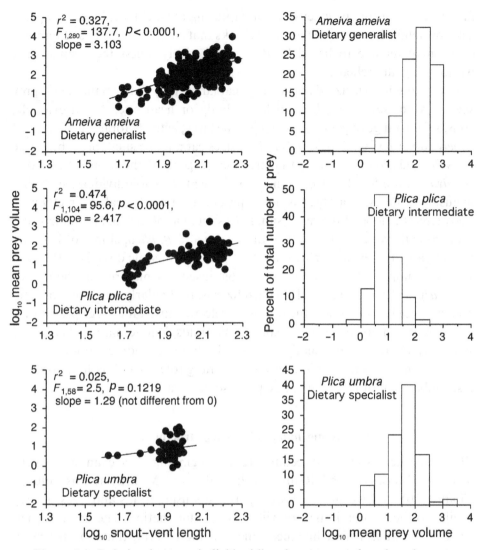

Figure 5.3. Relation between individual lizard snout–vent length and mean prey size for a dietary specialist (*Plica umbra*), a generalist (*Ameiva ameiva*), and a species with an intermediate diet (*Plica plica*).

morphological change in skull morphology results from specialization on ants (Montanucci, 1989). It might be particularly instructive to examine convergences (if any) in feeding mechanics of these unrelated lizards to determine whether apparent convergence in head morphology is associated with dietary specialization: why should a short, broad head be good for eating ants, and do the same skull modifications occur in ant-eating clades other than *Phrynosoma*? Most likely, underlying skull modifications are modifications in mechanics of tongue extension for capturing ants. If true, then one has to ask how teiids, lacertids, geckos, and skinks that eat large numbers of termites (also small relative to lizard head size) effectively ingest these small prey without lingual prehension.

Even though large lizards on average eat larger prey than smaller ones, prey size does not scale directly with lizard body or head size. As an example, consider the range of prey sizes eaten by the large, actively foraging teiid lizard *Ameiva ameiva* (Fig. 5.3). *Ameiva* of all sizes eat a lot of small prey, but large ones can and do eat a few prey that are much larger than those eaten by smaller *Ameiva*. Figure 5.3 shows only mean prey size for each individual, such that inclusion of many small prey is not apparent. If prey size scaled directly with body size, the slope of the regression lines should approach 3.0 (volume scales geometrically with length). For both *Ameiva* and *P. plica*, slopes of lines are considerably flatter than that. Failure of prey size to scale directly with lizard size may simply reflect the fact that most insects (a majority of the diet of *Ameiva* and *P. plica*) are small, so most lizards, particularly moderate to large-bodied species, do not have the option of selecting just a few large prey if they are to maintain positive energy balance. Varanids, helodermatids, *Lialis*, and snakes have shifted to unusually large prey by extreme modifications of the jaw and skull, modification or loss of the pectoral girdle (including limb loss), or dismemberment of unusually large prey (some varanids).

Biomechanics of feeding structures

The biomechanics of lizard feeding are just beginning to be examined experimentally (McBrayer and Reilly, 2002; Reilly and McBrayer, this volume, Chapter 10), and we suspect that experiments using kinds and sizes of prey similar to those eaten in nature will be most illuminating. The greatest challenge facing studies of feeding mechanics is demonstrating that morphological and functional variation affect observed differences in diets among species. To be convincing, such studies must be performed with natural prey of each lizard species. Differences among lizard species in the ability to manipulate and ingest prey that are well outside of the size range, of a different consistency, or of taxa different from those of naturally eaten prey are irrelevant.

As a group, squamates eat a remarkable diversity of prey types. Community ecologists divide prey into morphotypes to compare diets of species (see, for example, Pianka, 1973, 1986). Such categorization, useful in ecology, may not be best for studies of functional morphology. Lizards that use lingual prehension might be able to ingest any dry solid prey item of suitable size but unable to ingest a prey item with a highly flexible and damp body (e.g. an earthworm or pulmonate) (K. Schwenk, pers. comm.). Nevertheless, some Amazonian iguanians include earthworms in their diets (e.g. *Enyalioides* and some *Anolis* species). Lizards using jaw prehension can not only pick up such prey, they can use their snouts to root through surface debris and capture many hidden insect larvae as well as other invertebrates and small vertebrates (Vitt and Cooper, 1986). Moreover, jaw prehension allows a lizard to capture large prey, some of which are killed by rapid thrashing of the prey against rocks and other surface items. If jaw and tongue mechanics evolved in squamates in response to changes in diets (as opposed to diets changing in response to changes in jaw or lingual structure tied to something else), then diets of clades differing in jaw and tongue structure and use should differ in a predictable manner based on underlying jaw and tongue structure and function.

Dietary differences are apparent among major clades (Vitt and Pianka, 2005). The most obvious example of an association between feeding structure biomechanics and diet is in snakes, which have evolved a jaw structure accommodating ingestion of large prey relative to snake head size (Greene, 1997). Although members of one snake clade (Scolecophidia) have relatively rigid skulls with positioning of the quadrate and size of the supratemporal and mandible more or less similar to anguimorphan ancestors, higher snakes (alethinophidians) have various degrees of posterior migration of the quadrate, enlargement and extension of the supratemporal, loose mandibular symphysis, elongation of the mandible to accommodate ingestion of large prey, and independent movement of left and right maxillae (Cundall and Greene, 2000). These modifications, combined with loss of a pectoral girdle, accommodate a shift to vertebrates as primary prey. However, among lower-level clades (e.g. families within clades), dietary differences may not correlate with differences in feeding structures.

Within squamates typically referred to as "lizards," the most apparent difference in prey observed at the Iguania–Scleroglossa divergence is the drastic reduction in ants, other hymenopterans, and beetles in scleroglossan diets (Vitt and Pianka, 2005). Three non-exclusive hypotheses exist to explain this shift: (1) use of olfactory (Gekkota) and volmerofactory (Autarchoglossa) cues allowed scleroglossans to selectively eliminate prey producing noxious chemicals (particularly alkaloids) from their diets; (2) chemoreception in scleroglossans allowed them to discriminate prey quality, resulting in selection of prey of potentially higher energy content (Pianka and Vitt, 2003; Vitt *et al.*, 2003); and (3) the switch to jaw prehension resulted from a switch to larger prey (Schwenk, 2000b). Our data do not support hypothesis 3. Support exists for the first two hypotheses and they share common elements. Nevertheless, reasons for differences remain largely unexplored.

Thermoregulatory tactics and time of activity

Thermoregulatory tactics have not been examined with reference to squamate diets. Nevertheless, lizard species vary considerably in body temperatures while active, and different species are active at different times. The most extreme variation exists between diurnal and nocturnal squamates, which for lizards, breaks down phylogenetically (most gekkonids are nocturnal, but most other lizards are diurnal). However, considerable temperature variation exists between lizards that seek direct sun to elevate their body temperatures (thermoregulators, heliotherms) versus those that maintain most activity in shaded environments such that their body temperatures remain similar to low temperatures in shade (thermoconformers). Active body temperatures also vary between and within lizard clades. For example, among Teioidea, lacertids and teiids are active thermoregulators with relatively high body temperatures, whereas gymnophthalmids in rainforest have substantially lower body temperatures and appear to be thermoconformers (Vitt and Pianka, 2004).

Comparison of nocturnal versus diurnal lizard diets reveals the obvious: nocturnal lizards feed on more nocturnal insects, such as crickets, moths, and certain spiders, than do diurnal lizards (see Parker and Pianka, 1974; Pianka and Pianka, 1976; Avery, 1981; Doughty and Shine, 1995; Vitt, 1995; Vitt and Zani, 1997). Differences in resource availability between night and day offer the best explanation for such dietary differences. However, behavior of some nocturnal geckos suggest that maintenance of higher body temperatures during the day than those experienced at night may increase digestion rates and possibly facilitate other metabolic processes (Autumn *et al.*, 1999; Huey *et al.*, 1989). An Australian gecko, *Christinus marmoratus*, for example, thermoregulates by positioning itself in crevices and achieves relatively high body temperatures during the day (Kearney and Predavec, 2000). Similarly, behavioral thermoregulation increases growth rates in another gecko (Autumn and DeNardo, 1995).

Herbivorous lizards face the challenge of digesting plant materials that contain cellulose, often accomplished by microflora-induced gut fermentation (Troyer, 1984; Durtsche, 2000). Most studied herbivorous lizards are either active at higher body temperatures (e.g. *Dipsosaurus*; Pianka, 1971) or have extended activity to accommodate nutritional assimilation (Zimmerman and Tracy, 1989; van Marken Lichtenbelt, 1992; Vitt *et al.*, 2005). Small-bodied herbivorous lizards in the southern Andes appear to operate at lower body temperatures (Espinoza *et al.*, 2004), and it remains to be seen whether they compensate by increasing activity periods (see Vitt, 2004).

Sensory capabilities

Gekkotans use visual cues for detecting prey but also discriminate prey chemically via the nasal olfactory system. Autarchoglossans use both visual and chemical cues for locating prey, and they discriminate prey based on heavy non-volatile chemicals. The impact of these differences in sensory abilities on lizard diets is profound. For autarchoglossans, the ability to detect prey by means other than visual cues made available to them a huge diversity of insect taxa and life history stages not harvested to a significant degree by iguanians, including, but not limited to, termites, beetle larvae, cryptic insects and spiders, aquatic insects, mollusks, and hidden vertebrates (Vitt and Pianka, 2005). Coupled with jaw prehension (*sensu* Schwenk, 2000b), it also allowed some autarchoglossans to find, capture, and subdue large and in some instances potentially dangerous prey, such as those used by varanoid lizards and large teiids. Jaw prehension combined with serrated teeth in some varanids expanded dietary opportunities even more, allowing some species to kill prey much too large to ingest intact and later to dismember dead prey and swallow it in pieces (e.g. *Varanus komodoensis*; Auffenberg, 1981).

The most obvious effect of sensory capabilities on diet is the reduction of insects that use chemical defenses observed in the Scleroglossa. Ants, other hymenopterans, and beetles are less prevalent in scleroglossan diets. Development of an acute nasal olfactory system in gekkotans and an acute vomeronasal chemical sensing system in autarchoglossans provided opportunities for scleroglossans to discriminate among prey types based on chemical signals (see, for example, Cooper, 1994a,b, 1995b; Cooper and Hartdegen, 1999). Whether scleroglossans actively avoid beetles, ants, and other hymenopterans because they contain chemicals that might interfere with metabolic processes or whether they simply select prey with higher energy content remains uncertain. However, experiments on autarchoglossan responses to chemicals suggest the former (Cooper *et al.*, 2002a,b).

Physiological constraints

Physiological constraints on diet include phenomena such as differences between insectivorous and herbivorous lizards in relative stomach volume as well as gut length and anatomy (Ostrom, 1963; Iverson, 1982). However, without question, temperature is the primary physiological constraint on lizard diets. Body temperature has a major impact on all physiological processes because lizards are ectothermic poikilotherms (Huey, 1982). Not only do many lizards have relatively low body temperatures, but even those with high body temperatures experience hourly, daily, and seasonal variation in body temperatures. Effects of temperature related to feeding in squamates include foraging (Wilhoft, 1958; Ayers and Shine, 1997), hunger (Alexander _et al._, 2001), efficiency of capturing and handling prey (Greenwald, 1971; de Queiroz _et al._, 1987), efficiency of absorption (Harlow _et al._, 1976; Harwood, 1979; Beaupre _et al._, 1993), regulation of vitamin production (Ferguson _et al._, 2003), transport of nutrients to tissues via circulation, assimilation efficiency (Ballinger and Holscher, 1980; Troyer, 1987; Xue-Feng _et al._, 2001), specific dynamic action (Zaidan and Beaupre, 2003), overall metabolic rates (Zimmerman and Tracy, 1989; van Marken Lichtenbelt and Wesselingh, 1993; Spotila and Standora, 1985; but see Nussear _et al._, 1998 for a counter example), and assimilation of preformed water (Kaufmann and Pough, 1982; Clarke and Nicolson, 1994).

An as yet unexplored constraint to foraging and behavior in general is the potential effect of alkaloid intake in the diet as the result of hymenopteran and beetle consumption. These insects are well known to contain alkaloids (Blum, 1981). Reduction in intake of insects containing alkaloids associated with the shift from visual to chemosensory prey discrimination (see Pianka and Vitt, 2003; Vitt _et al._, 2003) at the Iguania–Scleroglossa transition released scleroglossans from effects of alkaloids on metabolic processes. Appropriate physiological studies to verify this are needed.

Foraging mode

As discussed above, foraging mode influences types of prey a lizard encounters. Sit-and-wait foragers hunt visually and hence only encounter mobile prey as they move past ambush stations, whereas widely foraging predators encounter more potential food items as well as a wider variety, because they search for hidden and sedentary prey (Huey and Pianka, 1981). Analyzing lizard diets strictly in the context of foraging mode presents a number of problems. First, foraging mode is strongly correlated with phylogeny, although some reversions have occurred. Almost without exception, iguanians ambush their prey, whereas most autarchoglossans forage widely (Cordylidae and a few Varanidae have reverted back to the ancestral sit-and-wait mode of foraging). Geckos have been variously classified, both as sit-and-wait ambush foragers, and as active widely foraging predators. Underlying the association of foraging mode with phylogeny are dramatic shifts in prey detection and handling between iguanians, gekkotans, and autarchoglossans. Some dietary differences associated with foraging mode are to be expected, but foraging mode may not be the primary cause of these differences (see below).

Resource availability

Effects of resource availability on lizard diets are complex. On the one hand, if competition accounts for some of the structure in lizard assemblages, then food resources may be limiting. However, whether resource levels measured by sticky traps, pitfalls, or any other methods bear directly on actual resource availability to lizards remains to be seen. In risky environments (high predator diversity or abundance), measured resources could be very high, but risks involved in acquiring those resources might be so high that effectively some resources are unavailable to lizards.

Many people naively assert that lizards are opportunistic feeders, eating whatever is available in their environment. If this were true, all lizards living together in an area would eat the same prey in the same proportions. This is manifestly not the case: dietary differences among species are the rule rather than the exception (see, for example, Pianka 1973, 1986; Vitt and Zani, 1996b). Nevertheless, for many species, diets vary with seasons as pointed out earlier. Within assemblages, dietary differences among species are maintained in spite of seasonal or annual variation in diets (see, for example, Vitt, 1991). Some species, however, are specialists, feeding on just a few prey categories to the exclusion of others, regardless of time of year or locality. Various species of lizard have specialized on ants, termites, scorpions, and large vertebrate prey. For example, ant specialization has evolved independently in agamids (*Phrynocephalus* and *Moloch*: Anderson, 1999; Pianka and Pianka, 1970; Pianka *et al.*, 1998), phrynosomatids (*Phrynosoma*: Pianka and Parker, 1975; Sherbrooke, 1981), and tropidurids (*Plica* and *Uracentron*: Vitt *et al.*, 1997; Vitt and Zani, 1996a) to mention a few. Termite specialization has evolved in scincids (*Typhlosaurus*: Huey *et al.*, 1974; *Ctenotus*: Pianka, 1969), lacertids (*Heliobolus, Pedioplanus*: Pianka, 1986), and gekkonids (*Diplodactylus*: Pianka and Pianka, 1976; *Pachydactylus*: Pianka and Huey, 1978; Pianka, 1986). The diurnal Kalahari lacertid *Nucras tessellata* and the nocturnal Australian pygopodid *Pygopus nigriceps* are scorpion specialists (Pianka, 1986). Another pygopodid, *Lialis burtonis*, preys almost exclusively on vertebrates, especially skinks. Monitor lizards of the genus *Varanus* and teiids in the genus *Tupinambis* also specialize on large vertebrate prey, in Australia and the neotropics, respectively.

As an example of interactions among foraging mode, time of activity, and resource availability, consider Australian termite specialist geckos and skinks (Huey *et al.*, 2001). Australian geckos are nocturnal sit-and-wait foragers, whereas sympatric skinks are diurnal and forage widely. These dietary specialists show striking variation in feeding success. More than 50% of all stomachs of three gecko dietary specialists were empty (*Diplodactylus conspicillatus*,

D. pulcher, Rhynchoedura ornata) (Pianka and Pianka, 1976; Pianka, 1986). Several sympatric Australian *Ctenotus* skink species that specialize on termites have substantially lower frequencies of empty stomachs: *C. ariadnae* 20%, *C. grandis* 6.4%, *C. pantherinus* 12.4% (Huey *et al.*, 2001). Diurnal WF termite specialists capture termites in their tunnels, in termitaria, or in open foraging trails, and would appear to have more reliable access to termites than nocturnal SW species, which must capture termites at night when these insects are active above ground. Termite activity at night appears to be unreliable (certainly termite swarms are).

Rather than there being a single vector of resource availability that applies to all species, instead each species experiences its own unique vector of resource availability, which is an outcome of the interaction between its perceptual abilities, body size, time of activity (daily and seasonal), use of space (microhabitat and habitat), as well as its foraging mode and thermoregulatory tactics (thermoconformer–thermoregulator). Because it would be difficult, if not impossible, to estimate such a unique resource availability vector for each species (utilization is much easier to estimate than "availability"), we have to seek other ways to analyze dietary differences among species. One useful way is to sum the diets of all lizards living together in a particular place and to use this as a bioassay of what foods are available to lizards at that locality (Winemiller and Pianka, 1990). Then, dietary utilization of each prey type by each species can be expressed as "electivities" (Ivlev, 1961) that reflect the degree to which each species uses each resource disproportionately to its relative abundance (as used by all species). Electivities can be scaled from -1 to $+1$, with -1 representing complete avoidance of a prey type and $+1$ complete specialization on a given prey type in which no other species eats that prey category. Alternatively, electivities can be scaled from 0 to 1, with 0.5 representing random utilization, numbers below 0.5 representing avoidance, and numbers above 0.5 representing positive selection of a given prey type.

Proportional utilization coefficients are heavily biased towards abundant resources, and thus tend to overestimate dietary similarity, whereas electivities give greater weight to scarce resources and better reflect dietary niche segregation (Winemiller and Pianka, 1990). These differences are rather dramatic when proportional utilization coefficients are compared with electivities in a tropical assemblage of Nicaraguan lizards (Fig. 5.4). The Winemiller and Pianka (1990) community analysis uses geometric means (g_i) of electivities (e_i) and proportional utilization coefficients (p_i) to reduce bias. Surprisingly, the "same" analysis in the ecological analytical software package Ecosim (Gotelli and Entsminger, 2004) uses only p_i data, thus producing results biased toward abundant resources.

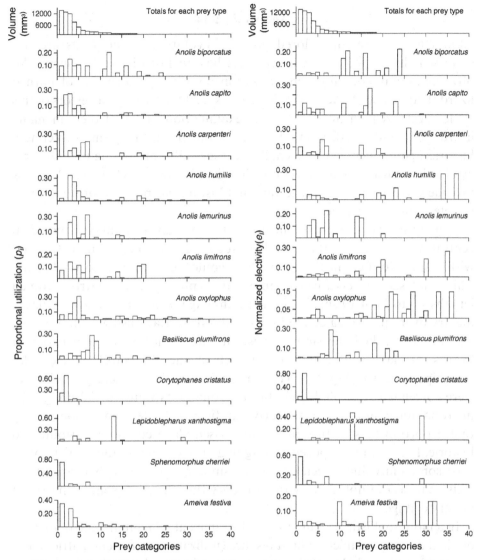

Figure 5.4. Differences between dietary data expressed as proportional utilization coefficients (p_i) and electivities (e_i). The 37 prey categories, from left to right, are: roaches, cicadas, grasshoppers and crickets, spiders, insect larvae, beetles, homopterans, fruit, hymenopterans, walking sticks, hemipterans, frogs, ants, lepidopterans, millipedes, lizards, lizard eggs, lizard shed skin, crustaceans, isopods, dragonflies, mantids, centipedes, phalangids, mollusks, flies, earthworms, scorpions, collembolans, termites, leeches, thysanurans, earwigs, pseudoscorpions, trichopterans, psocopterans, and mites (from Vitt and Zani, 1998).

Ancestral traits

Although relatively unexplored, modern evolutionary biologists are not surprised that a portion of lizard diets has an historical component. An obvious example of this is specialization on ants by 13 species of horned lizard (Pianka and Parker, 1975; Sherbrooke, 2003). Clearly, ancestors of this clade specialized on ants. However, examination of lizard diets in many lizard community studies reveals so much dietary diversity that detecting a historical component can be a challenge, especially considering that resource availability must have an effect on what lizards eat at a local level. We now provide several examples suggesting that history has played a profound role in determining lizard diets.

A quantitative analysis of lizard diets from an assemblage studied in central Amazonia by Vitt *et al.* (1999) suggested that a portion of structure with respect to diets in the assemblage could be attributed to phylogeny. This analysis conservatively compared a phylogenetic matrix (branch lengths set to 1) of 19 species in the assemblage with a matrix of calculated dietary overlaps and calculated microhabitat overlaps using a Mantel test. The analysis asked the simple question "are evolutionary similarities of lizards correlated with similarities in diets or microhabitats?" Species relationships appear in Fig. 5.5. This analysis showed that dietary similarities were correlated with phylogeny but microhabitats were not (although the microhabitat comparison was nearly significant). Aside from demonstrating that a portion of structure in a lizard assemblage could be attributed to historical effects (as opposed to ongoing species interactions), diets of at least some individual species might be similar to those of their ancestors. This analysis did not identify the source within the phylogenetic matrix of dietary change.

In a more sophisticated analysis of the same data, Giannini (2003) used the original dietary and microhabitat resource utilization coefficient matrices (not overlaps) and a phylogenetic matrix maintaining all monophyletic group structure (as opposed to similarities among species pairs) to tease out underlying historical effects on diets and microhabitat use. A canonical correspondence analysis (CCA) was applied to compare matrices using Monte Carlo methods to estimate statistical significance. The dietary portion of the analysis revealed significant phylogenetic effects for 7 of 15 comparisons (Table 5.2) shown as shaded circles on Fig. 5.5. Dietary divergence has occurred at several levels within the phylogeny, supporting the hypothesis that dietary differences among species might affect their relative abilities to exist in present-day communities and can have a history largely independent

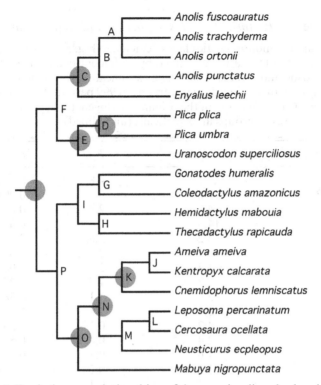

Figure 5.5. Evolutionary relationships of Amazonian lizards showing clades (groups) used in phylogenetic analysis of diets by Giannini (2003). Shaded circles with letters refer to significant phylogenetic effects on diets (see Table 5.2).

of present-day species interactions. Removing lower-level effects (species) produced two clades, *Plica* (= *Tropidurus*) and Teiidae, for which phylogeny explained 32.3% of the total dietary variation.

Because the analyses of Vitt *et al.* (1999) and Giannini (2003) identified *Plica*, owing to its use of large numbers of ants, as one source of underlying historical structure in dietary data for this Amazonian lizard assemblage, we take a closer look at its evolutionary history. Within the clade Tropiduridae, the Amazonian lizard *Uranoscodon superciliosus* is the sister taxon to the clade containing *Tropidurus, Plica,* and *Uracentron* (Frost, 1992) (Fig. 5.6). This lizard lives along streams and lagoons where it feeds on a variety of invertebrates, many of which are taken from waterway edges and may actually float up in the water (Howland *et al.*, 1990). The sister taxon contains a history of divergence centered in open cerrados of Brazil with reinvasion of arboreal microhabitats in rainforest of the Amazon and Orinoco River basins (Fig. 5.6). *Tropidurus* in open areas eat large numbers of ants, but volumetrically ants do not dominate their diets. *Plica* and *Uracentron* eat mostly ants, and at least

Table 5.2 *Historical effects on diet of Amazonian lizards*

Groups are monophyletic clades depicted in Fig. 5.5. Variation
is the percent variation explained by each identified clade with
F statistic and *p* values resulting a canonical correspondence
analysis with a reduced tree matrix from 999 permutations of a
Monte Carlo analysis on the original phylogenetic hypothesis.
The first seven groups are significant (underlined).

Group (s)	Variation	*F*	*p*
D	22.1	4.8	0.001
F/P	16.6	3.4	0.001
K	14.4	2.9	0.002
O	14.3	2.8	0.006
E	14.2	2.8	0.013
N	14.1	2.8	0.011
C	11.4	2.2	0.042
J	7.7	1.4	0.279
I	5.9	1.1	0.380
H	5.9	1.1	0.390
B	4.2	0.7	0.565
G	2.4	0.4	0.596
M	1.6	0.3	0.884
A	1.3	0.2	0.918
L	0.3	0.1	0.982

Source: Table 1 in Giannini (2003).

Uracentron flaviceps and *Plica umbra* are ant specialists. Thus the tropidurid
ancestor to the clade containing *Plica* and *Uracentron* likely had a diet consisting largely of ants. More importantly, the tendency toward ant-eating
evolved in ancestors living in open habitats (Brazilian cerrado), not
Amazonian rainforest, even though *Plica* and *Uracentron* are strictly rainforest species today.

To examine possible historical effects more closely, we performed a comparable canonical phylogenetic ordination on our combined neotropical and
desert lizard dataset with more prey categories (Vitt and Pianka, 2005). Diets
for 184 lizard species were summarized based on 27 prey categories: larvae/
eggs/pupae, vertebrates, ants, beetles, centipedes, earthworms, earwigs, flies,
grasshoppers/crickets, non-ant hymenopterans, isopods, lepidopterans, man-
tids/phasmids, millipedes, miscellaneous insects, mites, mollusks, odonates,
harvesters, plants, psocopterans, roaches, scorpions, spiders, springtails, termites, and bugs (Hemiptera + Homoptera). We used proportional utilization
data based on volumes of prey (electivities could not be used because they

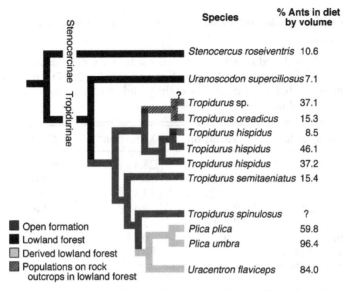

Figure 5.6. Relationships of tropidurid lizards showing the phylogenetic distribution of ant-eating in the clade.

must be based on lizard assemblages). We examined phylogenetic effects at the family and higher taxonomic category levels. Variation was significant in 14 of the 19 taxonomic groups (Vitt and Pianka, 2005). However, because residual variation changes with the inclusion of each clade, only 6 of the 14 remained significant in the final overall model (Table 5.3). Significant dietary shifts at these six major divergence points reduced variation in diets by a full 80.0%. Clearly, phylogenetic effects on lizard diets are profound. These results confirm the findings of Vitt *et al.* (2003) and identify numerous sources of dietary variation within the evolutionary history of lizards (Fig. 5.7). Unfortunately, there was a minor error in one line of the 184 in Vitt and Pianka (2005). Here, we present corrected results, which differ slightly from those reported earlier: Figure 5.8A is a biplot showing the position of each prey category in dietary niche space (prey types that are eaten together are close to each other on this plot, whereas those that are seldom eaten by the same lizard species are far apart). The origin at 0.0, 0.0 represents the lowest common denominator or the overall lizard diet summed across all 184 species. Vectors show positions of clades that significantly reduced residual variation in diet. Iguania and Scleroglossa vectors are diametrically opposed: iguanians prey heavily on ants, other Hymenoptera, beetles, and bugs. Agamids and iguanids are relatively close together in the lower left quadrant. However, scleroglossans are scattered around the other three quadrants. Scleroglossans consume a wide range of prey, with skinks feeding on termites and varanids on spiders

Table 5.3 *Results of canonical correspondence analysis after stepwise inclusion of significant clades*

Variation in diets is reduced by 80% by the six clades with significant F values. Correcting the error in our earlier dataset (Vitt and Pianka, 2005) strengthened the relationship between diet and phylogeny slightly and changed the rank order of Scincidae and Varanidae.

After inclusion of clades	Variation	Variation%	F value	p value
Iguania/Scleroglossa	0.176	28.16	9.223	0.0001
Varanidae	0.100	16.00	5.364	0.0001
Scincidae	0.082	13.12	4.505	0.0001
Gymnophthalmidae	0.057	9.12	3.140	0.0003
Teiidae	0.046	7.36	2.586	0.0034
Agamidae/Iguanidae	0.039	6.24	2.207	0.0168
Total of significant clades		80.00		

and vertebrates. Teiids and gymnophthalmids consume orthopterans and centipedes. This graph shows that scleroglossans eat many prey items rarely consumed by iguanians. Moreover, scleroglossans feed on prey types arranged at right angles to iguanian vectors (Fig. 5.8B). The shaded area in Fig. 5.8B corresponds to the area in Fig. 5.8A. Species are spread out in dietary niche space more than clades, which represent sets of species. Dietary generalists with broad food niches lie in the interior of the plot near the origin, and specialists are on peripheral areas of the diagram. We interpret these results as showing how acquisition of chemical prey discrimination, jaw prehension, and wide foraging opened up a new food resource base for scleroglossans, providing them access to sedentary and hidden prey that are unavailable to iguanians. Iguanians rely on visual prey detection, and are ambush predators that capture mobile prey moving past ambush sites via lingual prehension.

Up to this point, we have adhered to the traditional historical scenario. As such, evolutionary shifts in use of sensory systems for prey detection and discrimination (visual versus chemical), prey prehension (lingual versus jaw), and activity levels (particularly as they translate into differences in foraging behavior), appear to tie in well with associated shifts in prey types and microhabitat use of lizards (see, for example, Pianka and Vitt, 2003; Vitt *et al.*, 2003). However, phylogenetic hypotheses are just that, mere hypotheses. A recent re-evaluation of squamate evolutionary history based on a combination of nuclear (*RAG-1* and *c-mos*) and mitochondrial (*ND2* region) genes suggests a very different pattern, one that is only partly consistent with our former findings (Townsend *et al.*, 2004). This phylogenetic hypothesis places Iguania and Autarchoglossa as sister

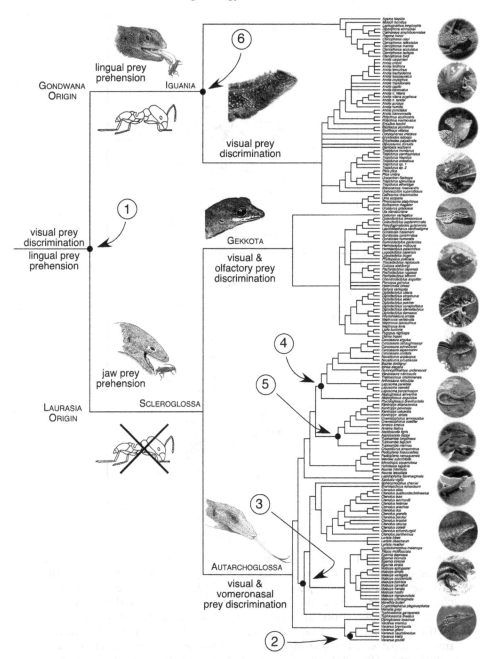

Figure 5.7. Phylogenetic hypothesis for 184 neotropical and desert lizard species. Solid circles indicate taxonomic groups that were significant in the CCA (1, Iguania/Scleroglossa; 2, Varanidae; 3, Scincidae; 4, Gymnophthalmidae; 5, Teiidae; 6, Iguanidae/Agamidae). Phylogenetic hypothesis is a composite based on published literature summarized by Vitt and Pianka (2005). Original version of figure, copyright 2005, National Academy of Sciences, USA.

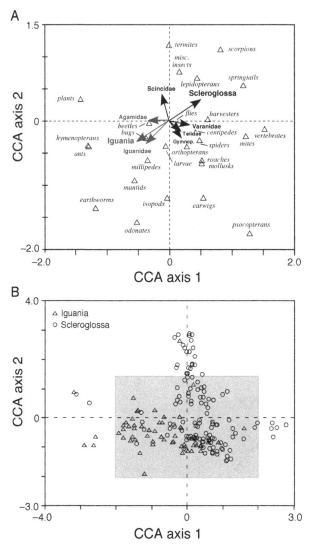

Figure 5.8. (A) Biplot showing corrected results of a canonical correspondence ordination analysis. Prey categories are plotted on the first two axes of dietary niche space; clades that significantly reduced variation are plotted with vectors radiating out from the origin (see Vitt and Pianka (2005) for further explanation, and Table 5.3 for a correction). The position of earthworms relative to lizard clades may partly be an artifact created by a small sample size for one highly unusual species of iguanian (*Enyalioides palpebralis*). (B) Plot showing positions of each species of iguanians (triangles) and scleroglossans (circles) in the first two CCA axes of dietary niche space. Original version of figures, copyright 2005, National Academy of Sciences, USA.

clades with Gekkota as the sister clade to them, thus eliminating Scleroglossa. We performed a CCA identical to that above, but with Iguania and Autarchoglossa as sister clades and Gekkota as sister to those (thus Scleroglossa no longer exists). The primary result obtained in the above analysis stands. Prey categories are distributed exactly as in Fig. 5.8 and vectors for significant clades are identical, with the single exception that no scleroglossan vector exists. The putative new clade of Townsend *et al.* (2004) (Iguania plus Autarchoglossa) did not achieve significance (nor did Gekkota) and was therefore not included in the final model. These dietary analyses thus lend support to the traditionally accepted phylogeny, but we recognize the circularity of this argument.

Conclusions

Based on relatively complete dietary data for 184 lizard species representing all major lizard clades and most minor ones, variation in lizard diets is reduced by 80% based on phylogeny alone. This result has major implications for under- standing the ecology, functional morphology, ecomorphology, behavior, and physiology of squamate reptiles. Differences reported here are nested deeply in lizard evolutionary history. A more realistic view of the evolution of lizard and likely squamate diets in general is a nested hierarchy in which key events at different points in deep history not only resulted in dramatic dietary shifts, but set the stage for potentially rapid diversification events. The traditional phy- logenetic hypothesis, involving the shift to chemical prey discrimination, jaw prehension, and a more active lifestyle including wide foraging to find prey (Schwenk, 2000b), appears to have made available a huge spectrum of prey that were previously unavailable to iguanians. Within scleroglossans, some clades (e.g. amphisbaenians, some anguids and skinks, dibamids, and some snakes) invaded a subterranean world that iguanians never had access to. Other clades (geckos) shifted to nocturnal activity, taking advantage of prey unavailable during the day. Yet others specialized on termites, scorpions, and vertebrates, prey rarely eaten by iguanians. The success of this shift is evident when comparing the number of iguanian versus scleroglossan species. About 1230 iguanians currently inhabit the planet, whereas there are nearly five times as many scleroglossans (more than 6000, about half of which are referred to as "lizards" and half as "snakes"). Considering that the traditional Scleroglossa and Iguania are equal ages (sister clades), this is a dramatic difference. However, because the divergence process is a Markovian process, the possi- bility exists that, dramatic as this difference is, it may have resulted from chance (see Vitt and Pianka, 2005). The striking ecological and morpho- logical shifts that have occurred among scleroglossans but not iguanians (nocturnality, multiple origins of limblessness and aquatic habits), adds support to the notion that historical traits of scleroglossans facilitated adaptive radiations and provided the basis for increased diversity.

Mechanistic studies of competition, evolutionary change in morphology in response to microhabitat structure, sexual selection, and thermoregulation and its correlates are best conducted among closely related species. Caribbean *Anolis* lizards, for example, have been ideal models for experimental field studies, likely because sympatric species share a vast majority of their evolutionary histories (Losos, 1992, 1994; Jackman *et al.*, 1999). Changes in response to other species or habitat change can be detected because species interactions between highly similar species are usually intense (see, for example, Losos *et al.*, 1997). Similar experiments performed on species embedded in different clades would be much less likely to detect measurable responses because deep historical differences mask characteristics of interest. Studies that seek to identify origins of differences among distantly related species or taxa require well-supported phylogenies and high-quality natural history data directed at the question of interest. Understanding foraging ecology of squamates, and likely of all organisms, requires identifying historical bases for differences among potentially interacting species in local assemblages if structure in natural communities is to be understood (see, for example, Losos, 1996; Webb *et al.*, 2002; Vitt *et al.*, 2003).

Clearly, predictions generated by early studies of foraging mode (e.g. Huey and Pianka, 1981) are not necessarily robust. Perhaps foraging mode should not be considered a paradigm, but rather an epiphenomenon that arose through the evolution of traits early in the history of squamate reptiles. Historical events need to be thoroughly examined in a phylogenetic context. Unfortunately, we do not have nearly the data necessary to do that. At best, we are just beginning to realize how important historical events are, yet the vast majority of research on lizards has been conducted on less than 10% of extant species. Recall that, for the Amazon assemblage of 19 species, 32.3% of the total dietary variation was explained by phylogeny. This assemblage is but a small subset of the 184 species. Poor taxon sampling strongly affects results. Considering that we can reduce variance by 32% with 19 species, but 80% with 184 species, imagine if we had all species! The "unexplained" variance in diets likely represents a combination of error due to poor taxon sampling, effects of species interactions, and other factors identified in this chapter.

Caveat

We have focused most of our analysis on 184 species from New and Old World deserts and New World tropics. Acquiring these data has taken both of our lifetimes (we are not dead yet!). As in almost all studies that attempt "global" hypothesis testing, our data suffer from grossly inadequate taxon sampling. One-hundred eighty-four species may sound impressive, but it represents only 4.25% of "lizards" and 2.25% of squamates. Although some interesting patterns have been identified, alternative partitions of our dataset are possible. For example, clade representation is strongly dependent on region (i.e. South America, Central America, North America, Australia, Kalahari) (Table 5.4). Resources undoubtedly differ between deserts and tropical forests or savanna, and effects of predators on effective resource availability for squamates is likely quite different among habitat types as well. Clearly, more complete and robust phylogenetic hypotheses and much more dietary data from natural populations are needed to fully understand evolution of diets in squamate reptiles.

Table 5.4 *Numbers of species in various clades and families by geographic region*

SA, South America; CA, Central America; NA, North America; AU, Australia; K, Kalahari.

Clade	SA	CA	NA	AU	K	Tot
Iguania	27	11	8	11	1	58
Scleroglossa	49	5	3	49	20	126
Gekkota	13	2	1	13	7	37
Autarchoglossa	34	4	2	36	13	89
Scincomorpha	37	4	1	27	13	82
Anguimorpha	1	—	—	6	—	7
Family						
Agamidae	—	—	—	11	1	12
Anguidae	1	—	—	—	—	1
Diplodactylidae	—	—	—	11	—	11
Eublepharidae	—	—	1	—	—	1
Gekkonidae	13	1	—	2	7	23
Sphaerodactyline	7	—	—	—	—	7
Gekkonine	6	1	—	2	7	16
Gymnophthalmidae	19	—	—	—	—	19
Iguanidae	27	11	8	—	—	46
Lacertidae	—	—	—	—	7	7
Pygopodidae	—	—	—	3	—	3
Scincidae	5	2	—	27	6	40
Teiidae	12	1	1	—	—	14
Varanidae	—	—	—	6	—	6
Xantusidae	—	1	1	—	—	2
Totals (by family)	76	16	11	60	21	184

Acknowledgments

Work in Brazil resulting in collection of lizard diet data was supported by NSF grants DEB-9200779 and DEB-9505518 to LJV and J. P. Caldwell. Brazilian agencies contributing to logistics include Instituto Nacional de Pesquisas da Amazonica (INPA), Conselho Nacional de Desenvolvimento Científico e Tecnológico (CNPq, Portaria MCT no. 170, de 28/09/94), Instituto Brasileiro do Meio Ambiente e dos Recursos Naturais Renováveis (IBAMA, permit no. 073/94-DIFAS), and Museu Paraense E. Goeldi in Belém. A research convenio between the Sam Noble Oklahoma Museum of Natural History and the Museu Paraense E. Goeldi in collaboration with Dr. T. C. S. Avila-Pires made this possible. All lizards were treated in accordance with federal, state, and university regulations (Animal Care Assurance 73-R-100, approved 8 November 1994). ERP's research has been supported by grants from the National Geographic Society, the John Simon Guggenheim Memorial Foundation, a senior Fulbright Research Scholarship, the Australian–American Educational Foundation, the University Research Institute of the Graduate School at the University of Texas at Austin, the Denton A. Cooley Centennial Professorship in Zoology at the University of Texas at Austin, the US National Science Foundation, and the US National Aeronautics and Space Administration. ERP also thanks the staffs of the Department of Zoology at the University of Western Australia and the Western Australian Museum plus the staff of the Department of Conservation and Land Management (CALM). LJV and ERP thank their respective universities for Big 12 faculty Fellowships. Guarino Colli and Alison Gainsbury generously gave us much needed assistance and guidance with canonical correspondence analysis. Vic Hutchison provided references on effects of temperature on various aspects of foraging and digestive biology. Last, and certainly not least, we thank the many colleagues who have helped shape our thinking on global ecology of lizards, often correcting our naïve mistakes along the way. In particular, we thank William E. Cooper Jr., Kurt Schwenk, and Aaron Bauer.

References

Alexander, G. J., van Der Heever, C. and Lazenby, S. L. (2001). Thermal dependence of appetite and digestive rate in the flat lizard, *Platysaurus intermedius wilhelmi. J. Herpetol.* **35**, 461–6.

Anderson, R. A. and Karasov, W. H. (1981). Contrasts in energy intake and expenditure in sit-and-wait and widely foraging lizards. *Oecologia* **49**, 67–72.

Anderson, R. A. and Vitt, L. J. (1990). Sexual selection versus alternative causes of sexual dimorphism in teiid lizards. *Oecologia* **84**, 145–57.

Anderson, S. C. (1999). *The Lizards of Iran*. Ithaca, NY: Society for the Study of Amphibians and Reptiles.

Auffenberg, W. (1978). Social feeding behavior in *Varanus komodoensis*. In *Behavior and Neurology of Lizards*, ed. N. Greenberg and P. D. MacLean, pp. 301–31. Poolesville, MD: National Institute of Mental Health.

Auffenberg, W. (1981). *The Behavioral Ecology of the Komodo dragon*. Gainesville, FL: University of Florida Press.

Autumn, K. and DeNardo, D. F. (1995). Behavioral thermoregulation increases growth rate in a nocturnal lizard. *J. Herpetol.* **29**, 157–62.

Autumn, K., Jindrich, D., DeNardo, D. F. and Mueller, R. (1999). Locomotor performance at low temperature and the evolution of nocturnality in geckos. *Evolution* **53**, 580–99.

Avery, R. A. (1981). Feeding ecology of the nocturnal gecko *Hemidactylus brookii* in Ghana. *Amph.-Rept.* **1**, 269–76.

Ayers, D. Y. and Shine, R. (1997). Thermal influences on foraging ability: body size, posture and cooling rate of an ambush predator, the python *Morelia spilota*. *Funct. Ecol.* **11**, 342–7.

Ballinger, R. E. and Holscher, V. L. (1980). Assimilation efficiency and nutritive state in the striped plateau lizard, *Sceloporus virgatus* (Sauria: Iguanidae). *Copeia* **1980**, 838–9.

Beaupre, S. J., Dunham, A. E. and Overall, K. L. (1993). The effects of consumption rate and temperature on apparent digestibility coefficient, urate production, metabolizable energy coefficient and passage time in canyon lizards (*Sceloporus merriami*) from two populations. *Funct. Ecol.* **7**, 273–80.

Beck, D. D. (1990). Ecology and behavior of the gila monster in southwestern Utah. *J. Herpetol.* **24**, 54–68.

Beck, D. D. and Lowe, C. H. (1991). Ecology of the beaded lizard, *Heloderma horridum*, in a tropical dry forest in Jalisco, Mexico. *J. Herpetol.* **25**, 395–406.

Blum, M. S. (1981). *Chemical Defenses of Arthropods*. New York: Academic Press.

Charnov, E. L. (1976). Optimal foraging: the marginal value theorem. *Theor. Pop. Biol.* **9**, 129–36.

Clarke, B. C. and Nicolson, S. W. (1994). Water, energy, and electrolyte balance in captive Namib sand-dune lizards (*Angolosaurus skoogi*). *Copeia* **1994**, 962–74.

Cooper, W. E. Jr. (1994a). Prey chemical discrimination, foraging mode, and phylogeny. In *Lizard Ecology: Historical and Experimental Perspectives*, ed. L. J. Vitt and E. R. Pianka, pp. 95–116. Princeton, NJ: Princeton University Press.

Cooper, W. E. Jr. (1994b). Chemical discrimination by tongue-flicking in lizards: a review with hypotheses on its origin and its ecological and phylogenetic relationships. *J. Chem. Ecol.* **20**, 439–87.

Cooper, W. E. Jr. (1995a). Foraging mode, prey chemical discrimination, and phylogeny in lizards. *Anim. Behav.* **50**, 973–85.

Cooper, W. E. Jr. (1995b). Prey chemical discrimination and foraging mode in gekkonoid lizards. *Herp. Monogr.* **9**, 120–9.

Cooper, W. E. Jr. (1997). Independent evolution of squamate olfaction and vomerolfaction and correlated evolution of vomerolfaction and lingual structure. *Amph.-Rept.* **18**, 85–105.

Cooper, W. E. Jr. and Hartdegen, R. (1999). Discriminative response to animal, but not plant, chemicals by an insectivorous, actively foraging lizard, *Scincella lateralis*, and differential response to surface and internal prey cues. *J. Chem. Ecol.* **25**, 1531–41.

Cooper, W. E. Jr. and Vitt, L. J. (2002). Distribution, extent, and evolution of plant consumption by lizards. *J. Zool. Lond.* **257**, 487–517.

Cooper, W. E. Jr., Caldwell, J. P., Vitt, L. J., Pérez-Mellado, V. and Baird, T. A. (2002a). Food chemical discriminations and correlated evolution between plant diet and plant chemical discrimination in lacertiform lizards. *Can. J. Zool.* **80**, 655–63.

Cooper, W. E. Jr., Pérez-Mellado, V., Vitt, L. J. and Budzinsky, B. (2002b). Behavioral responses to plant toxins by the omnivorous Balearic lizard, *Podarcis lilfordi*. *Physiol. Behav.* **76**, 297–303.

Cooper, W. E. Jr., Whiting, M. J. and Wyk, J. H. V. (1997). Foraging modes of cordyliform lizards. *S. Afr. J. Zool.* **32**, 9–13.

Cundall, D. and Greene, H. W. (2000). Feeding in snakes. In *Feeding*, ed. K. Schwenk, pp. 293–333. San Diego: Academic Press.

de Queiroz, A., Pough, F. H., Andrews, R. M. and Collazo, A. (1987). Thermal dependence of prey-handling costs for the scincid lizard, *Chalcides ocellatus*. *Physiol. Zool.* **60**, 492–8.

Doughty, P. and Shine, R. (1995). Life in two dimensions: natural history of the southern leaf-tailed gecko, *Phyllurus platurus*. *Herpetologica* **51**, 193–201.

Dunham, A. E. (1980). An experimental study of interspecific competition between the iguanid lizards *Sceloporus merriami* and *Urosaurus ornatus*. *Ecol. Monogr.* **50**, 304–30.

Durtsche, R. D. (2000). Ontogenetic plasticity of food habits in the Mexican spiny-tailed iguana, *Ctenosaura pectinata*. *Oecologia (Berlin)* **124**, 185–95.

Durtsche, R. D., Gier, P. J., Fuller, M. M. *et al.* (1997). Ontogenetic variation in the autecology of the greater earless lizard *Cophosaurus texanus*. *Ecography* **20**, 336–46.

Espinoza, R. E., Wiens, J. J. and Tracy, C. R. (2004). Recurrent evolution of herbivory in small, cold-climate lizards: breaking the ecophysiological rules of reptilian herbivory. *Proc. Natl. Acad. Sci. USA* **101**, 16819–24.

Ferguson, G. W., Gehrmann, W. H., Karsten, K. B. *et al.* (2003). Do panther chameleons bask to regulate endogenous vitamin d_3 production? *Physiol. Biochem. Zool.* **76**, 52–9.

Frost, D. R. (1992). Phylogenetic analysis and taxonomy of the *Tropidurus* group of lizards (Iguania: Tropiduridae). *Amer. Mus. Novit.* **3033**, 1–68.

Giannini, N. P. (2003). Canonical phylogenetic ordination. *Syst. Biol.* **52**, 684–95.

Gotelli, N. J. and Entsminger, G. L. (2004). *EcoSim: Null Models Software for Ecology*. Version 7. Burlington, VT: Acquired Intelligence Inc. & Kesey-Bear.

Greene, H. W. (1997). *Snakes: the Evolution of Mystery in Nature*. Berkeley, CA: University of California Press.

Greenwald, O. E. (1971). Thermal dependence of striking and prey capture by gopher snakes. *Copeia* **1971**, 141–8.

Harlow, H. J., Hillman, S. S. and Hoffman, M. (1976). The effect of temperature on digestive efficiency in the herbivorous lizard, *Dipsosaurus dorsalis*. *J. Comp. Physiol. B.* **111**, 1–6.

Harwood, R. H. (1979). The effect of temperature on the digestive efficiency of three species of lizards, *Cnemidophorus tigris, Gerrhonotus multicarinatus* and *Sceloporus occidentalis*. *Comp. Biochem. Physiol.* **63A**, 417–33.

Howland, J. M., Vitt, L. J. and Lopez, P. T. (1990). Life on the edge: the ecology and life history of the tropidurine iguanid lizard *Uranoscodon superciliosum*. *Can. J. Zool.* **68**, 1366–73.

Huey, R. B. (1982). Temperature, physiology, and the ecology of reptiles. In *Biology of the Reptilia*, vol. 12. *Physiology C*, ed. C. Gans and F. H. Pough, pp. 25–91. New York: Academic Press.

Huey, R. B. and Pianka, E. R. (1981). Ecological consequences of foraging mode. *Ecology* **62**, 991–9.

Huey, R. B., Pianka, E. R., Egan, M. E. and Coons, L. W. (1974). Ecological shifts in sympatry: Kalahari fossorial lizards (*Typhlosaurus*). *Ecology* **55**, 304–16.

Huey, R. B., Niewiarowski, P. H., Kaufmann, J. and Herron, J. C. (1989). Thermal biology of nocturnal ectotherms: is sprint performance of geckos maximal at low body temperatures? *Physiol. Zool.* **62**, 488–504.

Huey, R. B., Pianka, E. R. and Vitt, L. J. (2001). How often do lizards "run on empty?" *Ecology* **82**, 1–7.

Iverson, J. B. (1979). Behavior and ecology of the rock iguana *Cyclura carinata*. *Bull. Florida State Mus. Biol. Sci.* **24**, 175–358.

Iverson, J. B. (1982). Adaptations to herbivory in iguanine lizards. In *Iguanas of the World*, ed. G. M. Burghardt and A. S. Rand, pp. 60–76. Park Ridge, NJ: Noyes Publications.

Ivlev, V. S. (1961). *Experimental Feeding Ecology of Fishes*. New Haven, CT: Yale University Press.

Jackman, T. R., Larson, A., De Queiroz, K. and Losos, J. B. (1999). Phylogenetic relationships and tempo of early diversification in *Anolis* lizards. *Syst. Biol.* **48**, 254–85.

Kamil, A. C. (1983). Optimal foraging theory and the psychology of learning. *Amer. Zool.* **23**, 291–302.

Kaufmann, R. and Pough, F. H. (1982). The effect of temperature upon the efficiency of assimilation of preformed water by the desert iguana, *Dipsosaurus dorsalis*. *Comp. Biochem. Physiol.* **72A**, 221–4.

Kearney, M. and Predavec, M. (2000). Do nocturnal ectotherms thermoregulate? A study of the temperate gecko *Christinus marmoratus*. *Ecology* **81**, 2984–96.

Losos, J. B. (1992). The evolution of convergent structure in Caribbean *Anolis* communities. *Syst. Biol.* **41**, 403–20.

Losos, J. B. (1994). Historical contingency and lizard community ecology. In *Lizard Ecology: Historical and Experimental Perspectives*, ed. L. J. Vitt and E. R. Pianka, pp. 319–33. Princeton, NJ: Princeton University Press.

Losos, J. B. (1996). Phylogenetic perspectives on community ecology. *Ecology* **77**, 1344–54.

Losos, J. B., Warheit, K. I. and Schoener, T. W. (1997). Adaptive differentiation following experimental island colonization in *Anolis* lizards. *Nature* **387**, 70–3.

MacArthur, R. H. and Pianka, E. R. (1966). On optimal use of a patchy environment. *Amer. Nat.* **100**, 603–9.

Mautz, W. J. and Nagy, K. A. (1987). Ontogenetic changes in diet, field metabolic rate, and water flux in the herbivorous lizard *Dipsosaurus dorsalis*. *Physiol. Zool.* **60**, 640–58.

McBrayer, L. D. and Reilly, S. M. (2002). Prey processing in lizards: behavioral variation in sit-and-wait and widely foraging taxa. *Can. J. Zool.* **80**, 882–92.

Miranda, J. P. and Andrade, G. V. (2003). Seasonality in diet, perch use, and reproduction in the gecko *Gonatodes humeralis* from eastern Brazilian Amazon. *J. Herpetol.* **37**, 433–8.

Montanucci, R. R. (1989). The relationship of morphology to diet in the horned lizard genus *Phrynosoma*. *Herpetologica* **45**, 208–16.

Nussear, K. E., Espinoza, R. E., Gubbins, C. M., Field, K. J. and Hayes, J. P. (1998). Diet quality does not affect resting metabolic rate or body temperatures selected by an herbivorous lizard. *J. Comp. Physiol.* B **168**, 183–9.

Ostrom, J. H. (1963). Further comments on herbivorous lizards. *Evolution* **17**, 368–9.

Parker, W. S. and Pianka, E. R. (1974). Further ecological observations on the western banded gecko, *Coleonyx variegatus*. *Copeia* **1974**, 528–31.

Perry, G. (1999). The evolution of search modes: ecological versus phylogenetic perspectives. *Am. Nat.* **153**, 99–109.

Perry, G. and Pianka, E. R. (1997). Animal foraging: past, present, and future. *Trends Ecol. Evol.* **12**, 360–4.

Perry, G., Lampl, I., Lerner, A. *et al.* (1990). Foraging mode in lacertid lizards: variation and correlates. *Amph.-Rept.* **11**, 373–84.

Pianka, E. R. (1966). Convexity, desert lizards and spatial heterogeneity. *Ecology* **47**, 1055–9.

Pianka, E. R. (1969). Sympatry of desert lizards (*Ctenotus*) in western Australia. *Ecology* **50**, 1012–30.

Pianka, E. R. (1970). Comparative autecology of the lizard *Cnemidophorus tigris* in different parts of its geographic range. *Ecology* **51**, 703–20.

Pianka, E. R. (1971). Comparative ecology of two lizards. *Copeia* **1971**, 129–38.

Pianka, E. R. (1973). The structure of lizard communities. *Ann. Rev. Ecol. Syst.* **4**, 53–74.

Pianka, E. R. (1986). *Ecology and Natural History of Desert Lizards. Analyses of the Ecological Niche and Community Structure*. Princeton, NJ: Princeton University Press.

Pianka, E. R. (1994). Comparative ecology of *Varanus* in the Great Victoria desert. *Australian J. Ecol.* **19**, 395–408.

Pianka, E. R. and Huey, R. B. (1978). Comparative ecology, niche segregation, and resource utilization among gekkonid lizards in the southern Kalahari. *Copeia* **1978**, 691–701.

Pianka, E. R. and King, D. R., eds. (2004). *Varanoid Lizards of the World*. Bloomington, Indiana University Press.

Pianka, E. R. and Parker, W. S. (1975). Ecology of horned lizards: A review with special reference to *Phrynosoma platyrhinos*. *Copeia* **1975**, 141–62.

Pianka, E. R. and Pianka, H. D. (1970). The ecology of *Moloch horridus* (Lacertilia: Agamidae) in Western Australia. *Copeia* **1970**, 90–103.

Pianka, E. R. and Pianka, H. D. (1976). Comparative ecology of twelve species of nocturnal lizards (Gekkonidae) in the Western Australian desert. *Copeia* **1976**, 125–42.

Pianka, E. R. and Vitt, L. J. (2003). *Lizards: Windows to the Evolution of Diversity*. Berkeley, CA: University of California Press.

Pianka, G. A., Pianka, E. R. and Thompson, G. G. (1998). Natural history of thorny devils *Moloch horridus* (Lacertilia: Agamidae) in the Great Victoria Desert. *J. Roy. Soc. West. Australia* **81**, 183–90.

Rocha, C. F. D. (1996). Seasonal shift in lizard diet: the seasonality in food resources affecting the diet of *Liolaemus lutzae* (Tropiduridae). *Ciênc. Cult.* **48**, 264–9.

Schoener, T. W. (1971). Theory of feeding strategies. *Ann. Rev. Ecol. Syst.* **2**, 369–404.

Schwenk, K. (1995). Of tongues and noses: chemoreception in lizards and snakes. *Trends Ecol. Evol.* **10**, 7–12.

Schwenk, K. (2000a). Tetrapod feeding in the context of vertebrate morphology. In *Feeding*, ed. K. Schwenk, pp. 3–20. San Diego, CA: Academic Press.

Schwenk, K. (2000b). Feeding in lepidosaurs. In *Feeding*, ed. K. Schwenk, pp. 175–291. San Diego: Academic Press.

Sexton, O. J., Bauman, J. and Ortleb, E. (1972). Seasonal food habits of *Anolis limifrons. Ecology* **53**, 182–6.

Sherbrooke, W. C. (1981). *Horned lizards: unique reptiles of western North America.* Southwest Parks and Monuments Association.

Sherbrooke, W. C. (2003). *Introduction to Horned Lizards of North America.* Berkeley: University of California Press.

Shine, R. (1986). Food habits, habitats and reproductive biology of four sympatric species of varanid lizards in tropical Australia. *Herpetologica* **42**, 346–60.

Spotila, J. R. and Standora, E. A. (1985). Energy budgets of ectothermic vertebrates. *Amer. Zool.* **25**, 973–86.

Ter Braak, C. J. F. and Smilauer, P. (2002). *CANOCO Reference Manual and User's Guide to Canoco for Windows: Software for Canonical Community Ordination* (version 4.5). Ithaca, NY: Microcomputer Power (USA).

Townsend, T. M., Larson, A., Louis, E. and Macey, J. R. (2004). Molecular phylogenetics of Squamata: the position of snakes, amphisbaenians, and dibamids, and the root of the squamate tree. *Syst. Biol.* **33**, 735–57.

Troyer, K. (1984). Structure and function of the digestive tract of a herbivorous lizard *Iguana iguana. Physiol. Zool.* **57**, 1–8.

Troyer, K. (1987). Small differences in daytime body temperature affect digestion of natural food in a herbivorous lizard (*Iguana iguana*). *Comp. Biochem. Physiol.* **A87**, 623–6.

van Marken Lichtenbelt, W. D. (1992). Digestion in an ectothermic herbivore, the green iguana (*Iguana iguana*): effect of food composition on body temperature. *Physiol. Zool.* **65**, 649–73.

van Marken Lichtenbelt, W. D. and Wesselingh, R. A. (1993). Energy budgets in free-living green iguanas in a seasonal environment. *Ecology* **74**, 1157–72.

Van Wyk, J. H. (2000). Seasonal variation in stomach contents and diet composition in the large girdled lizard, *Cordylus giganteus* (Reptilia: Cordylidae) in the Highveld grasslands of the northeastern Free State, South Africa. *African Zool.* **35**, 9–27.

Vitt, L. J. (1991). Desert reptile communities. In *The Ecology of Desert Communities*, ed. G. A. Polis, pp. 250–76. Tucson, AZ: University of Arizona Press.

Vitt, L. J. (1995). The ecology of tropical lizards in the Caatinga of northeast Brazil. *Occ. Pap. Oklahoma Mus. Nat. Hist.* **1**, 1–29.

Vitt, L. J. (2004). Shifting paradigms: herbivory and body size in lizards. *Proc. Natl. Acad. Sci. USA* **101**, 16713–14.

Vitt, L. J. and Colli, G. R. (1994). Geographical ecology of a neotropical lizard: *Ameiva ameiva* (Teiidae) in Brazil. *Can. J. Zool.* **72**, 1986–2008.

Vitt, L. J. and Congdon, J. D. (1978). Body shape, reproductive effort, and relative clutch mass in lizards: resolution of a paradox. *Am. Nat.* **112**, 595–608.

Vitt L. J. and Cooper, W. E. Jr. (1986). Foraging and diet of a diurnal predator (*Eumeces laticeps*) feeding on hidden prey. *J. Herpetol.* **20**, 408–15.

Vitt, L. J. and Pianka, E. R. (2004). Historical patterns in lizard ecology: what teiids can tell us about lacertids. In *The Biology of Lacertids. Evolutionary and Ecological Perspectives*, ed. V. Perez-Mellado, N. Riera and A. Perera, Recerca 8, pp. 139–57. Menorca. Spain: Institut Menorquí d'Estudis.

Vitt, L. J. and Pianka, E. R. (2005). Deep history impacts present-day ecology and biodiversity. *Proc. Natl. Acad. Sci. USA* **102**, 7877–81.

Vitt, L. J. and Zani, P. A. (1996a). Ecology of the elusive tropical lizard *Tropidurus* [= *Uracentron*] *flaviceps* (Tropiduridae) in lowland rain forest of Ecuador. *Herpetologica* **52**, 121–32.

Vitt, L. J. and Zani, P. A. (1996b). Organization of a taxonomically diverse lizard assemblage in Amazonian Ecuador. *Can. J. Zool.* **74**, 1313–35.

Vitt, L. J. and Zani, P. A. (1997). Ecology of the nocturnal lizard *Thecadactylus rapicauda* (Sauria: Gekkonidae) in the Amazon region. *Herpetologica* **53**, 165–79.

Vitt, L. J. and Zani, P. A. (1998). Prey use among sympatric lizard species in lowland rain forest of Nicaragua. *J. Trop. Ecol.* **14**, 537–59.

Vitt, L. J., Caldwell, J. P., Sartorius, S. S. *et al.* (2005). Pushing the edge: extended activity as an alternative to risky body temperatures in an herbivorous teiid lizard (*Cnemidophorus murinus*: Squamata). *Funct. Ecol.* **19**, 152–8.

Vitt, L. J., Pianka, E. R., Cooper, W. E. Jr. and Schwenk, K. (2003). History and the global ecology of squamate reptiles. *Amer. Nat.* **162**, 44–60.

Vitt, L. J., Zani, P. A. and Avila-Pires, T. C. S. (1997). Ecology of the arboreal tropidurid lizard *Tropidurus* (= *Plica*) *umbra* in the Amazon region. *Can. J. Zool.* **75**, 1876–82.

Vitt, L. J., Zani, P. A. and Espósito, M. C. (1999). Historical ecology of Amazonian lizards: implications for community ecology. *Oikos* **87**, 286–94.

Wagner, G. P. and Schwenk, K. (2000). Evolutionarily stable configurations: functional integration and the evolution of phenotypic stability. In *Evolutionary Biology*, Vol. 31, ed. M. K. Hecht, R. J. McIntyre and M. T. Clegg, pp. 155–217. New York: Kluwer Academic/Plenum Publishers.

Webb, C. O., Ackerly, D. D., Peek, M. A. and Donoghue, M. J. (2002). Phylogenies and community ecology. *Ann. Rev. Ecol. Syst.* **33**, 475–505.

Wilhoft, D. C. (1958). Observations on preferred body temperature and feeding habits of some selected tropical iguanas. *Herpetologica* **14**, 161–4.

Winemiller, K. O. and Pianka, E. R. (1990). Organization in natural assemblages of desert lizards and tropical fishes. *Ecol. Monogr.* **60**, 27–55.

Xue-Feng, X. U., Xue-Jun, C. and Xiang, J. I. (2001). Selected body temperature, thermal tolerance and influence of temperature on food assimilation and locomotor performance in lacertid lizards, *Eremias brenchleyi*. *Zool. Res.* **22**, 443–5.

Zaidan, F. III and Beaupre, S. J. (2003). Effects of body mass, meal size, fast length, and temperature on specific dynamic action in the Timber Rattlesnake. *Physiol. Biochem. Zool.* **76**, 447–58.

Zimmerman, L. C. and Tracy, C. R. (1989). Interactions between the environment and ectothermy and herbivory in reptiles. *Physiol. Zool.* **62**, 374–409.

6

Why is intraspecific niche partitioning more common in snakes than in lizards?

RICHARD SHINE AND MICHAEL WALL

School of Biological Sciences, University of Sydney

Introduction

The scientific literature on foraging biology in lizards has tended to ignore intraspecific variation. Thus, the species is treated as the unit of analysis, under the implicit assumption that variation within a single species (and, to an even greater degree, within a single population) is trivial relative to variation among species. Many authors therefore talk of broad phylogenetic patterns in foraging mode, with all taxa within major lineages placed within the same major category (e.g. active or ambush). Such generalizations may have substantial value in pursuing broad issues, but at the population level they are simply wrong for many types of organism.

Groups of related species certainly share many distinctive features of foraging biology, and there is a genuine validity to statements about lineage-wide patterns. None the less, a detailed analysis of almost any single population (let alone one species) is likely to reveal diversity in trophic biology. For example, juveniles may feed in different ways, in different places and on different kinds and sizes of prey than do adults within the same population. Similarly, males and females may differ in prey utilization. For several reasons, snakes offer more dramatic examples of such intraspecific niche divergence than do lizards. In this chapter, we review published evidence for size and sex effects on foraging biology in snakes, consider underlying biological factors that generate such diversity, and attempt to explain why snakes and lizards – two very closely related groups of organisms – differ so dramatically in the occurrence of intraspecific niche partitioning.

Effects of body size on trophic biology in snakes

The body size of a snake exerts a pervasive influence on the types and sizes of prey items that it consumes. Although there are doubtless some species (and

Lizard Ecology: The Evolutionary Consequences of Foraging Mode, ed. S. M. Reilly, L. D. McBrayer and D. B. Miles. Published by Cambridge University Press. © Cambridge University Press 2007.

populations) of snakes in which an individual's body size does not affect the kinds of prey it eats, most detailed analyses have revealed significant links between predator body size and prey body size. Indeed, this phenomenon is so widespread that it has attracted reviews (Arnold, 1993) and experimental analyses in both the laboratory (Shine, 1991d) and the field (Shine and Sun, 2003). Table 6.1 reviews published examples of snake taxa in which individuals of different body sizes exhibit significant divergence in prey types. To reduce the information to a manageable level, we have adopted an arbitrary definition of "prey type": for example, we treated amphibians, reptiles, birds and mammals as different "types". A finer subdivision (e.g. with species of mammals as different "types") would generate a vastly larger number of cases of allometric dietary shifts (and may well be ecologically just as important). Similarly, we have ignored cases of subtle shifts in the relative proportion of different prey types, and the vast number of published cases supported only by anecdotal data. Importantly, in strong contrast to the situation in lizards, size-related dietary shifts are the rule and not the exception among snakes; many cases are apparent even after our exclusion of all these other kinds of evidence (Table 6.1). Equally important is the high degree of plasticity in this trait; although some lineages appear to show ontogenetic dietary shifts more often than others (see below), there are many cases where species within a single genus (and even populations within a single species, or sexes within a single population) differ in whether or not dietary composition shifts with body size (Table 6.1).

The causal mechanisms generating these size-dependent shifts are clear in many cases. The most important may be:

(1) *Ability to physically ingest prey changes with the predator's body size.* Gape-limitation may mean that the prey eaten by adult snakes are too large to be consumed by smaller conspecifics. In a few cases, larger snakes may actually consume prey killed by smaller snakes that were physically unable to ingest them (Shine et al., 2002b).

(2) *Ability to subdue prey changes with the predator's body size.* Pough (1977) argued that ontogenetic shifts in maximal aerobic performance in garter snakes (*Thamnophis sirtalis*) might be the reason why smaller snakes generally consumed relatively smaller prey items than did larger conspecifics.

(3) *Ability to capture prey changes with the predator's body size.* Smaller snakes may be slower or weaker, and thus less able to capture large vigorous prey. On the other hand, smaller snakes may be able to penetrate crevices used by prey (and hence search actively) whereas larger conspecifics cannot enter, and thus must rely on ambush. Water pythons (*Liasis fuscus*) in tropical Australia may show such a shift (T. Madsen and R. Shine, unpublished data). In this case, prey type and size may

Table 6.1 *Presence or absence of intraspecific shifts in prey type as a function of predator body size within snake species*

"Prey type" represents a subjective classification: for example, amphibians, reptiles, birds and mammals were regarded as separate "types". However, shifts from one type of mammal to another (e.g. mouse to pig) were not included. I, invertebrates; F, fishes; A, amphibians; R, reptiles; B, birds; M, mammals.

Family	Genus	Species	Shift or not?	Shift from	Shift to	Authority
Acrochordidae	*Acrochordus*	*arafurae*	no	F	F	Houston & Shine, 1993
Boidae	*Candoia*	*aspera, bibroni, carinata*	yes	R	M	Harlow & Shine, 1992
	Charina	*bottae*	yes	R	M	Rodríguez-Robles *et al.*, 1999b
	Corallus	*caninus*	no	M	M	Henderson, 1993b
	Corallus	*grenadensis*	yes	R	M	Henderson & Sajdak, 2002
	Corallus	*hortulanus*	yes	B	M	Henderson, 1993b
	Corallus	*ruschenbergerii*	yes	B	M	Henderson, 2002
	Epicrates	*striatus*	yes	R	B, M	Henderson *et al.*, 1987b
	Epicrates	*gracilis, monensis*	no	R	R	Henderson *et al.*, 1987b; Chandler & Tolson, 1990

Table 6.1 (cont.)

Family	Genus	Species	Shift or not?	Shift from	Shift to	Authority
Colubridae	Alsophis	vudii	no	A, R	A, R	Henderson & Sajdak, 1996
	Alsophis	cantherigerus	yes	A, R	A, R, M	Henderson & Sajdak, 1996
	Antillophis	parvifrons	no	A, R	A, R	Henderson et al., 1987a
	Arizona	elegans	yes	R	M, B	Rodriguez-Robles et al., 1999a
	Boiga	blandingi, dendrophila, irregularis (overall)	yes	R	M, B	Savidge, 1988; Greene, 1989b; Shine, 1991c; Luiselli et al., 1998
	Boiga	irregularis (New Guinea)	yes	R	M	Greene, 1989b
	Boiga	ceylonensis, ocracea	no	R	R	Greene, 1989b
	Boiga	cynodon, multomaculata, trigonata	yes	R	B	Greene, 1989b
	Boiga	pulverulenta	yes	R	R, M, B	Greene, 1989b
	Carphophis	vermis	no	I	I	Clark, 1970
	Cerberus	rynchops	no	F	F	Jayne et al., 1988
	Coluber	constrictor	yes	I, R	M, R	Fitch, 1999
	Coluber	hippocrepis	yes	R	M	Pleguezuelos & Moreno, 1990
	Coniophanes	fissidens	no	R, A	R, A	Seib, 1985
	Coronella	austriaca (Britain)	yes	R	M	Goddard, 1984
	Coronella	austriaca (Italy)	yes	I, R	M, R	Luiselli et al., 1996

Coronella	girondica	no	R	R	Luiselli et al., 2001
Darlingtonia	haetiana	no	A	A	Henderson & Schwartz, 1986
Dendrelaphis	punctulata (NT)	no	A	A	Shine, 1991a
Dendrelaphis	punctulata (QLD, NSW)	yes	R	A	Shine, 1991a
Diadophis	punctatus (Kansas)	no	I	I	Fitch, 1975
Elaphe	quatuorlineata	yes	R	M, B	Rugiero & Luiselli, 1996; Capizzi & Luiselli, 1997
Elaphe	situla	no	M	M	Rugiero et al., 1998
Fordonia	leucobalia	no	I	I	Shine, 1991c
Heterodon	nasicus	no	A	A	Platt, 1969
Heterodon	platyrhinos	no	A, R	A, R	Platt, 1969
Hypsiglena	torquata	no	R	R	Rodriguez-Robles et al., 1999c
Hypsirhynchus	ferox	no	R	R	Henderson, 1984a
Lampropeltis	zonata	yes	R	M, B	Greene & Rodriguez-Robles, 2003
Leptophis	mexicanus	no	A	A	Henderson, 1982
Liophis	lineatus, mossoroensis, poecilogyrus, viridis	no	A	A	Vitt & Vangilder, 1983
Macroprotodon	cucullatus	no	R	R	Pleguezuelos et al., 1994
Mehelya	capensis	no	A, R	A, R	Shine et al., 1996a
Mehelya	nyassae	no	R	R	Shine et al., 1996a
Natrix	maura	yes	I, A	F	Hailey & Davies, 1986; Santos & Llorente, 1998; Santos et al., 2000
Natrix	natrix (Italian Alp males)	no	A	A	Luiselli et al., 1997

Table 6.1 (*cont.*)

Family	Genus	Species	Shift or not?	Shift from	Shift to	Authority
	Natrix	*natrix* (Italian Alp females)	yes	A	A, M	Luiselli *et al.*, 1997
	Natrix	*natrix*	yes	A	A, F, R, M	Filippi *et al.*, 1996
	Natrix	*tessellata*	no	F	F	Filippi *et al.*, 1996
	Nerodia	*clarkii, cyclopion, harteri, rhombifera*	no	F	F	Mushinsky *et al.*, 1982; Plummer & Goy, 1984; Scott Jr. *et al.*, 1989; Miller & Mushinsky, 1990; Manjarrez & Garcia, 1991; Greene *et al.*, 1994
	Nerodia	*erythrogaster, fasciata*	yes	F	A	Mushinsky *et al.*, 1982; Miller & Mushinsky, 1990
	Oxybelis	*aeneus*	no	R	R	Henderson, 1982
	Oxyrhopus	*guibei*	yes	R	M	Andrade & Silvano, 1998
	Philodryas	*chamissonis*	yes	A, R	M, B	Greene & Jaksic, 1992
	Philodryas	*viridissimus*	yes	A, R	M	Martins & Oliveira, 1999
	Pituophis	*catenifer*	yes	R	M, B	Rodriguez-Robles, 2002
	Psammodynastes	*pulverulentus*	no	R, A	R, A	Greene, 1989a
	Pythonodipsas	*carinata*	yes	R	M	Branch *et al.*, 1997
	Regina	*alleni, grahamii, septemvittata*	no	I	I	Godley, 1980; Godley *et al.*, 1984
	Rhinobothryum	*lentiginosum*	no	R	R	Martins & Oliveira, 1999

	Genus	species				Reference
	Rhinocheilus	*lecontei*	no	M, R	M, R	Rodriguez-Robles & Greene, 1999
	Tachymenis	*chilensis*	no	A, R	A, R	Greene & Jaksic, 1992
	Thamnophis	*atratus*	yes	F, A	A	Lind & Welsh, 1994
	Thamnophis	*cyrtopsis*	yes	F	A	Jones, 1990
	Thamnophis	*eques*	yes	I	F, A	Macias Garcia & Drummond, 1988
	Thamnophis	*ordinoides*	no	I	I	Gregory, 1984
	Thamnophis	*sauritus*	no	A	A	Rowe et al., 2000
	Thamnophis	*sirtalis*	yes	I, A	A, M	Fitch, 1965
	Thamnophis	*validus*	yes	A	F	de Queiroz et al., 2001
	Thelotornis	*capensis*	yes	R	R, A	Shine et al., 1996c
	Thelotornis	*kirtlandii*	no	R	R	Akani et al., 2002
	Tripanurgos	*compressus*	no	R	R	Martins & Oliveira, 1999
	Tropidodryas	*serra, striaticeps*	yes	R	M	Sazima & Puorto, 1993
	Uromacer	*frenatus, oxyrhyncus*	no	R	R	Henderson, 1984b; Henderson et al., 1987c
	Waglerophis	*merremii*	no	A	A	Vitt & Vangilder, 1983
	Acanthophis	*antarcticus*	yes	R	M	Shine, 1980b
	Austrelaps	*labialis, ramsayi, superbus*	no	R, A	R, A	Shine, 1987b
	Cacophis	*harriettae, krefftii, squamulosus*	no	R	R	Shine, 1980a
Elapidae	*Dendroaspis*	*jamesoni*	yes	A, R, M	M, B	Luiselli et al., 2000b
	Denisonia	*devisii, maculata*	no	A	A	Shine, 1983
	Drysdalia	*coronoides*	no	R	R	Shine, 1981b
	Echiopsis	*curta*	yes	R, A	M, B	Shine, 1982
	Furina	*diadema*	no	R	R	Shine, 1981a
	Hemiaspis	*damelii*	no	A	A	Shine, 1987c
	Hoplocephalus	*bungaroides*	yes	R	M	Webb, 1996

Table 6.1 (*cont.*)

Family	Genus	Species	Shift or not?	Shift from	Shift to	Authority
	Micrurus	*corallinus, fulvius*	no	R	R	Marques & Sazima, 1997; Greene, 1984
	Naja	*melanoleuca*	no	R	R	Luiselli et al., 2002
	Naja	*nigricollis*	yes	R	B, M	Luiselli et al., 2002
	Notechis	*scutatus* (island)	no	B, M	B, M	Shine, 1987a
	Notechis	*scutatus* (mainland)	no	A	A	Shine, 1987a
	Oxyuranus	*microlepidota, scutellatus*	no	M	M	Shine & Covacevich, 1983
	Pseudechis	*porphyriacus*	no	R, A	R, A	Shine, 1977
	Pseudonaja	*affinis, ingrami, nuchalis, textilis*	yes	R	M	Shine, 1989
	Pseudonaja	*modesta*	no	R	R	Shine, 1989
	Rhinoplocephalus	*boschmai, nigrescens, pallidiceps*	no	R	R	Shine, 1984; Shine, 1988
	Simoselaps	*australis, bertholdi, bimaculatus, calonotus, fasciolatus, semifasciatus*	no	R	R	Shine, 1984
	Suta	*dwyeri, flagellum, gouldii, monachus, nigriceps, nigrostriatus, spectabilis*	no	R	R	Shine, 1988
	Vermicella	*annulata*	no	R	R	Shine, 1980c; Keogh & Smith, 1996

Family	Genus	species				Reference
Hydrophiidae	Enhydrina	schistosa	no	F	F	Voris et al., 1978
	Hydrophis	elegans	no	F	F	Fry et al., 2001
	Lapemis	hardwickii	no	F	F	Fry et al., 2001
	Pelamis	platurus	no	F	F	Kropach, 1975
Laticaudidae	Laticauda	colubrina, frontalis, laticauda	no	F	F	Shetty & Shine, 2002; Shine et al., 2002a
Leptotyphlopidae	Leptotyphlops	dulcis, humilis, scutifrons	no	I	I	Punzo, 1974; Webb et al., 2000
Pythonidae	Bothrochilus	boa	yes	R	M	Shine & Slip, 1990
	Liasis	fuscus, maculosus, stimsoni	yes	R	M	Shine & Slip, 1990
	Morelia	spilota, viridis	yes	R	M	Slip & Shine, 1988; Shine & Slip, 1990; Henderson, 1993b
	Python	regius	yes	B	M	Luiselli & Angelici, 1998
	Python	reticulatus	no	M	M	Shine et al., 1998c
Typhlopidae	Ramphotyphlops	australis, bituberculatus, hamatus, ligatus, nigrescens, pinguis, polygrammicus, proximus, subocularis, waitii, wiedii	no	I	I	Webb & Shine, 1993
Viperidae	Atheris	squamiger	yes	R, M	M, B	Luiselli et al., 2000a
	Bitis	caudalis	yes	R	M	Shine et al., 1998b
	Vipera	ammodytes	yes	R	M, B	Saint Girons, 1980; Luiselli, 1996
	Vipera	aspis, berus, kaznakovi, latastei latastei, seoanei	yes	R	M	Saint Girons, 1980; Luiselli & Agrimi, 1991; Luiselli & Anibaldi, 1991; Luiselli et al., 1995

Table 6.1 (*cont.*)

Family	Genus	Species	Shift or not?	Shift from	Shift to	Authority
	Vipera	*latastei monticola*	no	R	R	Saint Girons, 1980
	Vipera	*ursinii*	no	I	I	Saint Girons, 1980; Agrimi & Luiselli, 1992
	Agkistrodon	*bilineatus*	yes	R, A	M	Solorzano *et al.*, 1999
	Agkistrodon	*contortrix*	yes	A, R, I	M	Fitch, 1960
	Agkistrodon	*contortrix* (Texas)	yes	R	I, M	Lagesse & Ford, 1996
	Bothrops	*alcatraz*	no	I, R	I, R	Martins *et al.*, 2002
	Bothrops	*erythromelas, itapeniningae*	no	I, R, A, M	I, R, A, M	Martins *et al.*, 2002
	Bothrops	*mattogrossensis*	no	R, A	R, A	Martins *et al.*, 2002
	Bothrops	*asper, atrox, bilineatus, brazili, hyoprorus, jararaca, leucurus, jararacussu, pubescens, taeniatus*	yes	R, A	M	Martins *et al.*, 2002; Martins & Oliveira, 1999; Andrade & Abe, 1999; Sazima, 1992
	Bothrops	*alternatus, cotiara, fonsecai, neuwiedi*	no	M	M	Andrade & Abe, 1999; Martins *et al.*, 2002
	Bothrops	*moojeni*	yes	R, A	M, B	Andrade *et al.*, 1996; Martins *et al.*, 2002
	Bothrops	*pauloensis*	yes	R, A, I	M	Valdujo *et al.*, 2002; Martins *et al.*, 2002

Genus	species				Reference
Calloselasma	*rhodostoma*	yes	R, A, I	M, B	Daltry *et al.*, 1998
Cerrophidion	*godmani*	no	I, R, A, M	I, R, A, M	Campbell & Solorzano, 1992
Crotalus	*cerastes, enyo, helleri, willardi*	yes	R	M	Klauber, 1997; Secor, 1994; Taylor, 2001; Mackessy, 1988; Holycross *et al.*, 2002a
Crotalus	*durissus, horridus, oreganus*	no	M	M	Salomao *et al.*, 1995; Clark, 2002; Macartney, 1989; Wallace & Diller, 1990
Crotalus	*lepidus klauberi*	yes	I, R	R, M, B	Holycross *et al.*, 2002b
Crotalus	*pricei*	no	R, M	R, M	Prival *et al.*, 2002
Porthidium	*yucatanicum*	yes	R, I	R, M, B	McCoy & Censky, 1992; Martins *et al.*, 2002
Sistrurus	*catenatus*	yes	R	M	Holycross & Mackessy, 2002

remain the same for pythons across a wide range of body sizes, but foraging modes shift because of crevice accessibility. In the boa *Corallus grenadensis*, small animals can (actively) access sleeping anoles on thin branches. Larger snakes, however, are too bulky to do this and therefore switch to ambushing mammals closer to the ground (Henderson, 1993a).

A snake's body size may also influence thermal factors that affect its ability to forage in specific ways. For example, in aquatic garter snakes (*Thamnophis atratus*) that feed on amphibians and fishes in coldwater streams, smaller animals cool so quickly after entering the water that they cannot effectively chase prey; thus, they wait in ambush at the surface and strike when prey approach (Lind and Welsh, 1994). In contrast, the higher thermal inertia of larger conspecifics enables them to enter the water to locate and capture prey. Thermal factors may influence the feasibility of alternative foraging modes in terrestrial environments also. For example, ambush predators that rely upon very long periods of immobility to remain undetected by prey (as may be widespread in snakes: see, for example, Greene, 1997) cannot afford to shuttle between sun and shade. Larger body size increases thermal inertia (Grigg *et al.*, 1979) and thus may allow a snake to remain for longer in a suboptimally hot or cold site. The ability of snakes to reduce surface area by tight coiling also substantially expands the range of ambient thermal conditions under which ambush predation is possible. Ayers and Shine (1997) calculated that tightly coiled juvenile diamond pythons cooled from 33 to 12 °C (the ambient temperature) in 2 hours, whereas coiled adults took 8 hours to equilibrate to this temperature. This size-dependent thermal inertia allows adult pythons to keep hunting late into the evening (when mammalian prey are active), whereas the rapid heat loss of juveniles restricts them to diurnal foraging.

(4) *Ability to forage actively changes with the predator's body size.* In aquatic species, body size may determine the maximum duration of dives and hence the water depths that can be searched effectively (Heatwole, 1999). Body size also affects vulnerability to predators, possibly explaining why juvenile broad-headed snakes (*Hoplocephalus bungaroides*) rely on ambushing prey under rocks in outcrops year-round whereas adults switch to active searching for mammalian prey in the surrounding forests during summer (Webb, 1996). This kind of phenomenon may be common: juvenile snakes are notoriously under-represented in samples of active animals, suggesting much more limited movements than for conspecific adults (Bonnet *et al.*, 1999).

Importantly, these ontogenetic shifts in prey types and sizes (Table 6.1) also may influence where or when the animal forages, or (most relevant to the current volume) the foraging mode that it employs to obtain those prey. If the types of prey eaten by large and small snakes differ in aspects such as activity times or habitats, then the predators must also shift to forage effectively. For example, juvenile diamond pythons (*Morelia s. spilota*) consume diurnal lizards whereas adults of the same population take nocturnal mammals. Thus, foraging mode remains conservative (ambush, in both cases) but the time of foraging changes

(Table 6.1). Contrariwise, tree boas (*Corallus grenadensis*) of all size classes forage at night, but shift their foraging modes with body size. The sleeping lizards taken by juvenile boas are taken by active searching, whereas the active mammals eaten by adult boas are taken from ambush (Table 6.1). The larger reticulated python (*Python reticulatus*) shifts from dependence on rodents to much larger prey (monkeys, pangolins, etc.) with increasing size, and accordingly moves from disturbed habitats (villages) with high rodent abundance to forested habitats with larger prey (Table 6.1). Thus, all three species of constricting snake show ontogenetic shifts in prey types, but one achieves this dietary shift by changing its activity time, one by changing foraging mode, and the other by changing habitat use.

Sex differences in foraging biology in snakes

Males and females differ substantially in mean adult body sizes within many species of snake (from males 50% longer than females, to females 58% longer than males: Shine, 1994). Thus, the effects of body size on foraging biology (above) mean that diets of males and females will also differ in many taxa. Simple gape-limitation combined with sexual size dimorphism can engender massive disparities in diet. One spectacular example is the Western Australian carpet python (*Morelia spilota imbricata*) on Garden Island, where adult males (mean mass 300 g) eat mainly house-mice (*Mus domesticus*, up to 12 g) whereas adult females (mean mass 4 kg) feed primarily on wallabies (*Macropus eugenii*, up to 6 kg) (see Table 6.2). Similarly, female reticulated pythons (*Python reticulatus*) often grow large enough (to >30 kg) to take large prey (wild pigs, etc.) whereas most males remain within a size range (<10 kg) where rats are the most frequent dietary items (Table 6.2).

Although sexual size dimorphism may be the most general reason for the sexes to diverge in diets, it is not the only one. In several taxa, males and females eat different types or sizes of prey even at the same body sizes. The sexes may also differ in the frequency of feeding and the incidence of multiple meals; for example, female sea kraits (*Laticauda*) and acrochordids tend to contain single large prey items whereas males and juveniles contain multiple smaller items (Shetty and Shine, 2002; Houston and Shine, 1993). Such trophic divergence may reflect a trend for the sexes to forage in different habitats or seasons or at different times of day (e.g. *Acrochordus arafurae*: Houston and Shine, 1993). For example, male *Python regius* are small, relatively arboreal, and tend to eat birds rather than the mammals consumed by the larger terrestrial females (Luiselli and Angelici, 1998).

Sex divergences in foraging biology may have generated significant selection on the trophic morphology (relative head size, head shape, etc.) of snakes.

Table 6.2. *Sex-based differences in dietary habits within snake species*
SSD, sexual size dimorphism.

Family	Genus	Species	Larger sex	Males eat	Females eat	Authority
Boidae	*Eunectes*	*murinus*	F	reptiles, birds	mammals, reptiles, birds	Rivas, 1999
Pythonidae	*Morelia*	*spilota imbricata*	F	lizards, mice, birds	possums, wallabies	Pearson et al., 2002
	Python	*regius*	F	more birds	more mammals	Luiselli & Angelici, 1998
	Python	*reticulatus*	F	small mammals	large mammals	Shine et al., 1998c
Acrochordidae	*Acrochordus*	*arafurae*	F	small fishes	large fishes	Shine, 1986; Houston & Shine, 1993
Colubridae	*Boiga*	*irregularis*	M	more birds, mammals	more lizards	Savidge, 1988
	Coluber	*constrictor*	F	reptiles, invertebrates, mammals	more mammals	Fitch, 1999
	Coronella	*austriaca*	Same	lizards	mammals, snakes	Luiselli et al., 1996
	Geophis	*nasalis*	F	small worms	large worms	Seib, 1981
	Natrix	*maura*	F	small fishes	frogs, large fishes	Santos & Llorente, 1998; Santos et al., 2000
	Natrix	*natrix*	F	frogs, juvenile toads	frogs, adult toads, mice	Madsen, 1983; Luiselli et al., 1997
	Nerodia		F	small fishes	large fishes	

Family	Genus	species	Sex	Diet	Diet	Reference
	Nerodia	*cyclopion, rhombifera*		fishes	fishes, mudpuppies (*Necturus*)	Mushinsky et al., 1982
		sipedon	F			King, 1993
	Opheodrys	*aestivus*	F	more caterpillars	more odonates	Plummer, 1981
	Telescopus	*dhara*	F	lizards	birds	Zinner, 1985
	Thamnophis	*sirtalis*	F	frogs, dace	toads, suckers, mice	White & Kolb, 1974; Fitch, 1982
Elapidae	*Aspidelaps*	*scutatus*	F	more frogs	snakes, mammals	Shine et al., 1996b
	Pseudechis	*porphryriacus*	M	more frogs	more lizards	Shine, 1991d
Laticaudidae	*Laticauda*	*colubrina*	F	moray eels	conger eels	Shetty & Shine, 2002
	Laticauda	*frontalis*	F	small morays	large morays	Shine et al., 2002a
Viperidae	*Bitis*	*caudalis*	F	mostly lizards	more mammals	Shine et al., 1998b
	Vipera	*ursinii*	F	lizards, invertebrates mammals	mammals, lizards, invertebrates insects	Agrimi & Luiselli, 1992
	Agkistrodon	*contortrix*	M			Fitch, 1960, 1982

Males and females within a species often display different head sizes at the same body length (Shine, 1991b). These differences have been interpreted as adaptations to taking prey of different types and sizes (Shine, 1991b). In keeping with this adaptive hypothesis, neonatal snakes of several taxa display significantly different relative head sizes according to sex (Shine, 1991b; King et al., 1999). In other taxa, however, an individual's relative head size (and thus, sex differences in this trait) may be modified during ontogeny by its feeding experiences (Bonnet et al., 2001; Queral-Regil and King, 1998). The puzzling occurrence of sex differences in traits such as relative eye sizes and tongue lengths suggest that male and female snakes may diverge in many traits related to foraging biology (Shine, 1993; Shine et al., 1996a).

The selective forces responsible for intersexual niche divergence are likely diverse:

(1) *Simple sexual size dimorphism offers the simplest example*: males and females eat different prey types because they are of different sizes (or alternatively, males and females attain different sizes because they eat different prey). Intraspecific comparisons clearly show this causal link. For example, grass snakes (*Natrix natrix*) on the Swedish mainland show substantial sex divergence in body sizes and diets (females grow larger and eat toads as well as smaller prey) whereas on a nearby island where the larger prey type (toads) is absent both sexes eat newts and sexual size dimorphism is greatly reduced (Madsen and Shine, 1993b). Captive animals from both populations exhibit similar (and marked) sexual size dimorphism, suggesting that the lack of dimorphism in the island population is a phenotypically flexible response to the lack of large prey.

(2) *Reproduction is more energy-expensive for females than for males in most or all snake species*, so females may have evolved to feed more often, or on larger prey, than do conspecific males (see, for example, Houston and Shine, 1993; Shine 1993; Madsen and Shine, 1993a).

(3) *In many snake species, both sexes reduce or forego feeding during reproductive activities*: males during mate-searching and courtship, and females during gestation (see, for example, Madsen and Shine, 1993a; Daltry et al., 1998; Gregory et al., 1999; Lourdais et al., 2002). This seasonal anorexia may expose males and females to different spectra of prey availability, and thus favor divergent adaptations for foraging by the two sexes. Other selective forces on foraging may also differ seasonally; for example, post-parturient female *Vipera berus* experience high rates of mortality because they are so emaciated that they must resort to active foraging (and thus expose themselves to increased predation) (Madsen and Shine, 1993a).

(4) *If males move about more extensively or through different habitats than females in the course of mate-searching (as may often be true: Bonnet* et al., 1999*), they will encounter different kinds of potential prey*. This may not influence food intake for

ambush foragers, which will find it difficult to combine mate-searching with feeding (e.g. *Calloselasma rhodostoma*: Daltry *et al.*, 1998); for an active searcher like the Australian blacksnake (*Pseudechis porphyriacus*), however, there is no such conflict and hence adult males continue to feed throughout the mating season (Shine, 1977). Perhaps as a result, male blacksnakes consume different overall proportions of lizards versus frogs than do conspecific adult females (Shine, 1991b).

Other sources of intraspecific variation in foraging biology

Although our aim in this chapter is to highlight the effects of body size and sex on foraging biology in snakes, we note briefly that several other factors also generate variation in foraging traits within a single snake species. These include:

(1) *Color polymorphism* is widespread among snakes, especially ambush predators from multiple lineages (e.g. the boids *Corallus annulatus* and *C. hortulanus*, the pythonid *Morelia viridis*, the colubrids *Cerberus rynchops*, *Dispholidus typus*, and *Psammodynates pulverulentulus*, the elapids *Acanthophis antarcticus* and *Echiopsis curta*, the viperines *Atheris hispida*, *A. squamiger*, *Vipera aspis*, and *V. berus*, and the pit-vipers *Agkistrodon halys*, *Bothriechis schlegelii*, *Crotalus horridus*, *C. lepidus*, *Trimeresurus flavoviridis*, *Tropidolaemus mcgregori*, and *T. wagleri*) (Neill, 1963; Pitman, 1974; Rasmussen, 1975; Henderson, 1990; Ross and Marzec, 1990; Brown, 1991; Shine, 1991a; Garrett and Smith, 1994; Nilson *et al.*, 1994; Nobusaka *et al.*, 1994; Johnston, 1996; Stafford and Henderson, 1996; Greene, 1997; Cogger, 2000). This distribution suggests that polymorphism might evolve under frequency-dependent selection for background color matching in these sedentary snakes (Shine *et al.*, 1998a). In at least one case, different color morphs of a species from the same general area differ significantly in diets as well as several other attributes (*Python brongersmai*: Shine *et al.*, 1998a). In other cases, however, different color morphs appear to be broadly similar in most ecological attributes (e.g. *Thamnophis sirtalis*: King, 1988; *Vipera berus*: Forsman, 1995; *Emydocephalus annulatus*: Shine *et al.* 2003).

One foraging-related color polymorphism shows clear links both to a snake's sex and its body size. Several lineages of ambush predators use caudal luring (wriggling of the tail-tip) to encourage close approach by potential prey (see, for example, Neill, 1960; Henderson, 1970; Heatwole and Davison, 1976; Shine, 1980b; Sazima and Puorto, 1993). Modified colors on the tail-tip occur in several of these taxa, presumably enhancing the effectiveness of the lure. Interestingly, both the luring behavior and the modified tail colors are usually more common in juveniles than in conspecific adults, presumably because adult tails are too large to be effective lures, or because adults shift to other non-lurable prey (Sazima, 1991;

Sazima and Puorto, 1993). Further, among juveniles of both *Bothrops asper* and *B. atrox*, only males have colored tail-tips (Burger and Smith, 1950; Neill, 1960; Hoge and Federsoni, 1977; Tryon, 1985; Solorzano and Cerdas, 1989), perhaps reflecting the luring advantages of a longer tail in this respect (male snakes generally have longer tails than females: King, 1989; Shine, 1993).

(2) *Seasonal variation* affects food supply and thus diets. This variation may often be tied to life histories of major prey taxa. For example, a seasonal flush of meta-morphosing anurans, migratory wave of passerine birds or synchronized hatching of seabird eggs can generate massive but short-lived food abundance for snakes that consume these taxa. Similarly, seasonal shifts between orthopteran versus mammalian prey may be common in viperids (e.g. *Cerrophidion godmani*: Campbell and Solorzano, 1992; *Agkistrodon contortrix*: Gloyd and Conant, 1990). These different prey types may sometimes require alternative foraging tactics: for example, copperheads (*Agkistrodon contortrix*) actively forage for cicadas, often climbing trees to get them, but ambush most other prey (Gloyd and Conant, 1990). Sedentary eggs and hatchlings can only be obtained by mobile predators, whereas ambush may be the most effective foraging method at other times of year when only adult prey organisms are available. Even when prey types do not shift seasonally, prey behavior may do so and thus require a response from snake predators. For example, black-necked garter snakes (*Thamnophis cyrtopsis*) actively search for leopard frogs (*Rana yavapaiensis*) in spring when the frogs are concentrated along the banks of desert streams, but ambush them from beneath algal mats in summer when the frogs are more widely dispersed (Jones, 1990).

(3) *Year-to-year variation* can be considerable, either in prey availability *per se* or in weather conditions that influence a snake's foraging abilities either directly (e.g. because higher temperatures facilitate activity) or indirectly (e.g. by rainfall modifying vegetation density and thus the availability of ambush sites). Several long-term studies have detected significant year-to-year variation in traits such as feeding frequencies, prey types, and snake body condition. For example, Kephart and Arnold's (1982) seven-year study found that *Thamnophis elegans* preyed primarily on toads in years with normal or high rainfall; during drought years, however, toads stopped breeding, and the snakes ate proportionally more fishes.

(4) *Geographic variation* in dietary composition within wide-ranging snake species is likely to be the norm; even nearby populations sometimes differ substantially in this respect (e.g. garter snakes: Kephart, 1982; *Calloselasma rhodostoma*: Daltry *et al.*, 1998). Even where diets remain consistent, foraging modes may vary. For example, the homalopsine watersnake *Enhydris polylepis* in tropical Australia forages for fishes in shallow water. It does so by ambush in fast-moving streams (where the snakes anchor their tails into the substrate and intercept migrating fishes) but by active searching in still water (where the snakes move slowly along the margins of the pond in search of sleeping fishes). The two habitat types (and locations of different foraging tactics) may be only meters apart, and mark–recapture shows that individual snakes move between these habitats

(G. P. Brown and R. Shine, unpubl. data). Anthropogenically disturbed habitats offer another good example of such flexibility: for example, carpet pythons (*Morelia spilota mcdowelli*) that lie in wait for mammalian prey in natural habitats (Shine and Fitzgerald, 1996) shift facultatively to active pursuit of caged birds and domestic pets in suburban settings (Fearn *et al.*, 2001).

All of these sources of variation also interact in complex ways. For example, Shedao pit-vipers (*Gloydius shedaoensis*) occur both in island and mainland populations. All feed on migrating birds in spring and autumn, but the mainland snakes continue to feed (mostly on rodents) during summer, whereas the island snakes have little feeding opportunity through this season (S. Li, pers. comm.). Juvenile pit-vipers take invertebrate prey as well, so the geographic and seasonal shift in dietary composition also differs among size classes.

(5) *Flexibility within an individual through time* probably underlies much of the geographic and temporal variation described above (and even that correlated with color polymorphism, in species with ontogenetic color change such as *Morelia viridis*). Even quintessential ambush foragers such as Shedao pit-vipers (*Gloydius shedaoensis*) will consume dead birds (or live birds trapped in mist-nets) if they encounter them (Shine *et al.*, 2002b). Consumption of carrion has been reported for several snake taxa usually regarded as highly specialized ambush-foragers, especially viperids and piscivorous taxa (DeVault and Krochmal, 2002). This kind of flexibility presumably enhances overall feeding rates: if an ambush predator cannot find a site providing an opportunity to capture suitable prey, eventually it is likely to go looking for food. Similarly, a wide-forager that locates a site where prey are easily available in large numbers is unlikely to leave until that prey resource is depleted. The seasonal shift to active-searching arboreality in cicada-hunting copperheads (*Agkistrodon contortrix*: Gloyd and Conant, 1990) provides a clear example of this, as does the flexibility of the European water snake *Natrix maura*: adults usually ambush fishes, but switch to active foraging when the fish are concentrated in shallow pools (Hailey and Davies, 1986).

Thus, relatively few snakes can unambiguously be classed as strictly conforming to either extreme of the foraging-mode continuum (unlike organisms that construct and then rely upon static prey-capture devices like spider-webs or ant-lion pits). The most conservative (inflexible) snake taxa in this respect will be those that rely exclusively upon immobile prey such as eggs (e.g. most scolecophidians, *Dasypeltis*, *Emydocephalus*, some *Simoselaps*) because it is difficult to envisage a way for an ambush predator to capture such items. More commonly, snakes may use a variety of foraging tactics even within a single foraging bout (e.g. *Boiga irregularis*: Rodda, 1992). The same is presumably generally true among many kinds of animals. For example, lions ambush their prey during the wildebeest migration period, but may actively search for prey at other times (Schaller, 1972).

Why are intraspecific foraging-mode shifts more obvious in snakes than in lizards? Our compilation of published data (Tables 6.1 and 6.2) shows that individuals within a single snake population frequently differ in the types and sizes of prey that they consume, notably as a reflection of differences in predator body size. Foraging mode is intimately associated with such shifts, in two ways: (1) some of the ontogenetic and sex-based divergences reflect shifts in foraging modes (see above); and (2) snakes that rely upon ambush foraging often display such dietary shifts through ontogeny, whereas snakes that utilize active searching display ontogenetic shifts in prey type less frequently. We have not attempted a quantitative phylogenetically based test of the second of these patterns, but note that it appears in multiple lineages and thus is not a simple artifact of phylogenetic conservatism. The association between ontogenetic dietary shifts and ambush foraging is apparent in "basal" snakes (pythonids and boids vs. scolecophidians) as well as "advanced" snakes (viperids vs. colubrids); and even within one lineage, the Elapidae (where 4 of 6 ambush foragers show ontogenetic dietary shifts, compared with 5 of 39 active searchers: Table 6.1).

In general terms, what factors should increase the probability that a predator species will show intraspecific variation in prey type? We suggest that the simplest answer involves the range in absolute body masses of prey consumed by the predator population: a population that consumes prey over a very wide size range is likely to also take prey of more than one type (because each given prey type spans only a small part of the total range of prey sizes consumed). So the question becomes: what factors cause a predator population to consume a wide size range of prey? Important differences between lizards and snakes in this respect include:

(a) *The span of body sizes within a snake population is typically greater than within a lizard population.* On average, snakes grow larger than lizards. In Pough's (1980) compilation, 45% of 1592 snake species had estimated mean body masses >100 g, whereas this was true of less than 8% of 1780 lizards. Body lengths of neonatal snakes and lizards both average around 40%–50% of the size at maturation (Charnov et al., 1993). Snakes generally continue to grow throughout their lifetimes, whereas lizards display determinate growth (Andrews, 1982). Thus, the range of *absolute* body masses from hatching to maximum size is often much larger in snakes than in lizards.

(b) *Snakes tend to eat larger prey than do lizards of the same body mass,* because they consume their prey entire and because they often feed on vertebrates rather than invertebrates (Fig. 6.1). Thus, measures of relative prey mass (prey size divided by predator size, RPM) often exceed 0.50 for snakes (see, for example, Fitch and Twining, 1946; Luiselli et al., 1999; Martins et al., 2002) but rarely top 0.10 for lizards (see, for example, Losos and Greene, 1988). Greene (1984) found the

Figure 6.1. Snakes generally attain larger body sizes than lizards and take larger prey relative to their own size. These photographs show a viper (*Porthidium nasutum*) eating a frog (*Rana warszewitschii*) that is large relative to the predator's own body size, and a lizard (*Diporiphora winneckei*) eating a relatively small dipteran. Photographs by D. Warner and D. O'Connor.

average RPM for the elapid *Micrurus fulvius* to be 0.42, with a maximum of 1.31. Pit-vipers are capable of even more prodigious feats of ingestion. A neonate *Bothrops atrox*, for instance, contained a teiid lizard with an RPM of 1.56 (Greene, 1983) and a neonate *Crotalus cerastes* had eaten a *Cnemidophorus tigris* with an RPM of 1.72 (possibly leading to the snake's death: Mulcahy *et al.*, 2003).

On the other hand, a study of varanids (the lizards generally considered to be closest to snakes both phylogenetically and ecologically) found average RPMs of 26 species to range between 0.002 and 0.079 (Losos and Greene, 1988).

These two factors influence the size range of prey consumed by a predator population. If larger predators tend to eat larger prey, then the size range of prey consumed will depend upon (a) the span in absolute body sizes of the predators within a population, and (b) the rate at which prey size increases with predator size (which in turn depends upon relative prey mass). Thus, the range of prey sizes taken will be maximized if the predators themselves span a wide size range, and if they take prey items large relative to their own body size. The effect of the size range of predators is intuitive (for example, a tenfold range in predator size will generate a greater range in prey sizes than will a twofold range in predator size) but the other factor (relative prey mass, RPM) requires more explanation.

Figure 6.2 shows hypothetical relations between prey size and predator size for a series of reptile taxa differing in RPM. The critical dependent variable is the absolute range in prey sizes consumed (i.e. the vertical distance on the y-axis). If prey size does not increase with predator size, an increase in the size range of predators will not change the range in prey sizes consumed. However, any tendency for larger predators to take larger prey results in an increase in the size range of prey consumed with an increase in the size range of predators. The magnitude of the size range of prey is greatest, for any given range in predator size, if RPM is high (Fig. 6.2). For example, a span of predator body sizes from 100 g to 1 kg generates an increase in mean prey sizes of 45 g (from 5 to 50 g) if RPM is 0.05 (as may be true for many lizards: Losos and Greene, 1988). In contrast, the same increment in predator body size generates an increase of 450 g (from 50 to 500 g) if RPM is 0.50 (as may be true for some viperid snakes: see, for example, Fitch and Twining, 1946). All else being equal, a greater range in prey sizes makes it more likely that the predator population will consume prey of a diversity of types as well as sizes.

Thus, the reason why intraspecific niche partitioning is more common in snakes than in lizards may be twofold: snakes eat larger prey relative to their own size; and snake populations contain individuals over a wider range of absolute body sizes than do most lizards.

We can test this hypothesis by examining data in Tables 6.1 and 6.2. Our hypothesis predicts that shifts in prey type with the size or sex of the predator should be most frequent in:

(a) *species with large mean adult body size* (because these have the greatest range in absolute body size). In keeping with this prediction, species recorded to shift prey types were, on average, larger than closely related species that specialize on a single type of prey (see Table 6.1). For example, this pattern was evident within the

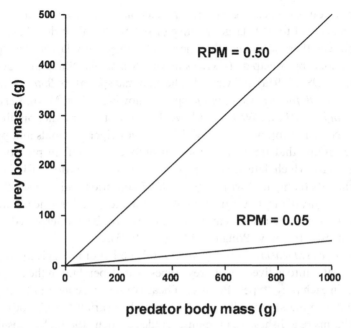

Figure 6.2. A simplified graphical model of predator body size and relative prey size as determinants of the total range of prey sizes consumed by a predator population. The range of prey sizes is shown by the vertical axis in this graph; it depends upon the absolute size range of predators (i.e. along the horizontal axis) and the slope of the relationship between mean prey size and predator body size (i.e. relative prey mass). The graph plots lines for relative prey masses (RPM, prey mass divided by predator mass) of 0.05 (typical of many lizard species) and 0.50 (seen in some snakes). The higher RPM translates into a wider range of prey sizes consumed.

Colubridae and Elapidae (especially in the latter group, where average SVL of shifters was 91 cm, of non-shifters 50 cm; $p = 0.002$). Greene (1989b) and Shine (1989) previously noted the same pattern among species within the colubrid genus *Boiga* and the elapid genus *Pseudonaja*, respectively. Interestingly, viperid species exhibiting a shift average about the same size as those that do not (roughly 65 cm); this result may reflect the fact that that small viperids (e.g. *Bitis caudalis*, *Crotalus lepidus*) are capable of swallowing relatively large endotherms as adults, and some large species (e.g. *Bothrops alternatus*, *Crotalus horridus*) produce young big enough to take mammals from birth (see below).

(b) *species with relatively small offspring relative to maximum adult size* (because these have the greatest range in absolute body sizes). In many snakes, the primary ontogenetic shift in diets occurs at a relatively small body size, often within the juvenile phase (Table 6.1). Thus, a decrease in offspring size might significantly decrease maximal ingestible prey size, and hence favor neonates eating prey that are smaller (and hence, likely to be of a different "type") than those consumed by conspecific adults. In keeping with this prediction, snake species that feed entirely on mammals tend to have larger relative offspring sizes than do related taxa that shift from ectothermic to endothermic prey as they grow larger. This pattern is evident among both elapids (*Oxyuranus* vs. *Pseudonaja*: Shine and Covacevich, 1983; Shine, 1989, 1991a) and viperids (the mammal specialists *Bothrops alternatus*, *B. cotiari*, and *B. fonsecai* vs. *Bothrops* species showing a shift: Martins *et al.*, 2002).

(c) *species with high RPMs.* We do not have data to test this prediction directly, but note that consumption of large prey characterizes viperids, boids and pythonids, the same groups that frequently show an ontogenetic shift in prey type. Of 45 viperid taxa for which data are available, 30 show a shift, as do 8 of 11 boids and 7 of 8 pythonids (compared with only 31 of 77 colubrids and 9 of 45 elapids: see Table 6.1). Many distinctive attributes of these snakes (e.g. large heads, long fangs, high numbers of scale-rows, hemotoxic venom) have been interpreted as adaptations to ingest large prey (Pough and Groves, 1983).

(d) *species that feed at least partly on endotherms.* Whether or not a given range in prey sizes translates into a diversity of prey "types" will depend upon the range of body sizes within each prey "type". Because of heat-conserving constraints, endotherms (birds and mammals) generally have larger and less variable adult body sizes than do ectotherms (e.g. fishes, amphibians, reptiles: Pough, 1980). As a result, even if adults of a snake species feed primarily upon mammals, there may be no mammals in the local area small enough to be ingested by a neonatal snake. Thus, juveniles necessarily rely upon ectothermic prey until they are large enough to utilize the endothermic prey resource.

Habitats differ in the availability of these different types of prey, and may thus differ in the incidence of ontogenetic shifts among prey "types". For example, aquatic habitats contain few ingestible-size birds or mammals, so that fishes form the majority of the diet for most highly aquatic snake species. Thus, even if a species shows immense ontogenetic and sexual divergence in prey sizes (as in laticaudids and acrochordids: see Tables 6.1 and 6.2), this may not translate into a shift among prey types. In contrast, terrestrial snakes will often encounter (small) ectotherms and (large) endotherms, and so an ontogenetic shift from dependence on ectothermic to endothermic prey is common (Table 6.1). One aquatic snake taxon that spans an extraordinary size range (the anaconda *Eunectes murinus*) grows large enough to consume aquatic mammals (capybaras) and hence shows an ontogenetic shift from ectotherms to endotherms similar to that observed in many terrestrial snake species (Rivas, 1999).

Our hypothesis on the causal basis for the scarcity of intraspecific dietary shifts in lizards compared with snakes can also be evaluated by reviewing data on lizards. Other chapters within this volume review the foraging biology of lizards in detail, so we consider this topic only briefly. Intraspecific diversity in foraging biology doubtless occurs in many lizard species, and many cases may even bear strong parallels to those we have documented above in snakes. For example, effects of body size on cooling rates allow large (male) Galapagos iguanas to forage for longer periods in the cold ocean, thus generating both size and sex effects on foraging tactics (Wikelski and Trillmich, 1994). Similarly, several varanid species forage widely for small prey but ambush larger items (King *et al.*, 2002). However, most lizards take almost exclusively invertebrate prey, often over a broad taxonomic range, and rarely show strong increases in prey size with predator size (because their prey are typically small, and even large items can be torn apart). The most dramatic intra-specific niche shift within lizards is that involving herbivory, whereby larger animals within a species take a higher proportion of plant food, with an obvious shift in foraging behavior (Chapter 7; Cooper and Vitt, 2002). However, such shifts are far from universal even in lizard species that take significant amounts of plant food (Cooper and Vitt, 2002). We predict that a detailed analysis would reveal intra-specific dietary shifts to be most common in lizard taxa displaying the traits we have identified above for snakes: for example, species with large mean adult body size, small offspring size, and high relative prey mass.

In summary, published studies reveal a remarkable dichotomy in the mag-nitude and frequency of intraspecific dietary shifts between two very closely related groups of organisms: snakes and lizards. The functional basis for that dichotomy relates to a series of other features of anatomy, physiology, ecology and behavior in which most snakes differ from most lizards. Notably, how-ever, some lizards (such as the pygopodid *Lialis burtonis*) have evolved to occupy traditional "snake" niches (infrequent ingestion of large vertebrate prey: see, for example, Patchell and Shine, 1986) whereas some snakes (e.g. the seasnake *Emydocephalus annulatus*) feed in a "lizard-like" manner on small immobile prey (fish eggs: Shine *et al.* 2004). This overlap in dietary niches between snakes and lizards offers considerable potential for robust tests on the selective forces that have shaped foraging biology in squamate reptiles.

Acknowledgements

We thank Dave O'Connor and Dan Warner for photographs. RS thanks the Australian Research Council for funding his research over many years.

References

Agrimi, U. and Luiselli, L. (1992). Feeding strategies of the viper *V. u. ursinii* (Reptilia: Viperidae) in the Apennines. *Herpetol. J.* **2**, 37–42.
Akani, G. C., Luiselli, L. and Angelici, F. M. (2002). Diet of *Thelotornis kirtlandii* (Serpentes: Colubridae: Dispholidini) from southern Nigeria. *Herpetol. J.* **12**, 179–82.

Andrade, D. V. and Abe, A. S. (1999). Relationship of venom ontogeny and diet in
    Bothrops. *Herpetologica* **55**, 200–4.

Andrade, R. O. and Silvano, R. A. M. (1998). Comportamento alimentar e dieta da
    "falsa coral" Oxyrhopus guibei Hoge & Romano (Serpentes, Colubridae). *Revta
    Bras. Zool.* **13**, 143–50.

Andrade, D. V., Abe, A. S. and dos Santos, M. C. (1996). Is the venom related to diet
    and tail color during *Bothrops moojeni* ontogeny? *J. Herpetol.* **30**, 285–8.

Andrews, R. M. (1982). Patterns of growth in reptiles. In *Biology of the Reptilia*, ed.
    C. Gans and F. H. Pough, pp. 273–320. London: Academic Press.

Arnold, S. J. (1993). Foraging theory and prey-size – predator-size relations in snakes.
    In *Snakes: Ecology and Behavior*, ed. R. A. Seigel and J. T. Collins, pp. 87–115.
    New York: McGraw-Hill.

Ayers, D. Y. and Shine, R. (1997). Thermal influences on foraging ability: body size,
    posture and cooling rate of an ambush predator, the python *Morelia spilota*.
    Funct. Ecol. **11**, 342–7.

Bonnet, X., Naulleau, G. and Shine, R. (1999). The dangers of leaving home: dispersal
    and mortality in snakes. *Biol. Conserv.* **89**, 39–50.

Bonnet, X., Shine, R., Naulleau, G. and Thiburce, C. (2001). Plastic vipers: influence
    of food intake on the size and shape of Gaboon vipers (*Bitis gabonica*). *J. Zool.
    Lond.* **255**, 341–51.

Branch, W. R., Shine, R., Harlow, P. S. and Webb, J. K. (1997). Sexual dimorphism,
    diet and aspects of reproduction of the western keeled
    snake, *Pythonodipsas carinata* (Serpentes: Colubridae). *Afr. J. Herpetol.* **46**,
    89–97.

Brown, W. S. (1991). Female reproductive ecology in a northern population of the
    timber rattlesnake, *Crotalus horridus*. *Herpetologica* **47**, 101–15.

Burger, W. L. and Smith, P. W. (1950). The coloration of the tail tip of young fer-
    de-lance: sexual dimorphism or adaptive coloration? *Science* **112**, 431–3.

Campbell, J. A. and Solorzano, A. (1992). Biology of the montane pitviper,
    Porthidium godmani. In *Biology of the Pitvipers*, ed. J. A. Campbell and
    E. D. Brodie Jr., pp. 223–50. Tyler, Tx: Selva.

Capizzi, D. and Luiselli, L. (1997). The diet of the four-lined snake (*Elaphe
    quatuorlineata*) in Mediterranean central Italy. *Herpetol. J.* **7**, 1–5.

Chandler, C. R. and Tolson, P. J. (1990). Habitat use by a boid snake, *Epicrates
    monensis*, and its anoline prey, *Anolis cristatellus*. *J. Herpetol.* **24**, 151–7.

Charnov, E. L., Berrigan, D. and Shine, R. (1993). The m/k ratio is the same for fish
    and reptiles. *Am. Nat.* **142**, 707–11.

Clark, D. R. Jr. (1970). Ecological study of the worm snake *Carphophis vermis*
    (Kennicott). *Univ. Kans. Pub. Mus. Nat. Hist.* **19**, 85–194.

Clark, R. W. (2002). Diet of the timber rattlesnake. *J. Herpetol.* **36**, 494–9.

Cogger, H. G. (2000). *Reptiles and Amphibians of Australia*, 6th edn. Sydney: Reed
    New Holland.

Cooper, W. E. and Vitt, L. J. (2002). Distribution, extent, and evolution of plant
    consumption by lizards. *J. Zool. Lond.* **257**, 487–517.

Daltry, J. C., Wüster, W. and Thorpe, R. S. (1998). Intraspecific variation in the
    feeding ecology of the crotaline snake *Calloselasma rhodostoma* in Southeast
    Asia. *J. Herpetol.* **32**, 198–205.

de Queiroz, A., Henke, C. and Smith, H. M.(2001). Geographic variation and
    ontogenetic change in the diet of the Mexican Pacific lowlands garter snake,
    Thamnophis validus. *Copeia* **2001**, 1034–42.

DeVault, T. L. and Krochmal, A. R. (2002). Scavenging by snakes: an examination of the literature. *Herpetologica* **58**, 429–36.

Fearn, S.,Robinson, B. Sambono, J. and Shine, R. (2001). Pythons in the pergola: the ecology of "nuisance" carpet pythons (*Morelia spilota*) from suburban habitats in south-eastern Queensland. *Wildl. Res.* **28**, 573–9.

Filippi, E., Capula, M., Luiselli, L. and Agrimi, U. (1996). The prey spectrum of *Natrix natrix* (Linnaeus, 1758) and *Natrix tesselata* (Laurenti, 1768) in sympatric populations. *Herpetozoa* **8**, 155–64.

Fitch, H. S. (1960). Autecology of the copperhead. *Univ. Kans. Pub. Mus. Nat. Hist.* **13**, 85–288.

Fitch, H. S. (1965). An ecological study of the garter snake, *Thamnophis sirtalis*. *Univ. Kans. Pub. Mus. Nat. Hist.* **15**, 493–564.

Fitch, H. S. (1975). A demographic study of the ringneck snake (*Diadophis punctatus*) in Kansas. *Univ. Kans. Pub. Mus. Nat. Hist.* **62**, 1–53.

Fitch, H. S. (1982). Resources of a snake community in prairie-woodland habitat of northeastern Kansas. In *Herpetological Communities*, ed. N. J. Scott, Jr., pp. 83–98. Washington, D. C.: U.S. Department of the Interior, Fish and Wildlife Service.

Fitch, H. S. (1999). *A Kansas Snake Community: Composition and Changes over 50 Years*. Malabar, FL: Krieger Publishing Company.

Fitch, H. S. and Twining, H. (1946). Feeding habits of the Pacific rattlesnake. *Copeia* **1946**, 64–71.

Forsman, A. (1995). Opposing fitness consequences of colour pattern in male and female snakes. *J. Evol. Biol.* **8**, 53–70.

Fry, G. C., Milton, D. A. and Wassenberg, T. J. (2001). The reproductive biology and diet of sea snake bycatch of prawn trawling in northern Australia: characteristics important for assessing the impacts on populations. *Pacif. Conserv. Biol.* **7**, 55–73.

Garrett, C. M. and Smith, B. E. (1994). Perch color preference in juvenile green tree pythons, *Chondropython viridis*. *Zoo Biol.* **13**, 45–50.

Gloyd, H. K. and Conant, R. (1990). *Snakes of the Agkistrodon Complex: A Monographic Review*. Society for the Study of Amphibians and Reptiles.

Goddard, P. (1984). Morphology, growth, food habits, and population characteristics of the smooth snake (*Coronella austriaca*) in southern Britain, UK. *J. Zool. Lond.* **204**, 241–58.

Godley, J. S. (1980). Foraging ecology of the striped swamp snake, *Regina alleni*, in southern Florida, USA. *Ecol. Monogr.* **50**, 411–36.

Godley, J. S., McDiarmid, R. W. and Rojas, N. N. (1984). Estimating prey size and number in crayfish-eating snakes (genus *Regina*). *Herpetologica* **40**, 82–8.

Greene, H. W. (1983). Dietary correlates of the origin and radiation of snakes. *Am. Zool.* **23**, 431–41.

Greene, H. W. (1984). Feeding behaviour and diet of the eastern coral snake, *Micrurus fulvius*. *Univ. Kans. Pub. Mus. Nat. Hist.* **10**, 147–62.

Greene, H. W. (1989a). Defensive behavior and feeding biology of the Asian mock viper, *Psammodynastes pulverulentus* (Colubridae), a specialized predator on scincid lizards. *Chinese Herpetol. Res.* **2**, 21–32.

Greene, H. W. (1989b). Ecological, evolutionary and conservation implications of feeding biology in Old World cat snakes, genus *Boiga* (Colubridae). *Proc. Calif. Acad. Sci.* **46**, 193–207.

Greene, H. W. (1997). *Snakes: The Evolution of Mystery in Nature*. Berkeley, CA: University of California Press.

Greene, H. W. and Jaksic, F. M. (1992). The feeding behavior and natural history of two Chilean snakes, *Philodryas chamissonis and Tachymenis chiliensis* (Colubridae). *Rev. Chil. Hist. Nat.* **65**, 485–93.

Greene, H. W. and Rodriguez-Robles, J. A. (2003). Feeding ecology of the California mountain kingsnake, *Lampropeltis zonata* (Colubridae). *Copeia* **2003**, 308–14.

Greene, B. D., Dixon, J. R., Mueller, J. M., Whiting, M. J. and Thornton, O. W. Jr. (1994). Feeding ecology of the Concho water snake, *Nerodia harteri paucimaculata*. *J. Herpetol.* **28**, 165–72.

Gregory, P. T. (1984). Habitat, diet, and composition of assemblages of garter snakes (*Thamnophis*) at eight sites on Vancouver Island. *Can. J. Zool.* **62**, 2013–22.

Gregory, P. T., Crampton, L. H. and Skebo, K. M. (1999). Conflicts and interactions among reproduction, thermoregulation and feeding in viviparous reptiles: are gravid snakes anorexic? *J. Zool. Lond.* **248**, 231–41.

Grigg, G. C., Drane, C. R. and Courtice, G. P. (1979). Time constants of heating and cooling in the eastern water dragon, *Physignathus leseurii*, and some generalizations about heating and cooling in reptiles. *J. Therm. Biol.* **4**, 95–103.

Hailey, A. and Davies, P. M. C. (1986). Diet and foraging behaviour of *Natrix maura*. *Herpetol. J.* **1**, 53–61.

Harlow, P. and Shine, R. (1992). Food habits and reproductive biology of the Pacific island boas (*Candoia*). *J. Herpetol.* **26**, 60–6.

Heatwole, H. F. (1999) *Sea Snakes*. Sydney, NSW: University of New South Wales Press.

Heatwole, H. and Davison, E. (1976). A review of caudal luring in snakes with notes on its occurrence in the Saharan sand viper, *Cerastes vipera*. *Herpetologica* **32**, 332–6.

Henderson, R. W. (1970). Caudal luring in a juvenile Russell's viper. *Herpetologica* **26**, 276–7.

Henderson, R. W. (1982). Trophic relationships and foraging strategies of some new world tree snakes (*Leptophis, Oxybelis, Uromacer*). *Amph.-Rept.* **3**, 71–80.

Henderson, R. W. (1984a). The diet of the Hispaniolan snake *Hypsirhynchus ferox* (Colubridae). *Amph.-Rept.* **5**, 367–71.

Henderson, R. W. (1984b). The diets of Hispaniolan colubrid snakes. I. Introduction and prey genera. *Oecologia* **62**, 234–9.

Henderson, R. W. (1990). Correlation of environmental variables and dorsal color in *Corallus enydris* (Serpentes: Boidae) on Grenada: some preliminary results. *Carib. J. Sci.* **26**, 166–70.

Henderson, R. W. (1993a). Foraging and diet in West Indian *Corallus enydris* (Serpentes: Boidae). *J. Herpetol.* **27**, 24–8.

Henderson, R. W. (1993b). On the diets of some arboreal boids. *Herpetol. Nat. Hist.* **1**, 91–6.

Henderson, R. W. (2002). *Neotropical Treeboas: Natural History of the* Corallus hortulanus *Complex*. Malabar FL: Krieger Publishing Company.

Henderson, R. W. and Sajdak, R. A. (1996). Diets of West Indian racers (Colubridae: *Alsophis*): composition and biogeographic implications. In *Contributions to West Indian Herpetology: A Tribute to Albert Schwartz*, ed. R. Powell and R. W. Henderson, pp. 327–38. Athens, OH: Society for the Study of Amphibians and Reptiles.

Henderson, R. W. and Sajdak, R. A. (2002). A preliminary look at the spatial distribution of treeboas at a site on Grenada. *Herpetol. Bull.* **81**, 5–7.

Henderson, R. W., Crother, B. I., Noeske-Hallin, T. A., Schwartz, A. and Dethloff, C. R. (1987a). The diet of the Hispaniolan snake *Antillophis parvifrons* (Colubridae). *J. Herpetol.* **21**, 330–4.

Henderson, R. W., Noeske-Hallin, T. A., Ottenwalder, J. A. and Schwartz, A. (1987b). On the diet of the boa *Epicrates striatus* on Hispaniola, Haiti with notes on *Epicrates fordi* and *Epicrates gracilis*. *Amph.-Rept.* **8**, 251–8.

Henderson, R. W. and Schwartz, A. (1986). Diet of *Darlingtonia haetiana*. *Copeia* **1986**, 529–31.

Henderson, R. W., Schwartz, A. and Noeske-Hallin, T. A. (1987c). Food habits of three colubrid tree snakes (genus *Uromacer*) on Hispaniola. *Herpetologica* **43**, 241–8.

Hoge, A. R. and Federsoni, P. A. Jr. (1977). Observations on a brood of *Bothrops atrox* (Linneaux, 1758): (Serpentes: Viperidae: Crotalinae). *Mems. Inst. But. (Sao Paulo)* **40/41**, 19–36.

Holycross, A. T. and Mackessy, S. P. (2002). Variation in the diet of *Sistrurus catenatus* (massasauga), with emphasis on *Sistrurus catenatus edwardsii* (desert massasauga). *J. Herpetol.* **36**, 454–64.

Holycross, A. T., Painter, C. W., Barker, D. G. and Douglas, M. E. (2002a). Foraging ecology of the threatened New Mexico ridge-nosed rattlesnake (*Crotalus willardi obscurus*). In *Biology of the Vipers*, ed. G. W. Schuett, M. Hoggren, M. E. Douglas and H. W. Greene, pp. 243–51. Eagle Mountain, UT: Eagle Mountain Publishing.

Holycross, A. T., Painter, C. W., Prival, D. B. *et al.* (2002b). Diet of *Crotalus lepidus klauberi* (banded rock rattlesnake). *J. Herpetol.* **36**, 589–97.

Houston, D. and Shine, R. (1993). Sexual dimorphism and niche divergence: feeding habits of the Arafura filesnake. *J. Anim. Ecol.* **62**, 737–48.

Jayne, B. C., Voris, H. K. and Kiew Bong, H. (1988). Diet, feeding behavior, growth, and numbers of a population of *Cerberus rynchops* (Serpentes: Homalopsinae) in Malaysia. *Field. Zool.* **50**, 1–15.

Johnston, G. R. (1996). Genetic and seasonal variation in body colour of the Australian death adder, *Acanthophis antarcticus* (Squamata: Elapidae). *J. Zool. Lond.* **239**, 187–96.

Jones, K. B. (1990). Habitat use and predatory behavior of *Thamnophis cyrtopsis* (Serpentes: Colubridae) in a seasonally variable aquatic environment. *Southwest. Nat.* **35**, 115–22.

Keogh, J. S. and Smith, S. A. (1996). Taxonomy and natural history of the Australian bandy-bandy snakes (Elapidae: *Vermicella*) with a description of two new species. *J. Zool. Lond.* **240**, 677–701.

Kephart, D. G. (1982). Microgeographic variation in the diets of garter snakes. *Oecologia* **52**, 287–91.

Kephart, D. G. and Arnold, S. J. (1982). Garter snake diets in a fluctuating environment: a 7-year study. *Ecology* **63**, 1232–6.

King, D. R., Pianka, E. R. and Green, B. (2002). Biology, ecology and evolution. In *Komodo Dragons. Biology and Conservation*, ed. J. B. Murphy, C. Ciofi, C. de la Panouse and T. Walsh, pp. 23–41. Washington, D.C.: Smithsonian Institution Press.

King, R. B. (1988). Polymorphic populations of the garter snake *Thamnophis sirtalis* near Lake Erie. *Herpetologica* **44**, 451–8.

King, R. B. (1989). Sexual dimorphism in snake tail length: sexual selection, natural selection, or morphological constraint? *Biol. J. Linn. Soc.* **38**, 133–54.

King, R. B. (1993). Microgeographic, historical, and size-correlated variation in water snake diet composition. *J. Herpetol.* **27**, 90–4.

King, R. B., Bittner, T. D., Queral-Regil, A. and Cline, J. H. (1999). Sexual dimorphism in neonate and adult snakes. *J. Zool. Lond.* **247**, 19–28.

Klauber, L. M. (1997). *Rattlesnakes: Their Habits, Life Histories, and Influence on Mankind*, 2nd edn. Los Angeles, CA: University of California Press.

Kropach, C. (1975). The yellow-bellied sea snake, *Pelamis* in the eastern Pacific. In *The Biology of Sea Snakes*, ed. W. A. Dunson, pp. 185–213. Baltimore, MD: University Park Press.

Lagesse, L. A. and Ford, N. B. (1996). Ontogenetic variation in the diet of the southern copperhead, *Agkistrodon contortrix*, in northeastern Texas. *Tex. J. Sci.* **48**, 48–54.

Lind, A. J. and Welsh, H. H. Jr. (1994). Ontogenetic changes in foraging behaviour and habitat use by the Oregon garter snake, *Thamnophis atratus hydrophilus*. *Anim. Behav.* **48**, 1261–73.

Losos, J. B. and Greene, H. W. (1988). Ecological and evolutionary implications of diet in monitor lizards. *Biol. J. Linn. Soc.* **35**, 379–407.

Lourdais, O., Bonnet, X. and Doughty, P. (2002). Costs of anorexia during pregnancy in a viviparous snake (*Vipera aspis*). *J. Exp. Zool.* **292**, 487–93.

Luiselli, L. (1996). Food habits of an Alpine population of the sand viper (*Vipera ammodytes*). *J. Herpetol.* **30**, 92–4.

Luiselli, L. and Agrimi, U. (1991). Composition and variation of the diet of *Vipera aspis francisciredi* in relation to age and reproductive stage. *Amph.-Rept.* **12**, 137–44.

Luiselli, L. and Angelici, F. M. (1998). Sexual size dimorphism and natural history traits are correlated with intersexual dietary divergence in royal pythons (*Python regius*) from the rainforests of southeastern Nigeria. *Ital. J. Zool.* **65**, 183–5.

Luiselli, L. and Anibaldi, C. (1991). The diet of the adder (*Vipera berus*) in two different alpine environments. *Amph.-Rept.* **12**, 214–17.

Luiselli, L., Akani, G. C. and Angelici, F. M. (2000a). Arboreal habits and viper biology in the African rainforest: the ecology of *Atheris squamiger*. *Isr. J. Zool.* **46**, 273–86.

Luiselli, L., Akani, G. C. and Barieenee, I. F. (1998). Observations of habitat, reproduction, and feeding of *Boiga blandingi* (Colubridae) in south-eastern Nigeria. *Amph.-Rept.* **19**, 430–6.

Luiselli, L., Akani, G. C., Otonye, L. D., Ekanem, J. S. and Capizzi, D. (1999). Additions to the knowledge of the natural history of *Bothrophthalmus lineatus* (Colubridae) from the Port Harcourt region of Nigeria. *Amph.-Rept.* **20**, 318–26.

Luiselli, L., Angelici, F. M. and Akani, G. C. (2000b). Large elapids and arboreality: the ecology of Jameson's green mamba (*Dendroaspis jamesoni*) in an Afrotropical forested region. *Contrib. Zool.* **69**, 147–55.

Luiselli, L., Angelici, F. M. and Akani, G. C. (2002). Comparative feeding strategies and dietary plasticity of the sympatric cobras *Naja melanoleuca* and *Naja nigricollis* in three diverging Afrotropical habitats. *Can. J. Zool.* **80**, 55–63.

Luiselli, L., Anibaldi, C. and Capula, M. (1995). The diet of juvenile adders, *Vipera berus*, in an alpine habitat. *Amph.-Rept.* **16**, 404–7.

Luiselli, L., Capula, M. and Shine, R. (1996). Reproduction output, costs of reproduction, and ecology of the smooth snake, *Coronella austriaca*, in the eastern Italian Alps. *Oecologia* **106**, 100–10.

Luiselli, L., Capula, M. and Shine, R. (1997). Food habits, growth rates, and reproductive biology of grass snakes, *Natrix natrix* (Colubridae) in the Italian Alps. *J. Zool. Lond.* **241**, 371–80.

Luiselli, L., Pleguezuelos, J. M., Capula, M. and Villafranca, C. (2001). Geographic variation in the diet composition of a secretive Mediterranean colubrid snake, *Coronella girondica*, from Spain and Italy. *Ital. J. Zool.* **68**, 57–60.

Macartney, J. M. (1989). Diet of the northern Pacific rattlesnake (*Crotalus viridis oreganus*) in British Columbia, Canada. *Herpetologica* **45**, 299–304.

Macias Garcia, C. and Drummond, H. (1988). Seasonal and ontogenetic variation in the diet of the Mexican garter snake (*Thamnophis eques*) in Lake Tecocomulco, Hidalgo, Mexico. *J. Herpetol.* **22**, 129–34.

Mackessy, S. P. (1988). Venom ontogeny in the Pacific rattlesnakes *Crotalus viridis helleri* and *Crotalus viridis oreganus*. *Copeia* **1988**, 92–101.

Madsen, T. (1983). Growth rates, maturation, and sexual size dimorphism in a population of grass snakes, *Natrix natrix*, in southern Sweden. *Oikos* **40**, 277–82.

Madsen, T. and Shine, R. (1993a). Costs of reproduction in a population of European adders. *Oecologia* **94**, 488–95.

Madsen, T. and Shine, R. (1993b). Phenotypic plasticity in body sizes and sexual size dimorphism in European grass snakes. *Evolution* **47**, 321–5.

Manjarrez, J. and Garcia, C. M. (1991). Feeding ecology of *Nerodia rhombifera* in a Veracruz swamp. *J. Herpetol.* **25**, 499–502.

Marques, O. A. V. and Sazima, I. (1997). Diet and feeding behavior of the coral snake, *Micrurus corallinus*, from the Atlantic forest of Brazil. *Herpetol. Nat. Hist.* **5**, 88–93.

Martins, M. and Oliveira, M. E. (1999). Natural history of snakes in forests of the Manaus region, Central Amazonia, Brazil. *Herpetol. Nat. Hist.* **2**, 78–150.

Martins, M., Marques, O. A. V. and Sazima, I. (2002). Ecological and phylogenetic correlates of feeding habits in neotropical pitvipers of the genus *Bothrops*. In *Biology of the Vipers*, ed. G. W. Schuett, M. Hoggren, M. E. Douglas and H. W. Greene, pp. 307–28. Eagle Mountain, UT: Eagle Mountain Publishing.

Miller, D. E. and Mushinsky, H. R. (1990). Foraging ecology and prey size in the mangrove water snake, *Nerodia fasciata compressicauda*. *Copeia* **4**, 1099–106.

Mushinsky, H., Hebrard, J. J. and Vodopich, D. S. (1982). Ontogeny of water snake foraging ecology. *Ecology* **63**, 1624–9.

McCoy, C. J. and Censky, E. J. (1992). Biology of the Yucatan hognosed pitviper, *Porthidium yucatanicum*. In *Biology of the Pitvipers*, ed. J. A. Campbell and E. D. Brodie Jr., pp. 217–22. Tyler, TX: Selva.

Mulcahy, D. G., Mendelson III, J. R. and Setser, K. W. (2003). *Crotalus cerastes*: prey/predator weight ratio. *Herpetol. Rev.* **34**, 64.

Neill, W. T. (1960). The caudal lure of various juvenile snakes. *Quart. J. Fl. Acad. Sci.* **23**, 173–200.

Neill, W. T. (1963). Polychromatism in snakes. *Quart. J. Fl. Acad. Sci.* **26**, 194–216.

Nilson, G., Hoggren, M., Tuniyev, B. S., Orlov, N. L. and Andren, C. (1994). Phylogeny of the vipers of the Caucasus (Reptilia, Viperidae). *Zool. Script.* **23**, 353–360.

Nobusaka, R., Nakamoto, E. and Sawai, Y. (1994). Studies on the reproduction of habu (*Trimeresurus flavoviridis*) and coloration of the offsprings. *Snake* **26**, 1–9.

Patchell, F. C. and Shine, R. (1986) Food habits and reproductive biology of the Australian legless lizards (Pygopodidae). *Copeia* **1986**, 30–9.

Pearson, D., Shine, R. and How, R. (2002). Sex-specific niche partitioning and sexual size dimorphism in Australian pythons (*Morelia spilota imbricata*). *Biol. J. Linn. Soc.* **77**, 113–25.

Pitman, C. R. S. (1974). *A Guide to the Snakes of Uganda*, revised edn. Codicote, England: Wheldon and Wesley.

Platt, D. R. (1969). Natural history of the hognose snakes *Heterodon platyrhinos* and *Heterodon nasicus*. *Univ. Kans. Pub. Mus. Nat. Hist.* **18**, 253–420.

Pleguezuelos, J. M., Honrubia, S. and Castillo, S. (1994). Diet of the false smooth snake, *Macroprotodon cucullatus* (Serpentes, Colubridae), in the western Mediterranean area. *Herpetol. J.* **4**, 98–105.

Pleguezuelos, J. M. and Moreno, M. (1990). Nutrition of *Coluber hippocrepis* in the southeastern Iberian peninsula. *Amph.-Rept.* **11**, 325–38.

Plummer, M. V. (1981). Habitat utilization, diet, and movements of a temperate arboreal snake (*Opheodrys aestivus*). *J. Herpetol.* **15**, 425–32.

Plummer, M. V. and Goy, J. M. (1984). Ontogenetic dietary shift of water snake (*Nerodia rhombifera*) in a fish hatchery. *Copeia* **1984**, 550–2.

Pough, F. H. (1977). Ontogenetic change in blood oxygen capacity and maximum activity in garter snakes (*Thamnophis sirtalis*). *J. Comp. Physiol.* **B116**, 337–45.

Pough, F. H. (1980). The advantages of ectothermy for tetrapods. *Am. Nat.* **115**, 92–112.

Pough, F. H. and Groves, J. D. (1983). Specializations of the body form and food habits of snakes. *Am. Zool.* **23**, 443–54.

Prival, D. B., Goode, M. J., Swann, D. E., Schwalbe, C. R. and Schroff, M. J. (2002). Natural history of a northern population of twin-spotted rattlesnakes, *Crotalus pricei*. *J. Herpetol.* **36**, 598–607.

Punzo, F. (1974). Comparative analysis of the feeding habits of two species of Arizona blind snakes, *Leptotyphlops h. humilis* and *Leptotyphlops d. dulcis*. *J. Herpetol.* **8**, 153–6.

Queral-Regil, A. and King, R. B. (1998). Evidence for phenotypic plasticity in snake body size and relative head dimensions in response to amount and size of prey. *Copeia* **1998**, 423–9.

Rasmussen, J. B. (1975). Geographical variation, including an evolutionary trend, in *Psammodynastes pulverulentus* (Boie, 1827) (Boiginae, Homalopsidae, Serpentes). *Viden. Med. Dansk Nat. For.* **138**, 39–64.

Rivas, J. (1999) The life history of the green anaconda (*Eunectes murinus*), with emphasis on its reproductive biology. Ph.D. dissertation, University of Tennessee.

Rodda, G. H. (1992). Foraging behavior of the brown tree snake, *Boiga irregularis*. *Herpetol. J.* **2**, 110–14.

Rodriguez-Robles, J. A. (2002). Feeding ecology of the North American gopher snake (*Pituophis catenifer*, Colubridae). *Biol. J. Linn. Soc.* **77**, 165–83.

Rodriguez-Robles, J. A. and Greene, H. W. (1999). Food habits of the long-nosed snake (*Rhinocheilus lecontei*), a 'specialist' predator? *J. Zool. Lond.* **248**, 489–99.

Rodriguez-Robles, J. A., Bell, C. J. and Greene, H. W. (1999a). Food habits of the glossy snake, *Arizona elegans*, with comparisons to the diet of sympatric long-nosed snakes, *Rhinocheilus lecontei*. *J. Herpetol.* **33**, 87–92.

Rodriguez-Robles, J. A., Bell, C. J. and Greene, H. W. (1999b). Gape size and evolution of diet in snakes: feeding ecology of erycine boas. *J. Zool. Lond.* **248**, 49–58.

Rodriguez-Robles, J. A., Mulcahy, D. G. and Greene, H. W. (1999c). Feeding ecology of the desert nightsnake, *Hypsiglena torquata* (Colubridae). *Copeia* **1999**, 93–100.

Ross, R. A. and Marzec, G. (1990). *The Reproductive Husbandry of Pythons and Boas*. Stanford, CA: Institute for Herpetological Research.

Rowe, J. W., Campbell, K. C. and Gillingham, J. C. (2000). Diet of the ribbon snake on Beaver Island, Michigan: temporal variation and the relationship of prey size to predator size. *Herpetol. Nat. Hist.* **7**, 145–52.

Rugiero, L., Capizzi, D. and Luiselli, L. (1998). Aspects of the ecology of the leopard snake, *Elaphe situla*, in southeastern Italy. *J. Herpetol.* **32**, 626–30.

Rugiero, L. and Luiselli, L. (1996). Ecological notes on an isolated population of the snake *Elaphe quatuorlineata*. *Herpetol. J.* **6**, 53–5.

Saint Girons, H. (1980). Modifications sélectives du régime des vipères (Reptilia: Viperidae) lors de la croissance. *Amph.-Rept.* **1**, 127–36.

Salomao, M. D. G., Santos, S. M. A. and Puorto, G. (1995). Activity pattern of *Crotalus durissus* (Viperidae, Crotalinae): feeding, reproduction and snakebite. *Stud. Neotrop. Fauna Env.* **30**, 101–6.

Santos, X. and Llorente, G. A. (1998). Sexual and size-related differences in the diet of the snake *Natrix maura* from the Ebro Delta, Spain. *Herpetol. J.* **8**, 161–5.

Santos, X., González-Solís, J. and Llorente, G. A. (2000). Variation in the diet of the viperine snake, *Natrix maura*, in relation to prey availability. *Ecography* **23**, 185–92.

Savidge, J. A. (1988). Food habits of *Boiga irregularis*, an introduced predator of Guam, west Pacific. *J. Herpetol.* **22**, 275–82.

Sazima, I. (1991). Caudal luring in two neotropical pitvipers, *Bothrops jararaca* and *B. jararacussu*. *Copeia* **1991**, 245–8.

Sazima, I. (1992). Natural history of the Jararaca pitviper, *Bothrops jararaca*, in southeastern Brazil. In *Biology of the Pitvipers*, ed. J. A. Campbell and E. D. Brodie Jr., pp. 199–216. Tyler, TX: Selva.

Sazima, I. and Puorto, G. (1993). Feeding technique of juvenile *Tropidodryas striaticeps*: probable caudal luring in a colubrid snake. *Copeia* **1993**, 222–6.

Schaller, G. B. (1972) *The Serengeti Lion*. Chicago, IL: Chicago University Press.

Scott, N. J., Maxwell, T. C., Thornton, O. W. Jr., Fitzgerald, L. A. and Flury, J. W. (1989). Distribution, habitat, and future of Harter's water snake, *Nerodia harteri*, in Texas. *J. Herpetol.* **23**, 373–89.

Secor, S. M. (1994). Natural history of the sidewinder, *Crotalus cerastes*. In *Herpetology of the North American Deserts: Proceedings of a Symposium*, ed. P. R. Brown and J. W. Wright, pp. 281–301. Southwestern Herpetologists Society.

Seib, R. L. (1981). Size and shape in a neotropical burrowing colubrid snake, *Geophis nasalis*, and its prey. *Am. Zool.* **21**, 933.

Seib, R. L. (1985). Euryphagia in a tropical snake, *Coniophanes fissidens*. *Biotropica* **17**, 57–64.

Shetty, S. and Shine, R. (2002). Sexual divergence in diets and morphology in Fijian sea snakes *Laticauda colubrina* (Laticaudinae). *Austral Ecol.* **27**, 77–84.

Shine, R. (1977). Habitats, diet, and sympatry in snakes: a study from Australia. *Can. J. Zool.* **55**, 1118–28.

Shine, R. (1980a). Comparative ecology of three Australian snake species of the genus *Cacophis* (Serpentes: Elapidae). *Copeia* **1980**, 831–8.

Shine, R. (1980b). Ecology of the Australian death adder (*Acanthophis antarcticus*: Elapidae): evidence for convergence with the Viperidae. *Herpetologica* **36**, 281–9.

Shine, R. (1980c). Reproduction, feeding, and growth in the Australian burrowing snake *Vermicella annulata. J. Herpetol.* **14**, 71–8.

Shine, R. (1981a). Ecology of Australian elapid snakes of the genera *Furina* and *Glyphodon. J. Herpetol.* **15**, 219–24.

Shine, R. (1981b). Venomous snakes in cold climates: ecology of the Australian genus *Drysdalia* (Serpentes: Elapidae). *Copeia* **1981**, 14–25.

Shine, R. (1982). Ecology of the Australian elapid snake *Echiopsis curta. J. Herpetol.* **16**, 388–93.

Shine, R. (1983). Food habits and reproductive biology of Australian elapid snakes of the genus *Denisonia. J. Herpetol.* **17**, 171–5. .

Shine, R. (1984). Reproductive biology and food habits of the Australian elapid snakes of the genus *Cryptophis. J. Herpetol.* **18**, 33–9.

Shine, R. (1986). Sexual differences in morphology and niche utilization in an aquatic snake, *Acrochordus arafurae. Oecologia* **69**, 260–7.

Shine, R. (1987a). Ecological comparisons of island and mainland populations of Australian tiger snakes (*Notechis*: Elapidae). *Herpetologica* **43**, 233–40.

Shine, R. (1987b). Ecological ramifications of prey size: food habits and reproductive biology of Australian copperhead snakes, *Austrelaps* (Elapidae). *J. Herpetol.* **21**, 21–8.

Shine, R. (1987c). Food habits and reproductive biology of Australian snakes of the genus *Hemiaspis* (Elapidae). *J. Herpetol.* **21**, 71–4.

Shine, R. (1988). Food habits and reproductive biology of small Australian snakes of the genera *Unechis* and *Suta* (Elapidae). *J. Herpetol.* **22**, 307–15.

Shine, R. (1989). Constraints, allometry, and adaptation: food habits and reproductive biology of Australian brownsnakes (*Pseudonaja*: Elapidae). *Herpetologica* **45**, 195–207.

Shine, R. (1991a). *Australian Snakes: A Natural History*. Ithaca, NY: Cornell University Press.

Shine, R. (1991b). Intersexual dietary divergence and the evolution of sexual dimorphism in snakes. *Am. Nat.* **138**, 103–22.

Shine, R. (1991c). Strangers in a strange land: ecology of the Australian colubrid snakes. *Copeia* **1991**, 120–31.

Shine, R. (1991d). Why do larger snakes eat larger prey items? *Funct. Ecol.* **5**, 493–502.

Shine, R. (1993). Sexual dimorphism in snakes. In *Snakes: Ecology and Behavior*, ed. R. A. Seigel and J. T. Collins, pp. 49–86. New York: McGraw-Hill.

Shine, R. (1994). Sexual size dimorphism in snakes revisited. *Copeia* **1994**, 326–46.

Shine, R. and Covacevich, J. (1983). Ecology of highly venomous snakes: the Australian genus *Oxyuranus* (Elapidae). *J. Herpetol.* **17**, 60–9.

Shine, R. and Fitzgerald, M. (1996). Large snakes in a mosaic rural landscape: the ecology of carpet pythons, *Morelia spilota* (Serpentes: Pythonidae), in coastal eastern Australia. *Biol. Conserv.* **76**, 113–22.

Shine, R. and Slip, D. J. (1990). Biological aspects of the adaptive radiation of Australasian pythons (Serpentes: Boidae). *Herpetologica* **46**, 283–90.

Shine, R. and Sun, L. X. (2003). Attack strategy of an ambush predator: which attributes of the prey trigger a pit-viper's strike? *Funct. Ecol.* **17**, 340–8.

Shine, R., Ambariyanto, P., Harlow, S. and Mumpuni. (1998a). Ecological divergence among sympatric colour morphs in blood pythons, *Python brongersmai. Oecologia* **116**, 113–9.

Shine, R., Branch, W. R., Harlow, P. S. and Webb, J. K. (1996a). Sexual dimorphism, reproductive biology, and food habits of two species of African filesnakes (*Mehelya*: Colubridae). *J. Zool. Lond.* **240**, 327–40.

Shine, R., Branch, W. R., Harlow, P. S. and Webb, J. K. (1998b). Reproductive biology and food habits of horned adders, *Bitis caudalis* (Viperidae), from southern Africa. *Copeia* **1998**, 391–401.

Shine, R., Bonnet, X., Elphick, M. and Barrott, E. (2004). A novel foraging mode in snakes: browsing by the sea snake *Emydocephalus annulatus* (Serpentes, Hydrophiidae). *Funct. Ecol.*, **18**, 16–24.

Shine, R., Haagner, G. V., Branch, W. R., Harlow, P. S. and Webb, J. K. (1996b). Natural history of the African shieldnose snake *Aspidelaps scutatus* (Serpentes: Elapidae). *J. Herpetol.* **30**, 361–6.

Shine, R., Harlow, P. S., Branch, W. R. and Webb, J. K. (1996c). Life on the lowest branch: sexual dimorphism, diet, and reproductive biology of an African twig snake, *Thelotornis capensis* (Serpentes, Colubridae). *Copeia* **1996**, 290–9.

Shine, R., Harlow, P. S., Keogh, J. S. and Boeadi. (1998c). The influence of sex and body size on food habits of a giant tropical snake, *Python reticulatus*. *Funct. Ecol.* **12**, 248–58.

Shine, R., Reed, R. N., Shetty, S. and Cogger, H. G. (2002a). Relationships between sexual dimorphism and niche partitioning within a clade of sea-snakes (Laticaudinae). *Oecologia* **133**, 45–53.

Shine, R., Shine, T. and Shine, B. (2003). Intraspecific habitat partitioning by the sea snake *Emydocephalus annulatus* (Serpentes, Hydrophiidae): the effects of sex, body size, and color pattern. *Biol. J. Linn. Soc.* **80**, 1–10.

Shine, R., Sun, L. X., FitzGerald, M. and Kearney, M. (2002b). Accidental altruism in insular pit-vipers (*Gloydius shedaoensis*, Viperidae). *Evol. Ecol.* **16**, 541–8.

Slip, D. J. and Shine, R. (1988). Feeding habits of the diamond python, *Morelia s. spilota*: ambush predation by a boid snake. *J. Herpetol.* **22**, 323–30.

Solorzano, A. and Cerdas, L. (1989). Reproductive biology and distribution of the terciopelo, *Bothrops asper* Garman (Serpentes: Viperidae) in Costa Rica. *Herpetologica* **45**, 444–50.

Solorzano, A., Romero, M., Gutierrez, J. M. and Sasa, M. (1999). Venom composition and diet of the cantil, *Agkistrodon bilineatus howardgloydi* (Serpentes: Viperidae). *Southwest. Nat.* **44**, 478–83.

Stafford, P. J. and Henderson, R. W. (1996). *Kaleidoscopic Tree Boas: The Genus Corallus of Tropical America.* Malabar, FL: Krieger.

Taylor, E. N. (2001). Diet of the Baja California rattlesnake, *Crotalus enyo* (Viperidae). *Copeia* **2001**, 553–5.

Tryon, B. (1985). *Bothrops asper* (terciopelo): reproduction. *Herpetol. Rev.* **16**, 28.

Valdujo, P. H., Nogueira, C. and Martins, M. (2002). Ecology of *Bothrops neuwiedi pauloensis* (Serpentes: Viperidae: Crotalinae) in the Brazilian Cerrado. *J. Herpetol.* **36**, 169–76.

Vitt, L. J. and Vangilder, L. D. (1983). Ecology of a snake community in northeastern Brazil. *Amph.-Rept.* **4**, 273–96.

Voris, H. K., Voris, H. H. and Boo-Liat, L. (1978). The food and feeding behavior of a marine snake, *Enhydrina schistosa* (Hydrophiidae). *Copeia* **1978**, 134–46.

Wallace, R. L. and Diller, L. V. (1990). Feeding ecology of the rattlesnake *Crotalus viridis oreganus* in northern Idaho, USA. *J. Herpetol.* **24**, 246–53.

Webb, J. K. (1996). Ecology and conservation of the threatened broad-headed snake, *Hoplocephalus bungaroides*. Ph.D. dissertation, University of Sydney, Australia.

Webb, J. K. and Shine, R. (1993). Dietary habits of Australian blindsnakes (Typhlopidae). *Copeia* **1993**, 762–70.

Webb, J. K., Shine, R., Branch, W. R. and Harlow, P. S. (2000). Life-history strategies
  in basal snakes: reproduction and dietary habits of the African thread snake
  Leptotyphlops scutifrons (Serpentes: Leptotyphlopidae). *J. Zool. Lond.*
  250, 321–7.
White, M. and Kolb, J. A. (1974). A preliminary study of *Thamnophis* near Sagehagen
  Creek, California. *Copeia* **1974**, 126–36.
Wikelski, M. and Trillmich, F. (1994). Foraging strategies of the Galapagos marine
  iguana (*Amblyrhynchus cristatus*): adapting behavioral rules to ontogenic size
  change. *Behaviour* **128**, 255–79.
Zinner, H. (1985). On behavioral and sexual dimorphism of *Telescopus dhara* Forscal
  1776 (Reptilia: Serpentes, Colubridae). *J. Herpetol. Assoc. Afr.* **31**, 5–6.

7

Herbivory and foraging mode in lizards

ANTHONY HERREL

Department of Biology, University of Antwerp

Introduction

Active foraging and sit-and-wait predation are often considered as two very disparate foraging strategies (Huey and Pianka, 1981). As implied by the name, active foragers actively search for food and, in the process, will cover large distances. Typical sit-and-wait predators on the other hand will wait motionless for prey to pass by, at which they will dart with a sudden burst of movement. These animals are often cryptically colored and do not move around for prolonged periods of time (see Table 7.1). Herbivores, however, need to forage actively for food, but have radiated extensively within a group of lizards seemingly predisposed to a sit-and-wait strategy.

Differences in foraging mode in lizards seem to be roughly associated with a deep split within the lizard phylogeny: the transition from "fleshy-tongued" Iguania to the "scaly-tongued" Scleroglossa. It has been suggested that this split is tightly associated with evolution of the use of the tongue for chemo-receptive purposes either at the level of the scleroglossans (Schwenk, 1993, 2000; Cooper, 1994) or independently in each of the scleroglossan radiations (Reilly and McBrayer, this volume, Chapter 10). The development of a tongue that allowed animals to detect and assess sedentary prey is thought to have allowed lizards to search extensive areas for prey that would otherwise remain undetected. As a consequence, active foragers tend to consume unpredictably distributed and clumped prey (e.g. termites). Sit-and-wait predators, on the other hand, tend to eat more active and larger prey (Pough *et al.*, 2001). Despite the general consensus that most iguanians are indeed sit-and-wait foragers, Perry (1999) totally rejected the notion that Iguania and Autarchoglossa are dichotomous in foraging mode, and did not detect bimodality between or within either clade.

Surprisingly, herbivores, which are active foragers by definition (waiting motionless for plants to pass by is an unviable evolutionary strategy), are

Lizard Ecology: The Evolutionary Consequences of Foraging Mode, ed. S. M. Reilly, L. D. McBrayer and D. B. Miles. Published by Cambridge University Press. © Cambridge University Press 2007.

Table 7.1 *Ecological and morphological characteristics of sit-and-wait predators, active foragers, and herbivores*

Predictions for herbivores clearly have a number of characteristics in common with both sit-and-wait and active foraging lizards. Whereas behaviorally and ecologically herbivores are more like active foragers, morphologically herbivores are expected to be more similar to sit-and-wait predators.

	Sit-and-wait predators	Active foragers	Herbivores
Ecology and behavior			
Prey type	mobile, large, selective, occasional	small, sedentary, clumped, opportunistic, often	sedentary, selective, seasonal
Prey mass/day	low	high	high
Energy requirement	low	high	low
Tongue flicking	no	yes	yes
Escape	camouflage, burst speed	endurance, speed	size/aggression
Home range	small	large	intermediate
Performance			
Endurance	low	high	high
Sprint speed	high	low	low
Bite force	high	low	high
Morphology			
Body size	medium to large	small to medium	large
Body shape	stocky	narrow, elongated	stocky, large body volume
Head shape	large, broad	elongate, narrow	large, broad, short
Tooth structure	robust	narrow, elongate, curved	broad, cusped, resistant to wear
Tongue structure	fleshy	narrow, bifurcated	fleshy, bifurcated
Gut structure	plastic, rapidly upregulated, long	continuous function, short	continuous function, long, partitioned colon

Table modified after Huey and Pianka (1981); Pough *et al.* (2001).

predominantly found among the Iguania. Indeed, roughly 60% of all lizards considered herbivores *sensu stricto* (*s.s.*) are iguanians (Pough, 1973; King, 1996; Pough *et al.*, 2001). This creates an interesting paradox: the radiation within a, by definition, actively foraging niche (herbivory) by members of the clade that seems most constrained to sit-and-wait foraging. However, in a recent paper, Cooper and Vitt (2002) suggested that in contrast to herbivory *s.s.* the number of origins of omnivory (i.e. partial consumption of plant matter) might in fact be more common among scleroglossans and associated with their ability to assess potential food items by using chemoreception. In this chapter I examine in more detail which characters are traditionally considered to be associated with the two foraging strategies, and compare them with the traits typically associated with a herbivorous life-style. Characters analyzed were chosen because clear predictions could be made regarding their polarity for either foraging mode or dietary groups. I chose to analyze ecological, performance and morphological traits to evaluate the differences between dietary groups and foraging mode on all different levels. Through my examination of characters associated with foraging mode and/or dietary groups, I try to evaluate why so many herbivores *s.s.* have arisen within the Iguania and whether, and how, herbivores transcend the evolutionary impasse of both sides of the foraging paradigm.

Methods

Terminology

In this chapter I use the term omnivore as an indicator of any lizard incorporating plant matter into its diet on a regular basis. I use the term herbivore *s.s.* for those lizards eating only, or predominantly (e.g. over 80% of the diet) plant matter. The term herbivore *s.l.* (*sensu lato*) will be used to define any lizard including plant matter into its diet (i.e. omnivores + herbivores *s.s.*)

Bite forces and head shape Bite force and head shape data were collected for a wide range of lizards including representatives from most major radiations (Table 7.2, Fig. 7.1). Most animals were measured in the field and released at the exact site of capture immediately after the experiments. Data for *Varanus exanthematicus* and *Uromastix acanthinurus* were obtained for captive animals housed at the Antwerp Zoo. Data for *Xenosaurus grandis* were taken from Herrel *et al.* (2001a).

Bite forces were measured by using a Kistler piezo-electric force transducer with portable amplifier as described in Herrel *et al.*, 1999a, 2001a,c. Head dimensions were determined immediately after the bite force trials by using a pair of digital Mitutoyo calipers. Head length was determined from the back of the parietal bone to the tip of the premaxilla; head width was measured at the widest point of the head, just posterior to the orbits, and includes the potential bulging of jaw muscles; head height was measured at highest part of the head, at the level of the frontoparietal suture, and includes the height of the lower jaw and associated muscles; lower jaw length was determined from the back of the retro-articular process to the tip of the lower jaw.

Table 7.2 *Head shape, bite forces, and ecology in lizards*

Data are means (in millimetres, with the exception of bite force, which is expressed in newtons) from adult male individuals. Classification of species into dietary and foraging categories is based on literature data. The number in parentheses after a species name indicates the sample size.

Species	SVL	Head length	Head width	Head height	Bite force (N)	Diet	Foraging mode
Dipsosaurus dorsalis (10)	125.74 ± 5.7	22.38 ± 0.6	19.15 ± 1.31	14.17 ± 0.87	18.99 ± 2.21	herbivore	active forager
Callisaurus draconoides (12)	83.04 ± 8.18	15.66 ± 0.93	12.59 ± 0.84	8.80 ± 0.48	4.79 ± 1.46	insectivore	sit-and-wait
Sceloporus undulatus (16)	59.64 ± 4.35	12.80 ± 1.11	10.70 ± 1.01	7.17 ± 0.73	3.57 ± 2.54	insectivore	sit-and-wait
Pogona vitticeps (8)	217.76 ± 28.75	38.1 ± 3.00	48.37 ± 6.27	29.06 ± 3.71	112.39 ± 18.32	omnivore	sit-and-wait
Plocederma stellio (4)	95.79 ± 7.28	23.64 ± 0.79	23.42 ± 0.84	12.93 ± 0.75	22.23 ± 5.01	insectivore	sit-and-wait
Uromastix acanthinurus (4)	159.46 ± 26.11	31.79 ± 4.91	32.69 ± 4.58	18.81 ± 2.56	68.39 ± 23.74	herbivore	active forager
Cordylus mossambicus (5)	109.03 ± 3.17	30.29 ± 0.56	24.93 ± 0.34	13.92 ± 0.45	19.31 ± 0.88	insectivore	sit-and-wait
Eulamprus heathwoli (3)	109.17 ± 10.70	21.37 ± 0.43	13.70 ± 0.83	11.24 ± 0.23	16.66 ± 1.70	insectivore	active forager
Corucia zebrata (2)	275.00 ± 7.07	50.60 ± 4.53	48.88 ± 2.37	36.03 ± 3.01	206.85 ± 69.15	herbivore	active forager
Tiliqua rugosa (13)	282.44 ± 17.07	45.90 ± 3.08	51.64 ± 3.80	33.21 ± 2.54	161.64 ± 30.11	omnivore	active forager
Gallotia galloti (15)	112.03 ± 10.24	30.22 ± 3.19	19.90 ± 2.14	16.31 ± 2.70	107.63 ± 32.24	herbivore	active forager
Psammodromus algirus (9)	74.84 ± 3.33	18.34 ± 0.78	11.61 ± 0.72	8.77 ± 0.54	11.52 ± 1.65	insectivore	active forager
Podarcis lilfordi (8)	62.08 ± 2.57	16.44 ± 1.34	9.63 ± 0.84	7.71 ± 0.83	10.71 ± 0.81	omnivore	active forager
Cnemidophorus tigris (5)	77.26 ± 9.89	18.97 ± 2.07	10.28 ± 1.75	8.87 ± 1.15	6.69 ± 4.38	insectivore	active forager
Xenosaurus grandis (9)	110.81 ± 7.97	26.55 ± 2.58	21.73 ± 1.80	13.20 ± 0.64	19.91 ± 5.16	insectivore	sit-and-wait
Varanus exanthematicus (7)	237.90 ± 24.38	48.01 ± 4.65	33.58 ± 3.07	23.99 ± 2.94	86.61 ± 21.00	insectivore	active forager

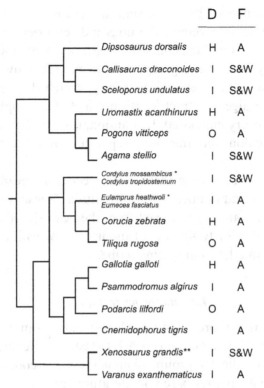

	D	F
Dipsosaurus dorsalis	H	A
Callisaurus draconoides	I	S&W
Sceloporus undulatus	I	S&W
Uromastix acanthinurus	H	A
Pogona vitticeps	O	A
Agama stellio	I	S&W
Cordylus mossambicus * *Cordylus tropidosternum*	I	S&W
Eulamprus heathwoli * *Eumeces fasciatus*	I	A
Corucia zebrata	H	A
Tiliqua rugosa	O	A
Gallotia galloti	H	A
Psammodromus algirus	I	A
Podarcis lilfordi	O	A
Cnemidophorus tigris	I	A
*Xenosaurus grandis***	I	S&W
Varanus exanthematicus	I	A

Figure 7.1. Phylogeny depicting the relationships between the species in the analysis. Interfamilial relationships based on Pough *et al.* (2001). Phrynosomatid relationships based on Reeder and Wiens (1996); scincid relationships based on Hutchinson (1980); lacertid relationships based on Arnold (1989). *Xenosaurus grandis* was classified as a sit-and-wait predator based on data in Ballinger *et al.* (1995); *Cordylus tropidosternum* and *C. mossambicus* based on data in Cooper *et al.* (1997). A, active forager; D, dietary classification; F, foraging mode; H, herbivore; I, insectivore; O, omnivore; S&W, sit-and-wait forager. * Species for which only bite force and head shape data could be obtained. For intestinal tract length data the closely related species below was used (see methods). ** Only included in the bite force and head shape data set as no preserved specimens could be obtained for this species or any close relatives.

Intestinal tract and teeth

Preserved adult male individuals of the species included in the bite force and head shape dataset were dissected to extract the digestive tract. As no preserved individuals of the species *Cordylus mossambicus* or *Eulamprus heathwoli* were available, I used two closely related species with similar dietary preferences and foraging mode (*Eumeces fasciatus* and *Cordylus tropidosternum*). First, snout–vent length and head dimensions were determined for each animal.

Animals were then opened by a mid-ventral incision and the entire digestive tract, lungs, and heart were removed. Lungs and heart were carefully removed from the digestive tract and the connective tissue associated with the digestive tract was cut away. Next, the digestive tract from each individual was pinned at its anterior end, straightened, and gently stretched to a fully extended position. The total digestive tract length not including esophagus or rectum was measured for every individual. In addition, the length of the stomach, the intestine and the colon were measured separately by using Mitutoyo digital calipers.

To inspect tooth structure in these same species, the heads were removed from the specimens and cleaned. To do so, muscle, sensory organs and connective tissue were manually removed from the head skeleton. Pictures were taken from the posterior tooth row of the upper jaw with a Nikon Coolpix camera mounted on a Leica dissecting scope.

Locomotor performance

I decided to re-analyze previously published data from Van Damme and Vanhooydonck (2001) and Garland (1999) to test for differences in locomotor performance among dietary groups. Animals were re-coded as being either herbivores *s.l.* or insectivores based on literature data.

Statistical analyses As species are not independent data points, but related through their evolutionary history, I analyzed the data taking into account the phylogenetic relationships between species. I used the same phylogenetic tree as given in Van Damme & Vanhooydonck (2001) and Garland (1999) to analyze the locomotor performance data. To analyze the bite force, head shape, and intestinal tract data, I compiled a tree based on published data (see caption of Figure 7.1 for sources). All data were $\log_{10}$ transformed before analysis, and species means were entered as tip-node values. All analyses were run with the PDAP package (Garland *et al.*, 1999). Branch lengths were set to unity as no data on actual branch lengths could be obtained for all the species in the analysis. I selected Brownian motion as the model of evolutionary change. A thousand simulations were run and analyzed with the PDSIMUL and PDANOVA routines in the PDAP package. F values obtained from the simulation analysis (F_{phyl}) were compared with the F values obtained by doing regular ANCOVAs (F_{95}) in SPSS (v.10). If the F_{95} values were larger than the F_{phyl} values, I considered the results significant. Exact p values were determined by comparing the F_{95} values with the F-table generated by the PDANOVA routine.

Results

Ecology and behavior

Prey type and size (see also Chapter 5)

The original predictions by Huey and Pianka (1981) were that sit-and-wait predators are expected to eat more mobile prey, and that widely foraging lizards should catch more sedentary and unpredictable prey. Additionally, Huey and Pianka (1981) made the argument that active foragers often eat clumped prey such as termites, or large prey such as scorpions (e.g. *Nucras tesselata*). In some cases large prey including small mammals are also captured by actively foraging lizards (i.e. varanids). Later, other authors (see summary table in Pough *et al.*, 2001, p. 443) have added an additional component to this scheme that seems contradictory to the original predictions: sit-and-wait predators should also consume, on average, larger prey (see also Andrews, 1979). However, this seems a reasonable assumption as animals that tend to eat infrequently should take the largest possible prey available. This is most exemplary in snakes such as vipers that are extreme sit-and-wait predators and eat large, bulky prey (Cundall and Greene, 2000). Unfortunately, to my knowledge, no quantitative analyses of prey size have been performed to test these hypotheses.

Plants clearly cannot be considered a mobile or active food source. Rather, plants are sedentary with a clumped distribution and often of unpredictable availability (e.g. perennial plants or plant parts such as flowers and fruits). These characteristics are generally in accordance with those of food types eaten by active foragers. Yet most plants are large food items that need to be reduced in size before they can be consumed. Although some herbivores will swallow rather big pieces of plants (e.g. *Corucia zebrata*: pers. obs.), most herbivorous lizards effectively crop pieces from plants, thus reducing plant matter to bite-sized pieces (Fig. 7.2). Whereas herbivores typically eat a variety of plants and plant parts (see, for example, Dubuis *et al.*, 1971; Sylber, 1988; Valido & Nogales, 1994), most are selective and select those plants that allow them to obtain a balanced diet (Dearing & Schall, 1992; Rocha, 2000). Omnivores eat plants in variable proportions that may change seasonally, or according to local environmental conditions (Sadek, 1981; Dubas and Bull, 1991; Valido and Nogales, 1994).

Prey detection (see also Chapter 8)

A clear and strong dichotomy has been demonstrated with respect to foraging mode and prey detection capacity in lizards. Whereas active foragers are capable of detecting prey, and even assessing prey quality (i.e. nutrient composition) by using chemical cues, sit-and-wait predators typically are not

Figure 7.2. Stomach contents of an iguanian herbivore (*D. dorsalis*) obtained after stomach flushing. Note how the leaves are often cut into smaller parts. Note also how the edges of the cut are clean, as if made by a pair of scissors (arrow). Scale bar, 5 mm.

(Cooper, 1994; Cooper *et al.*, 2002). Although there is limited evidence suggesting that some sit-and-wait foragers are capable of recognizing immobile items as potential food by using gustation rather than chemoreception (Herrel *et al.*, 1998b), most sit-and-wait foragers likely rely on visual cues to detect prey (Cooper, 1994; Cooper *et al.*, 2002).

Herbivores, even when belonging to otherwise strictly sit-and-wait foraging clades, are always capable of detecting food items by using chemical cues (Cooper, 1994; Cooper and Vitt, 2002). Because of the tight association between chemosensory prey detection and herbivory, Cooper and Vitt (2002) proposed that the evolution of omnivory would be facilitated in clades of actively foraging lizards. Yet, whereas many actively foraging lizards have made the switch to omnivory (see Cooper and Vitt, 2002), only few have become completely herbivorous (King, 1996).

Energy requirements (see also Chapter 4)

As sit-and-wait predators spend most of their foraging time waiting motionless for prey to pass by, they will obviously be spending less energy than actively foraging lizards (e.g. roughly one third less in lacertid lizards: Huey and Pianka, 1981). Despite their need to forage actively, herbivores are also thought to spend little energy during foraging. The rationale behind this argument is simple: as plants do not escape from a potential forager, the forager will not need to pursue its food (at high speed) and will thus utilize less energy (Pough, 1973). Body-mass-specific energy requirements are low, as herbivores are generally less active and spend considerably less time foraging and thermoregulating (Wilson and Lee, 1974; Gleeson, 1979). Yet, if the cost of transport is similar in herbivores and non-herbivores (which it presumably is; see Gleeson 1979), and if herbivores need to amass more food per unit body mass (as presumably less energy can be extracted per unit mass from plant material; see King, 1996), then the cost of foraging might still be higher in herbivorous lizards. Unfortunately, not enough comparative data on the energetic cost of foraging in lizards are available to test the above mentioned hypothesis.

Home ranges

Despite the results of previous studies (Stamps, 1977, 1983), a recent analysis of home range sizes indicated no differences associated with foraging mode (Perry and Garland, 2002). Rather, home range size seems to scale directly to energetic requirements (Perry and Garland, 2002). Not unexpectedly, home range size did differ significantly between herbivores, omnivores, and carnivores, with omnivores having the smallest, herbivores somewhat bigger, and carnivores having the largest home range sizes.

Performance

Locomotion (see Chapters 2 and 3)

Although it is generally assumed that representatives of the two foraging modes differ in sprint speed and endurance capacity (with active foragers having lower speeds but greater stamina; see Garland, 1994; Miles *et al.*, this volume, Chapter 2), this has rarely been tested. However, a recent comparative analysis of sprint speed data by Van Damme and Vanhooydonck (2001) was unable to show differences in sprint speed between active foragers and sit-and-wait predators. Endurance capacity, on the other hand, was shown to be significantly related to quantitative measures of lizard movement patterns under field conditions (Garland, 1999). Consequently, actively foraging lizards are expected to have greater endurance than typical sit-and-wait predators.

To my knowledge, no study has explicitly investigated whether differences in locomotor performance exist between herbivores *s.l.* and non-herbivores. Thus, I decided to re-analyze previously published data from Van Damme and Vanhooydonck (2001) and Garland (1999) to test for differences in locomotor performance. The results from these analyses indicate that there are no significant differences in sprint speed (phylogenetic ANCOVA: $F_{1,97} = 2.57$; $p = 0.5$) or endurance capacity (phylogenetic ANCOVA: $F_{1,22} = 0.02$; $p = 0.96$) when comparing herbivores *s.l.* with other lizards. Although herbivores have to forage actively, they do not show greater endurance capacity than non-herbivores. Moreover, herbivores also show no reduced sprint performance capacity despite the fact that they do not need to pursue prey at high speed. This is all the more surprising given that herbivores are also thought to rely predominantly on their larger size and aggression (for example, *Uromastix* lizards will readily use their spiny tails when confronted with predators) as anti-predator strategies (see Table 7.1).

Bite forces (see also Chapter 9)

Although no explicit predictions exist for differences in bite forces among representatives of the two foraging modes, bite forces can be expected to be higher in sit-and-wait foragers. Not only would high bite forces allow these animals to crush and kill the relatively large prey they feed on (Pough *et al.*, 2001); in addition, high bite forces would ensure that a broad spectrum of potential prey with varying exoskeleton thickness can be handled efficiently. Indeed, high bite forces allow for a faster processing of hard-shelled prey, thus considerably reducing handling time and saving energy (Pough *et al.*, 1997; Verwaijen *et al.*, 2002). However, an analysis of bite forces among a range of lizard taxa (see Figs. 7.1, 7.3; Tables 7.2, 7.3) indicates that sit-and wait predators do not bite harder than closely related active foragers (irrespective of dietary preference). This result holds when taking into account the evolutionary relationships among species. Thus, it appears that, at least for the species included in the present analysis, sit-and-wait predators do not bite harder than active foragers.

As most plant parts are typically tough and fibrous (with the exception of fruits, buds and young shoots), herbivores are expected to have high bite forces. This would allow them to efficiently crop smaller, bite-size pieces from larger plants (Herrel *et al.*, 1998a). The ingestion of small pieces of plant matter is considered important as this allows for a more efficient digestion (Bjorndal *et al.*, 1990; Bjorndal & Bolten, 1992). A comparison of *in vivo* bite forces among the lizard species in our analysis (Fig. 7.3, Table 7.2) indicates that lizards including plant matter into their diet (herbivores *s.l.*) do indeed bite harder than insectivores (Table 7.3). Even when taking into account the evolutionary relationships among species, herbivores *s.l.* still bite significantly harder than insectivores (Fig. 7.3, Table 7.3). However, no differences could be demonstrated between herbivorous *s.s.* and omnivorous lizards (see Table 7.2), indicating that the increase in bite force is likely associated with the initial transition from insectivory to omnivory.

Table 7.3 *Results of phylogenetic analyses of covariance testing for differences in head shape and bite force*

Variable	F_{phyl}	F_{95}	p_{95}	p_{phyl}
Test for differences between active and sit-and-wait foragers				
Head length	6.44	0.02	0.90	0.95
Head width	6.27	11.16	0.005	0.009
Head height	6.07	0.72	0.41	0.53
Bite force	6.80	1.14	0.30	0.43
SVL[a]	7.73	0.84	0.38	0.59
Test for differences between herbivores s.l. and other lizards				
Head length	3.86	0.09	0.77	0.75
Head width	3.45	0.13	0.73	0.70
Head height	3.65	3.85	0.07	0.045
Bite force	3.75	5.39	0.04	0.023
SVL[a]	3.67	4.00	0.07	0.041

[a] Tests for differences in snout–vent length (SVL) are phylogenetic analyses of variance. All other tests are phylogenetic analyses of covariance with SVL as covariate.

Feeding behavior (see also Chapters 9 and 10)

In a recent paper, McBrayer and Reilly (2002) demonstrated that prey processing behavior differs significantly between iguanian and autarchoglossan lizards. Moreover, they showed how certain aspects of prey handling appear to covary with foraging mode in lizards (see Chapter 10). Whereas active foragers use more puncture crushing, sit-and-wait predators used fewer processing cycles on average. Given that sit-and-wait foragers are expected to eat large prey this is rather surprising, especially as bite forces do not differ between members of both foraging modes (see above). Prey capture in sit-and-wait predators also defies the prediction that these animals will eat larger prey on average. Given that adhesive forces are proportional to the contact area between tongue and prey, jaw capture would be expected for animals eating large prey. This is indeed counterintuitive, as most iguanians use their tongue to capture prey. Even more remarkable is the fact that the only scleroglossans known to use their tongues during prey capture have secondarily evolved a sit-and-wait foraging strategy (e.g. *Zonosaurus laticaudatus*: Urbani and Bels, 1995; *Gerrhosaurus major*: McBrayer and Reilly, 2002; *Eumeces schneideri*: McBrayer and Reilly, 2002; Chapter 10).

As mentioned earlier, herbivores need to reduce plant matter to optimize digestive efficiency. Interestingly, the one herbivore *s.s.* (*Ctenosaurus*

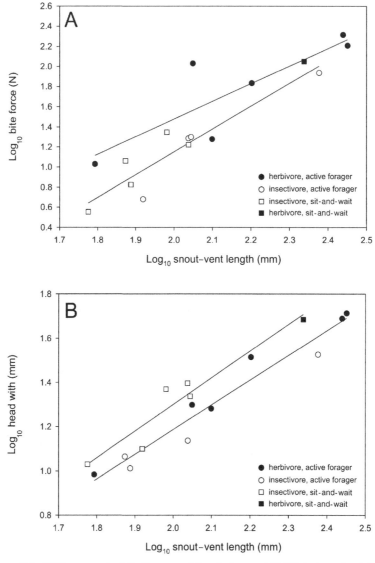

Figure 7.3. Differences in bite force and head shape. (A) Note how herbivores (filled symbols) bite significantly harder than non-herbivores (open symbols) for a given body size. However, no differences are apparent among sit-and-wait (squares) or active foragers (circles). (B) Sit-and-wait predators have significantly wider heads than active foragers for a given snout–vent length. No differences in head width are present among dietary categories. Symbols as in (A).

quinquecarinata) included in the dataset by McBrayer and Reilly (2002) uses the most puncture crushing of all iguanids studied. Personal observations of feeding behavior in other herbivores such as *Uromastix acanthinurus* and *Corucia zebrata* indicate that these, too, chew plant matter extensively before swallowing. Although it has been previously suggested that cranial kinesis would be beneficial for herbivorous lizards (see, for example, King, 1996), data based on cineradiographic recordings of animals during feeding do not support this (Throckmorton, 1976; Herrel *et al.*, 1998a). Given that omnivores eat a wide variety of prey differing in size and other physical attributes (e.g. hardness), I expect extensive modulation capacity. Moreover, well-developed feedback pathways are also likely to be important for omnivorous lizards (Herrel *et al.*, 2001b). The evolution towards 'true' herbivory, on the other hand, might be associated with the loss of functional versatility and flexibility of the feeding system (i.e. a lack of modulatory capacity when confronted with different prey types; see Ralston and Wainwright, 1997) as indicated by data for *Uromastix acanthinurus* (Herrel and De Vree, 2000; pers. obs.). However, it should be noted that this is definitely not the case for all herbivores: *Corucia zebrata* shows no decrease in its modulatory capacity associated with its herbivorous diet (Herrel *et al.*, 1999b).

Morphology

Body size and shape (see also Chapter 2)

Sit-and wait predators are expected to be rather stocky when compared with active predators, as they move less and rely mostly on crypsis as an antipredatory strategy (Huey and Pianka, 1981; Vitt and Congdon, 1978). Aditionally, a large body and stomach volume is presumably beneficial when eating occasional large prey. Active foragers, on the other hand, are expected to have slimmer and more streamlined bodies. As these animals move about a lot and need to be agile while searching for food in a variety of microhabitats, a lower body mass and smaller overall body size are likely to be beneficial. An analysis of snout–vent length (SVL) data (see Tables 7.3 and 7.5) indicates no systematic differences in body size between active foragers and sit-and-wait predators. However, rather than snout–vent length *per se*, body width and relative body mass (relative to SVL) would be expected to differ between foraging modes.

A classic paper by Pough (1973) predicts that larger-bodied lizards should be herbivorous as they are presumably unable to meet their (absolutely higher) caloric demands based on a diet of insects alone. Herbivory is thus considered a neccesary byproduct of large body size in the absence of distinct morphological, physiological, or ecological specializations (Pough, 1973). On the other hand, larger body size presumably also facilitates herbivory, as a large body allows for a large gastrointestinal volume. This is probably important, as herbivores often consume large amounts of plant matter of potentially low nutritional quality in one sitting. For the species included in our analysis

(Table 7.2), differences in snout–vent length between herbivores and insecti-vores are non-significant. However, when taking into account the evolutionary relationships among species, herbivores *s.l.* are indeed significantly larger than insectivores (Table 7.3). This implies that the evolution towards an omnivo-rous or herbivorous diet in lizards has occurred in tandem with an increase in body size. Although it should be noted that the list of species included in this analysis is by no means exhaustive, these results confirm earlier observations by Van Damme (1999) for lacertid lizards.

Head size and shape (see also Chapter 9)

One of the interesting predictions of the paradigm is that representatives of the two foraging modes will differ in head shape. As sit-and-wait predators are thought to eat large prey, the prediction is that they should have wide mouths, large gapes, and wide passages to the esophagus, and thus essentially be big-headed. Active foragers, on the other hand, are expected to have rather slender, light, and elongate heads to allow them to move about for prolonged periods of time. An analysis of the external head shape data in Table 7.2 indicates that sit-and-wait predators and active foragers do indeed have differ-ently shaped heads (MANCOVA with SVL as covariate, Wilk's $\lambda = 0.251$; $F_{4,10} = 7.468$; $p = 0.005$). As indicated by subsequent univariate F-tests, this difference is mainly due to differences in relative head width (Fig. 7.3, Table 7.3). When taking into account the evolutionary relationships among the different species in the analysis, differences in relative head width are still highly significant (Table 7.3).

Lizards that include plant matter in their diet can also expected to have differently shaped heads, given that they tend to bite harder than insectivorous lizards (see above) and that head shape has been shown to be correlated with bite force in lizards (Herrel *et al.*, 1999b, 2001a, c). An analysis of the dataset in Table 7.2 indicates that differences in head shape do indeed exist (MANCOVA with SVL as covariate, Wilk's $\lambda = 0.374$; $F_{4,10} = 4.18$; $p = 0.03$) between herbi-vores and other lizards. Subsequent univariate F-tests show that this difference is largely due to differences in head height, with herbivores having taller heads than insectivores (Table 7.3). Phylogenetic analysis indicate that the evolution of herbivory is indeed associated with an evolutionary increase in head height independent of the group to which lizards belong (Table 7.3).

Tooth shape

Although no clear predictions exist with respect to tooth structure, one might expect sit-and-wait predators to have more robust teeth that would allow them to crush large or thick-shelled prey. Active foragers, on the other hand, are expected to be more agile and would be expected to have light skulls with slender, pointed teeth for puncturing smaller prey. However, even a cursory overview of literature data indicates that this dichotomous view is overly simplistic. Rather, more specific functional adaptations of dentition to diet are observed, irrespective of foraging mode. For example, whereas ant specialists have numerous, short, peg-like teeth (Hotton, 1955; Montanucci, 1989), durophagous specialists typically have blunt, rounded teeth (Dalrymple, 1979; Rieppel, 1979; Rieppel and Labhardt, 1979).

The plant matter that creates most problems for herbivorous animals consists of leaves, stems and grasses, as these are composed of both tough and fibrous materials (Lucas and Luke, 1984). Tough materials need a lot of mechanical work before fracture occurs as they are generally ductile and resist crack propagation. Similarly, fibrous materials are notch-insensitive (Jeronimidis, 1991) and cracks are readily absorbed at the matrix–fiber interface (Sibbing, 1991). To fracture such materials, tooth penetration is essential. To increase local stresses, a set of carefully aligned opposing blades that pass close to each other (shearing) is presumably optimal (Lucas, 1979, 1982; Lucas and Luke, 1984; Sibbing, 1991). The teeth of herbivorous and even omnivorous lizards examined here (Fig. 7.4) have some of the above characteristics in common: they are mediolaterally flattened, generally robust, and appear blade-like (see also Hotton, 1955; Montanucci, 1968). However, in contrast to data gathered by Hotton (1955) and Montanucci (1968), non-iguanid herbivores do not always show increased degree of cuspation of the teeth (Fig. 7.4). Acrodont herbivores differ from this general pattern: species like *Uromastix* have highly specialized but less mediolaterally flattened teeth. Still, the tooth structure as observed in *Uromastix* is highly suited for cropping and cutting up plant matter (Cooper and Poole, 1973; Robinson, 1976; Throckmorton, 1979). Although not related to either group of foraging strategies, the acrodont tooth type appears mammalian-like in some aspects (Cooper *et al.*, 1970) and seems generally predisposed for herbivory owing to its flattened and robust nature. Interestingly, natural history observations indicate that several agamid lizards, although considered typical insectivores, will include plant matter in their diets in times of food scarcity (Greer, 1989; Disi *et al.*, 2001; Spawls *et al.*, 2002) suggesting that the tooth structure observed in acrodonts might indeed facilitate the switch to omnivory.

Digestive tract length and morphology

I was unable to find clear predictions as to what degree differences in digestive tract length and morphology should be expected between actively foraging lizards and sit-and-wait predators. However, if sit-and-wait predators do indeed

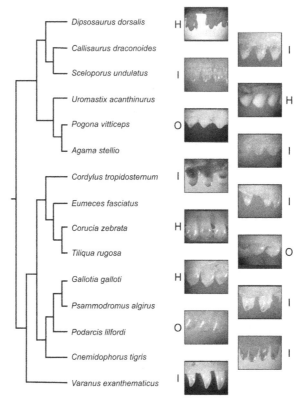

Figure 7.4. Phylogeny depicting the relationships between the species in the analysis. Photographs to the right of the phylogeny show a lateral view of the posterior teeth of the upper jaw for each species (not to scale). Note that herbivores have generally flattened and blade-like teeth. However, only the inguanid herbivore *Dipsosaurus dorsalis* has well-developed excentric cusps on its teeth.

consume larger prey on average, one could predict that these animals would have larger stomachs to allow the digestion of large prey. Additionally, as sit-and wait predators are expected to eat infrequently, the physiologically active tissues should be up- and down-regulated depending on the presence of food (Piersma and Lindstrom, 1997). Many snakes use this strategy; some, such as pythons, are able to increase the size of the active tissues of the small intestine to up to three times the fasting level (Starck and Beese, 2001). However, an analysis of differences in total digestive tract length as well as the length of specific compartments within the digestive tract did not indicate any differences between the active foragers and sit-and-wait predators included in our dataset (phylogenetic ANCOVA, all $p > 0.05$; Table 7.4). This is not completely unexpected, as one might expect differences to be situated in the volume rather than the length of these elements. It would be interesting to test for differences in volume by

Table 7.4 *Intestinal tract and snout–vent length in representative lizard species*

Data represent means (in millimetres) from adult male individuals. Classification of species into dietary and foraging categories is based on literature data. Total intestinal tract length does not include the esophagus or the rectum as these parts could not always be removed intact in all specimens. The number in parentheses after a species name indicates the sample size.

Species	SVL	Stomach	Intestine	Colon	Total length	Diet	Foraging mode
Dipsosaurus dorsalis (1)	110.25	23.15	105.08	19.18	147.41	herbivore	active forager
Callisaurus draconoides (3)	81.59 ± 7.55	27.98 ± 4.36	65.65 ± 10.52	9.98 ± 1.33	103.60 ± 14.86	insectivore	sit-and-wait
Sceloporus undulatus (3)	61.08 ± 3.59	18.89 ± 2.44	45.23 ± 3.50	10.14 ± 3.76	74.26 ± 5.41	insectivore	sit-and-wait
Pogona vitticeps (1)	173.06	49.14	163.36	39.73	252.23	omnivore	sit-and-wait
Plocederma stellio (2)	104.15 ± 0.05	37.46 ± 0.45	98.28 ± 22.02	21.34 ± 3.27	157.08 ± 18.30	insectivore	sit-and-wait
Uromastix acanthinurus (3)	171.81 ± 4.11	61.74 ± 5.27	126.64 ± 12.96	50.66 ± 2.93	239.04 ± 18.90	herbivore	active forager
Cordylus tropidosternon (2)	87.50 ± 0.60	26.58 ± 1.35	45.52 ± 7.8	15.35 ± 2.89	87.44 ± 9.34	insectivore	sit-and-wait
Eumeces fasciatus (3)	71.35 ± 1.06	26.96 ± 0.78	42.62 ± 5.20	12.63 ± 0.63	82.21 ± 5.41	insectivore	active forager
Corucia zebrata (2)	240.00 ± 0.50	73.41 ± 1.69	164.72 ± 14.72	72.50 ± 3.08	310.62 ± 16.11	herbivore	active forager
Tiliqua rugosa (2)	244.00 ± 29.70	72.66 ± 12.54	214.76 ± 4.99	56.32 ± 2.58	343.73 ± 4.98	omnivore	active forager
Gallotia galloti (4)	116.88 ± 6.79	33.64 ± 2.35	103.03 ± 19.84	28.25 ± 3.37	164.92 ± 17.81	herbivore	active forager
Psammodromus algirus (4)	68.77 ± 3.17	20.53 ± 5.00	40.91 ± 11.64	13.65 ± 4.28	75.09 ± 18.78	insectivore	active forager
Podarcis lilfordi (3)	71.01 ± 6.29	24.36 ± 4.57	73.35 ± 14.21	21.61 ± 4.01	119.32 ± 22.30	omnivore	active forager
Cnemidophorus tigris (3)	83.98 ± 8.48	30.62 ± 8.12	75.73 ± 16.62	16.72 ± 3.03	123.07 ± 25.84	insectivore	active forager
Varanus exanthematicus (2)	261.36 ± 21.79	65.82 ± 1.30	257.04 ± 10.88	57.06 ± 8.42	379.91 ± 20.61	insectivore	active forager

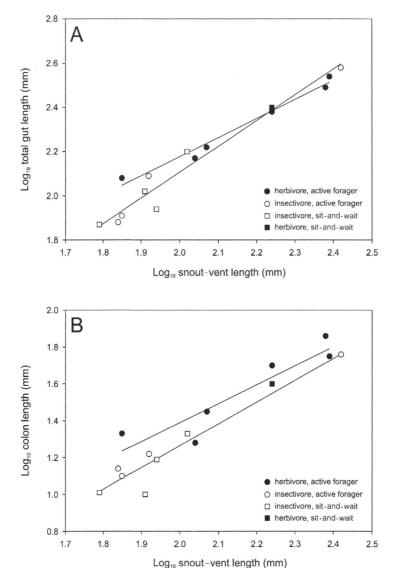

Figure 7.5. Differences in intestinal tract length. Despite the absence of differences in total intestinal tract length (A), herbivores do have longer colons for a given body length (B). In addition, the colon in herbivores has often been modified into a regular cecum with cecal ridges or valves that help slow down the passage of food.

using silicone casts of fresh specimens. Additionally, differences in digestive physiology between active and sit-and-wait foragers are likely to prove an interesting avenue for further research.

As vertebrates rely heavily on commensal microorganisms to digest the cellulose in the cell walls of plants, having a large digestive tract that can be

Table 7.5 *Results of phylogenetic analyses of covariance testing for differences in intestinal tract length*

Variable	F_{phyl}	F_{95}	p_{95}	p_{phyl}
Test for differences between active and sit-and-wait foragers				
stomach	3.35	0.32	0.58	0.52
intestine	3.44	0.92	0.36	0.28
colon	3.59	5.42	0.04	0.03
total length	3.43	1.78	0.21	0.15
SVL[a]	3.00	4.29	0.059	0.02
Test for differences between herbivores s.l. and other lizards				
stomach	8.08	0.01	0.92	0.94
intestine	7.29	0.02	0.90	0.93
colon	7.36	3.31	0.10	0.18
total length	7.72	0.30	0.60	0.71
SVL[a]	7.47	0.10	0.34	0.53

[a] Tests for differences in snout–vent length (SVL) are phylogenetic analyses of variance. All other tests are phylogenetic analyses of covariance with SVL as covariate. Differences in esophageal and rectal length were not tested as it was impossible to take out the entire intestinal tract intact in all specimens. Total length was measured from the beginning of the stomach to the end of the colon.

partly modified as a fermenting chamber is expected for any herbivore. Indeed, in many herbivorous *s.s.* lizards part of the colon has been modified into a well-developed cecum, often with cecal valves that help slow down the passage of food (Iverson, 1980, 1982). In addition, an increase in the surface area of the digestive tract to allow for maximal resorption of nutrients is probably important. In some herbivorous mammals that feed on bulky, low caloric foods, this is achieved by a dramatic increase in the length of the intestine (up to ten times longer) (see Liem *et al.*, 2001). Data for herbivorous whiptail lizards (*Cnemidophorus murinus*) and lacertid lizards suggest that an increase in intestinal tract length may also be an important evolutionary strategy in omnivorous lizards (Dearing, 1993; A. Herrel *et al.*, unpublished data). An analysis for the species included in Table 7.4 indicates that although no differences in total intestinal tract length exist (phylogenetic ANCOVA, $p = 0.151$; see Fig. 7.5 and Table 7.5), herbivores *s.l.* do differ from other lizards in having a well-developed and significantly longer colon (phylogenetic ANCOVA, $p = 0.026$, see Fig. 7.5 and Table 7.5). This suggests that, rather than a general increase in intestinal tract length, the length of specific physiologically important components may increase as a response to a herbivorous diet. Potentially, changes in overall intestinal tract length may only be important when constraints on cecal mass and dimensions are present, as may be the case in actively moving lizards such as whiptails. All 'true' herbivores examined in the present paper, including the lacertid lizard *Gallotia galloti*, showed well-developed ceca with ridges or valves that probably function to slow down the passage of food. In addition, the omnivorous lacertid lizard *Podarcis lilfordi* as well as the omnivorous scincid *Tiliqua rugosa* showed distinct modifications of the colon, including cecal ridges that presumably also function to slow down the passage of food. The modifications in gut structure in herbivorous lizards probably play an important role in assuring that their digestive efficiency is comparable to that of insectivores (Throckmorton, 1973; Nagy, 1977; Johnson and Lillywhite, 1979).

Tongue structure (see also Chapters 8 and 10)

Among lizards using their tongue to capture and/or transport prey, sit-and-wait foragers would probably benefit from having big tongues, given that the adhesive force exerted by the tongue upon the prey is proportional to the surface area of the tongue itself. In additional, we would expect the tongues of sit-and-wait predators to have large numbers of well-developed papillae to ensure maximal adhesion (i.e. maximize the potential for interlocking). However, this prediction relies strongly on the fact that sit-and-wait foragers are expected to eat large prey. For very large prey, tongue transport is no longer a functional strategy, owing to the disproportionate increase of prey mass to surface area. Active foragers, on the other hand, are likely to benefit from having a long, bifurcated tongue that can be used for chemoreceptive purposes. Actively foraging scleroglossans often also have scaly tongues, which might be better for transport of chemicals to the sublingual plicae and ultimately the vomeronasal organ. As active foragers eat light, small prey they do not need large, fleshy, papillose tongues for transport. In varanid lizards (*Varanus*) and some tegus (*Tupinambis*) that eat very large prey, inertial rather than lingual transport is used (Gans, 1969; Elias *et al.*, 2000; Schwenk, 2000).

Herbivores presumably do not need a large and fleshy tongue as they tend to eat rather light food items. In addition, as the tongue is used for chemoreceptive purposes, one would expect herbivores to have longer and more deeply bifurcated tongues. However, as in most lizards, the tongue is still of prime importance in transporting the food to the esophagus. As iguanians rely strongly on gustation to assess prey chemicals, herbivores are also expected to have a larger number of taste buds on the tongue and associated tissues. As herbivorous iguanians also use vomerolfactive cues (Cooper and Alberts, 1991; Cooper, Chapter 8), one would also expect a more strongly developed vomeronasal organ. It is noteworthy to mention that, at least in one iguanian herbivore for which data on tongue structure are available, the papillae on the tongue were strongly reduced and resembled the more scaly tongue structure of scleroglossan lizards (see Fig. 3 in Herrel *et al.*, 1998b). Omnivores would, in contrast to herbivores, be expected to have rather big and fleshy tongues to pick up and/or transport a wide range of vegetable and animal food items.

Discussion

How similar are herbivores to sit-and-wait predators and active foragers? Although herbivores are, by definition, active foragers, the data presented here indicate that they are unique and highly specialized, possessing an array of both sit-and-wait and active foraging traits. Although herbivores need to forage actively to find their food, the food itself must be reduced extensively before being swallowed. The clumped and often unpredictable nature of certain plant foods is, however, reminiscent of foods eaten by active foragers. One aspect of foraging where herbivores have clearly adopted a typical active forager trait is prey detection. All herbivores tested to date, whether nested within sit-and-wait clades or not, use prey chemical cues to identify and assess food items. Owing to the sessile nature of their food, herbivores have relatively low energy requirements despite their active mode of foraging (see above). As a result of their relatively low energy requirements, herbivores also have home ranges that are smaller than those of insectivores (Perry and Garland, 2002).

The analysis of locomotor performance data indicates that herbivores are again different from active foragers in that they do not have an increased endurance capacity. Although herbivores need to forage actively, they presumably cover smaller distances (smaller home ranges) and do not need to sustain fairly high locomotor speeds, as the food does not move. Herbivores differ from both active foragers and sit-and-wait predators in having larger bite forces. This seems to be a trait essential for the evolution of a herbivorous lifestyle in lizards. The feeding behavior, on the other hand, is more similar to that of an active forager, with an emphasis on prolonged chewing. Obviously, this is important as prolonged chewing increases the amount of plant cells directly broken (which frees the content to be digested), and increases the surface area of the plant food for digestion by commensal organisms.

Morphologically, herbivores are a mixed bag, with some traits more like those of sit-and-wait predators, but also with a whole suite of derived and highly specialized characteristics. Although herbivores have the wide, stocky body and head in common with sit-and-wait predators, they have uniquely tall heads and specialized, mediolaterally flattened and blade-like teeth. The digestive tract is also uniquely specialized in herbivores, with an elongated colon that is often modified into a regular cecum inclusive of cecal ridges and valves. The tongue of herbivores is often broad and fleshy, which may aid in the transport of plant matter, but presumably possesses a specialized epithelium and more taste buds than is typical for sit-and-wait predators. Unfortunately, quantitative data on tongue structure are largely lacking, thereby making tests for differences in tongue structure impossible.

One of the major issues related to the foraging paradigm is the seemingly rigid association of morphological and ecological traits with foraging mode. Because foraging mode is tightly associated with phylogeny (with iguanians being mostly sit-and-wait predators and scleroglossans being mostly active foragers) it is hard to test whether the invariance of character states associated with foraging mode is due to phylogenetic inertia or the result of true adaptive evolution. However, herbivores may offer some unique insights as they have arisen independently in both Iguania and Scleroglossa and appear to possess a number of very specific adaptations, unrelated to foraging mode as well as characters that are probably associated specifically with foraging mode. The results presented here clearly indicate that most of the characters typically tightly associated with foraging mode do change when evolving a herbivorous lifestyle (morphology, locomotor performance, energy requirements, etc.). In addition, some of the characters typically associated with a certain foraging mode show reversals in herbivores. For example, iguanian herbivores that arose within a sit-and-wait clade all evolved the use of chemosensory prey detection typically associated with an actively foraging lifestyle. Scleroglossan herbivores, on the other hand, evolve large stocky bodies and big heads typically associated with iguanian sit-and-wait foragers. Clearly, many of these traits are not constrained by phylogeny, suggesting that the ecological and morphological character suite associated with foraging mode is a true adaptive phenomenon.

Given that many of the traits associated with herbivory are uniquely derived, independent of the ancestral condition, the question of why there are so many 'true' herbivores among the Iguania becomes even more pertinent. What are the differences between omnivores and 'true' herbivores as judged from the previous overview? Apparently, very few if any quantitative differences can be discerned in traits examined here for omnivores and true herbivores. The one character uniquely associated with true herbivores seems to be the high degree of specialization of the colon with very prominent ridges or valves. Yet this trait too seems unconstrained by phylogeny, as at least two scleroglossan herbivores *s.s.* possess this derived morphology (*Corucia zebrata* and *Gallotia galloti*). Could it be that, rather than being a real phenomenon, the apparent lack of scleroglossan herbivores *s.s.* is due to our limited knowledge of herbivory among scleroglossans? An overview of literature data indicates that data for iguanian herbivores are indeed much more common than for scleroglossans. Yet most of the new herbivores that are being 'discovered' are small scleroglossans, suggesting that our current understanding might indeed be taxonomically biased.

Conclusions

What can we conclude regarding foraging strategy and herbivory in lizards? Is there really a paradox? Where do herbivores fit in? How do they support or fail to support the supposed dichotomy between active foragers and sit-and-wait predators? Although the data presented here raise as many questions as can be answered, one thing the analyses clearly show is that there is no evidence of lizards being constrained or predisposed with respect to dietary mode because of their ancestry. For iguanians, sit-and-wait predators readily evolve the use of chemosensory cues to recognize and judge the quality of plant matter, and on the scleroglossan side active foragers grow bigger and develop broader and higher heads. Regardless of foraging mode, the change to a herbivorous diet universally associated with the evolution of flat blade-like teeth, high bite forces, larger body size, and a longer colon. As most of these traits are already observed in omnivorous lizards, this suggests that it is the transition to an omnivorous diet that may be the crucial step in the evolution of herbivory. It may then be surprising that, even though independent origins of omnivory are likely to be more common among scleroglossans (Cooper and Vitt, 2002), most herbivores *s.s.* are Iguania. If the crucial transitions indeed take place going from an insectivorous to an omnivorous diet, why are there not more scleroglossan herbivores *s.s.*? The current knowledge of the distribution of herbivory within lizards and of the associated ecological and morphological traits is unfortunately insufficient to answer this question.

Unresolved questions, future directions

The often heard, but often justified, cry for more quantitative data is applicable here too. Clearly, more data are needed on all aspects of herbivory, especially in scleroglossan lizards. Cordylid (e.g. *Angolosaurus skoogi*), anguid (which species if any might be herbivores?) and xantusiid (e.g. *Xantusia riversiana*) herbivores are largely unstudied, but could provide significant insights into the evolution of herbivory within scleroglossan taxa. More quantitative data on foraging behavior (i.e. quantifying movement during foraging) and associated morphological traits would be equally beneficial to our understanding of the co-variation and co-evolution of dietary specialization and foraging mode in lizards. Wider sampling of herbivorous and non-herbivorous taxa is thus essential to allow a quantitative test for associations between foraging mode and herbivory (see Cooper and Vitt, 2002). In addition, I think it would be most important to identify current, and ongoing, transitions from insectivory to omnivory, and from omnivory to herbivory. Clearly, we also need to document the ecological, behavioral and functional changes that accompany these transitions. Lizards showing large amounts of population-level variation in the degree of herbivory (e.g. lacertid lizards of the genus *Gallotia*) would be especially suited for these types of analyses and might offer truly exciting windows into adaptive evolution in action.

Acknowledgements

I thank Jay Meyers for sharing unpublished data. Bieke Vanhooydonck, Raoul Van Damme, and Jay Meyers are gratefully acknowledged for stimulating discussions on aspects of this work. I also thank the Antwerp Zoo for allowing me to use the animals in their care. A. H. is a postdoctoral fellow of the Fund for Scientific Research, Flanders, Belgium (FWO-Vl).

References

Andrews, R. M. (1979). The lizard *Corytophanes cristatus*: an extreme "sit-and-wait" predator. *Biotropica* **11**, 136–9.

Arnold, N. E. (1989). Towards a phylogeny and biogeography of the Lacertidae: relationships within an Old World family of lizards derived from morphology. *Bull. Brit. Mus. Nat. Hist. (Zool.)* **44**, 291–339.

Ballinger, R. E., Lemos-Espinal, J., Sanoja-Sarabia, S. and Coady, N. R. (1995). Ecological observations of the lizard *Xenosaurus grandis* in Cautlapan, Veracruz, Mexico. *Biotropica* **27**, 128–32.

Bjorndal, K. A., Bolten, A. B. and Moore, J. E. (1990). Digestive fermentation in herbivores: effects of food particle size. *Physiol. Zool.* **63**, 710–21.

Bjorndal, K. A. and Bolten, A. B. (1992). Body size and digestive efficiency in a herbivorous freshwater turtle: Advantages of a small bite size. *Physiol Zool.* **65**, 1028–39.

Cooper, J. S. and Poole, D. F. G. (1973). The dentition and dental tissues of the agamid lizard *Uromastyx*. *J. Zool. Lond.* **169**, 85–100.

Cooper, J. S., Poole, D. F. G. and Lawson, R. (1970). The dentition of agamid lizards with special reference to tooth replacement. *J. Zool. Lond.* **162**, 85–98.

Cooper, W. E. Jr. (1994). Prey chemical discrimination, foraging mode, and phylogeny. In *Lizard Ecology*, ed. L. J. Vitt and E. R. Pianka, pp. 95–116. Princeton, NJ: Princeton University Press.

Cooper, W. E. Jr. and Alberts, A. C. (1991). Tongue flicking and biting in response to chemical food stimuli by an iguanid lizard (*Dipsosaurus dorsalis*) having sealed vomeronasal ducts: vomerolfaction may mediate these behavioural responses. *J. Chem. Ecol.* **17**, 135–46.

Cooper, W. E. Jr. and Vitt, L. J. (2002). Distribution, extent, and evolution of plant consumption by lizards. *J. Zool. Lond.* **257**, 487–517.

Cooper, W. E. Jr., Caldwell, J. P., Vitt, L. J., Perez-Mellado, V. and Baird, T. A. (2002). Food-chemical discrimination and correlated evolution between plant diet and plant-chemical discrimination in lacertiform lizards. *Can. J. Zool.* **80**, 655–63.

Cooper, W. E. Jr., Whiting, M. J. and Van Wyk, J. H. (1997). Foraging modes of cordyliform lizards. *S. Afr. J. Zool.* **32**(1), 9–13.

Cundall, D. and Greene, H. W. (2000). Feeding in snakes. In *Feeding*, ed. K. Schwenk, pp. 293–336. San Diego: Academic Press.

Dalrymple, G. H. (1979). On the jaw mechanism of the snail-crushing lizards, *Dracaena* Daudin, 1802 (Reptilia, Lacertilia, Teiidae). *J. Herpetol.* **13**, 303–11.

Dearing, M. D. (1993). An alimentary specialization for herbivory in the tropical whiptail lizard, *Cnemidophorus murinus*. *J. Herpetol.* **27**, 111–14.

Dearing, M. D. and Schall, J. J. (1992). Testing models of optimal diet assembly by the generalist herbivorous lizard *Cnemidophorus murinus*. *Ecology* **73**, 845–58.

Disi, A. M., Modry, D., Necas, P. and Rifai, L. (2001). *Amphibians and Reptiles of the Hashemite Kingdom of Jordan*. Frankfurt: Edition Chimaira.

Dubas, G. and Bull, C. M. (1991). Diet choice and food availability in the omnivorous lizard *Trachydosaurus rugosus*. *Wildl. Res.* **18**, 147–55.

Dubuis, A., Faurel, L., Grenot, C. and Vernet, R. (1971). Sur le regime alimentaire du lezard saharien *Uromastix acanthinurus* Bell. *C. R. Acad. Sci.* D **273**, 500–3.

Elias, J. A., McBrayer, L. D. and Reilly, S. M. (2000). Prey transport kinematics in *Tupinambis teguixin* and *Varanus exanthematicus*: conservation of feeding behavior in 'chemosensory-tongued' lizards. *J. Exp. Biol.* **203**, 791–801.

Gans, C. (1969). Comments on inertial feeding. *Copeia* **1969**, 855–7.

Garland, T. Jr. (1994). Phylogenetic analyses of lizard endurance capacity in relation to body size and body temperature. In *Lizard Ecology*, ed. L. J. Vitt and E. R. Pianka, pp. 237–85. Princeton, NJ: Princeton University Press.

Garland, T. Jr. (1999). Laboratory endurance capacity predicts variation in field locomotor behaviour among lizard species. *Anim. Behav.* **58**, 77–83.

Garland, T. Jr., Midford, P. E. and Ives, A. R. (1999). An introduction to phylogenetically based statistical methods, with a new method for confidence intervals on ancestral states. *Am. Zool.* **39**, 374–88.

Gleeson, T. T. (1979). Foraging and transport costs in the Galapagos marine iguana, *Amblyrhynchus cristatus*. *Physiol. Zool.* **52**, 549–57.

Greer, A. E. (1989). *Biology and Evolution of Australian Lizards*. New York: Surrey Beatty and Sons.

Herrel, A. and De Vree, F. (2000). Kinematics of intraoral transport and swallowing in the herbivorous lizard *Uromastix acanthinurus*. *J. Exp. Biol.* **202**, 1127–37.

Herrel, A., Aerts, P. and De Vree, F. (1998a). Ecomorphology of the lizard feeding apparatus: a modelling approach. *Neth. J. Zool.* **48**, 1–25.

Herrel, A., De Grauw, E. and Lemos-Espinal, J. A. (2001a). Head shape and bite performance in xenosaurid lizards. *J. Exp. Zool.* **290**, 101–7.

Herrel, A., Meyers, J. J., Nishikawa, K. C. and De Vree, F. (2001b). The evolution of feeding motor patterns in lizards: modulatory complexity and possible constraints. *Am. Zool.* **41**, 1311–20.

Herrel, A., Spithoven, L., Van Damme, R. and De Vree, F. (1999a). Sexual dimorphism of head size in *Gallotia galloti*: testing the niche divergence hypothesis by functional analyses. *Funct. Ecol.* **13**, 289–97.

Herrel, A., Timmermans, J.-P. and De Vree, F. (1998b). Tongue flicking in agamid lizards: morphology, kinematics, and muscle activity patterns. *Anat. Rec.* **252**, 102–16.

Herrel, A., Van Damme, R., Vanhooydonck, B. and De Vree, F. (2001c). The implications of bite performance for diet in two species of lacertid lizards. *Can J. Zool.* **79**, 662–70.

Herrel, A., Verstappen, M. and De Vree, F. (1999b). Modulatory complexity of the feeding repertoire in scincid lizards. *J. Comp. Physiol.* A**184**, 501–18.

Hotton, N. III (1955). A survey of adaptive relationships of dentition to diet in the North American Iguanidae. *Am. Midl. Nat.* **53**, 88–114.

Huey, R. B. and Pianka, E. R. (1981). Ecological consequences of foraging mode. *Ecology* **62**, 991–9.

Hutchinson, M. N. (1980). The systematic relationships of the genera *Egernia* and *Tiliqua* (Lacertilia: Scincidae): a review and immunological reassessment. In *Proceedings of the Melbourne Herpetological Symposium*, ed. C. B. Banks and A. A. Martin, pp. 176–93. Victoria: Royal Melbourne Zoological Gardens.

Iverson, J. B. (1980). Colic modifications in iguanine lizards. *J. Morphol.* **163**, 79–93.

Iverson, J. B. (1982). Adaptations to herbivory in iguanine lizards. In *Iguanas of the World: their Behavior, Ecology, and Conservation*, ed. G. M. Burghardt and A. S. Rand, pp. 60–76. Park Ridge, NJ: Noyes.

Jeronimidis, G. (1991). Mechanical and fracture properties of cellular and fibrous materials. In *Feeding and the Texture of Food*, ed. J. F. V. Vincent and P. J. Lillford, pp. 1–17. Cambridge: Cambridge University Press.

Johnson, R. N. and Lillywhite, H. B. (1979). Digestive efficiency of the omnivorous lizard *Klauberina riversiana*. *Copeia* **1979**, 431–7.

King, G. (1996). *Reptiles and Herbivory*. London: Chapman and Hall.

Liem, K. F., Bemis, W. E., Walker, W. F. Jr. and Grande, L. (2001). *Functional Anatomy of the Vertebrates*. San Diego, CA: Harcourt.

Lucas, P. W. (1979). The dental-dietary adaptations of mammals. *N. Jb. Geol. Paleont. Mh.* **1979**, 486–512.

Lucas, P. W. (1982). Basic principles of tooth design. In *Teeth: Form, Function and Evolution*, ed. B. Kurten, pp. 154–62. New York: Columbia University Press.

Lucas, P. W. and Luke, D. A. (1984). Basic principles of food breakdown. In *Food Acquisition and Processing in Primates*, ed. D. A. Chivers, B. A. Woods and A. Bilsborough, pp. 283–301. New York: Plenum Press.

McBrayer, L. D. and Reilly, S. M. (2002). Prey processing in lizards: behavioral variation in sit-and-wait and widely foraging taxa. *Can. J. Zool.* **80**, 882–92.

Montanucci, R. R. (1968). Comparative dentition in four iguanid lizards. *Herpetologica* **24**, 305–15.

Montanucci, R. R. (1989). The relationship of morphology to diet in the horned lizard genus Phrynosoma. *Herpetologica* **45**, 208–16.

Nagy, K. A. (1977). Cellulose digestion and nutrient assimilation in *Sauromalus obesus*, a plant-eating lizard. *Copeia* **1977**, 355–62.

Perry, G. (1999). The evolution of search modes: ecological versus phylogenetic perspectives. *Am. Nat.* **153**, 98–109.

Perry, G. and Garland, T. Jr. (2002). Lizard home ranges revisited: effects of sex, body size, diet, habitat and phylogeny. *Ecology* **83**, 1870–85.

Piersma, T. and Lindstrom, A. (1997). Rapid reversible changes in organ size as a component of adaptive behavior. *Trends Ecol. Evol.* **12**, 134–8.

Pough, F. H. (1973). Lizard energetics and diet. *Ecology* **54**, 837–44.

Pough, F. H., Preest, M. R. and Fusari, M. H. (1997). Prey-handling and the evolutionary ecology of sand-swimming lizards (Lerista: Scincidae). *Oecologia* **112**, 351–61.

Pough, F. H., Andrews, R. M., Cadle, J. E., *et al.* (2001). *Herpetology*, 2nd edn. New Jersey: Prentice Hall.

Ralston, K. R. and Wainwright, P. C. (1997). Functional consequences of trophic specialisation in pufferfishes. *Funct. Ecol.* **11**, 43–52.

Reeder, T. W. and Wiens, J. J. (1996). Evolution of the lizard family Phrynosomatidae as inferred from diverse types of data. *Herp. Monogr.* **10**, 43–84.

Rieppel, O. (1979). A functional interpretation of the varanid dentition (Reptilia, Lacertilia, Varanidae). *Gegenbauers Morph. Jahrb.* **125**, 797–817.

Rieppel, O. and Labhardt, L. (1979). Mandibular mechanics in *Varanus niloticus* (Reptilia: Lacertilia). *Herpetologica* **35**, 158–63.

Robinson, P. L. (1976). How *Sphenodon* and *Uromastix* grow their teeth and use them. *Linn. Soc. Symp. Ser.* **3**, 43–64.

Rocha, C. F. D. (2000). Selectivity in plant food consumption in the lizard *Liolaemus lutzae* from southeastern Brazil. *Stud. Neotrop. Fauna Environm.* **35**, 14–18.

Sadek, R. A. (1981). The diet of the Madeiran lizard *Lacerta dugesii*. *Zool. J. Linn. Soc.* **73**, 313–41.

Schwenk, K. (1993). The evolution of chemoreception in squamate reptiles: a phylogenetic approach. *Brain. Behav. Evol.* **41**, 124–37.

Schwenk, K. (2000). *Feeding*. San Diego, CA: Academic Press.

Sibbing, F. A. (1991). Food processing by mastication in cyprinid fish. *Symp. Soc. Exp. Biol.* **43**, 57–92.

Spawls, S., Howell, K., Drewes, R. and Ashe, J. (2002). *A Field Guide to the Reptiles of East Africa*. San Diego, CA: Academic Press.

Stamps, J. A. (1977). Social behavior and spacing patterns in lizards. In *Biology of the Reptilia*, vol. 7, ed. C. Gans and D. W. Tinkle, pp. 265–334. London: Academic Press.

Stamps, J. A. (1983). Sexual selection, sexual dimorphism, and territoriality. In *Lizard Ecology: Studies of a Model Organism*, ed. R. B. Huey, E. R. Pianka and T. W. Schoener, pp. 169–204. Cambridge, MA: Harvard University Press.

Starck, J. M. and Beese, K. (2001). Structural flexibility of the intestine of burmese python in response to feeding. *J. Exp. Biol.* **204**, 325–35.

Sylber, C. K. (1988). Feeding habits of the lizards *Sauromalus varius* and *S. hispidus* in the Gulf of California. *J. Herpetol.* **22**, 413–24.

Throckmorton, G. S. (1973). Digestive efficiency in the herbivorous lizard *Ctenosaura pectinata*. *Copeia* **1973**, 431–5.

Throckmorton, G. S. (1976). Oral food processing in two herbivorous lizards, *Iguana iguana* (Iguanidae) and *Uromastix aegyptius* (Agamidae). *J. Morphol.* **148**, 363–90.

Throckmorton, G. S. (1979). The effect of wear on the cheek teeth and associated dental tissues of the lizard *Uromastix aegyptius* (Agamidae). *J. Morphol.* **160**, 195–208.

Urbani, J. M. and Bels, V. L. (1995). Feeding behavior in two scleroglossan lizards: *Lacerta viridis* (Lacertidae) and *Zonosaurus laticaudatus* (Cordylidae). *J. Zool. Lond.* **236**, 265–90.

Valido, A. and Nogales, M. (1994). Frugivory and seed dispersal by the lizard *Gallotia galloti* (Lacertidae) in a xeric habitat of the Canary Islands. *Oikos* **70**, 403–11.

Van Damme, R. (1999). Evolution of herbivory in lacertid lizards: effects of insularity and body size. *J. Herpetol.* **33**, 663–74.

Van Damme, R. and Vanhooydonck, B. (2001). Origins of interspecific variation in lizard sprint capacity. *Funct. Ecol.* **15**, 186–202.

Verwaijen, D., Van Damme, R. and Herrel, A. (2002). Relationships between head size, bite force, prey handling efficiency and diet in two sympatric lacertid lizards. *Funct. Ecol.* **16**, 842–50.

Vitt, L. J. and Congdon, J. D. (1978). Body shape, reproductive effort and relative clutch mass in lizards: resolution of a paradox. *Am. Nat.* **112**, 595–608.

Wilson, K. J. and Lee, A. K. (1974). Energy expenditure of a large herbivorous lizard. *Copeia* **1974**, 338–48.

8

Lizard chemical senses, chemosensory behavior, and foraging mode

WILLIAM E. COOPER, JR.

Department of Biology, Indiana University–Purdue University at Ft. Wayne

Introduction

Even to casual observers, one of the most obvious things about lizards is tongue-flicking. The form and frequency of tongue-flicking vary greatly among lizard species, but all do it. No other animals do. Even tuataras, the closest living relatives of squamate reptiles and sole survivors of Rhynchocephalia, do not tongue-flick (Schwenk, 1986, 1993a; Cooper *et al.*, 2001a). Biologists have suggested several functions for tongue-flicking in lizards and snakes, including sensory functions such as gustation (Schwenk, 1985), touch (Klauber, 1956; Bodnar *et al.*, 1975), detection of airborne vibration (Ditmars, 1937; Klauber, 1956), and chemoreception. Some lizards extend the tongue as part of antipredatory displays (Greene, 1988), and the author F. Scott Fitzgerald (1945), undoubtedly based on frequent tongue-flicking during courtship by some species, has likened tongue-flicks to a kiss:

The kiss originated when the first male reptile licked the first female, implying in a subtle, complimentary way that she was as succulent as the small reptile he had for dinner the night before.

All of the available evidence suggests that the primary function of tongue-flicking is chemosensory. It has been amply demonstrated that lizards can detect food, predators, and conspecifics by tongue-flicking (Halpern, 1992; Mason, 1992; Cooper, 1997a; Downes and Shine, 1998). The tongue bears variable numbers of taste buds in lizards, but not in snakes, suggesting possible use of tongue-flicking to taste potential foods and other stimulus sources (Schwenk, 1985). Although direct demonstrations are lacking, the tongue presumably is sensitive to tactile stimuli because it is supplied with abundant tactile receptors (Moncrieff, 1967). Whatever the literary merits of Fitzgerald's suggestion, the relationship between tongue-flicking and kissing is whimsical,

Lizard Ecology: The Evolutionary Consequences of Foraging Mode, ed. S. M. Reilly, L. D. McBrayer and D. B. Miles. Published by Cambridge University Press. © Cambridge University Press 2007.

although mutual tactile stimulation undoubtedly occurs during lingual contact with body surfaces during tongue-flicking. The focus here will be entirely on chemosensory functions of tongue-flicking and their relationships to foraging behavior.

An early indication that tongue-flicking might be related to foraging mode was a set of field observations by L. T. Evans (1961), who noted that lizard species that hunted by moving through the habitat searching for prey tongue-flicked frequently, whereas species that waited immobile for prey tongue-flicked infrequently. Quantitative data on tongue-flicking by zoo specimens from six lizard families reinforced Evan's impression that tongue-flick rates vary greatly among taxa (Bissinger and Simon, 1979). The authors noted a possible correlation between lingual forking and tongue-flick rate. When my own observations suggested that tongue-flick rates were higher in families that forage more actively, in agreement with Evans' comments, I conducted a few experiments which showed that actively foraging species were capable of discriminating between chemicals from prey and those from other sources (Cooper, 1989a, 1990a). This ability is hereafter called prey chemical discrimination. When further preliminary experiments revealed no similar ability in ambush foragers (Cooper, 1989b), over 15 years ago I began a long-term research project to investigate the relationships among tongue-flicking, lingual morphology, the cranial chemical senses, use of the tongue to locate and identify food, diet, and foraging behavior in lizards.

A remarkable set of interrelationships emerged from data on these variables. A simplified overview omitting many important details follows. Use of the tongue to locate and identify food evolved in tandem with increased abundance of chemoreceptors in the vomeronasal organs and increased elongation and forking of the tongue. The advantages of being able to find hidden prey by chemoreception favored evolution of active foraging for animal prey from ancestral ambush foragers and chemosensory evaluation of plant food by herbivores. Active foraging has in turn favored more frequent incorporation of plants into the diet than has ambush foraging.

This chapter begins with descriptions of the morphology and physiology of lizard chemosensory systems, experimental evidence for prey chemical discrimination (and its absence), and a brief consideration of the discreteness of lizard foraging modes and its relevance to the relationships to be examined. The bulk of the chapter presents evidence that correlated evolution has occurred among the lingual–vomeronasal system, food chemical discrimination, and foraging mode, and that foraging mode has influenced the evolution of diet, which in turn has affected responsiveness to chemicals from specific food types. Finally, I speculate about the influences of prey chemical discrimination mediated by lingual chemical sampling and foraging mode on the evolutionary diversification of lizards.

Lizard chemosensory systems

Lizards have three major cephalic chemosensory systems. The nasal chemical senses are olfaction and vomerolfaction, the latter sense mediated by the vomeronasal organs (Fig. 8.1) (Cooper and Burghardt, 1990a). The third

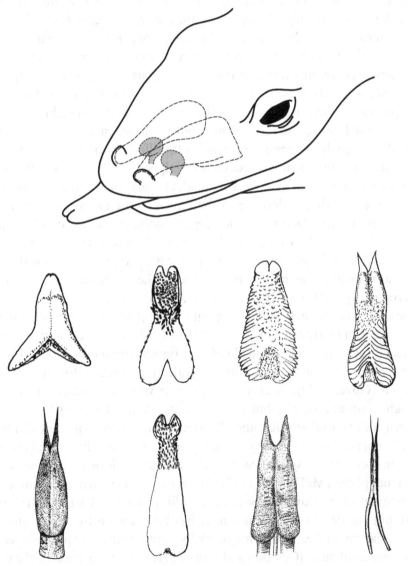

Fig. 8.1. Lizard chemosensory systems and variation in lingual morphology.

cephalic chemical sense is gustation. Olfaction senses airborne volatile molecules, vomerolfaction detects larger, non-volatile molecules as well as some smaller ones (Halpern, 1992; Inouchi *et al.*, 1993), and gustation detects limited classes of compounds dissolved in water. The absolute and relative degrees of development of these systems vary among lizard taxa (Parsons, 1970; Gabe and Saint Girons, 1976; Schwenk, 1985).

In lizards the air entering the external nares passes through a tubular vestibulum nasi to an enlarged chamber, the cavum nasi proprium, where the olfactory epithelium containing chemoreceptor cells is located (Parsons, 1970). Air passes through the cavum to the nasopharyngeal duct, which leads to the internal nares (choanae) and the respiratory tree. Volatile molecules in air passing through the cavum stimulate chemoreceptors on its wall.

Vomeronasal organs are chemosensory organs widespread among vertebrates, but reach their peak expression in squamate reptiles (Bertmar, 1981). The squamate vomeronasal organs or Jacobson's organs are mushroom-shaped bodies situated above the roof of the mouth, one on each side (Fig. 8.1). Air inspired through the external nares reaches the olfactory epithelium, but not the vomerolfactory epithelium, of squamates (Halpern, 1992). Access to the vomeronasal organs is through the vomeronasal ducts, one opening into the roof of the mouth on each side and leading to the lumen of the ipsilateral vomeronasal organ where vomerolfactory chemoreceptors occur in the surrounding epithelium (Halpern, 1992). Radioactively labeled chemicals sampled by tongue-flicking and, to a lesser degree, those that come into contact with the labial scales surrounding the mouth pass through the vomeronasal ducts to the vomerolfactory epithelium (Graves and Halpern, 1989).

The tongue is crucial for efficient chemical sampling by tongue-flicking, but the mechanism of transfer of chemicals from the tongue to the vomeronasal organs is not fully understood. During a typical tongue-flick, the tongue is protruded on an upward and outward trajectory, sweeps downward, and is retracted (Gove, 1979). It may contact some volatile chemicals as it passes through a volume of air, but chemicals are sampled primarily by contact between an external surface and the ventral tip of the tongue (McDowell, 1972; Toubeau *et al.*, 1994). When the tongue is retracted, its ventral tip comes to rest in contact with specialized raised areas on the floor of the mouth called sublingual plicae (McDowell, 1972). These may be elevated into contact with the openings of the vomeronasal ducts (Gillingham and Clark, 1981; Young, 1993), but the details are unknown. It has been shown by X-ray cinematography that the forked tips of tongues of varanid lizards are not inserted into the vomeronasal ducts (Oelofsen and Van den Heever, 1979). Because the tongue serves as the sampling device for the vomeronasal organs, the relationship between these structures is often emphasized by reference to the tongue–vomeronasal or lingual–vomeronasal system.

Although the olfactory and vomerolfactory organs are adjacent to each other above the roof of the mouth, they are distinct sensory systems that lack connections between their lumina and differ in access and structure as discussed, as well as in neural pathways. Olfactory receptors send messages through the olfactory nerves to the primary olfactory bulbs, whereas the vomerolfactory receptors communicate via separate vomeronasal nerves with the lateral (accessory) olfactory bulbs (Halpern, 1992).

Taste buds occur on the tongues of most lizards, but not on those of varanid lizards or snakes (Schwenk, 1985). However, even in the latter groups, taste buds are located in epithelia of the oral cavity (Schwenk, 1985). Chemicals may come into contact with the taste buds when brought into the mouth or when contacted by the protruded tongue during tongue-flicking or licking, which is distinct from tongue-flicking in that the dorsal surface of the tongue contacts a substrate.

Roles of chemosensory systems in prey discrimination

Olfactory receptors of squamates are sensitive to airborne volatile molecules, as indicated by electro-olfactograms, recordings made directly from the olfactory epithelium (Inouchi *et al.*, 1993). However, behavioral roles of olfaction in foraging and feeding are not well understood. The Cowles and Phelan (1958) hypothesis, first presented for rattlesnakes, proposes that detection of airborne molecules by olfaction elicits tongue-flicking for vomerolfactory sampling. Lesions of the olfactory nerves prevent the increase in tongue-flick rate normally elicited by airborne odorants, confirming the Cowles–Phelan hypothesis (Halpern *et al.*, 1997).

Olfaction is inadequate for food chemical discriminations by garter snakes (*Thamnophis sirtalis*) having severed vomeronasal nerves (Halpern and Frumin, 1979) and desert iguanas (*Dipsosaurus dorsalis*) having sealed vomeronasal ducts (Table 8.1) (Cooper and Alberts, 1991). Because the desert iguanas lost the ability to discriminate both prey chemicals and plant food (carrot) chemicals, olfaction appears not to be important for discriminations of prey and plant foods that are not highly odorous. The possibilities remain that olfaction is important for food chemical discriminations by geckos, which have more highly developed olfactory systems than other lizards (Gabe and Saint Girons, 1976; Schwenk, 1993b), and for location of highly odorous foods such as flowers and fruits.

One species of gecko, *Cosymburus platyurus*, suffered reduced ability to capture fruit flies after its olfactory nerves were transected (Chou *et al.*, 1988),

Table 8.1 *Effect of sealing vomeronasal ducts on response strength (tongue-flick attack score) to food chemicals (carrot) and control stimuli in the desert iguana,* Dipsosaurus dorsalis

	Treatment			
	ducts open		ducts sealed	
Stimulus	$\bar{x}$	SE	$\bar{x}$	SE
carrot	49.5	5.5	0.9	0.7
cologne	5.5	1.7	1.1	1.0
distilled water	1.0	0.4	0.5	0.5

Source: Cooper and Alberts (1991).

suggesting that olfactory cues may aid in locating or identifying prey. Omnivorous geckos, lacertids, and teiids may locate flowers and fruit by olfactory cues, but there is no direct evidence. The most suggestive evidence is that the lacertid *Podarcis lilfordi* rapidly locates pieces of fruit hidden under cups in the field (Cooper and Pérez-Mellado, 2001a) and that this species, the lacertids *Gallotia simonyi* and *G. caesaris*, and the teiid *Cnemidophorus murinus* are readily captured in traps baited with fruit or tomatoes (Cooper and Pérez-Mellado, 2001a,b; Cooper *et al.*, 2002a). Some nocturnal geckos in New Zealand appear to aggregate at flowering plants, suggesting that they are attracted by floral scents (Whitaker, 1987). Experimental studies blocking olfaction are needed to ascertain what roles are played by olfaction in such cases.

Even less is known about roles of gustation in lizards, but a few studies suggest its involvement in feeding decisions. *Cnemidophorus arubensis* and the polychrotid *Anolis carolinensis* reject food adulterated with a low concentration of quinine (Schall, 1990; Stanger-Hall *et al.*, 2001) and *Podarcis lilfordi* exhibits decreased tongue-flicking rate in response to cotton swabs bearing 1 g quinine/200 g distilled water (Cooper *et al.*, 2002c). These responses presumably reflect aversion to the bitter taste of the alkaloid. Olfaction is unlikely to be involved in responses to adulterated food because the foods were taken into the mouth rather than rejected without oral sampling. In *A. carolinensis*, which lacks prey chemical discrimination mediated by tongue-flicking (Cooper, 1989b), bitter and sweet substances elicit responses even when vomerolfaction is blocked, strongly suggesting a role for gustation, but not excluding olfaction. In the lacertids *P. lilfordi* and *Gallotia caesaris*, cotton swabs bearing sucrose solutions initially elicit exploratory tongue-flicks, which are quickly replaced by licks that bring the dorsal lingual surface bearing taste buds into contact with the swabs (Cooper *et al.*, 2002b; Cooper and Pérez-Mellado, 2001b). The licking rate increases greatly at the highest sucrose concentration in *P. lilfordi* (Cooper *et al.*, 2002b), suggesting a feeding response to the sweet taste. Desert iguanas having sealed vomeronasal ducts failed to discriminate among prey chemicals, plant chemicals, and control stimuli despite having intact taste buds, showing that food chemical discriminations, at least for crickets and carrots, are not mediated by gustation (Cooper and Alberts, 1991).

Vomerolfaction appears to be the primary sense responsible for food chemical discriminations in squamate reptiles, based primarily on studies of garter snakes of the genus *Thamnophis*. Ingestively naive hatchling garter snakes normally discriminate between prey chemicals and control stimuli, but fail to do so after removal of the tips of the forked tongue (Burghardt and Pruitt, 1975). Following lesions of the vomeronasal nerves, adult garter snakes do not attack prey extracts (Halpern and Frumin, 1979) and lose the ability to follow prey scent trails (Kubie and Halpern, 1979).

Data for other lizards are scanty. After its vomeronasal ducts are sealed with tissue adhesive, the scincid lizard *Chalcides ocellatus* shows a great decline in prey ingestion and exhibits no increase in tongue-flicking rate in novel environments, a species-typical exploratory behavior (Graves and Halpern, 1990). As noted above, desert iguanas lose the ability to discriminate between food and control chemicals when their vomeronasal ducts are sealed, blocking access of chemicals to the vomeronasal organs (Cooper and Alberts, 1991). Although experimental data for lizards other than snakes are limited, the two species studied are representatives of the two major lizard clades, Iguania and Scleroglossa, suggesting that vomerolfaction is necessary for food chemical discriminations in a wide taxonomic range of lizards, with possible exceptions for geckos and other taxa that utilize highly odorous foods that can be located by using olfaction and/or recognized as palatable by their sweet taste.

Relationships among chemosensory anatomy and prey chemical discrimination

If vomerolfaction mediates prey chemical discrimination, correlated evolution may be expected to have occurred between prey chemical discrimination and both abundance of vomerolfactory chemoreceptors and aspects of lingual anatomy that enhance chemical sampling. Similar relationships would not be expected between prey chemical discrimination and the other chemical senses. Vomerolfaction would be predicted to have evolved independently of other

chemical senses unless optimizing lingual sampling for vomerolfaction inter-
feres with gustatory functions of the tongue.

Methods

To assess relative development of the chemical senses among lizard taxa, I used
data from two comparative studies, one of the abundance of olfactory and
vomerolfactory chemoreceptors (Gabe and Saint Girons, 1976) and the other
of the distribution and abundance of taste buds (Schwenk, 1985). Information
on lingual morphology was obtained from McDowell's (1972) survey, but the
primary data were from my own comparative measurements of lingual elongation and depth of forking. Data on the size of the region sampled by the
tongue during a single tongue-flick cycle relative to lizard size were obtained
from Gove's (1979) work on the form and variation of tongue-flicks.

I experimentally studied prey chemical discrimination in numerous species
representing all of the major lizard clades by using the swab method pioneered
by Wilde (1938) and Burghardt (1967) and modifications of it (Cooper and
Burghardt, 1990b; Cooper, 1998; Cooper *et al.*, 2003). In this technique, a
moistened cotton swab bearing control stimuli or food chemicals is placed
before a lizard's snout and the lizard's responses are recorded for 60 s beginning with the first tongue-flick. Typical stimuli tested in an experiment include
food chemicals, an odorless control (deionized or distilled water), a pungency
control for responses to a trophically irrelevant, yet odorous substance
(usually cologne), and sometimes a plant control for insectivores or a non-
preferred prey type for prey specialists. The number of tongue-flicks and
latency to bite the swab (60 s if no bite) are recorded for each trial. Numbers
of tongue-flicks, latency to bite or proportion of individuals that bite, and a
composite measure of response strength called tongue-flick attack score are
analyzed by parametric or non-parametric ANOVA as needed. Prey (or plant)
chemical discrimination is considered to be present when responses are significantly greater to food chemicals than to control stimuli. For quantitative
comparative assessments of the strength of discrimination, the variable analyzed was the ratio of the tongue-flick attack score for prey to that for distilled
or deionized water (Cooper, 1997a). For some species that did not respond to
chemical cues on swabs, I presented the chemical stimuli on ceramic tiles and
observed responses from a blind to eliminate intimidating effects of experi-
menter presence (Cooper, 1998).

The evolutionary histories of traits were reconstructed by using the TRACE
routine of MacClade (Maddison and Maddison, 1992). The basic phylogeny
assumed was that of Estes *et al.* (1988), supplemented by modifications based

Table 8.2 *Rank correlations among cephalic chemical senses and aspects of lingual structure*

	ORA	TBA	DLF	LEL	PAL
vomerolfactory receptor abundance	0.21	−0.82*	0.81*	0.65*	0.74*
olfactory receptor abundance (ORA)		−0.04*	−0.20	0.00	−0.06
taste bud abundance (TBA)			−0.78*	−0.54*	−0.69*
depth of lingual forking (DLF)				0.66*	−0.55*
lingual elongation (LEL)					0.60*

*Asterisks, statistically significant.
Source: Cooper (1996, 1997b).

on more recent phylogenetic studies (as cited in the individual works discussed below). Squamata is divided into two major subclades, Iguania (iguanas and their relatives) and Scleroglossa, which is further divided into Gekkota (geckos) and Autarchoglossa (skinks, monitor lizards, etc.) (Estes *et al.*, 1988). Several alternative phylogenies were used for analyses in some studies to examine effects of different likely phylogenetic branching patterns for cases in which relationships among some lizard taxa are uncertain.

To assess whether correlated evolution has occurred between pairs of morphological and behavioral variables, I used Felsenstein's (1985) method of phylogenetically independent contrasts for continuous variables and concentrated changes tests (Maddison, 1990) and/or Pagel's (1994) tests for binary variables. Both methods take into account phylogenetic relationships among the taxa studied. Felsenstein's and Pagel's methods additionally include estimates of branch lengths.

Correlated evolution among sensory variables

The abundance of olfactory receptors is uncorrelated with abundance of either vomerolfactory receptors or lingual taste buds (Table 8.2) (Cooper, 1997b), suggesting that olfactory sensitivity has evolved independently of vomerolfactory and lingual gustatory sensitivity (Cooper, 1997b). Abundance of vomerolfactory receptors was negatively correlated with abundance of lingual taste buds (Table 8.2). The utility of the tongue for making gustatory discriminations appears to decrease as the importance of vomerolfaction increases among lizard taxa.

Abundance of vomerolfactory receptor cells is correlated with several aspects of lingual morphology (Table 8.2), which is highly variable among lizard taxa

(Fig. 8.1). Receptor abundance increases with the degree of elongation of the tongue (Cooper, 1995a) and the depth of forking (Cooper, 1996, 1997b). Increasing elongation presumably enhances the tongue's ability to sample substrates in the external environment (Cooper, 1995a). Forking is hypothesized to permit scent-trailing by tropotaxis, a form of orientation in which an animal compares concentrations at two simultaneously sampled points and can proceed in the direction of greater concentration to locate the source (Schwenk, 1994), and might also affect ability to access stimuli in narrow crevices. Depth of forks is correlated with the area sampled during tongue-flicking (Cooper, 1996), suggesting that tongues used for scent-trailing are highly extensible. Lingual shapes run the gamut from stout, fleshy forms with shallow notches at their tips characteristic of iguanas and their relatives to the very elongated, deeply forked types found in snakes and varanid and teiid lizards (Fig. 8.1) (McDowell, 1972; Schwenk, 1994; Cooper, 1996, 1997b). Ventral pallets are the ventral surfaces at the tip of tongue that contact substrates during tongue-flicking (McDowell, 1972). They are largest in taxa having low abundance of vomerolfactory receptors, decrease in size as receptor abundance increases, and are lost in snakes and varanid lizards, the taxa having the highest vomerolfactory receptor abundance (Cooper, 1997b). Pallet condition is negatively correlated with degree of forking, suggesting that the ventral pallets diminish and disappear as the tongue splits into separate tines to improve tropotaxis or other aspects of vomerolfactory sampling (Cooper, 1996, 1997b).

Prey chemical discrimination and chemical senses

Tests for responses to food chemicals have been conducted for about 100 species of lizard (Appendix 8.1). Prey chemical discrimination occurs in many lizards, but is absent among insectivores in some large taxa (Fig. 8.2) (Cooper, 1995b, 1997a; Cooper *et al.*, 2001a). It is absent among insectivorous/carnivorous species in Iguania (Cooper, 1995b, 1997a; Cooper *et al.*, 2001a), in gekkonid geckos (Cooper and Habegger, 2000), and in cordylid lizards (Cooper and Van Wyk, 1994). Initial studies revealed the presence of prey chemical discrimination in representatives of one herbivorous iguanian family, in eublepharid geckos, and in all studied families of Autarchoglossa other than Cordylidae (Cooper, 1994a,b, 1995b). Subsequent studies have established that intrafamilial variation in presence of prey chemical discrimination occurs among insectivores in several families and even some genera, and that such variation is related to diet and foraging mode (Cooper, 1999, 2000, 2002, 2003a).

In a study including representatives of ten lizard families, the degree of prey chemical discrimination was highly variable, the ratios of tongue-flick attack

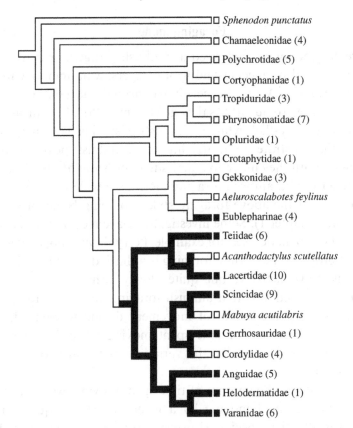

Fig. 8.2. Phylogenies of prey chemical discrimination and foraging mode are identical in insectivorous lizards. White, absent; black, present. Numbers adjacent to names of higher taxa indicate numbers of species for which prey chemical discrimination has been studied experimentally.

scores for prey chemicals to those for water ranging from 1.00 to 15.66 (Cooper, 1997a). Strength of discrimination was significantly correlated with the abundance of vomerolfactory receptors, and with lingual variables correlated with vomerolfactory receptor abundance, i.e. elongation and degree of forking, and negatively correlated with the size/presence of ventral pallets. The degree of prey chemical discrimination increased with area sampled relative to head size during tongue-flicking (Cooper, 1997a). Strength of discrimination was negatively correlated with abundance of taste buds, and uncorrelated with abundance of olfactory receptors.

The relative area sampled during tongue-flicking increased as the degree of prey chemical discrimination increased. These correlations suggest linkage between sensitivity of vomerolfaction, efficiency of lingual vomerolfactory sampling, and strength of prey chemical discrimination.

Foraging mode

Lizards are commonly characterized as ambush foragers or active foragers. Ambush foragers, also called sit-and-wait foragers, wait immobile for prey to approach, then attack. Active foragers move through the habitat searching for prey, tongue-flicking frequently as they go (Evans, 1961). In contrast, ambush foragers rarely tongue-flick while at ambush posts; instead, they sometimes tongue-flick the substrate upon arriving at an ambush post and then stop tongue-flicking until they move to a new site (Cooper et al., 1994b).

Although active and ambush foraging modes are widely accepted as short-hand descriptors of the above foraging styles, foraging behavior of lizards is highly variable (Chapter 1). Some investigators have recognized that variation exists within the two modes (see, for example, Perry, 1999; Cooper et al., 2001a). Regal (1978) proposed that cruise foraging is a third, distinct foraging mode in which lizards move actively, but quite slowly, while searching for prey. A common interpretation is that ambush foraging and active foraging lie on a spectrum reflecting the degree of movement during foraging. If the only relevant variable is percent of time spent moving while foraging, a simple spectrum is adequate to describe the variation, but lizard foraging is more complex than this.

By focusing only on the amount of time spent moving, we assume that the proportion of time spent moving by a given species is equivalent to the proportion for all other species. However, foraging behaviors might differ greatly among species in other ways. For example, this approach leaves us unable to distinguish between a species that forages by making movements of long duration between ambush posts, but hunting only while immobile, from another species that spends the same relative amount of time moving, but hunts as it moves. One alternative measure of foraging mode is the percent of attacks initiated while moving (versus while immobile), but data on this variable are very limited (Cooper and Whiting, 1999; Cooper et al., 2001b).

In my studies of lizard foraging behavior and its relationships to other variables, I have taken two approaches. For comparisons involving binary variables such as presence/absence of prey chemical discrimination, I have treated foraging mode as a discrete variable by defining ambush foragers to be those that spend less than 10% of the time moving during foraging hours and active foragers to be those that spend $\geq 10\%$ of the time moving (following Perry, 1995). For comparisons with continuous variables, I have used percent time moving as the index of degree of foraging activity.

When lizards are considered to be either ambush foragers or active foragers, foraging mode is a conservative trait among insectivores (Chapter 1). As far as is known, all insectivores in all iguanian families are ambush foragers (Perry, 1999; Cooper *et al.*, 2001b). Foraging mode is more variable among scleroglossans (Chapters 1, 10). In Gekkota, gekkonids appear to be ambush foragers with some possible exceptions, but the foraging behavior of these primarily nocturnal lizards may differ qualitatively from that of other lizards (Arnold, 1984; Cooper, 1994a; Perry, 1999; Werner *et al.*, 1997; Bauer, this volume, Chapter 12). Eublepharid geckos are active foragers with the exception of the basal genus *Aeluroscalabotes*, which are ambush foragers (Cooper, 1995b; unpublished observations). Among major families of autarchoglossans, only the Cordylidae consists entirely of ambush foragers (Cooper *et al.*, 1997). The vast majority of species in the other autarchoglossan families are active foragers, but a few ambush foragers occur in Lacertidae and Scincidae (Cooper, 1994a,b; Perry, 1999).

Relationships between foraging mode and prey chemical discrimination

From the foregoing, it is clear that prey chemical discrimination has undergone correlated evolution with the lingual–vomeronasal system and that active foragers tongue-flick at much higher rates than ambush foragers while foraging. It can be predicted from their frequent tongue-flicking that active foragers are capable of prey chemical discrimination. On the other hand, it is difficult to imagine that tongue-flicking while at an ambush post can help an ambush forger locate prey. In agreement with these ideas, active foragers exhibit prey chemical discrimination, but ambush foragers do not (Appendix 8.1a). However, because most species retain the character states of their ancestors, the striking correspondence alone is not adequate evidence that correlated evolution has occurred between foraging mode and prey chemical discrimination.

The first test of the relationship between foraging mode and prey chemical discrimination examined binary data on both variables for representatives of 14 lizard families (Cooper, 1995b). The distributions of the two traits on lizard phylogeny were identical and the relationship between foraging mode and prey chemical discrimination was significant in tests using four alternative phylogenies (Cooper, 1995b). In the preferred phylogeny, active foraging and prey chemical discrimination had three joint origins (i.e. on the same branch) and one joint loss (Cooper, 1995b). However, in that test one of the joint gains was based on data for the desert iguana, *Dipsosaurus dorsalis*. At the time, I characterized desert iguanas as active foragers because their herbivorous diet requires them to move to plants rather than ambush them. In retrospect, it would have been preferable to limit the test to insectivorous/carnivorous taxa because herbivores and actively foraging insectivores may have different foraging requirements and styles.

The next test of the relationship between prey chemical discrimination and foraging mode (1) used continuous variables indicating the strength of prey chemical discrimination and the degree of foraging activity and (2) excluded data on herbivores (Cooper, 1997a). Strength of prey chemical discrimination was highly correlated with percent time spent moving, confirming the relationship between foraging mode and prey chemical discrimination (Cooper, 1997a).

Since these tests were done, data have accumulated for numerous additional species of insectivores, including representatives of additional families, and for cases in which foraging mode has shifted within genera in the Scincidae and Lacertidae. The skink *Mabuya acutilabris* is an ambush forager that lacks prey chemical discrimination, in contrast to its congeners, which forage actively and exhibit prey chemical discrimination (Cooper, 2000). Typical lacertids are active foragers, but *Acanthodactylus scutellatus* is an ambush forager. However, it moves 7.7% of the time, placing it near the upper limit for ambush foraging (Perry *et al.*, 1990). Its congener *A. boskianus* is an active forager (Perry *et al.*, 1990). Both species are capable of prey chemical discrimination, but the discrimination was much weaker in the ambush forager (Cooper, 1999). These findings suggest that prey chemical discrimination may be in the process of being lost in the ambush forager, or possibly that its importance is stable, but has diminished importance owing to lesser reliance on tongue-flicking during active search (Cooper, 1999).

These additional cases of changes in foraging mode and the data on additional ambush foragers and active foragers (in Appendix 8.1a) firmly establish the relationship between foraging mode and prey chemical discrimination based on lingually sampled chemicals analyzed by the vomeronasal system. I also had the opportunity to observe two *Aeluroscalabotes feylinus*, which had been assumed to be ambushers lacking prey chemical discrimination based on descriptions of their behavior in captivity by others (Cooper, 1995b). In my laboratory *A. feylinus* moved little after exploration on the day of their arrival. In tests for prey chemical discrimination, they did not tongue-flick or bite swabs in multiple tests. These observations strengthen the previous assumptions, but more extensive tests and field observations are needed for confirmation.

In a new analysis of correlated evolution between prey chemical discrimination and foraging mode, I have excluded herbivores and omnivores (Appendix 8.1b). Herbivores such as iguanids cannot ambush plants, and virtually all exhibit prey chemical discrimination regardless of the foraging behavior of related insectivores (Cooper, 2002, 2003a). With herbivores excluded, all iguanians in the analysis are ambush foragers. On the phylogeny in Fig. 8.2 (iguanian phylogeny based on Macey *et al.*, 1997) there are two simultaneous gains and three simultaneous losses of both traits. Significant correlated evolution has occurred between the traits (concentrated changes test, $p < 0.014$). In related tests conducted dividing lizards into ambush foragers versus non-ambushers (active foragers, omnivores, and herbivores), the relationship is even stronger (concentrated changes test, $p \ll 0.001$).

The advantages to active foragers of using chemical cues to locate hidden or immobile prey and to follow scent trails are obvious. The helodermatid *Heloderma suspectum* (Bogert and Martín del Campo, 1956), the Bengal monitor, *Varanus bengalensis* (Auffenberg, 1984), and the teiid *Cnemidophorus uniparens* use chemical cues to detect prey hidden beneath soil (Cooper *et al.*, 2001b), and the Komodo dragon, *V. komodoensis*, follows prey scent trails (Auffenberg, 1981).

It is less obvious that prey chemical discrimination has little or no utility for ambush foragers. Once an ambush forager has settled upon an ambush post and becomes immobile, it cannot locate prey by tongue-flicking at that site. Worse, because ambush foragers rely on immobility to avoid detection by prey and predators (Vitt and Congdon, 1978; Vitt and Price, 1982), the motion of tongue-flicking might reveal their presence (Cooper, 1994b, 1995b). However, because ambush foragers tongue-flick substrates most frequently when they first arrive at new sites (Simon *et al.*, 1981; Cooper *et al.*, 1994b), they might use the presence of prey chemical cues to select ambush sites.

Two recent studies suggest that ambush foragers do not use chemical cues on substrates to select ambush posts, i.e. to enhance foraging efficiency by staying longer at sites where prey chemical cues occur. In a laboratory experiment the phrynosomatid *Sceloporus malachiticus* was offered a choice between two ambush posts having minimally overlapping views, one bearing prey chemicals, the other not (Cooper, 2003b). The lizards occupied posts randomly with respect to prey scent. When the cordylid *Platysaurus broadleyi* was presented with tiles labeled with prey chemicals or control substances in the field, prey chemicals had no effect on giving-up time, i.e. the time spent at an ambush post (Cooper and Whiting, 2003). Thus, it appears that ambush foragers not only do not use chemical cues to search for prey, but further do not use prey chemical cues to select ambush posts. In many species that select elevated ambush posts, but attack prey moving below, presence of chemical cues at the elevated site give little or no indication about the presence of prey in the foraging area. Thus, all available evidence suggests that ambush foragers do not use chemical cues to find prey or to select ambush sites. However, the possibility remains that they can identify prey by using chemical cues, but give no observable indication of doing so because search for prey is visual and attack is triggered by visual cues.

Strike-induced chemosensory searching and foraging mode

Strike-induced chemosensory searching (SICS) is a method of relocating prey that has been bitten or struck and then voluntarily released, dropped accidentally, or lost owing to escape efforts by the prey. It was first described in rattlesnakes and other venomous snakes (Chiszar and Scudder, 1980) that strike to envenomate potentially dangerous prey, then release the prey and wait for a brief interval, during which the prey may wander some distance before the venom overcomes it. Then the snake relocates the prey by scent-trailing and consumes it without threat of retaliation. Some non-venomous snakes and lizards also exhibit SICS (Cooper, 1989c; Cooper *et al.*, 1989), but lack the strike–release–trail strategy of viperids. After bitten prey are removed from their mouths experimentally, non-venomous snakes show an increase in tongue-flick rate and initiate searching movements (Cooper, 2003c). In most lizards the duration of the post-strike elevation in tongue-flicking rate is only a minute or two, in contrast to up to two hours or even more in rattlesnakes. This difference is undoubtedly due in part to the high probability that an envenomated prey has died and remains available nearby and the low probability that a mobile insect does so following its escape. The greater abundance of smaller prey and lower energetic content per item also favor shorter giving-up time for searching. Some of the most active foragers among lizards also eat larger prey and exhibit greater durations of searching (Cooper, 1993; Cooper *et al.*, 1994a).

I conducted a comparative experimental study of SICS in representatives of 16 families of lizards and the tuatara as an outgroup to assess the relationship between SICS and both foraging behavior and prey chemical discrimination. The phylogenetic distribution of SICS after removal of bitten prey was identical to those of prey chemical discrimination and foraging mode (Fig. 8.3); (Cooper, 2003c). Only active foragers and the herbivorous desert iguana exhibited SICS, which was absent in the common ancestor of lizards and all iguanians except the representative of Iguanidae. It evolved independently in Eublepharidae and the common ancestor of Autarchoglossa. The sole reversion to ambush foraging (in Cordylidae) was accompanied by loss of SICS. Significant correlated evolution has occurred between SICS and foraging mode in lizards (Cooper, 2003c). Thus, ambush foragers, which attack prey

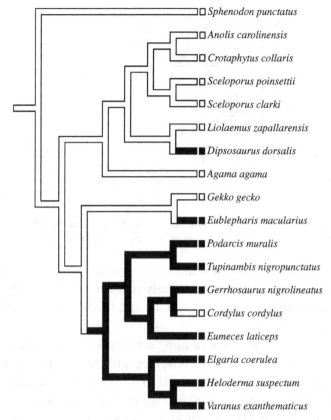

Fig. 8.3. Phylogenetic distributions of strike-induced chemosensory searching and foraging mode coincide. White, absent; black, present.

based on visual cues, do not use chemical cues to search for lost prey and may lack the ability to do so. In active foragers SICS represents use of their typical active search and chemical sampling to relocate lost prey known to be nearby or perhaps to find other individuals of the same prey type in prey species that aggregate. During SICS active search is likely to be more area-concentrated than during typical active foraging, but this remains to be studied.

Diet, food chemical discrimination, and foraging mode

For active foragers that locate and identify food by using chemical cues, responsiveness to prey chemicals must correspond to diet to be efficient. This correspondence is typical for snakes (see, for example, Arnold, 1981; Burghardt, 1969; Cooper *et al.*, 1990; Cooper *et al.*, 2000). However, because most lizards are generalist predators of small animals, they may be expected to

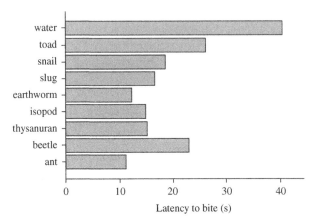

Fig. 8.4. Latency to bite swabs bearing chemical stimuli from potential prey types and deionized water, an odorless control (Cooper and Pérez-Mellado, 2002a).

respond to chemical cues from a wide range of potential prey. For example, the lacertid *Podarcis lilfordi* responds strongly to chemicals from three orders of insects, isopods, earthworms, slugs, snails, and toads (Fig. 8.4); (Cooper and Pérez-Mellado, 2002a).

Few prey specialists are known among actively foraging lizards, but the gila monster (*Heloderma suspectum*) consumes primarily small mammals and birds, including eggs (Beck, 1990). In tests of prey chemical discrimination, gila monsters bit more cotton balls bearing chemical cues from mice than from earthworms and fish, and tongue-flicked at higher rates to mouse cues than to earthworms, fish, another lizard species, and crickets (Cooper and Arnett, 2003). This finding suggests that responsiveness to chemical cues has been selected to correspond to diet, but comparative data are needed to determine whether correlated evolution has occurred between diet and response strength to chemical cues in insectivorous/carnivorous species.

Insectivorous/carnivorous lizards tested informally early in my investigations typically did not respond strongly to plant chemicals, but desert iguanas, which are primarily herbivorous, did so despite being derived from ancestral insectivorous ambush foragers (Cooper and Alberts, 1990). This suggested that any correspondence between diet and responsiveness to food chemical might be detectable in a comparative survey of responses to plant chemicals in insectivores and in plant eaters. Even herbivorous lizards whose immediate ancestors were ambush foragers may be expected to have evolved plant chemical discrimination. They are not constrained from moving during foraging, must move to reach plant food, and could benefit from abilities to identify nutritionally beneficial plant parts and stages of maturity and to avoid consuming chemically defended species.

I conducted experiments to determine whether prey and plant chemical discrimination are present in a series of insectivores, omnivores, and herbivores from a wide range of lizard taxa and in the tuatara, *Sphenodon punctatus*, as an outgroup. In both Iguania (16 species from 6 families) and Scleroglossa (21 species from 11 families), omnivorous and herbivorous species exhibited plant chemical discrimination (Cooper, 2002, 2003a). Only one marginally omnivorous iguanian species did not respond preferentially to plant chemicals (Cooper *et al.*, 2001a). Another possible exception is the omnivorous tropidurid *Leiocephalus inaguae*. Prey chemical discrimination was not detected in this species, but the method of testing was different from currently used methods and responses to plant chemicals were not examined (Noble and Kumpf, 1936). Herbivorous iguanians additionally acquired prey chemical discrimination (Cooper, 2003a), perhaps because they consume some prey or perhaps as a side effect of similarities between prey chemicals and chemicals that indicate desirable nutritional qualities in plant food.

A large majority of ambushing insectivores did not exhibit discriminative responses to either plant or prey chemicals, whereas actively foraging insectivores exhibited prey chemical discrimination, but not plant chemical discrimination (Appendix 8.1b). Two exceptions have been discovered, both involving fig fruits. The cordylid *Platysaurus broadleyi*, which usually ambushes insects but eats substantial quantities of fig fruits when available, responds strongly to fig chemicals, yet does not exhibit prey chemical discrimination (Whiting, 1999). The actively foraging lacertid *Lacerta perspicillata* also is primarily insectivorous, but consumes fig fruits in season. This species responds strongly to both prey and fig chemicals (Cooper and Pérez-Mellado, 2002b; A. Perera, unpublished data). Because some lacertids respond strongly to sugar, responses to figs might be gustatory rather than vomerolfactory.

Significant correlated evolution has occurred between plant diet and plant chemical discrimination in both Iguania (Fig. 8.5) and Scleroglossa (Fig. 8.6), as well as within ambush foragers and active foragers (Cooper, 2002, 2003a). Data for lacertids not used in Cooper (2002) are incorporated in Fig. 8.6. Based on these data, significant correlated evolution has occurred in Scleroglossa (concentrated changes test, $p \ll 0.001$) between plant diet and plant chemical discrimination. These findings for the two major lizard taxa strongly confirm the hypothesis that responsiveness to chemical cues is evolutionarily adjusted to match changes in diet.

Foraging mode influences not only lingual–vomeronasal structure and function and food chemical discrimination, but diet selection (Vitt and Pianka, this volume, Chapter 5). Huey and Pianka (1981) proposed that active

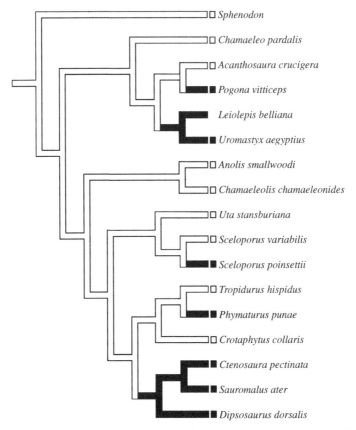

Fig. 8.5. In Iguania the distributions of plant chemical and prey chemical discrimination are identical to that shown for substantial plant consumption (omnivory and herbivory) with the exception that discriminations are absent in *Sceloporus poinsettii*. White, absent; black, present.

foragers are more likely than active foragers to eat patchily distributed prey and that active foragers should be more likely to consume ambushing prey and *vice versa*. Because active foragers move much more extensively than ambush foragers, they might be preadapted for wide searching for plant food when conditions favoring plant consumption arise. To become omnivorous or herbivorous, ambush foragers probably would have to evolve search movements *de novo*. A survey of diets of hundreds of lizard species revealed 10 independent origins of omnivory or herbivory in 221 ambush foragers and 22 independent origins in 231 species of active foragers (Cooper and Vitt, 2002). Evolution of plant consumption has been significantly favored by active foraging (Cooper and Vitt, 2002). Chapter 7 discusses consequences of herbivory for foraging behavior and feeding physiology and morphology.

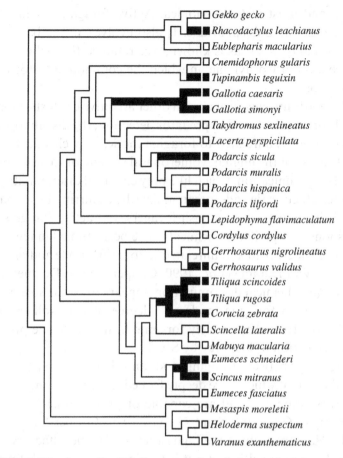

Fig. 8.6. In Scleroglossa the distributions of plant chemical and prey chemical discrimination are identical to that shown for substantial plant consumption (omnivory and herbivory). White, absent, black, present.

Foraging mode, food chemical discrimination, lingual–vomeronasal system, and lizard evolution

Foraging behavior may have had a profound impact on the evolution of lizard taxonomic diversity, as well as foraging and dietary diversity. The strong relationships between degree of specialization of the lingual–vomeronasal system and prey chemical discrimination on the one hand and between prey chemical discrimination and foraging behavior on the other suggest that foraging might have influenced evolution of lingual morphology and vomer-olfactory receptor abundance. Relatively few shifts in foraging mode have occurred in lizards, and some are linked to the origins of major taxa. The ancestral ambush foraging was retained in Iguania, the basal branch of

Scleroglossa, and most of Gekkonoidea. Active foraging originated within, but near the base of, Eublepharidae. It also evolved in the basal branch of Autarchoglossa and, so far as is known, was retained throughout that taxon with the exception of Cordylidae, in which the common ancestor reverted to ambush foraging.

Although lizards use chemical cues in other adaptive contexts, most notably pheromonal communication and predator detection, the distributions of these chemosensory roles is unrelated to foraging mode, prey chemical discrimination, specialization of the lingual–vomeronasal system, or to lizard phylogeny. Pheromonal communication, which like prey chemical discrimination is mediated by vomerolfaction, is widespread in lizards, occurring in all major lizard taxa studied, including both ambush foragers and active foragers (Cooper, 1994a). Responses to predator chemicals have been studied in fewer taxa, but also seem to be widespread and unrelated to foraging mode and prey chemical discrimination (Van Damme et al., 1990; Cooper, 1994a; Downes and Shine, 1998). Thus, food chemical discrimination appears to be the sole basis for the relationship between foraging mode and degree of vomerolfactory specialization in lizards. Its use by actively foraging lizards to locate prey has been extended to relocation of lost prey by SICS.

The relationships between foraging mode and chemosensory behavior break down in snakes, all of which have deeply forked, elongated tongues, abundant vomerolfactory receptors, and are capable of prey chemical discrimination (see, for example, Gabe and Saint Girons, 1976; Cooper, 1990b, 1996; see Chapter 11). Having attained the highest level of vomerolfactory specialization, snakes have retained it and put it to use even in ambush foragers. For example, SICS reaches its peak development in venomous rattlesnakes and other viperid snakes, which are ambush foragers (Chiszar and Scudder, 1980). Rattlesnakes also use chemical cues to locate profitable foraging areas, but then lie in wait for prey (Duvall et al., 1985).

Since a seminal paper by Huey and Pianka (1981) and a later chapter by Huey and Bennett (1986) predicted that foraging mode affects numerous aspects of lizard ecology and presented empirical evidence supporting some of the predictions, it has been customary to view correlates of foraging mode as consequences of foraging mode. My studies of relationships among foraging mode, chemosensory structure, and prey chemical discrimination support this view. However, differences in foraging mode were not intended to be, and do not provide, a comprehensive explanation for the ecological and taxonomic diversity of lizards.

Adaptations for feeding and foraging account for a great deal of lizard diversity. Loss of the ancestral lingual prey prehension by scleroglossans

(Schwenk and Throckmorton, 1989) freed the tongue to be molded evolutionarily for other tasks, especially chemical sampling for vomerolfaction. Loss of lingual prey prehension was thus a key innovation that separated Iguania and Scleroglossa (Schwenk, 2000; Vitt *et al.*, 2003) and allowed further diversification within Scleroglossa as active foraging and improved vomerolfaction co-evolved to enhance ability to find hidden food by tongue-flicking. Another important factor in the evolution of lizard diversity that is not a direct consequence of foraging mode is the shift to nocturnality by geckos (Vitt *et al.*, 2003). However, geckos may have become nocturnal to avoid competition for food with competitively superior, actively foraging autarchoglossans (Vitt *et al.*, 2003). Other important innovations are discussed by Vitt *et al.* (2003).

Tongue-flicking for vomerolfactory sampling presumably arose shortly after the evolution of vomeronasal ducts opening into the roof of the mouth in the common ancestor of Squamata (Cooper, 1994a). However, selection favoring use of the lingual–vomeronasal system by insectivorous/carnivorous taxa for prey chemical discrimination to locate and identify food was delayed until the advent of active foraging. Conditions that favor active foraging are poorly known, but may include competition among ambushers, availability of abundant food resources not active on the surface, ability to detect immobile prey on the surface, and other improvements in ability to detect prey not as readily accessible to ambush foragers.

The advent of wide foraging in lizards already employing tongue-flicking for chemical sampling in contexts other than foraging very likely spurred the evolution of prey chemical discrimination. The wide distribution of pheromonal communication and perhaps use of chemical cues to detect predators suggests that early lizards tongue-flicked substrates, perhaps as do modern ambush foragers upon reaching new locations (Cooper *et al.*, 1994b). Once lizards began to move more frequently while foraging, lingual contact with prey chemicals on substrates would have provided the opportunity for natural selection to promote evolution of prey chemical discrimination.

Use of prey chemical discrimination in turn favored selection for increased vomeronasal discriminatory power, accounting for the increasing abundance of vomerolfactory receptors in active foragers. Similarly, selection for enhanced sampling ability accounts for the greater elongation of the tongue and deepening of notch at its tip in active foragers. Evolution of progressively active foraging in some autarchoglossan taxa led to increasing vomerolfactory receptor abundance and to increasingly elongated and more deeply notched tongues, as well as to progressive decrease in abundance of taste buds. In the most highly active foragers such as varanids and teiids, this process has culminated in the most abundant vomerolfactory receptors and in highly elongated, forked tongues having few or no taste buds.

Although adoption of a plant diet leads to plant chemical discrimination, prey chemical discrimination has had a much greater impact on lizard diversification than has plant chemical discrimination. About 88% of lizard species consume little or no plant matter; only 11% are omnivores and 1% are herbivores (Cooper and Vitt, 2002). As described, several changes in lingual–vomeronasal morphology are associated with prey chemical discrimination and active foraging. Whether or not plant chemical discrimination promotes further lingual–vomeronasal enhancements in families of active foragers or has any effects on chemosensory morphology in plant eaters derived from ambush foragers is unknown. However, any such effects must be minor. Species that consume significant quantities of plant are widely distributed among lizard taxa and do not exhibit marked morphological differences in lingual or vomeronasal morphology from congeneric and confamilial insectivores. That some effect may occur is suggested by the greater abundance of vomerolfactory receptors in iguanids, which are herbivores, than in insectivorous species in other iguanian families (Gabe and Saint Girons, 1976).

Thus, differences in degree of foraging activity among insectivores may have driven a substantial part of the evolutionary diversification of lizards via alterations in chemosensory behavior and morphology. Once loss of lingual prey prehension separated Scleroglossa and Iguania, progressive changes occurred in scleroglossans, especially within Autarchoglossa. This scenario can account for the close resemblance between a classification of lizards based solely on chemosensory characters (Schwenk, 1993a) and the current classification based on a wide range of characters, few of which are chemosensory (Estes *et al.*, 1988). Despite the existence of other key innovations affecting lizard evolution, differences in foraging behavior have played a central role in generation of diversity.

References

Arnold, S. J. (1981). Behavioral variation in natural populations. I. Phenotypic, genetic and environmental correlations between chemoreceptive responses to prey in the garter snake, *Thamnophis elegans. Evolution* **35**, 489–509.

Arnold, E. N. (1984). Ecology of lowland lizards in the eastern United Arab Emirates. *J. Zool. Lond.* **204**, 329–54.

Auffenberg, W. (1981). *The Behavioral Ecology of the Komodo Monitor.* Gainesville, FL: University Presses of Florida.

Auffenberg, W. (1984). Notes on the feeding behaviour of *Varanus bengalensis* (Sauria: Varanidae). *J. Bombay Nat. Hist. Soc.* **80**, 286–302.

Beck, D. D. (1990). Ecology and behavior of the gila monster in southwestern Utah. *J. Herpetol.* **24**, 54–68.

Bertmar, G. (1981). Evolution of vomeronasal organs in vertebrates. *Evolution* **35**, 359–66.

Bissinger, B. E. and Simon, C. A. (1979). Comparison of tongue extrusions in representatives of six families of lizards. *J. Herpetol.* **13**, 133–9.

Bodnar, J. D., Poulos, D. A. and Tapper, D. N. (1975). A survey of the peripheral trigeminal system in the common boa constrictor. *Neurosci. Abstr.* **1(874)**, 561.

Bogert, C. M. and Martin del Campo, R. (1956). The gila monster and its allies: the relationships, habits, and behavior of the lizards of the family Helodermatidae. *Bull. Amer. Mus. Nat. Hist.* **109**, 1–238.

Burghardt, G. M. (1967). Chemical-cue preferences of inexperienced snakes: comparative aspects. *Science* **157**, 718–21.

Burghardt, G. M. (1969). Comparative prey-attack studies in newborn snakes of the genus *Thamnophis*. *Behaviour* **33**, 77–114.

Burghardt, G. M. and Pruitt, C. H. (1975). The role of the tongue and senses in feeding of newborn garter snakes. *Physiol. Behav.* **14**, 185–94.

Chiszar, D. and Scudder, K. M. (1980). Chemosensory searching by rattlesnakes during predatory episodes. In *Chemical Signals in Vertebrates and Aquatic Invertebrates*, ed. D. Muller-Schwarze and R. M. Silverstein, pp. 125–39. New York: Plenum Press.

Chou, L. M., Leong, C. F. and Choo, B. L. (1988). The role of optic, auditory and olfactory senses in prey hunting by two species of geckos. *J. Herpetol.* **22**, 349–51.

Cooper, W. E. Jr. (1989a). Prey odor discrimination in the varanoid lizards *Heloderma suspectum* and *Varanus exanthematicus*. *Ethology* **81**, 250–8.

Cooper, W. E. Jr. (1989b). Absence of prey odor discrimination by iguanid and agamid lizards in applicator tests. *Copeia* **1989**, 472–8.

Cooper, W. E. Jr. (1989c). Strike-induced chemosensory searching occurs in lizards. *J. Chem. Ecol.* **15**, 1311–20.

Cooper, W. E. Jr. (1990a). Prey odor detection by teiid and lacertid lizards and the relationship of prey odor detection to foraging mode in lizard families. *Copeia* **1990**, 237–42.

Cooper, W. E. Jr. (1990b). Prey odour discrimination by lizards and snakes. In *Chemical Signals in Vertebrates*, ed. D. W. Macdonald, D. Muller-Schwarze and S. E. Natynczuk, pp. 533–8. Oxford: Oxford University Press.

Cooper, W. E. Jr. (1993). Duration of poststrike elevation in tongue-flicking rate in the savannah monitor lizard, *Varanus exanthematicus*. *Ethol. Ecol. Evol.* **5**, 1–18.

Cooper, W. E. Jr. (1994a). Chemical discrimination by tongue-flicking in lizards: a review with hypotheses on its origin and its ecological and phylogenetic relationships. *J. Chem. Ecol.* **20**, 439–87.

Cooper, W. E. Jr. (1994b). Prey chemical discrimination, foraging mode, and phylogeny. In *Lizard Ecology: Historical and Experimental Perspectives*, ed. L. J. Vitt and E. R. Pianka, pp. 95–116. Princeton, NJ: Princeton University Press.

Cooper, W. E. Jr. (1995a). Evolution and function of lingual shape in lizards, with emphasis on elongation, extensibility, and chemical sampling. *J. Chem. Ecol.* **21**, 477–505.

Cooper, W. E. Jr. (1995b). Foraging mode, prey chemical discrimination, and phylogeny in lizards. *Anim. Behav.* **50**, 973–85.

Cooper, W. E. Jr. (1996). Variation and evolution of forked tongues in squamate reptiles. *Herp. Nat. Hist.* **4**, 135–50.

Cooper, W. E. Jr. (1997a). Correlated evolution of prey chemical discrimination with foraging, lingual morphology, and vomeronasal chemoreceptor abundance in lizards. *Behav. Ecol. Sociobiol.* **41**, 257–65.

Cooper, W. E. Jr. (1997b). Independent evolution of squamate olfaction and vomerolfaction and correlated evolution of vomerolfaction and lingual structure. *Amph.-Rept.* **18**, 85–105.

Cooper, W. E. Jr. (1998). Evaluation of swab and related tests as a bioassay for assessing responses by squamate reptiles to chemical stimuli. *J. Chem. Ecol.* **24**, 841–66.

Cooper, W. E. Jr. (1999). Supplementation of phylogenetically correct data by two species comparison: support for correlated evolution of foraging mode and prey chemical discrimination in lizards extended by first intrageneric evidence. *Oikos* **86**, 97–104.

Cooper, W. E. Jr. (2000). An adaptive difference in the relationship between foraging mode and responses to prey chemicals in two congeneric scincid lizards. *Ethology* **106**, 193–206.

Cooper, W. E. Jr. (2002). Convergent evolution of plant chemical discrimination by omnivorous and herbivorous scleroglossan lizards. *J. Zool. Lond.* **257**, 53–66.

Cooper, W. E. Jr. (2003a). Correlated evolution of herbivory and food chemical discrimination in iguanian and ambush foraging lizards. *Behav. Ecol.* **14**, 409–6.

Cooper, W. E. Jr. (2003b). Prey chemicals do not affect perch choice by an ambushing lizard, *Sceloporus malachiticus*. *J. Herpetol.* **37**, 197–9.

Cooper, W. E. Jr. (2003c). Foraging mode and evolution of strike-induced chemosensory searching in lizards. *J. Chem. Ecol.* **29**, 995–1008.

Cooper, W. E. Jr. and Alberts, A. C. (1990). Responses to chemical food stimuli by an herbivorous actively foraging lizard, *Dipsosaurus dorsalis*. *Herpetologica* **46**, 259–6.

Cooper, W. E. Jr. and Alberts, A. C. (1991). Tongue-flicking and biting in response to chemical food stimuli by an iguanid lizard (*Dipsosaurus dorsalis*) having sealed vomeronasal ducts: Vomerolfaction may mediate these behavioral responses. *J. Chem. Ecol.* **17**, 135–46.

Cooper, W. E. Jr. and Arnett, J. (2003). Correspondence between diet and chemosensory responsiveness by helodermatid lizards. *Amph.-Rept.* **24**, 86–91.

Cooper, W. E. Jr. and Burghardt, G. M. (1990a). Volmerolfaction and vomodor. *J. Chem. Ecol.* **16**, 103–5.

Cooper, W. E. Jr. and Burghardt, G. M. (1990b). A comparative analysis of scoring methods for chemical discrimination of prey by squamate reptiles. *J. Chem. Ecol.* **16**, 45–65.

Cooper, W. E. Jr. and Habegger, J. J. (2000). Lingual and biting responses to food chemicals by some eublepharid and gekkonid geckos. *J. Herpetol.* **34**, 360–8.

Cooper, W. E. Jr. and Pérez-Mellado, V. (2001a). Location of fruit using only airborne odor cues by a lizard. *Physiol. Behav.* **74**, 339–2.

Cooper, W. E. Jr. and Pérez-Mellado, V. (2001b). Chemosensory responses to sugar and fat by the omnivorous lizard *Gallotia caesaris* with behavioral evidence suggesting a role for gustation. *Physiol. Behav.* **73**, 509–6.

Cooper, W. E. Jr. and Pérez-Mellado, V. (2002a). Responses by a generalist predator, the Balearic lizard *Podarcis lilfordi*, to chemical cues from taxonomically diverse prey. *Acta Ethol.* **4**, 119–24.

Cooper, W. E. Jr. and Perez-Mellado, V. (2002b). Responses to food chemicals by two insectivorous and one omnivorous species of lacertid lizards. *Neth. J. Zool.* **52**, 11–28.

Cooper, W. E. Jr. and Van Wyk, J. H. (1994). Absence of prey chemical discrimination by tongue-flicking in an ambush-foraging lizard having actively foraging ancestors. *Ethology* **97**, 317–328.

Cooper, W. E. Jr. and Vitt, L. J. (2002). Distribution, extent, and evolution of plant consumption by lizards. *J. Zool. Lond.* **257**, 487–517.

Cooper, W. E. Jr. and Whiting, M. J. (1999). Foraging modes in lacertid lizards from southern Africa. *Amph.-Rept.* **20**, 299–311.

Cooper, W. E. Jr. and Whiting, M. J. (2003). Prey chemicals do not affect giving-up-time at ambush posts by the cordylid lizard *Platysaurus broadleyi*. *Herpetologica* **59**, 455–8.

Cooper, W. E. Jr., Burghardt, G. M. and Brown, W. S. (2000). Behavioural responses by hatchling racers (*Coluber constrictor*) from two geographically distinct populations to chemical stimuli from prey and predators. *Amph.-Rept.* **21**, 103–15.

Cooper, W. E. Jr., Buth, D. G. and Vitt, L. J. (1990). Prey odor discrimination by ingestively naive coachwhip snakes (*Masticophis flagellum*). *Chemoecology* **1**, 86–91.

Cooper, W. E. Jr., Caldwell, J. P., Vitt, L. J., Pérez-Mellado, V. and Baird, T. A. (2002a). Food-chemical discrimination and correlated evolution between plant diet and plant-chemical discrimination in lacertiform lizards. *Can. J. Zool.* **80**, 655–63.

Cooper, W. E. Jr., Deperno, C. S. and Arnett, J. (1994a). Prolonged poststrike elevation in tongue-flicking rate with rapid onset in gila monster, *Heloderma suspectum*: relation to diet and foraging and implications for evolution of chemosensory searching. *J. Chem. Ecol.* **20**, 2867–81.

Cooper, W. E. Jr., Ferguson, G. W. and Habegger, J. J. (2001a). Responses to animal and plant chemicals by several iguanian insectivores and the tuatara, *Sphenodon punctatus*. *J. Herpetol.* **35**, 255–63.

Cooper, W. E. Jr., McDowell, S. G. and Ruffer, J. (1989). Strike-induced chemosensory searching in the colubrid snakes *Elaphe g. guttata* and *Thamnophis sirtalis*. *Ethology* **81**, 19–28.

Cooper, W. E. Jr., Pérez-Mellado, V. and Vitt, L. J. (2002b). Responses to major categories of food chemicals by the lizard *Podarcis lilfordi*. *J. Chem. Ecol.* **28**, 689–700.

Cooper, W. E. Jr., Pérez-Mellado, V. and Vitt, L. J. (2003). Cologne as a pungency control in tests of lizard chemical discrimination: effects of concentration, brand, and simultaneous and sequential presentation. *J. Ethol.* **21**, 101–6.

Cooper, W. E. Jr., Pérez-Mellado, V., Vitt, L. J. and Budzynski, B. (2002c). Behavioral responses to plant toxins by two omnivorous lizard species. *Physiol. Behav.* **76**, 297–303.

Cooper, W. E. Jr., Vitt, L. J. and Caldwell, J. P. (1994b). Movement and substrate tongue flicks in Phrynosomatid lizards. *Copeia* **1994**, 234–7.

Cooper, W. E. Jr., Vitt, L. J., Caldwell, J. P. and Fox, S. F. (2001b). Foraging modes of some American lizards: relationships among measurement variables and discreteness of modes. *Herpetologica* **57**, 65–76.

Cooper, W. E. Jr., Whiting, M. J. and Van Wyk, J. H. (1997). Foraging modes of cordyliform lizards. *S. Afr. J. Zool.* **32**, 9–13.

Cowles, R. B. and Phelan, R. L. (1958). Olfaction in rattlesnakes. *Copeia* **1958**, 77–83.

Ditmars, R. L. (1937). *Snakes of the World*. New York: Macmillan.

Downes, S. and Shine, R. (1998). Sedentary snakes and gullible geckos: predator-prey coevolution in nocturnal rock-dwelling reptiles. *Anim. Behav.* **55**, 1373–85.

Du Toit, A., Mouton, P. le F. N., Geertsema, H. and Fleming, A. F. (2002). Foraging mode of serpentiform, grass-living cordylid lizards: a case study of *Cordylus anguinus*. *Afr. Zool.* **37**, 141–9.

Duvall, D., King, M. B. and Gutzwiller, K. J. (1985). Behavioral ecology and ethology of the prairie rattlesnake. *Nat. Geogr. Res.* **1985**, 80–111.

Estes, R., De Queiroz, K. and Gauthier, J. (1988). Phylogenetic relationships within Squamata. In *Phylogenetic Relationships of the Lizard Families*, ed. R. Estes and G. Pregill, pp. 119–281. Stanford, CA: Stanford University Press.

Evans, L. T. (1961). Structure as related to behavior in the organization of populations of reptiles. In *Vertebrate Speciation*, ed. W. F. Blair, pp. 148–78. Houston, TX: University of Texas Press.

Felsenstein, J. (1985). Phylogenies and the comparative method. *Am. Nat.* **125**, 1–15.

Fitzgerald, F. S. (1945). The note-books. In *The Crack-Up*. ed. E. Wilson. New York: New Directions, pp. 127–128.

Gabe, M. and Saint Girons, H. (1976). Contribution à la morphologie comparée des fosses nasales et de leurs annexes chez lepidosauriens. *Mem. Mus. Natl. Hist. Nat., Nouv. Ser.* **A98**, 1–87 + 49 figs., 10 pl.

Garrett, C. M. and Card, W. C. (1993). Chemical discrimination of prey by naive neonate Gould's monitors *Varanus gouldii*. *J. Chem. Ecol.* **19**, 2599–604.

Gillingham, J. C. and Clark, D. L. (1981). Snake tongue-flicking: transfer mechanics to Jacobson's organ. *Can. J. Zool.* **59**, 1651–7.

Gove, D. (1979). A comparative study of snake and lizard tongue-flicking, with an evolutionary hypothesis. *Z. Tierpsychol.* **51**, 58–76.

Graves, B. M. and Halpern, M. (1989). Chemical access to the vomeronasal organs of the lizard *Chalcides ocellatus*. *J. Exp. Zool.* **249**, 150–7.

Graves, B. M. and Halpern, M. (1990). Roles of vomeronasal organ chemoreception in tongue flicking, exploratory, and feeding behaviour of the lizard, *Chalcides ocellatus*. *Anim. Behav.* **39**, 692–8.

Greene, H. W. (1988). Antipredator mechanisms in reptiles. In *Biology of the Reptilia*, vol. 16., ed. C. Gans and R. B. Huey, pp. 1–152. New York: Alan R. Liss.

Halpern, M. (1992). Nasal chemical senses in reptiles: structure and function. In *Biology of the Reptilia*, vol. 18., ed. C. Gans and D. Crews, pp. 423–523. Chicago, IL: University of Chicago Press.

Halpern, M. and Frumin, N. (1979). Roles of the vomeronasal and olfactory systems in prey attack and feeding in adult garter snakes. *Physiol. Behav.* **22**, 1183–9.

Halpern, M., Halpern, J., Erichsen, E. and Borghjid, S. (1997). The role of nasal chemical senses in garter snake responses to airborne odor cues from prey. *J. Comp. Psychol.* **111**, 251–60.

Huey, R. B. and Bennett, A. F. (1986). A comparative approach to field and laboratory studies in evolutionary biology. In *Predator Prey Relationships: perspectives and approaches from the study of lower vertebrates*. ed. M. E. Feder and G. V. Lauder. Chicago: University of Chicago Press, pp. 82–98.

Huey, R. B. and Pianka, E. R. (1981). Ecological consequences of foraging mode. *Ecology* **62**, 991–9.

Inouchi, J., Wang, D., Jiang, X. C., Kubie, J. L. and Halpern, M. (1993). Electrophysiological analysis of the nasal chemical senses in garter snakes. *Brain Behav. Evol.* **41**, 171–82.

Kaufman, J. D., Burghardt, G. M. and Phillips, J. A. (1996). Sensory cues and foraging decisions in a large carnivorous lizard, *Varanus albigularis*. *Anim. Behav.* **52**, 727–6.

Klauber, L. M. (1956). *Rattlesnakes, Their Habits, Life Histories, and Influence on Mankind*. Los Angeles, CA: University of California Press.

Kubie, J. L. and Halpern, M. (1979). Chemical senses involved in garter snake prey trailing. *J. Comp. Physiol. Psychol.* **93**, 648–67.

López, P., Martín, J. and Cooper, W. E., Jr. (2002). Chemosensory responses to plant chemicals by the amphisbaenian *Blanus cinereus*. *Amph.-Rept.* **23**, 348–53.

Macey, J. R., Larson, A., Ananjeva, N. B. and Papenfuss, T. J. (1997). Evolutionary shifts in three major structural features of the mitochondrial genome among iguanian lizards. *J. Molec. Evol.* **44**, 660–74.

Maddison, W. P. (1990). A method for testing the correlated evolution of two binary characters: Are gains or losses concentrated on certain branches of a phylogenetic tree? *Evolution* **44**, 539–57.

Maddison, W. P. and Maddison, D. R. (1992). *Macclade: Analysis of Phylogeny and Character Evolution*, version 3.0. Sunderland, MA: Sinauer Associates.

Marcos-Leon, M. B. (1999) Ecolofisiologia y estrategias de obtencion de alimento en cuatro especies de Lacertidae. Doctoral dissertation, University of Salamanca.

Mason, R. T. (1992). Reptilian pheromones. In *Biology of the Reptilia*, vol. 18, ed. C. Gans and D. Crews, pp. 114–228. Chicago, IL: University of Chicago Press.

McDowell, S. B. (1972). The evolution of the tongue of snakes, and its bearing on snake origins. In *Evolutionary Biology*, ed. T. Dobzhansky, M. K. Hecht and W. C. Steere, pp. 191–273. New York: Appleton-Century-Crofts.

Moncrieff, P. W. (1967). *The Chemical Senses*, 3rd edn. London: Hill.

Nelling, C. (1996) Responses to prey odors by three species of lacertid lizards: prey odor discrimination, aged odor detection, strike-induced chemosensory searching, and chemosensory search images. M. S. thesis, Shippensburg University, Shippensburg, Pennsylvania.

Noble, G. K. and Krumpf, K. F. (1936). The function of Jacobsen's organ in lizards. *J. Genet. Psychol.* **48**, 371–82.

Oelofsen, B. W. and Van Den Heever, J. A. (1979). Role of the tongue during olfaction in varanids and snakes. *S. Afr. J. Sci.* **75**, 365–6.

Pagel, M. (1994). Detecting correlated evolution on phylogenies: a general method for the comparative analysis of discrete characters. *Proc. R. Soc. Lond.* B**255**, 37–45.

Parsons, T. S. (1970). The nose and Jacobson's organ. In *Biology of the Reptilia*, vol. 2, ed. C. Gans and T. S. Parsons, pp. 99–191. London: Academic Press.

Perry, G. (1995). The evolutionary ecology of lizard foraging: a comparative study. Ph.D. thesis, University of Texas.

Perry, G. (1999). The evolution of search modes: ecological versus phylogenetic perspectives. *Am. Nat.* **153**, 99–109.

Perry, G., Lampl, I., Lerner, A. *et al.* (1990). Foraging mode in lacertid lizards: variation and correlates. *Amph.-Rept.* **11**, 373–84.

Regal, P. J. (1978). Behavioral differences between reptiles and mammals: an analysis of activity and mental capabilities. In *Behavior and Neurology of Lizards*, ed. N. Greenberg, pp. 183–202. Rockville, MD: National Institute of Mental Health.

Schall, J. J. (1990). Aversion of whiptail lizards (*Cnemidophorus*) to a model alkaloid. *Herpetologica* **46**, 34–9.

Schwenk, K. (1985). Occurrence, distribution and functional significance of taste buds in lizards. *Copeia* **1985**, 91–101.

Schwenk, K. (1986). Morphology of the tongue in the tuatara, *Sphenodon punctatus* (Reptilia: Lepidosauria), with comments on function and phylogeny. *J. Morphol.* **188**, 129–6.

Schwenk, K. (1993a). The evolution of chemoreception in squamate reptiles: a phylogenetic approach. *Brain Behav. Evol.* **41**, 124–37.

Schwenk, K. (1993b). Are geckos olfactory specialists? *J. Zool. Lond.* **229**, 289–302.

Schwenk, K. (1994). Why snakes have forked tongues. *Science* **263**, 1573–7.

Schwenk, K. (2000). Feeding in Lepidosaurs. In *Feeding: Form, Function and Evolution in Tetrapod Vertebrates*, ed. K. Schwenk, pp. 175–291. New York: Academic Press.

Schwenk, K. and Throckmorton, G. S. (1989). Functional and evolutionary morphology of lingual feeding in squamate reptiles: phylogenetics and kinematics. *J. Zool. Lond.* **219**, 153–5.

Simon, C. A., Gravelle, K., Bissinger, B. E., Eiss, I. and Ruibal, R. (1981). The role of chemoreception in the iguanid lizard *Sceloporus jarrovi*. *Anim. Behav.* **29**, 46–54.

Stanger-Hall, K. F., Zelmer, D. A., Bergren, C. and Burns, S. A. (2001). Taste discrimination in a lizard (*Anolis carolinensis*, Polychrotidae). *Copeia* **2001**, 490–8.

Toubeau, G., Cotman, C. and Bels, V. (1994). Morphological and kinematic study of the tongue and buccal cavity in the lizard *Anguis fragilis* (Reptilia: Anguidae). *Anat. Rec.* **240**, 423–33.

Traeholt, C. (1994). Notes on the water monitor *Varanus salvator* as a scavenger. *Malay. Nat. J.* **47**, 345–53.

Van Damme, R., Bauwens, D., Vanderstighelen, D. and Verheyen, R. F. (1990). Responses of the lizard *Lacerta vivipara* to predator chemical cues: the effects of temperature. *Anim. Behav.* **40**, 298–305.

Vitt, L. J. and Congdon, J. D. (1978). Body shape, reproductive effort, and relative clutch mass in lizards: resolution of a paradox. *Am. Nat.* **112**, 595–608.

Vitt, L. J. and Price, H. J. (1982). Ecological and evolutionary determinants of relative clutch mass in lizards. *Herpetologica* **38**, 237–5.

Vitt, L. J., Pianka, E. R., Cooper, W. E. and Schwenk, K. (2003). History and global ecology of squamate reptiles. *Am. Nat.* **162**, 44–60.

Werner, Y. L., Okada, S., Ota, H., Perry, G. and Tokunaga, S. (1997). Varied and fluctuating foraging modes in nocturnal lizards of the family Gekkonidae. *Asiatic Herp. Rev.* **7**, 153–65.

Whitaker, A. H. (1987). The roles of lizards in New Zealand plant reproductive strategies. *New Zealand J. Bot.* **25**, 315–28.

Whiting, M. J. (1999). When to be neighbourly: differential agonistic responses in the lizard *Platysaurus broadleyi*. *Behav. Ecol. Sociobiol.* **46**, 210–14.

Wilde, W. S. (1938). The role of Jacobson's organ in the feeding reaction of the common garter snake (*Thamnophis sirtalis sirtalis*). *J. Exp. Zool.* **77**, 445–65.

Wymann, M. A. and Whiting, M. J. (2002). Foraging ecology of rainbow skinks (*Mabuya margaritifer*) in southern Africa. *Copeia* **2002**, 943–57.

Young, B. A. (1993). Evaluating hypotheses for the transfer of stimulus particles to Jacobson's organ in snakes. *Brain Behav. Evol.* **41**, 203–9.

Appendix 8.1 *Chemosensory and foraging mode data for lizards and tuataras that have been tested for food chemical discrimination*

References not cited elsewhere follow the data. AC, active forager; AM, ambush forager; FM, foraging mode; PlantCD, plant chemical discrimination; PreyCD, prey chemical discrimination.

(a) Insectivores–carnivores

Taxon		PreyCD	PlantCD	FM
Rhynchocephalia				
	Sphenodon punctatus	—	—	AM
Squamata				
Iguania	Agamidae			
	Acanthosaura crucigera	—	?	AM
	Agama agama	—	?	AM
	Calotes mystaceus	—	?	AM
	Chamaeleonidae			
	Chamaeleo pardalis	—	?	AM
	Corytophanidae			
	Corytophanes cristatus	—	?	AM
	Crotaphytidae			
	Crotaphytus collaris	—	?	AM
	Iguanidae			
	Ctenosaura pectinata hatchlings	—	?	AM
	Opluridae			
	Oplurus cuvieri	—	?	AM
	Phrynosomatidae			
	Sceloporus clarkii	—	?	AM
	S. jarrovii	—	?	AM
	S. malachiticus	—	?	AM
	S. poinsettii	—	?	AM
	S. undulatus	—	?	AM
	S. variabilis	—	?	AM
	Uta stansburiana	—	?	AM
	Polychrotidae			
	Anolis carolinensis	—	?	AM
	A. lineatopus	—	?	AM
	A. smallwoodi	—	?	AM
	Chamaeleolis smallwoodi	—	?	AM
	Polychrus acutirostris	—	?	AM
	Tropiduridae			
	Liolaemus zapallarensis	—	?	AM
	Tropidurus hispidus	—	?	AM
	T. torquatus	—	?	AM
Scleroglossa				
Gekkota				
	Gekkonidae			
	Gekko gecko	—	—	AM

Appendix 8.1 *(cont.)*

Taxon		PreyCD	PlantCD	FM
	Pachydactylus turneri	—	?	AM
	Thecadactylus rapicauda	—	?	AM
	Eublepharidae			
	Coleonyx brevis	+	?	AC
	C. variegatus	+	?	AC
	Eublepharis macularius	+	—	AC
	Goniourosaurus luii	+	?	AC
Autarchoglossa				
Scincomorpha				
	Lacertidae			
	Acanthodactylus boskianus	+	?	AC
	A. scutellatus	+[a]	?	AM
	Lacerta agilis[b]	+	?	AC
	L. monticola[c]	+	?	AC
	L. perspicillata	+	—	AC
	Podarcis bocagei[c]	+	?	AC
	P. hispanica	+	—	AC
	P. muralis	+	—	AC
	Psammodromus algirus[c]	+	?	AC
	Takydromus septentrionalis	+	?	AC
	T. sexlineatus[b]	+	—	AC
	Teiidae			
	Ameiva exsul	+	?	AC
	A. saurimanensis	+	?	AC
	A. undulata	+	?	AC
	Cnemidophorus gularis	+	–	AC
	Tupinambis nigropunctatus	+	?	AC
	T. rufescens	+	?	AC
	Xantusiidae			
	Lepidophyma flavimaculatum	+	—	?
	Amphisbaenidae			
	Blanus cinereus	+	–[d]	?
	Scincidae			
	Chalcides ocellatus	+	?	AC
	Eumeces fasciatus	+	—	AC
	E. inexpectatus	+	?	AC
	E. laticeps	+	?	AC
	Mabuya acutilabris	-	?	AM
	M. macularia	+	—	AC
	M. margaritifer[e]	-	?	AM
	M. quinquetaeniata	+	?	AC
	M. striata	+	?	AC
	Scincella lateralis	+	—	AC

Appendix 8.1 *(cont.)*

Taxon		PreyCD	PlantCD	FM
	Cordylidae			
	Cordylus anguina[f]	-	?	AM
	C. cordylus	—	—	AM
	Platysaurus broadleyi	—	?	AM
	P. pungweensis	—	?	AM
	Gerrhosauridae			
	Gerrhosaurus nigrolineatus	+	—	AC
Anguimorpha				
	Anguidae			
	Abronia graminea[g]	+	?	AC
	Elgaria coerulea	+	?	AC
	E. multicarinatus	+	?	AC
	Mesaspis moreletii	+	—	AC
	Ophisaurus apodus	+	?	AC
	O. attenuatus	+	?	AC
	Xenosauridae			
	Xenosaurus platyceps	+	?	?
	Helodermatidae			
	Heloderma suspectum	+	—	AC
	Varanidae			
	Varanus albigularis[h]	+	?	AC
	V. bengalensis	+	?	AC
	V. exanthematicus	+	—	AC
	V. gouldii[i]	+	?	AC
	V. komodoensis	+	?	AC
	V. salvator[j]	+	?	AC

(b). Omnivores–herbivores

Taxon		PreyCD	PlantCD
Squamata			
Iguania			
	Chamaeleonidae		
	Leiolepis belliana	+	+
	Pogona vitticeps	+	+
	Uromastyx aegyptius	+	+
	Iguanidae		
	Ctenosaura pectinata (adults)	+	+
	Dipsosaurus dorsalis	+	+
	Sauromalus ater	+	+
	Phrynosomatidae		
	Sceloporus poinsettii	–	–
	Tropiduridae		
	Phymaturus punae	+	+

Appendix 8.1 *(cont.)*

Taxon	PreyCD	PlantCD
Scleroglossa		
Gekkota		
Gekkonidae		
Rhacodactylus leachianus	+	+
R. leachianus	+	+
Autarchoglossa		
Scincomorpha		
Lacertidae		
Gallotia caesaris	+	+
G. simonyi	+	+
Podarcis lilfordi	+	+
P. sicula	+	-
Teiidae		
Cnemidophorus murinus	+	+
T. teguixin	+	+
Scincidae		
Corucia zebrata	+	+
Novoeumeces schneideri	+	+
Scincus mitranus	+	+
Tiliqua rugosa	+	+
T. scincoides	+	+
Gerrhosauridae		
Gerrhosaurus validus	+	+

[a] Strength of prey chemical discrimination less than that of actively foraging congener.

[b] Nelling (1996).

[c] Marcos-León (1999).

[d] López *et al.* (2002).

[e] Wymann and Whiting (2002).

[f] Du Toit *et al.* (2002).

[g] E. De Grauw, unpublished data.

[h] Kaufman *et al.* (1996).

[i] Garrett and Card (1993).

[j] Traeholt (1994).

9

Patterns of head shape variation in lizards: morphological correlates of foraging mode

LANCE D. McBRAYER

Department of Biology, Georgia Southern University

CLAY E. CORBIN

Department of Biological and Allied Health Sciences, Bloomsburg University

Introduction

The relationship between cranial morphology, diet, and feeding performance has been explored in most vertebrate classes. In fact, key biomechanical elements and regions of the skull are known to be associated with various prey types in a wide range of species (Radinsky, 1981; Kiltie, 1982; Lauder, 1991; Zweers *et al.*, 1994; Perez-Barberia and Gordon, 1999). Numerous examples in teleosts have linked form, function, and diet (Lauder, 1991; Turingan *et al.*, 1995; Wainwright, 1996); in birds, beak morphology and lever mechanics have been correlated with various dietary patterns (Beecher, 1962; James, 1982; Barbosa and Moreno, 1999). In mammals, the rostrum (snout) often becomes narrower and incisor tooth structure changes as dietary selectivity increases (Radinsky, 1981; Solounias, 1988; Gordon and Illius, 1994; Biknevicius, 1996).

In lizards (non-ophidian squamates), there are relatively few quantitative and comparative studies relating diet to skull morphology, especially with regard to foraging modes (McBrayer, 2004). Classic works provide descriptions of lizard skull and muscle morphology (see, for example, Haas, 1973; Gomes, 1974). Some functional morphological studies have detailed particularly interesting forms such as the outgroup to lizards, *Sphenodon* (Gorniak *et al.*, 1982), durophagous species (Wineski and Gans, 1984; Gans *et al.*, 1985; Gans and De Vree, 1986, 1987), carnivorous species (Smith, 1982, 1984; Throckmorton and Saubert, 1982), ovophagous species (Herrel *et al.*, 1997b), and herbivorous species (Throckmorton, 1976, 1978, 1980; Herrel and De Vree, 1999; Herrel *et al.*, 1999a). However, these works are far from comprehensive because the limited functional data that do exist come from focal species with distinctive diets.

The most comprehensive body of literature on lizard skull function is on the origin and adaptive significance of cranial kinesis (Rieppel, 1978; Smith, 1980;

Lizard Ecology: The Evolutionary Consequences of Foraging Mode, ed. S. M. Reilly, L. D. McBrayer and D. B. Miles. Published by Cambridge University Press. © Cambridge University Press 2007.

Frazzetta, 1983; Smith and Hylander, 1985; Condon, 1987; De Vree and Gans, 1987; Iordansky, 1990, 1996; Arnold, 1998; Herrel *et al.*, 2000; Metzger, 2002). However, some works have suggested that skull form and muscle architecture are important in shaping the range of dietary specializations in lizards. Gans *et al.* (1985) provided a detailed analysis of jaw adductor muscle architecture and motor patterns to demonstrate how blue-tongued skinks are able to feed on snails. In addition, Herrel *et al.* (1999a) showed that motor patterns change with different types of insect prey. Other works investigated motor patterns and kinematics (Schwenk and Throckmorton, 1989; Herrel *et al.*, 1997a), and one quantified motor patterns and biting performance simultaneously (McBrayer and White, 2002). Herrel and colleagues (2001a, c) quantified how variation in head shape corresponds to biting performance in lacertids and in xenosaurids. However, no one has yet tried to place these functional and morphological data in the specific context of foraging mode and diet (but see Chapter 7).

Most extant lizards are small insectivores and show remarkable diversity in skull morphology, jaw adductor anatomy, and dietary preferences (Schwenk, 2000). Yet considering their extreme morphological and ecological diversity, scant data are available demonstrating the functional consequences (e.g. dietary variation) of variation in skull morphology and feeding performance (Schwenk, 2000). Understanding the relationship between morphology and performance with the broader perspective of the lizard skull as a form–function complex is critical to describing the ecological and evolutionary consequences of the morphological variation observed among lizard clades.

As detailed in this volume, the sit-and-wait (SW) and wide-foraging (WF) paradigm makes specific predictions about prey type and other ecological/morphological factors (Huey and Pianka, 1981; Pianka and Vitt, 2003). Sit-and-wait and WF lizards differ in their exposure to different types of prey. Theoretically, SW lizards should encounter active and highly mobile prey items such as grasshoppers and beetles. On the other hand, WF should encounter more sedentary, cached prey, such as termites and larvae (Huey and Pianka, 1981; Pianka, 1986). Although some overlap of prey taxa exists between SW and WF (Perry *et al.*, 1990), the relative amounts (proportionately and volumetrically) of sedentary and active prey consistently differ for each foraging mode (Huey and Pianka, 1981; Pianka, 1981). Although 'active' and 'sedentary' are broad categorizations of prey types, arthropods such as crickets (an active prey item) and spiders (sedentary) have been shown to differ in hardness (Herrel *et al.*, 2001c). Furthermore, there is growing evidence for dietary transitions across the Iguania–Scleroglossa transition, which would correlate well with many traits associated with foraging mode

(Vitt and Pianka, this volume, Chapter 5; Reilly and McBrayer, this volume, Chapter 10; Vitt *et al.*, 2003).

This chapter reviews the ecological consequences of head shape and jaw biomechanics in SW and WF lizards. First, patterns of morphological variation are quantified among species to investigate how species varying in foraging mode differ in size-corrected head shape. Second, initial steps are taken to quantify the role of history (i.e. phylogeny) in shaping these morphological patterns so that we may examine evolutionary transitions in head shape. Third, functional explanations for variation in head length and width are explored via a biomechanical model proposed by Greaves (1988). The model makes predictions regarding how bite force may be maximized and therefore provides testable hypotheses for future work. The primary objective of this comparative study is to generate hypotheses regarding the trade-offs between head shape, biting performance, and foraging mode. The principal questions to be addressed are: (1) what features of head shape vary among foraging modes and in what direction? (2) Do these features vary in predictable directions functionally or phylogenetically? and (3) When examined in a comparative context, do WF lizards consistently have longer, narrower heads than SW lizards and vice versa?

Materials and methods

Linear measurements of the head were chosen to reflect overall variation in head size and shape as well as biomechanically informative aspects of the jaw. Using dial calipers, external measurements were taken to the nearest 0.1 mm from individuals in the field and from preserved specimens loaned from the California Academy of Science (CAS), the Natural History Museum of Los Angeles County (LACNHM), or the Vertebrate Natural History Museum at Stephen F. Austin State University (Appendix 9.1). Head length (HL), depth (HD), and width (HW) represent general measures of head size and shape; tooth row length (TRL) and quadrate to coronoid length (QCL) characterize the jaw-closing lever mechanics (Fig. 9.1). We also estimated the approximate position of the muscle resultant in the occlusal plane (r) for the force lever (dashed line in Fig. 9.1B). The r value represents the perpendicular distance from a line connecting the jaw joints to the point where the resultant force of the jaw muscles act. The farther anteriorly the muscle resultant force acts, the greater the average biting force along the length of the lower jaw. Thus, larger r values indicate higher potential average bite force along the length of the jaw, all else being equal (Druzinsky and Greaves, 1979; Greaves, 1988).

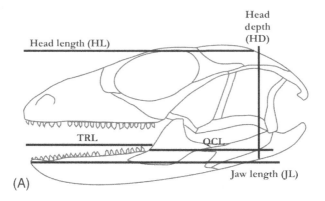

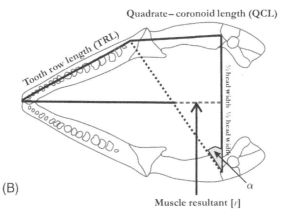

Figure 9.1. (A) Morphological measurements taken from 248 lizard specimens from 11 families. (B) The muscle resultant was calculated by solving the equation $r = \frac{1}{2} \, HW \, \tan \, \alpha$, where $\alpha = \arctan(QC/HW)$. See text for explanation.

Twenty-two species of lizard were used to quantify morphological and functional variation. We chose species based on their foraging mode such that they represented a broad phylogenetic distribution across the squamates (excluding snakes). However, because foraging mode has a strong phylogenetic signal (Perry, 1999), we also chose several species from clades that contain both SW and WF species. Although these 22 species are not representative of all of the morphological variation present in lizards, we feel that they serve as an adequate starting point because they span a range of body sizes, dietary preferences, head shapes, foraging modes, and lizard clades.

The phylogenetic relationships of the species in this study (based on the hypothesis of Estes *et al.*, 1988) are depicted in Fig. 9.2. Within-family relationships were estimated (Iguanidae: Schulte *et al.*, 2003) or taken from the literature (Lacertidae: Arnold, 1991; Cordyliformes: Lang, 1991). Foraging

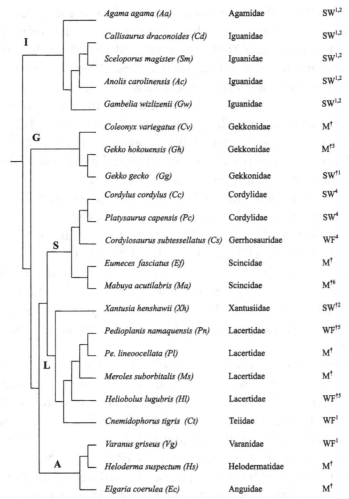

Figure 9.2. Foraging mode designations, families, and phylogenetic relationships among the species included in this study. The phylogeny is based on those of Estes *et al.* (1988), Arnold (1991), and Lang (1991). Iguanid relationships were estimated from the results of Schulte *et al.* (2003). Foraging mode designations with † were made by using the criteria of Reilly and McBrayer (this volume) and based on data from the following published accounts: 1, Cooper (1994b, table 5.1); 2, Dunham *et al.* (1988); 3, Werner *et al.* (1997); 4, Cooper *et al.* (1997); 5, Huey and Pianka (1981); 6, described as SW in Costanzo and Bauer (1993) or mixed by Reilly and McBrayer (this volume). Major clade abbreviations: A, Anguimorpha; L, Lacertoidea; S, Scincomorpha; I, Iguania; G, Gekkota.

mode assignments (SW, M (mixed or intermediate foraging mode), and WF) were made based upon either published accounts or the following criteria for mean percent time moving (PTM): SW < 10%; M 10–25%; WF > 30% (see Reilly and McBrayer, this volume, Chapter 10). These 22 species vary considerably in body size. Hence, the morphological data were size corrected by using the technique of Mosimann and James (1979). Briefly, the raw morphological measures of each individual were log_{10} transformed, summed, and divided by the total number of measurements. This quotient represents the log size component. The log size component of each observation was then subtracted from each log-transformed morphological measurement.

We used multivariate techniques to investigate patterns of variation within the morphological space. Principal components (PCs) were calculated (PROC PRINCOMP, SAS Institute, 2001) from a covariance matrix of the six morphological variables (HL, JL, HW, HD, TRL, QCL) using all observations. The scores from the PCs were then used to characterize the species positions in morphological space. Because the muscle resultant (*r*) was derived from other variables (see Fig. 9.1) in the analysis, this variable was size-corrected, but to reduce redundancy the variable was not included in the mulitvariate analyses.

Most ecomorphological studies are based on the assumption that morphology predicts ecology (Bock and Von Wahlert, 1965; Ricklefs and Travis, 1980; Arnold, 1983; Bock, 1994). If this is true for head morphology and foraging mode, we predict that morphological differences will be apparent between those lizards that sit and wait for prey and wide-foraging species. Hence, a canonical variates analysis (PROC CANDISC) (SAS Institute, 2001) was performed to (1) test for differences in the group centroids in multivariate space using foraging mode as a class variable, (2) determine which of the six morphological variables were primarily responsible for separating the groups, and (3) assess the amount of morphological misclassification of each mode into others (i.e. M → SW, SW → WF, etc.). Because the groups are different sizes and because this could lead to among-class departures from multivariate normality, we conducted a classification and regression tree analysis (CART) using learning samples (Breiman *et al.*, 1984; Steinberg and Colla, 1997) to verify the results of the canonical variates analysis (Kerels *et al.*, 2004). To analyze the results of the canonical variates analysis in a historical context, we layered the Estes *et al.* (1998) phylogeny of our species into the canonical variates space. Ancestral character values for both canonical variates scores and foraging mode were estimated by using an unordered parsimony model (Maddison and Maddison, 2003, 2004).

An unweighted pairwise group methods average (UPGMA) cluster analysis was performed by using MSVP (MVSP, 2000) as an initial investigation into how phylogeny influences the position of species in morphological space. The goal of the cluster analysis in this study is to construct a tree diagram of the species based on mean values for all size-corrected head shape variables (HL, HW, HD, JL, TRL, QCL) and then interpret the resulting tree diagram in light of their foraging mode and taxonomic affiliation. In this diagram, if phylogeny has an overpowering effect on the species they will cluster according to taxonomic affiliation (Iguania, Gekkota, Scincomorpha, and Anguimorpha); if morphology responds to changes in foraging regardless of phylogenetic membership, the species will cluster according to foraging mode. Moreover, a strict consensus tree between the UPGMA and our phylogeny was computed to see how well the morphological data "fit" the Estes *et al.* phylogeny. A Consensus Fork Index (CFI) was calculated to objectively analyze this fit (Swofford, 2002). The CFI measures the similarity between trees by calculating the number of shared clades divided by the total number possible ($k - 2$, where $k =$ no. of taxa) and this value is normalized (see Colless, 1980).

We tested for differences in the average HL, HW, and muscle resultant (r; see Fig. 9.1) between SW, M, and WF species with ANOVA and subsequent Bonferroni correction. Each variable for all observations was entered as a dependent variable and foraging mode as the independent. We predicted that HL would be greater in WF and HW would be greater in SW. The WF species should have a longer head to increase the velocity of closure of the jaws, whereas the SW species should have a wider head to increase gape and, potentially, bite force. Given this, we also predicted that the mean values for r would be different among foraging types owing to variation in head length and head width.

Results

Four PCs explained 95% of the total variation in the morphological data; however, upon visual analysis of a scree plot, we decided to interpret three axes (Table 9.1).

The first axis explains 53.2% of the variation. Variables important along the first axis are TRL, QCL, and HD. Hence lizards with long QCL and short TRL and low HD are at the left of the origin and lizards with short QCL, long TRL and high HD are to the right of the origin (Fig. 9.3). Along the second PC (explaining 19.3% of the variation), lizards with long narrow heads and long jaws are above the origin and species with the opposite characteristics are below the origin. The third axis (explaining 16.6% of the variation) is a head

Table 9.1 *Variable names with abbreviations, eigenvalues, variance metrics and loadings of the size-corrected skull morphological measurements and principal components (PCs)*

Variables	PC1	PC2	PC3
Head length (HL)	27	77	−22
Jaw length (JL)	58	69	21
Head width (HW)	20	−75	24
Head depth (HD)	65	−05	74
Tooth row length (TRL)	89	28	−09
Quadrate–coronoid length (QCL)	−88	29	34
Eigenvalue	0.013	0.005	0.004
% Variance	53.2	19.3	16.6
Cumulative % variance	53.2	72.5	89.1

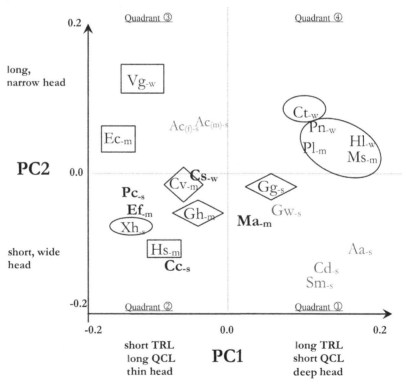

Figure 9.3. Positions of species in morphological space resulting from a principal components analysis on size-adjusted data. Species and family names can be found in Fig. 9.2. Abbreviations: s, sit-and-wait; m, mixed; w, wide forager. Symbols: circles, Lacertoidae; squares, Anguimorpha; diamonds, Gekkota; gray, Iguania; bold, Scincomorpha.

Table 9.2 *Loadings obtained from a canonical variates analysis of head shape and foraging mode*

The CART column indicates the variable importance in determining the placement of species into each foraging category. This importance score is a normalized value that is an index to their importance in splitting species into foraging classes.

Variables	Can1	Can2	CART
Head length (HL)	0.74	0.4	82.93
Jaw length (JL)	0.63	0.6	70.11
Head width (HW)	− 0.85	0.28	100.00
Head depth (HD)	0.059	0.009	57.48
Tooth row length (TRL)	0.194	0.66	43.24
Quadrate–coronoid length (QCL)	0.046	− 0.287	64.26
% Variance	83.3	16.7	—

depth axis. The PC analysis shows that WF have longer, narrower heads than SW species and that lacertoidean and anguimorphan WF species have reached these morphologies in different ways (see quadrant 3 vs. 4).

The canonical variates analysis reveals significant differences in the skull morphologies based on foraging mode (Wilk's $\Lambda = 0.43$, $F_{12,480} = 20.88$, $p < 0.0001$). Two canonical axes explained 83.3% and 16.7% of the variation, respectively (Table 9.2). In canonical space, the species are separated along an axis largely explained by HL and HW, where sit-and-wait foragers have short, wide heads and wide foragers tend to have long narrow skull morphologies (Fig. 9.4). The mixed foraging species as well as many of the estimated ancestral values are occupying an intermediate position in the canonical space. The class means (excluding the ancestral values) for the two canonical variates (Can1 and Can2) were SW $= -1.05$ and 0.255, M $= 0.19$ and -0.578, WF $= 1.39$ and 0.42, respectively. These mean values were significantly different from one another after accounting for within-class variability (Mahalanobis distances, 'D^2'). The statistics for these tests were, M vs. SW, $D^2 = 2.24$, $F_{6,240} = 16.93$, $p < 0.0001$; WF vs. SW, $D^2 = 6.01$, $F_{6,240} = 37.24$, $p < 0.0001$; M vs. WF, $D^2 = 2.44$, $F_{6,240} = 14.47$, $p < 0.0001$. The map of the Estes *et al.* (1988) tree and the estimated ancestral states reveals a combination of phylogenetic (short branch lengths within clades) and idiosyncratic adaptive signaling (long branching) within the morphological space.

We detected only infrequent misclassification among the three groups which reflects differences in foraging mode. Using a generalized square distance function (see SAS Institute, 2001) and setting prior probabilities proportional to

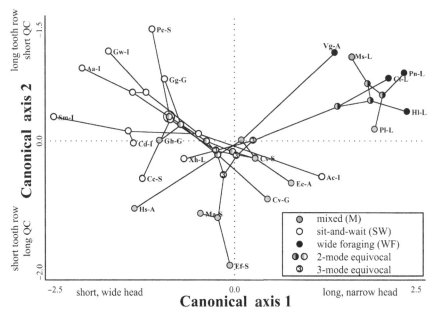

Figure 9.4. The canonical variates space in context of a phylogeny. The first canonical axis explains 83.3% of the variance and demonstrates a gradient in head shape. Branches of the phylogenetic tree (black lines) were mapped onto the species values for morphology in canonical variates space. Maximum parsimony criteria were used to estimate ancestral character values (unlabeled spots) for the canonical variates scores (position in space) and foraging mode (shading of spots). If the spot is divided, it is equivocal with respect to two foraging modes; "3" indicates that all three modes are equally likely at that node. *Anolis carolinensis* scores were averaged across sexes. Species abbreviations (two letters) and major clades (one letter) are same as in Fig. 9.2.

sample size, the analysis shows little morphological overlap between the SW, M, and WF groups (Table 9.3). The CART analysis showed very similar results; hence, we report the classification percentages for each analysis in Table 9.3.

In the principal components space, species seem to cluster together according to either phylogenetic affinity (i.e. *Callisaurus draconoides* (Cd) and *Sceloporus magister* (Sm) are phrynosomatids) or foraging mode (e.g. *Gecko gecko* (Gg) and *Gambelia wislizenii* (Gw)). Hence, we ran a cluster analysis as an initial effort to determine whether phylogeny or foraging mode has greater importance in determining their juxtaposition in morphological space. Not surprisingly, the cluster analysis mirrors the results of the PCA. The resulting patterns suggest a division of the morphologies into two clusters along a QCL–TRL dimension (compare PC1, Fig. 9.3, and the tree diagram resulting from the UPGMA, Fig. 9.5). In Fig. 9.5, the eleven species in the upper cluster are those with a short TRL and long QCL; the lower eleven species tend to

Table 9.3 *Number of observations and percent classified into either sit-and-wait (SW), mixed (M) or wide-foraging (WF) modes*

Prior probabilities were set proportional to group sample size. Numbers in parentheses represent results from a non-parametric analysis (CART).

To	From			Total
	M	SW	WF	
M	41 (75)	23 (1)	24 (12)	88 (88)
	46.6% (85.2)	26.1%	27.3%	100%
SW	19 (1)	78 (94)	1 (3)	98 (98)
	19.39%	**79.6% (96)**	1.02%	100%
WF	12 (2)	6 (1)	44 (59)	62 (62)
	19.35%	9.68%	**71.0% (95.1)**	100%
Prior probabilities	0.35	0.40	0.25	—

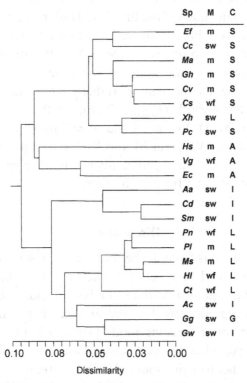

Figure 9.5. UPGMA cluster analysis on size-adjusted morphological data. See text for explanation. Abbreviations: Sp, species; M, foraging mode; C, major clade (abbreviations as in Fig. 9.2.)

have opposite characteristics. This similarity dendrogram is different from the phylogeny derived by Estes *et al.* (1988) and the consensus fork index (CFI) is relatively low (normalized CFI from a strict consensus $= 0.263$). Hence, the Estes *et al.* (1988) phylogenetic hypothesis is a weak predictor of our dendrogram based upon the similarity of head shape among these species.

The results of the ANOVA reveal that the muscle resultant (r) is not significantly different among SW, M, and WF species ($F_{2,245} = 2.20$; $p < 0.11$). This is likely due to the overlaps between the foraging types ($\bar{x} \pm SE$; SW $= 0.51 \pm 0.001$; M $= 0.48 \pm 0.012$; WF $= 0.50 \pm 0.010$). Thus we ran an additional one-way ANOVA (with Bonferroni correction) and grouped the data by their higher clade (e.g. Iguania, Anguimorpha, etc.). Of the four clades with derived foraging mode states, two clades showed significant differences (Lacertoidea (M–WF): $F_{1,58} = 5.28$; $p < 0.03$; Anguimorpha (M–WF): $F_{1,30} = 15.48$; $p < 0.001$), while another approached significance (Scincomorpha (SW–M–WF): $F_{2,69} = 2.86$; $p < 0.06$) and the fourth was not significant (Gekkota (SW–M): $F_{1,26} = 0.02$; $p < 0.90$). Because r seemed to be longer and is a derivative of relative head length and width, and these two variables seem to explain much of the functional difference among the foraging modes across several vertebrate groups, these two variables were explored in greater detail. Again using a one-way ANOVA (with Bonferroni correction), we tested for differences in mean values for head width and length, with foraging mode as the independent variable. Significant differences were observed for both head width ($F_{2,245} = 70.76$; $p < 0.0001$) and head length ($F_{2,245} = 44.07$; $p < 0.0001$). For head width, SW foragers were significantly greater than each M and WF, but M and WF were not different from each other. For head length, WF was significantly greater than each M and SW, but M and SW were not different from each other.

Discussion

Given that cranial shape is likely to be under intense selection, we predicted that foraging mode would be a better predictor of morphological variation than phylogeny. The canonical variates analysis demonstrated that SW, M, and WF species are significantly different in head shape and the differences were reliable predictors of foraging mode (Table 9.3, Fig. 9.4). In fact, the classification rates exceeded our expectations and thus indicate that cranial morphology can predict foraging mode. Furthermore, this study supports the findings of Huey and Pianka (1981) and McBrayer (2004). Hence, the relationship between foraging mode and skull morphology is a generalized pattern across squamates rather than only in lacertids.

In principal components space, SW, M, and WF lizards show interesting patterns of variability in head shape. The SW species are distributed along PC1 such that they grade into the head shapes of the mixed foragers (i.e. the cloud of species along PC1 in Fig. 9.3; species with –s or –m subscript). The SW and M species are similar on PC2 in having short, wide heads (Q1, Q2). On the other hand, the heads of most wide foragers (Q3, Q4) are long and narrow (species with –w subscript; positive end of PC2 in Fig. 9.3). This general head shape has been described elsewhere (Huey and Pianka, 1981; McBrayer, 2004). Yet, it appears that WF species are divergent (Q3 vs. Q4) along PCs 1 and 2. *Varanus griseus* (Vg) loads on the negative end of PC1 (short TRL, long QCL, thin head) whereas *Cnemidophorus tigris* (Ct), *Pedioplanis namaquensis*, and *Heliobolus lugubris* (Hl) each load that axis positively (long TRL, short QCL, deep head). Thus this lacertoidean group of species is similar to SW species in terms of TRL, QCL and HD, but is different from them in terms of HL and HW (PC2). The variation in TRL and QCL among wide foragers suggests that the Anguimorpha (Q3) and Lacertoidea (Q4) attained their long heads in fundamentally different ways. From their SW or M ancestors, the WF lacertoideans (teiids and lacertids) have lengthened the tooth row relative to the distance between the quadrate and coronoid process (QCL). The anguimorphs, however, have attained their long heads independently via an increase in the quadrate–coronoid length, not in tooth row length (TRL). Changes in TRL and QCL have a potentially important impact on the biomechanics of the lower jaw (see below).

Note that one species, Ac (*Anolis carolinensis*), occupies a unique region of the space compared with other SW and M foragers and especially with other iguanians (Gw, Cd, Sm, Aa). This species is sexually dimorphic in head dimensions (Preest, 1994). *A posteriori*, we tested the six male and six female individuals included here and found that they exhibited dramatic differences in head length, width, and depth. Given that sexual dimorphism in this species is probably the result of sexual selection (Preest, 1994), the inclusion of this species may not be appropriate to examine more general patterns of association between skull morphology and foraging mode. To explore the impact of sexual dimorphism, male and female anoles were treated separately (Fig. 9.3, Acm/Acf); however, their position was relatively unchanged. Thus, anoles have longer, more pointed heads than other iguanian or SW lizards studied to date. Sexual dimorphism was not observed in the other species included in the dataset.

Patterns within clades (e.g. family) with diverse foraging modes mirrored the larger-scale patterns of ecomorphology. Within the Gekkota, the mixed foraging *Coleonyx variegatus* (Cv) is morphologically similar to another mixed foraging gecko, *Gecko hokouensis* (Gh) (note the corresponding branch lengths in Fig. 9.4); however, the SW foraging *Gecko gecko* (Gg) is placed toward the other SW taxa (longer branch lengths in Fig. 9.4). Similarly, in the two skink species the SW *Mabuya acutilabris* is morphologically similar to the other SW species (PC1; long TRL, short QCL, high HD) whereas the WF *Eumeces fasciata* is more similar to the mixed and WF species on the negative end of PC1. In lizards in the family Lacertidae (Hl, Ms, Pn, Pl) all four species are similar in TRL and QCL (PC1). However, the M species (Pl, Ms) are positioned downward toward the other M and SW taxa on PC2. The opposite pattern is observed in the cordyliforms (SW) where the secondarily WF *Cordylosaurus subtessellatus* (Cs) is positioned upward toward the other WF species. Similarly, among the anguimorphs, the WF *Varanus griseus* has shifted upward from its M sister taxa. This shift is quite dramatic with respect to the phylogeny, as indicated by the very long estimated branch between the desert monitor and its ancestry in this dataset. These patterns show that shifts in foraging mode appears to be consistently associated with similar changes in cranial shape. Although there is a continuum of head shapes (especially in SW and M foragers) that likely covary with a continuum of foraging modes, it appears that shifts in foraging mode have predictable effects on head shape (or vice versa). In Fig. 9.4, notice the relatively long branch lengths between the SW lizards and their estimated ancestral values and their "finger-like" projection into the upper left corner of the morpho-logical space. If phylogeny were playing a significant role rather than a adaptive response, then one would expect those species to be more clustered with their relatives (e.g. the star-like arrangement among SW and M foraging lacertids).

However, the role of phylogeny in foraging mode evolution has been well established (Cooper, 1994a,b; 1995b; Schwenk, 1995; Perry, 1999). Like pre-vious analyses, our results show that clades tend to occupy common regions of morphospace (Fig. 9.4). For example, the members of the Iguanidae grouped together morphologically. The same could be said of the teiid and lacertid species and gekkotans. Thus, these data further support the fact that phylo-genetic history also plays a role in foraging mode evolution (Perry, 1999; this volume, Chapter 1). However given the morphological diversity of the 3000 + lizard species, the present analysis has but scratched the surface of potential relationships that might exist between morphology and ecology. Regardless, given the consistency of the relationships between foraging mode and head shape, we feel that the analysis indicates that SW and WF foragers differ in four morphological measures of the head (HL, HW, QCL, TRL). This pattern is exhibited both among clades (e.g. iguanian–scleroglossan transition) and within clades (e.g. gekkonids or lacertids). Below we comment as to why these variables are important.

The cluster analysis demonstrates that iguanian taxa (I) tend to cluster together (and have short QCL and long TRL) although they occupy two distinct regions within the major cluster. Likewise, the scleroglossan taxa (S, L, and A) cluster together (and have long QCL and short TRL) (Fig. 9.5). The gekkotan taxa (G) are evenly split into two groups with the M foraging *Coleonyx variegatus* (Cv) and *Gekko hokouensis* (Gh) clustering with the other scleroglossan taxa and the SW *Gekko gecko* (Gg) clustering with iguanian SW species. Most of the autarchoglossan taxa that have either mixed or second-arily derived foraging modes cluster according to taxonomic grouping. Thus, even though there is phylogenetic signal in these data, foraging and head shape have co-evolved throughout the history of these lineages.

These interpretations can be further explored in a more direct evolutionary framework by mapping the mean of each size-corrected variable onto the phylogeny of the study taxa (Maddison and Maddison, 2003). We mapped the assigned foraging mode states onto the phylogeny of the study species by using a maximum parsimony criterion. This reconstruction of foraging mode evolution found fourteen equally parsimonious trees each with nine evolu-tionary "steps" (Fig. 9.6). By examining head length (HL) (tabulated data at top of Fig. 9.6), it is clear that iguanians and most other SW species (except *Anolis*, no.1) have shorter heads (range $= 0.13–0.21$) than WF ($0.23–0.28$), whereas M foragers have head lengths spanning the entire range ($0.18–0.27$). The reverse is seen in head width (HW): SW species typically have the wider heads whether examined across the Iguania–Autarchoglossa transition or within geckos, cordyliforms, or lacertids (ranges: $SW = 0.06–0.14$; $M = -0.11–0.05$; $WF = 0.01–0.04$). Like many other variables associated with foraging mode (e.g. VNO, tongue morphology, feeding behavior), higher anguimorphan and lacertoidean taxa have independently evolved similarities in head shape (long, narrow: Fig. 9.6, nos. 3–5, and see quadrants 3 and 4 of Fig. 9.3) from more generalized forms seen in scincomorphan, gekkotan, and iguanian lizards (Fig. 9.6, nos. 1, 2 and tabulated data). In addition, Fig. 9.6 shows that for those species that shift foraging modes within particular clades (e.g. M foragers, Ms, Pl; Lacertidae), species are displaced toward the derived foraging mode head shape space (Fig. 9.6, tabulated data for HL, HW). The same is true in the cordyliforms, where a secondarily derived WF species (Cs) retains a more generalized head shape but is divergent from other species of the clade toward the head shapes seen in the other WF taxa. Thus when foraging mode and head shape are mapped on the phylogeny the patterns of covariation appear impressive.

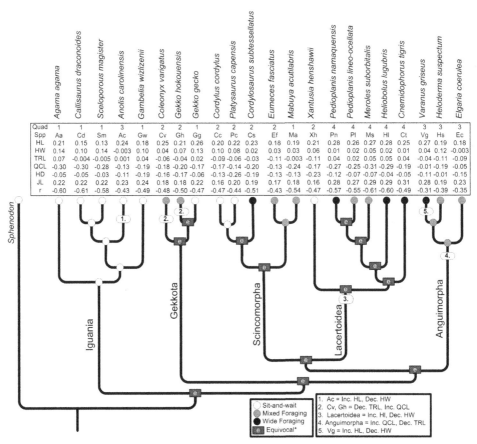

Figure 9.6. Foraging mode and morphology with respect to phylogeny. This figure illustrates the phylogenetic relationships among 22 representative lizard species from five clades. The phylogenetic hypothesis is based on Estes *et al.* (1988), Arnold (1991), and Lang (1991). Foraging mode states were mapped onto the phylogeny and are depicted by differently shaded circles at the tips and at the nodes. Nodal (ancestral) states were estimated from a maximum parsimony algorithm (Maddison and Maddison, 2003). A total of 14 equally parsimonious trees were produced, all with 9 evolutionary steps. Tabulated data at the top are the size corrected species' mean values for head shape variables (abbreviations as in Fig. 9.2, species codes (spp)); location of each species in principal components space ("Quad" corresponds to the four quadrants in Fig. 9.3). Five evolutionary trends observed along particular lineages are also placed within the tree, showing that head shape, particularly head length and width, change concominently with foraging mode. These trends are explained in the legend (1–5) and further in the text. Hatched circles are nodal states where there is an equivocal solution to the algorithm below the node; these nodes would be further resolved with the addition of species to these lineages. Abbreviations: Dec., decrease in morphological trait X; Inc., increase in morphological trait X.

Why head length and head width?

The six skull variables measured here are likely part of an integrated suit of characters, some of which are closely associated with foraging mode (HL, HW), others of which are less so. Tongue morphology and chemoreceptive ability are undoubtedly coupled to head shape and foraging mode evolution (Schwenk, 1994; Cooper, this volume, Chapter 8; Reilly and McBrayer, this volume, Chapter 10). Wagner and Schwenk (2000) have argued that the tongue and lingual prehension mechanism of iguanians represents an integrated character complex such that the coupling of these characters is internally stable and represents an evolutionarily stable configuration. An evolutionarily stable configuration (ESC) is a character complex that performs a particular set of functions and is closed to substantial modification owing to selection. Our analysis demonstrates that iguanians share many similarities in head shape compared with scleroglossans (Figs. 9.3, 9.5). This may be reflective of an ESC in iguanians in that the internally stable character complex of the hyolingual apparatus used for lingual prey prehension prevents major morphological modification (i.e. to a long, narrow head). The ESC was probably dissolved within the scleroglossans and thereby allowed for the transition of the tongue from primarily a prey capture device to a prey detection device (Wagner and Schwenk, 2000). Although we are limited to only five iguanian species, our data show a greater diversity of head shapes in scleroglossans and as such may provide qualitative support for the idea that the dissolution of the ESC led to increased morphological diversity in scleroglossans.

One potential scenario for the transition between a short, wide head (good for lingual prey capture) to a longer, narrower skull is that once the prey capture ESC was broken, head size increased and thereby allowed for increases in tongue length and, by default, prey size. Increases in tongue length would facilitate better scent trailing and chemoreception (Cooper, 1994a, 1995a). A small shift in prey size is detectable at the Iguania–Scleroglossa transition (Vitt *et al.*, 2003), and tongue length, surface area, taste buds and prey chemical discrimination all covary phylogenetically with foraging mode (Cooper, 1994a,b, 1995a,b, 1997).

Another plausible scenario could be that as chemoreception abilities directed foraging mode evolution, the change in total head length was a consequence of selection on correlated characters for better scent trailing. It seems highly likely that traits "tongue length" and "head length" are genetically linked as a highly integrated character suite. If true, then it is parsimonious to assume that selection would act on both traits simultaneously rather than independently or that selection on one trait forced a correlated response in another trait, given there was a genetic correlation between the two. Therefore the "evolutionary line of least resistance" (Schluter, 1996; Marriog and Cheverud, 2005) to an increase in tongue length and improved scent trailing would be one in which head size, particularly head length, increased, and with it tongue length. This may explain why wide foragers (Fig. 9.3) have the longer skulls (this study; Huey and Pianka, 1981; McBrayer, 2004) and longer tongues (Cooper, 1995a). This hypothesis can be tested via comparative studies of the scaling patterns of head length and tongue length in relation to foraging mode.

The existence of a relationship between head length and width does nothing to explain the functional significance of this pattern or why it exists. In most predators, prey size selection is likely to be strongly correlated with gape size. Vitt *et al.* (2003) show that a small but detectable shift in prey size exists at the Iguania–Scleroglossa transition. They hypothesize that the increase in prey size may have been due to increases in gape due to morphological evolution in the skull that allowed for cranial kinesis (specifically meso-kinesis, where the maxillary unit of the skull may be lifted [see below]). Although this is plausible, it is important to note that gape size may be measured in many ways. Relative gape is defined by the maximum size of a prey item that might enter the esophagus, whereas absolute gape is the max-imum distance between the tips of the jaws when fully opened. Increases in head width increase relative gape; increases in head length increase absolute gape (Emerson, 1985). From a functional perspective, then, an increase in prey size could be related to the addition of cranial kinesis (especially if head length and width remained unchanged). A less complex evolutionary scenario might be increases in gape via an increase in head length and/or head width through peramorphosis.

A functional trade-off in bite force may exist as absolute and/or relative gape changes. A model to predict how bite force would change along the length of the jaw in reptiles was proposed by Druzinsky and Greaves (1979) and modified by Greaves (1988). Central to the Greaves model is the estimation of the muscle resultant. The muscle resultant is the point at which the resultant force vector acts anterior of the jaw joints due to the action of the jaw musculature. The model assumes that output force is maximized along the length of the entire jaw, that the physiological cross sectional areas of the muscle is equal, and that each side of the jaw musculature is activated equally. This final assumption seems plausible based on the electromyographic studies of lizard jaw muscles conducted to date, which show bilateral activation (Smith, 1982; Gans *et al.*, 1985; Herrel *et al.*, 1997a,b, 2001b; McBrayer and White, 2002). Greaves' (1988) work demonstrated that, as the muscle resultant moves anteriorly (i.e. gets longer), bite force along the length of the jaw should increase (Fig. 4, p. 301). Using trigonometry and external head measurements only, we estimated the anterior or posterior position of the muscle resultant for each specimen in this study and found mixed results. Across all clades, the three foraging modes were not significantly different. This is probably due to the similarity of head shape among M and SW foragers (see Fig. 9.3). However, when analyzed within each clade by using ANOVA, some interesting differences were observed. Gekkotans and scincomorphans have a more generalized head shape (Fig. 9.3), and hence it is not surprising that their muscle resultant values are not significantly different. However, the WF species in the Lacertoidea and the Anguimorpha have significantly longer muscle resultant values than their M foraging sister taxa. This observation suggests that as species change to higher degrees of wide foraging (Fig. 9.3, Q3, Q4), the skull becomes longer in a manner that would biomechanically serve to maintain (or increase?) bite force along the length of the jaw. The reverse would be true for those species that are mixed or sit-and-wait; head width (and relative gape) might be optimized rather than the position of the force resultant.

The implication of this finding is that functional trade-offs in gape, biting performance, and jaw speed occur as head shape changes with foraging modes (particularly true for wide foraging). These changes occur at a similar phylogenetic level as many other cranial characters, e.g. tongue morphology, chemoreceptive ability, lingual prey capture, and prey processing behavior (Reilly and McBrayer, this volume, Chapter 10). All else being equal, a short wide skull will increase relative gape and not dramatically effect the muscle resultant position (and thus bite force along the length of the lower jaw) (Fig. 9.7) (Greaves, 1988). On the other hand, a longer skull, particularly the distance between the quadrate and coronoid (QCL), for a given width could increase bite force and increase absolute gape (but not necessarily relative gape) (Fig. 9.7). Having greater average bite force along the length of the jaw might help in consuming larger prey items, as wide-foraging scleroglossans apparently do. However, long jaws of wide foragers would also serve to increase the velocity of movement of the jaws during prey processing. Increased mouth opening or closing velocity would be beneficial in rapidly processing many tiny prey items such as termites or insect larvae. (Jaw length, HW, QCL, etc. will also influence the functional significance of cranial kinesis, if present; see below.) Thus, having longer jaws could be important in both prey capture and processing, whether capturing a hard prey item or quickly processing several small prey items. As we gain more dietary and morphological data from more taxa, these functional hypotheses may be tested and

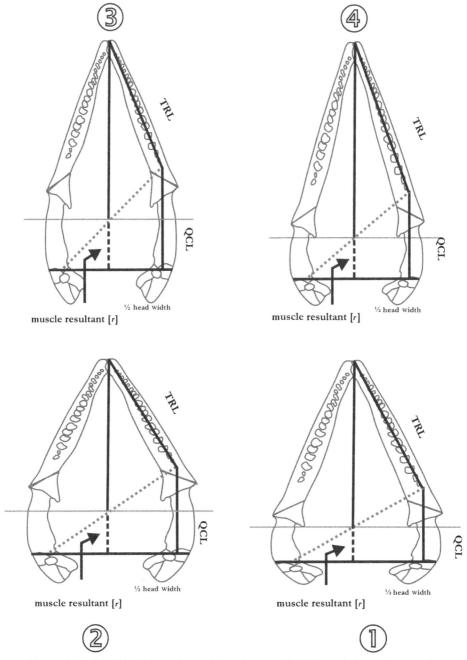

Figure 9.7. Graphic illustration of the Greaves (1988) model to estimate the muscle resultant vector for a reptile. The four head forms (represented by the lower jaw) correspond to the morphospace indicated in the four quadrants in Fig. 9.3. On each head form, the solid line is drawn for comparison of the locations of the approximate anterior–posterior position of the muscle resultant vector in the occlusal plane (see Greaves, 1988). Wide foragers

further refined. However, currently, the level of integration, and potentially correlated evolution, between the head, hyolingual apparatus, vomeronasal system, and foraging mode seems remarkable.

Cranial kinesis

Cranial kinesis is a property of some vertebrate skulls in which joints exist among the cranial bones and thus permit movement of one segment of the skull relative to another. Obviously, changes to the cranial links (i.e. HL, HW, etc.) will affect the functional properties of a kinetic skull. Lizards may have joints at the frontoparietal (mesokinesic), parietal–supraoccipital (metakinesic), and/or quadrate–squamosal (streptostylic) (Frazetta, 1983; Smith, 1982). The phylogenetic distribution of cranial kinesis is somewhat varied. Although many lizard clades contain species with kinetic joints, most lizards do not eat exceptionally large prey, as macrostomatan snakes do. Extreme cranial kinesis as in snakes has been observed in some fossorial lizards such as *Lialis* (Patchell and Shine, 1986) but is not ubiquitously distributed among fossorial taxa. Unfortunately, specific phylogenetic patterns for cranial kinesis are unknown in most lizard clades owing to lack of sufficient comparative and/or experimental data (Metzger, 2002). Particularly for metakinetic and mesokinetic skulls, cranial kinesis is described via the manipulation of dead specimens rather than through direct observation of movement (e.g. cineradiography). Thus, the presence and function of cranial kinesis *in vivo* remains elusive for most species. Metzger (2002) and Schwenk (2000) provide excellent reviews of cranial kinesis in lizards, and readers are encouraged to seek detailed information there. The topic is addressed here to explore the potential for cranial kinesis to be an important characteristic for SW and WF and because each of the variables studied here will influence the functional properties of kinetic skulls.

Caption for Figure 9.7 (*cont.*)
  typically have long narrow heads (3,4) whereas the heads of mixed and sit-and-wait foragers are shorter and wider (1,2). Within foraging types, the relative length of the tooth row and position of the estimated resultant vector is relatively shorter (vector estimate more posterior) (1,4) or longer (vector estimate more anterior) (2,3) with corresponding differences in the rear of the skull and depth of the head. Accordingly there is a range of head shapes among SW and M forms but the wide foraging head, though long and narrow, differs in having relatively shorter (3, varanids) or longer (4, teiids, lacertids) tooth rows and associated changes in the approximate position of the vector. See text for details.

Several hypotheses exist for the function of cranial kinesis. Streptostyly has been hypothesized to potentially increase bite force, increase gape, aid in food transport, and/or facilitate shearing movements during prey processing (see Metzger, 2002). Mesokinesis has been hypothesized to potentially increase gape, clamp the prey between pterygoid bones, decrease the duration of the gape cycle, provide shock absorption during prey capture, and increase bite force (Metzger, 2002). Metakinesis has proven hard to define anatomically and is thought to involve relative small excursions and thus be of limited functional significance (Metzger, 2002).

Current data suggest the following taxonomic patterns in cranial kinesis. No iguanian lizards are reported as having mesokinetic skulls; however, iguanids and some agamids do have streptostylic skulls (Schwenk, 2000; Metzger, 2002; Vitt *et al.*, 2003). Geckos have both streptostylic and mesokinetic kinetic skulls (Herrel *et al.*, 1999c). Among the scleroglossans, scincomorphans and lacertoideans are either yet undiagnosed or have only streptostyly, and anguimorphs (anguids and varanids) have both mesokinesis and streptostyly. The remaining families of anguimorphs are either unknown (xenosaurids) or possess no cranial kinesis (helodermatids) (Metzger, 2002).

Cranial kinesis may be important to SW and/or WF if it increases gape size (and therefore prey size), increases capture success of prey, or provides efficient prey processing. Vitt *et al.* (2003) hypothesized that scleroglossans improved prey capture and manipulation capabilities by the addition of a mesokinetic joint and that cranial kinesis may have facilitated a shift to larger prey in the Scleroglossa. Both streptostyly and mesokinesis are kinetic mechanisms thought to increase absolute gape, which in turn might increase capture success for large prey (i.e. absolute gape). In the few iguanian species (and thus most SW foragers) measured to date, only streptosyly, not mesokinesis, is found. However, in scleroglossans (and thus M and WF), streptostyly and mesokinesis are found in some taxa (Schwenk, 2000). This phylogenetic distribution of mesokinesis correlates well with the observation that scleroglossans eat slightly larger prey (Vitt *et al.*, 2003). Thus, despite having access to a limited amount of quantitative data (Metzger, 2002), some evidence that indicates cranial kinesis is of potential importance to the evolution of head shape in lizards.

We propose that for cranial kinesis to be of any significance in feeding and potentially foraging mode decisions, two things must be demonstrated in a wide variety of SW, M, and WF species. First, diets must be shown to differ among SW, M, and WF species that also vary in degrees of cranial kinesis (see Vitt *et al.*, 2003; Vitt and Pianka, this volume, Chapter 5). Second, direct measurement of kinesis must be demonstrated in a variety of SW and WF species and the competing purposes of cranial kinesis must be explained on a functional basis. Only on knowing this could we then begin to explore how variation in general head shape (e.g. HL, HW, QCL, etc.) influences the functional properties of the kinetic skull in SW or WF species. Unfortunately, we currently lack enough functional data to provide causal links between cranial kinesis and aspects of feeding ecology. However with more data, it may become apparent that cranial kinesis played an important role in the evolution of foraging mode and lizard head shape.

Conclusions and future directions

Our analysis contained taxonomically broad, yet somewhat limited, sampling (22 of 3000 + non-ophidian squamate species). Stayton (2005) conducted a survey of skull shape evolution in lizards using a much larger sample (147 genera) and landmark-based methods, which are more sensitive to shape variation than our linear distance measures. Interestingly, however, he observed patterns similar to ours, although he did not specifically probe his data on the basis of foraging mode. Cranial variables related to the length of the head explained most of the variation in skull shape (head length, eye length, pterygoid length, maxilla length), just as our analysis did (HL, QCL, and TRL). Stayton (2005) did not include head width in his study; however, given these other similarities, we would expect that it would be a major explanatory factor if included. As in our study, he found that taxonomic groups cluster together in morphospace (i.e. iguanians together, teiids and varanids together). However, most intriguing was his finding that iguanids and agamids typically occupy opposite regions of the morphospace compared with teiids and varanids. Furthermore, the positions of these groups in morphospace largely resemble ours, such that in multivariate space a gradient of species is formed by iguanians occupying one end of the axis (shorter, blunter skulls), skinks and geckos occupy the middle, and teiids and varanids occupy the other end (longer, narrower skulls) (compare Figs. 3–5, 8 in Stayton, 2005 to our Figs. 9.3–9.4).

The evolutionary patterns suggested here are based upon the Estes *et al.* (1988) hypothesis of squamate relationships. However, other phylogenies have been proposed. Interpreting our data on the Lee (1998) phylogeny changes our conclusions very little as it places the Scincidae and Cordyliformes between the Lacertoidea and Anguimorpha. However, another radically different hypothesis has been proposed where the Iguania and most Autarchoglossa are sister clades, the Scincoidea are basal to this clade, and the Gekkota are the most basal group (Townsend *et al.*, 2004). Under this hypothesis, the short wide heads of iguanians and the long narrow heads of teiids and varanids are all derived from a generalized skink-like head. In our analysis, a basal and generalized head shape would now be near the origin and in Quadrant 2 of Fig. 9.3 (geckos and skinks) whereas the short wide iguanian head would be derived (Quadrant 1) as would the two types of long narrow head (Quadrants 3, 4; varanids, teiids, lacertids).

In light of this drastically different phylogenetic hypothesis, we repeated two of our analyses using the Townsend *et al.* (2004) hypothesis. First, we plotted the scores from the canonical variates analysis in the context of the Townsend phylogeny instead of the Estes phylogeny (see Fig. 9.4) and estimated ancestral states for both canonical variates scores and foraging mode using methods identical to those given above. Interestingly, the Townsend phylogeny shows fewer equivocal ancestral nodes (12 in the Estes phylogeny compared with 7 in Townsend) and a tendency for ancestral nodes to move away from the origin, and it provides a more parsimonious view of evolution of foraging mode (9 steps in the Estes phylogeny compared with 8 in Townsend). In addition, the Townsend phylogeny provides a better fit to the UPGMA based on head morphology (normalized CFI with "strict consensus" $= 0.737$). Hence, in light of the Townsend *et al.* (2004) hypothesis it seems that phylogeny may play an even greater role in the morphological variability of these lizards than what we can infer from the Estes phylogeny. However, the fork index of the UPGMA and the Townsend phylogeny is not unity; given the existence of peripherally oriented long branches in the CV space, we also conclude that the relationship between head morphology and behavior, at least in part, is due to adaptive evolution among and within the clades of this study.

One major outcome of our study was to identify reliable morphological characters that co-evolve with foraging mode. It is unquestionable that head length and head width are related to foraging mode evolution. However, much remains to be understood about the precise ecological and functional reasons for this relationship. Head length and jaw length certainly affect the jaw opening speed, absolute gape, and the muscle resultant; head width can influence relative gape and the muscle resultant. The current analysis should be expanded to test these effects specifically, especially within clades with multiple foraging modes (e.g. Gekkonidae, Lacertidae, cordyliforms, Scincidae). These clades will provide the best resolution to test whether lizard skull morphologies are changing in tandem with foraging mode alone or with other traits simultaneously. Inclusion of more species within each of these families will help develop stronger support for our hypothesis of an association between foraging mode and head shape. We endeavored to include species with derived foraging mode states such that the within-clade resolution of the foraging mode – head shape relationship might be maximized. However, it is clear that the crown Anguimorpha and crown Lacertoidea (with WF) have independently evolved morphologically different "pointed heads," and due to this, have modified jaw mechanics. In addition, we find it striking that consistent patterns of directional change in head shape within gekkotans, scincomorphs, and lacertids with changes in foraging mode. Thus, we predict that as additional species are added to this dataset head shape will covary with foraging mode. In any case, the hypothesis that skull morphology changes in concert with foraging modes should be further tested by using a comparative method such as independent contrasts or phylogenetic autocorrelation. An examination of the correlation between "phylogenetically corrected data" for head shape and a continuous metric of foraging mode (e.g. percent time moving [PTM]) would be highly beneficial in teasing apart the role of phylogeny in shaping our observed patterns, which undoubtedly contain some phylogenetic signal. Most importantly, other variables like diet (Vitt and Pianka, this volume, Chapter 5), chemoreceptive ability (Cooper, this volume, Chapter 8), and sexual selection (especially in iguanians) should be related to changes in head shape as those data become available in these species.

Acknowledgements

We gratefully acknowledge John Hranitz, Keith Metzger, and Steve Reilly for discussion and helpful insight on this project. Two anonymous reviewers helped to improve this manuscript. The work was supported by Bloomsburg University College of Science and Technology, Department of Biological and Allied Health Sciences (CEC) and the Department of Biology, Georgia Southern University (LDM).

References

Arnold, S. J. (1983). Morphology, performance and fitness. *Am. Zool.* **23**, 347–61.

Arnold, E. N. (1991). Relationships of the South African lizards assigned to *Aporosaura, Meroles*, and *Pedioplanis* (Reptilian: Lacertidae) *J. Nat. Hist.* **25**, 783–807.

Arnold, E. N. (1998). Cranial kinesis in lizards: variations, uses, and origins. *Evol. Biol.* **30**, 323–57.

Barbosa, A., and Moreno, E. (1999). Evolution of foraging strategies in shorebirds: an ecomorphological approach. *Auk* **116**, 712–25.

Beecher, W. J. (1962). The bio-mechanics of the bird skull. *Bull. Chic. Acad. Sci.* **11**, 10–33.

Biknevicius, A. R. (1996). Functional discrimination in the masticatory apparatus of juvenile and adult cougars *(Puma concolor)* and spotted hyenas *(Crocuta crocuta)*. *Can. J. Zool.* **74**, 1934–42.

Bock, W. J. (1994). Concepts and methods in ecomorphology. *J. Biosci.* **19**, 403–13.

Bock, W. J. and Von Wahlert, G. (1965). Adaptation and the form-function complex. *Evolution* **19**, 269–99.

Breiman, L., Friedman, J., Olshen, R. and Stone, C. (1984). *Classification and Regression Trees.* Pacific Grove, CA: Wadsworth Publishing.

Condon, K. (1987). A kinematic analysis of mesokinesis in the Nile monitor *(Varanus niloticus). J. Exp. Biol.* **47**, 73–87.

Colless, D. H. (1980). Congruence between morphometric and allozyme data for *Menidia* species: a reappraisal. *Syst. Zool.* **29**, 288–99.

Cooper, W. E. Jr. (1994a). Chemical discrimination by tongue-flicking in lizards: A review with hypotheses on its origin and its ecological and phylogenetic relationships. *J. Chem. Ecol.* **20**, 439–87.

Cooper, W. E. Jr. (1994b). Prey chemical discrimination, foraging mode, and phylogeny. In *Lizard Ecology: Historical and Experimental Perspectives*, ed. L. J. Vitt and E. R. Pianka, pp. 97–116. Princeton, NJ: Princeton University Press.

Cooper, W. E. Jr. (1995a). Evolution and function of lingual shape in lizards, with emphasis on elongation, extensibility, and chemical sampling. *J. Chem. Ecol.* **21**, 477–505.

Cooper, W. E. Jr. (1995b). Foraging mode, prey chemical discrimination, and phylogeny in lizards. *Anim. Behav.* **50**, 973–85.

Cooper, W. E. Jr. (1997). Correlated evolution of prey chemical discrimination with foraging, lingual morphology and vomeronasal chemoreceptor abundance in lizards. *Behav. Ecol. Sociobiol.* **41**, 257–65.

Cooper, W. E. Jr., Whiting, M. J. and Van Wyk, J. H. (1997). Foraging modes of cordyliform lizards. *S. Afr. J. Zool.* **32**, 9–13.

Costanzo, R. A. and Bauer, A. M. (1993). Diet and activity of *Mabuya acutilabris* (Reptilia: Scincidae) in Namibia. *Herpetol. J.* **3**, 130–5.

De Vree, F. and Gans, C. (1987.) Kinetic movements in the skull of adult *Trachydosaurus rugosus. An. Hist. Emb.* **16**, 206–9.

Druzinsky, R. E. and Greaves, W. S. (1979). A model to explain the posterior limit of the bite point in reptiles. *J. Morphol.* **160**, 165–8.

Dunham, A. E., Miles, D. B. and Reznick, D. N. (1988). Life history patterns in squamate reptiles. In *Biology of the Reptilia*, vol. 16, ed. C. Gans and R. Huey, pp. 441–519. New York: A. R. Liss.

Emerson, S. (1985). Skull shape in frogs – correlations with diet. *Herpetologica* **41**, 177–88.

Estes, R., de Queiroz, K. and Gauthier, J. (1988) Phylogenetic relationships within Squamata. In *Phylogenetic Relationships of the Lizard Families: Essays Commemorating Charles L. Camp*, ed. R. Estes and G. Pregill, pp. 119–281. Stanford: Stanford University Press.

Frazzetta, T. (1983). Adaptation and function of cranial kinesis in reptiles: a time-motion analysis of feeding in alligator lizards. In *Advances in Herpetology and Evolutionary Biology*, ed. A. Rhodin and K. Miyata, pp. 222–44. Cambridge: Harvard University Museum of Comparative Zoology.

Gans, C. and De Vree, F. (1986). Shingle-back lizards crush snail shells using temporal summation (tetanus) to increase the force of the adductor muscles. *Experientia* **42**, 387–9.

Gans, C. and De Vree, F. (1987). Functional bases of fiber length and angulation in muscle. *J. Morphol.* **192**, 63–85.

Gans, C., De Vree, F. and Carrier, D. (1985). Usage pattern of the complex masticatory muscles in the shingleback lizard, *Trachydosaurus rugosus*: a model for muscle placement. *Amer. J. Anat.* **173**, 219–40.

Gomes, N. M. B. (1974). Antomie comparée de la musculature trigeminale des lacertiliens. *Mem. Mus. Nat. Hist. Nat.* **A90**, 1–107.

Gordon, I. and Illius, A. (1994). The functional significance of the browser-grazer dichotomy in African ruminants. *Oecologia* **98**, 167–75.

Gorniak, G. C., Rosenberg, H. I. and Gans, C. (1982). Mastication in the tuatara, *Sphenodon punctatus* (Reptilia: Rhynchocephalia): structure and activity of the motor system. *J. Morphol.* **171**, 321–53.

Greaves, W. S. (1988). The maximum average bite force for a given jaw length. *J. Zool. Lond.* **214**, 295–306.

Haas, G. (1973). Muscles of the jaws and associated structure in the Rynchocephalia and Squamata. In *Biology of the Reptilia*, ed. C. Gans and T. Parsons, pp. 285–490. London: Academic Press.

Herrel, A. and De Vree, F. (1999). Kinematics of intraoral transport and swallowing in the herbivorous lizard *Uromastix acanthinurus*. *J. Exp. Biol.* **202**, 1127–37.

Herrel, A., Aerts, P. and De Vree, F. (2000). Cranial kinesis in geckoes: Functional implications. *J. Exp. Biol.* **203**, 1415–23.

Herrel, A., Aerts, P., Fret, J. and De Vree, F. (1999a). Morphology of the feeding system in agamid lizards: ecological correlates. *Anat. Rec.* **254**, 496–507.

Herrel, A., Cleuren, J. and De Vree, F. (1997a). Quantitative analysis of jaw and hyolingual muscle activity during feeding in the lizard *Agama stellio*. *J. Exp. Biol.* **200**, 101–115.

Herrel, A., De Grauw, E. and Lemos-Espinal, J. A. (2001a). Head shape and bite performance in xenosaurid lizards. *J. Exp. Zool.* **290**, 101–7.

Herrel, A., De Vree, F., Delheusy, V. and Gans, C. (1999b). Cranial kinesis in gekkonid lizards. *J. Exp. Biol.* **202**, 387–98.

Herrel, A., Meyers, J., Nishikawa, K. and De Vree, F. (2001b). The evolution of feeding motor patterns in lizards: modulatory complexity and possible constraints. *Amer. Zool.* **41**, 1311–20.

Herrel, A., Spithoven, L., Van Damme, R. and de Vree, F. (1999c). Sexual dimorphism of head size in *Gallotia galloti*: Testing the niche divergence hypothesis by functional analyses. *Funct. Ecol.* **13**, 289–97.

Herrel, A., Van Damme, R., Vanhooydonck, B. and De Vree, F. (2001c). The implications of bite performance for diet in two species of lacertid lizards. *Can. J. Zool.* **79**, 662–70.

Herrel, A., Wauters, I., Aerts, P. and De Vree, F. (1997b). The mechanics of ovophagy in the beaded lizards *(Heloderma horridum)*. *J. Herpetol.* **31**, 383–393.

Huey, R. and Pianka, E. R. (1981). Ecological consequences of foraging mode. *Ecology* **62**, 991–9.

Iordansky, N. (1990). Evolution of cranial kinesis in lower tetrapods. *Neth. J. Zool.* **40**, 32–54.

Iordansky, N. (1996). The temporal ligaments and their bearing on cranial kinesis in lizards. *J. Zool. Lond.* **239**, 167–75.

James, F. (1982). The ecological morphology of birds: a review. *Ann. Zool. Fenn.* **19**, 265–75.

Kerels, T. J., Bryant, A. A. and Hick, D. S. (2004). Comparison of discriminant functions and classification tree analyses for age classification of marmots. *Oikos* **105**, 575–87.

Kiltie, R. A. (1982). Bite force as a basis for niche differentiation between rain forest peccaries (*Tayassu tajacu* and *T. pecari*). *Biotropica* **14**, 188–95.

Lang, M. (1991). Generic relationships within Cordyliformes (Reptilia: Squamata). *Biologie* **61**, 121–88.

Lauder, G. (1991). Biomechanics and evolution: integrating physical and historical biology in the study of complex systems. In *Biomechanics in Evolution*, ed. J. Rayner and R. Wootton, pp. 1–19. Cambridge: Cambridge University Press.

Lee, M. S. Y. (1998). Convergent evolution and character correlation in burrowing reptiles: towards a resolution of squamate relationships. *Biol. J. Linn. Soc.* **65**, 369–453.

Maddison, D. R. and Maddison, W. P. (2003). *MacClade 4: Analysis of Phylogeny and Character Evolution*. Version 4.06. Sunderland, MA: Sinauer Associates.

Maddison, W. P. and Maddison, D. R. (2004). *Mesquite: A Modular System for Evolutionary Analysis*. Version 1.05. http://mesquiteproject.org.

Marriog, G. and Cheverud, J. M. (2005). Size as a line of least evolutionary resistance: diet and adaptive morphological radiation in new world monkeys. *Evolution* **59**, 1128–42.

McBrayer, L. D. (2004). The relationship between skull morphology, biting performance and foraging mode in Kalahari lacertid lizards. *Zool. J. Linn. Soc.* **140**, 403–16.

McBrayer, L. D. and White T. D. (2002). Bite force, behavior, and electromyography in the teiid lizard, *Tupinambis teguixin*. *Copeia* **2002**, 111–19.

Metzger, K. (2002). Cranial kinesis in lepidosaurs: skulls in motion. In *Topics in Functional and Ecological Vertebrate Morphology*, ed. P. Aerts, K. D'Aout, A. Herrel and R. Van Damme, pp. 15–46. Maastricht, The Netherlands: Shaker Publishing.

Mosimann, J. E. and James, F. C. (1979). New statistical methods for allometry with application to Florida red-winged blackbirds. *Evolution* **33**, 444–59.

MVSP (2000). *Multivariate Statistical Package, Version 3.11f*. Anglesey, Wales: Kovach Computing Services.

Patchell, F. C. and Shine, R. (1986). Feeding mechanisms in pygopodid lizards: how can *Lialis* swallow such large prey? *J. Herpetol.* **20**, 59–64.

Perez-Barberia, F. J. and Gordon, I. J. (1999). The functional relationship between feeding type and jaw and cranial morphology in ungulates. *Oecologia* **118**, 157–65.

Perry, G., Lampl, I., Lerner, A. *et al.* (1990). Foraging mode in lacertid lizards: variation and correlates. *Amph.-Rept.* **11**, 373–84.

Perry, G. (1999). The evolution of search modes: ecological versus phylogenetic perspectives. *Am. Nat.* **153**, 98–109.

Pianka, E. R. (1981). Resource acquisition and allocation among animals. In *Physiological Ecology: An Evolutionary Approach to Resource Use*, ed. C. Townsend and P. Calow, pp. 300–14. Sunderland: Sinauer Associates.

Pianka, E. R. (1986). *Ecology and Natural History of Desert Lizards: Analyses of the Ecological Niche and Community Structure*. Princeton, NJ: Princeton University Press.

Pianka, E. R. and Vitt, L. J. (2003). *Lizards: Windows to the Evolution of Diversity*. Berkeley, CA: University of California Press.

Preest, M. R. (1994). Sexual size dimorphism and feeding energetics in *Anolis carolinensis:* why do females take smaller prey than males? *J. Herpetol.* **28**, 292–8.

Radinsky, L. (1981). Evolution of skull shape in carnivores. 1. Representative modern carnivores. *Biol. J. Linn. Soc.* **15**, 369–88.

Rieppel, O. (1978). Streptostyly and muscle function in lizards. *Experientia* **34**, 776–7.

Ricklefs, R. E. and Travis, J. (1980). A morphological approach to the study of avian community organization. *Auk* **97**, 321–38.

SAS Institute (2001). *SAS. Version 8.02*. Cary, NC: SAS Institute.

Schulte, J. A., Valladares, J. P. and Larson, A. (2003). Phylogenetic relationships within Iguanidae inferred using molecular and morphological data and a phylogenetic taxonomy of iguanian lizards. *Herpetologica* **59**, 399–419.

Schluter, D. (1996). Adaptive radiation along genetic lines of least resistance. *Evolution* **50**, 1766–74.

Schwenk, K. (1994). Why snakes have forked tongues. *Science* **263**, 1573–7.

Schwenk, K. (1995). Of tongues and noses: chemoreception in lizards and snakes. *Trends Ecol. Evol.* **10**, 7–12.

Schwenk, K. (2000). Feeding in lepidosaurs. In *Feeding*, ed. K. Schwenk, pp. 175–291. San Diego, CA: Academic Press.

Schwenk, K. and Throckmorton, G. S. (1989). Functional and evolutionary morphology of lingual feeding in squamate reptiles: phylogenetics and kinematics. *J. Zool. Lond.* **219**, 153–75.

Smith, K. K. (1980). Mechanical significance of streptostyly in lizards. *Nature* **283**, 778–9.

Smith, K. K. (1982). An electromyographic study of the function of jaw adducting muscle in *Varanus exanthematicus* (Varanidae). *J. Morphol.* **173**, 137–58.

Smith, K. K. (1984). The use of the tongue and hyoid apparatus during feeding in lizards (*Ctenosaura similis* and *Tupinambis nigropunctatus*). *J. Zool. Lond.* **202**, 115–3.

Smith, K. K. and Hylander, W. L. (1985). Strain gauge measurement of mesokinetic movement in the lizard *Varanus exanthematicus*. *J. Exp. Biol.* **114**, 53–70.

Solounias, N. S. (1988). Interpreting the diet of extinct ruminants: the case of a non-browsing giraffid. *Paleobiology* **14**, 287–300.

Stayton, C. T. (2005). Morphological evolution of the lizard skull: a geometric morphometrics survey. *J. Morphol.* **263**, 47–59.

Steinberg, D. and Colla, P. (1977). *CART – Classification and Regression Trees*. San Diego, CA: Salford Systems.

Swofford, D. L. (2002). *PAUP* Version 4.0: Phylogenetic Analysis Using Parsimony*. Sunderland, MA: Sinauer Associates.

Throckmorton, G. S. (1976). Oral food processing in two herbivorous lizards, *Iguana iguana* (Iguanidae) and *Uromastix aegyptius* (Agamidae). *J. Morphol.* **148**, 363–90.

Throckmorton, G. S. (1978). Action of the pterygoideus muscle during feeding in the lizard *Uromastix aegyptius* (Agamidae). *Anat. Rec.* **190**, 217–22.

Throckmorton, G. S. (1980). The chewing cycle in the herbivorous lizard *Uromastix aegyptius* (Agamidae). *Arch. Oral Biol.* **25**, 225–33.

Throckmorton, G. S. and Saubert, C. W. (1982). Histochemical properties of some jaw muscles of the lizard *Tupinambis nigropunctatus* (Teiidae). *J. Morphol.* **203**, 345–52.

Townsend, T. M., Larson, A., Louis, E. and Macey, J. R. (2004). Molecular phylogenetics of squamata: the position of snakes, amphisbaenians, and dibamids, and the root of the squamate tree. *Syst. Biol.* **53**, 735–57.

Turingan, R. G., Wainwright, P. C. and Hensley, D. A. (1995). Interpopulation variation in prey use and feeding biomechanics in Caribbean triggerfishes. *Oecologia* **102**, 296–304.

Vitt, L. J., Pianka, E. R., Cooper, W. E. and Schwenk, K. (2003). History and global
    ecology of squamate reptiles. *Am. Nat.* **162**, 44–60.
Wagner, G. P. and Schwenk, K. (2000). Evolutionarily stable configurations:
    functional integration and the evolution of phenotypic stability. In *Evolutionary
    Biology*, vol. 31, ed. M. Hecht, pp. 155–217. New York: Kluwer Academic/
    Plenum.
Wainwright, P. C. (1996). Ecological explanation through functional morphology: the
    feeding biology of sunfishes. *Ecology* **77**, 1336–43.
Werner, Y. L., Okada, S., Ota, H., Perry, G. and Tonkunaga, S. (1997). Varied and
    fluctuating foraging modes in nocturnal lizards of the family Gekkonidae. *As.
    Herpetol. Res.* **7**, 153–65.
Wineski, L. E. and Gans, C. (1984). Morphological basis of the feeding mechanics in
    the shingleback lizard *Trachydosaurus rugosus* (Scincidae, Reptilia). *J. Morphol.*
    181, 271–95.
Zweers, G. A., Berkhoudt, H. and Vanden Berge, J. C. (1994). Behavioral mechanisms
    of avian feeding. In *Advances in Comparative and Environmental Physiology*, vol.
    18, ed. V. L. Bels, M. Chardon and P. Vandewalle, pp. 241–79. Berlin: Springer-
    Verlag.

Appendix 9.1. Catalog numbers of the specimens used in this study

California Academy of Science (CAS)
Agama agama: 103162, 103163, 103498, 103499, 103500, 103501, 103502, 103503,
    103504, 103505, 125605, 125606.
Coleonyx variegatus: 178506, 188553, 188554, 188555, 188557, 188560, 188561,
    188562, 188563, 188564, 188565, 188568.
Gekko gecko: 7593, 7594, 7595, 18491, 18492, 19595, 19596, 142067, 204948, 204949,
    204950, 214046.
Gekko hokouensis: 18041, 18042, 18046, 18047, 21730, 21731, 21732, 21734, 21736,
    21738, 21739, 21743.
Cordylus cordylus: 165665, 165666, 165667, 165668, 165669, 165677, 165678, 165679,
    180361, 180363, 180366, 180377.
Platysaurus capensis: 193456, 193458, 193459, 193460, 193461, 193462, 193463,
    193464, 193465, 193466, 193467, 193468.
Cordylosaurus subtessellatus: 165627, 173424, 193373, 193629, 206906, 206965,
    223968, 223967, 214798, 206966, 173384, 165628.
Mabuya acutilabris: 76310, 176312, 176314, 176315, 176319, 176330, 176341, 176347,
    176349, 176350, 176351, 176353.
Xantusia henshawi henshawi: 64297, 64298, 64301, 64302, 64304, 64305, 64308, 64309,
    64312, 64314, 64318, 64319.
Heloderma suspectum suspectum: 34283, 34285, 34286, 34290, 34291, 34292, 34293,
    35301, 190164, 190167.
Heloderma suspectum: 190164, 190165, 190166.
Varanus griseus caspius: 143291.
Varanus griseus koniecznyi: 99940.
Varanus griseus: 84267, 84300, 84301, 84473, 86630, 138853, 140475.

Natural History Museum of Los Angeles County (NHMLAC):
Heliobolus lugubris: 79900, 79917, 79918, 80049, 80053, 80063, 80068, 80074, 80081,
    80093, 80094, 80100.

Meroles suborbitalis: 81858, 81875, 81877, 81889, 81929, 81944, 81946, 81956, 81967, 81974, 81980, 81990.
Pedioplanis lineoocellata: 78757, 78760, 78772, 78773, 78775, 78791, 78793, 78798, 78802, 78821, 78827, 78904.
Pedioplanis namaquensis: 80186, 80192, 80199, 80200, 80238, 80256, 80262, 80443, 80448, 80454, 80460, 80484.

Stephen F. Austin State University Vertebrate Natural History Museum:
Eumeces fasciatus: not cataloged.
Anolis carolinensis: 1182, 1945, 2717, 2749, 1184, 3084; others not cataloged.

Personal and field collections (LDM)
Sceloporus magister, Callisaurus draconoides, Gambelia wizlizenii, Cnemidophorus (Aspidoscelis) tigris, Elgaria coerulea.

10

Prey capture and prey processing behavior and the evolution of lingual and sensory characteristics: divergences and convergences in lizard feeding biology

STEPHEN M. REILLY
Program in Ecology and Evolutionary Biology, Department of Biological Sciences, Ohio University

LANCE D. McBRAYER
Department of Biology, Georgia Southern University

Introduction

Prey location, capture, and subsequent processing are fundamentally important behaviors critical to the assimilation of food resources. All three of these behaviors involve movements of the tongue and jaws and it is well known that both tongue movements and tongue morphology vary widely among lizards (Schwenk, 2000). A central element of the sit-and-wait (ambush) vs. wide foraging paradigm involves the trade-off between prey capture function and chemosensory acuity. In general, ambush feeders are thought to use the tongue primarily to capture prey located visually, whereas wide foragers are thought to have traded tongue-based prey capture for tongue-flicking, which is critical to locating widely dispersed prey by using chemoreception (Pianka and Vitt, 2003; Cooper, 1997a). The switch to chemosensory tongue function among sclero-glossan lizards is certainly linked to their wide-foraging strategy; in fact, this transition has enabled wide foragers to dominate lizard communities worldwide (Vitt *et al.*, 2003). In this chapter we examine the trade-off between feeding behaviors (prey capture and subsequent prey processing) and chemosensory function in lizards with data available to date. First, we present new data and a review of kinematic patterns of "prey capture" behaviors. This analysis illustrates three basic prey capture modes used by lizards. Next, we review patterns of post-capture prey processing behavior that reveal three evolutionary transitions in lizard "chewing" behavior. Finally, we compare changes in lizard feeding behavior with quantified characteristics of the vomeronasal system, tongue morphology, prey discrimination ability, and foraging behavior from the literature to examine how changes in feeding function correlate with changes in chemosensory function. The behavioral and morphological data show that, contrary to the prevailing hypotheses, the Iguania have shifted to a novel tongue

Lizard Ecology: The Evolutionary Consequences of Foraging Mode, ed. S. M. Reilly, L. D. McBrayer and D. B. Miles. Published by Cambridge University Press. © Cambridge University Press 2007.

protrusion strategy, and that the advent of chemosensory dedicated tongue function is not a scleroglossan trait but has independently evolved with wide foraging twice, once within the Lacertoidea and once within the Anguimorpha.

Methods

Data for prey capture behaviors (Figs. 10.1 and 10.2) and kinematics (Fig. 10.3A) were obtained from the literature or from gape profiles for the following species we studied by using high-speed video (200 or 250 images per second): *Anolis carolinensis, Ctenosaurus quinquecarinata, Gambelia wislizenii, Oplurus cuvieri, Sceloporus clarkii, Agama agama, Trapelus savignii, Eublepharis macularis, Eumeces schneideri, Scincus scincus, Gerrhosaurus major, Ameiva ameiva, Cnemidophorus sexlineatus, Tupinambis teguixin, Takydromus sexlineatus, Lepidophyma flavimaculatum, Elgaria coerulea,* and *Varanus exanthematicus.* Numbers of individuals per species included in the study and the phylogeny for these species are presented in Fig. 10.4.

Patterns among data for prey processing behaviors were quantified with independent contrasts analysis (Felsenstein, 1985) expanding on the study of McBrayer and Reilly (2002) with the addition of data from six new species (number of feeding bouts per individual): *Anolis carolinensis* (10, 6, 5), *Gambelia wislizenii* (5, 5, 5), *Trapelus savignii* (5), *Eublepharis macularis* (7, 5, 3), *Scincus scincus* (8, 7, 5), and *Lepidophyma flavimaculatum* (6, 5, 5). Prey processing behavior in *Lepidophyma flavimaculatum* was scored visually under red light "night" conditions. The independent contrast analysis was conducted at the level of family, because key aspects of lizard feeding biology (Cooper, 1994a; Schwenk, 1988) are highly conserved at this taxonomic level and because phylogenies with family branch lengths are available. Branch lengths (My BP) for the independent contrasts analysis (from Estes *et al.*, 1988; Cooper, 1997a) that differ from those in McBrayer and Reilly (2002) are due to the addition of the six new species: Gekkota (166), Teiidae (141), Lacertidae (141), Xantusiidae (148), Teiidae + Lacertidae (7), and Lacertoidea (7). Taxonomy and the phylogenetic topology for mapping evolutionary patterns follow Estes *et al.* (1988).

To control for relative prey size during processing, lizards were offered prey (mealworms, crickets, or mice of varying size) selected to be approximately the length of the head of each individual. These prey were used because they are prey shapes commonly eaten by lizards and can easily be obtained in a range of sizes. Additional prey capture behavior was obtained while the animals fed on pieces of prey that fell to the substrate. Strike behavior and gape cycle kinematics were observed in each species. Post-capture prey processing behaviors were scored for the six new species from the strike (prey capture) until the first swallowing cycle (i.e. when prey was no longer seen in the oral cavity but had entered the pharynx). Following McBrayer and Reilly (2002), we scored the occurrence to the following post-capture prey processing behaviors. *Transport* (**T**): any behavior during which the prey item was moved anteroposteriorly within the mouth (inertial transports occurred in three species; their occurrence rates are indicated in parentheses in Fig. 10.4). Two post-capture chewing behaviors (*sensu* Reilly *et al.* 2001) were defined as any behavior during which the size, shape, and/or structural integrity of the prey item was changed via contact with the tongue, palate, jaws and/or teeth without anteroposterior or side to side movement of the prey item. These were *puncture crushing* (**PC**) when the jaws close on the prey, puncturing the prey with the teeth, and *palatal crushing* (**PLC**) when the prey is crushed within the oral cavity between the palate and the tongue. The *total number of processing cycles* (**TOT**) was also scored (side-to-side prey movements were scored separately but not included as a variable in the independent contrasts analysis, except for their inclusion in the total and interspersibility, because they were of low and relatively uniform occurrence across the families). Variation in the sequence of chewing and transport behaviors used during a bout was quantified by calculating another variable, *interspersibility* (**INT**) from the ethogram for each feeding trial. Interspersibility is the number of times a lingual transport behavior (T or the occasional side-to-side transport) was followed by a chewing behavior (either PC or PLC), indicating how often prey items are repositioned for further chewing. This was used as an additional measure of an increase in chewing complexity. The raw occurrence data for each behavior for each bout were pooled across individuals within species to calculate species means for use in the independent contrasts analysis.

Prey capture behavior in lizards

Prey capture is the initial behavior during food acquisition that results in the movement of the prey item into the oral cavity. Prey capture has been studied in *Sphenodon* (Gorniak *et al.*, 1982), about a dozen iguanians, and an equal number of scleroglossans (reviewed in Bels *et al.*, 1994; Schwenk, 2000; Bels, 2003) (Figs. 10.1 and 10.2). Obviously, kinematic and morphological studies of feeding are difficult and more are needed for many more taxa, but this is the most complete dataset and synthesis of prey capture and processing behavior to date and it provides sufficient pattern to illuminate some clear general patterns among the Lepidosauria. *Sphenodon* and most non-iguanian lizards (iguanians have a novel tongue protrusion system, described below) use both

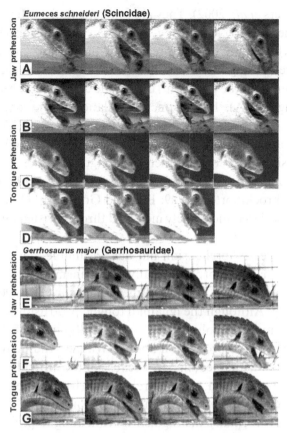

Figure 10.1. Jaw prehension (A, E) and tongue prehension (B, C, D, F, G), characteristic of non-iguanian lizards and *Sphenodon*, illustrated in scincid and gerrhosaurid species. Both behaviors are characterized by "peaked" gape cycles with only fast opening (FO) and closing (C) phases (Fig. 10.3), but the tongue is retracted during the FO phase in jaw prehension and protracted during FO in tongue prehension. Neither has a slow opening (SO) phase. (B, F) Tongue prehension on crickets; tongue prehension used on cricket legs (C, F) and tarsi (D, G) shows the range of tongue prehension skills of these scincomorphan scleroglossans.

"jaw prehension" (as in Fig. 10.1A, E; Fig. 10.2A) and "tongue prehension" (as in Fig. 10.1B, C, D, F, G) to capture prey (Gorniak *et al.*, 1982; Herrel *et al.*, 1995; Schwenk, 2000) and we will use these terms to describe these two ancestral capture behaviors of lizards. In both jaw prehension and tongue prehension the gape cycle involves a peaked gape cycle profile (Fig. 10.3A: black profiles) consisting of a fast opening (FO) and a closing (C) phase (Goosse and Bels, 1992; Delheusy *et al.*, 1994, 1995; Urbani and Bels, 1995; Bels *et al.*, 1994; Herrel *et al.*, 1995; Delheusy and Bels, 2000; Schwenk, 2000;

Bels, 2003; Meyers *et al.*, 2002). They are fundamentally different, however, in that the tongue is retracted during FO in jaw prehension (Fig. 10.1A, E, Fig. 10.2A), whereas in tongue prehension the tongue moves forward to contract the prey during FO and is then retracted (with the prey item, and/or the jaws continue over the prey item) during the closing (C) phase (Fig. 10.1 B, C, D, F, G, Fig. 10.3A). This distinct modulation of the tongue movements relative to the gape cycle controls the difference between the two capture behaviors that are identical in gape profile (FO, C) (Bels, 2003). There tends to be a prey size effect on whether the tongue is used or not, with jaw prehension dominating on large prey items (Fig. 10.1) (Smith, 1984; Schwenk and Throckmorton, 1989; Bels and Goosse, 1990; Schwenk, 2000). Thus, the modulation is most likely mediated through visual input prior to the strike.

The Iguania use jaw prehension but have a unique, derived tongue protrusion system that is morphologically and kinematically different from tongue prehension (Smith, 1988; Wagner and Schwenk, 2000; Bels *et al.*, 1994; Bels, 2003). Although the iguanian tongue protrusion system has been heralded as a unique "evolutionary stable configuration" facilitating the radiation of the tongue protrusion ability (Wagner and Schwenk, 2000), it has not been formally named, in part because of the difficulty of distinguishing it in a few words from tongue prehension (for example, Schwenk and Wagner [2001] called it "whole-tongue protrusion coupled to hyobranchial protraction"). To formally distinguish it from tongue prehension, we will term tongue based prey capture in the Iguania "*translational tongue protrusion*". This is based on the unique features that distinguish it from tongue prehension: contraction of specialized tongue muscles (hyoglossus, genioglossus, verticalis, circular fibers) pushes (*translates*) the tongue along a long (>50% tongue length), tapered lingual process (hyobranchial rod) to *protrude* the entire tongue beyond the jaws (Wagner and Schwenk, 2000). Translational tongue protrusion in iguanians produces a diagnostic novel kinematic phase in the jaw opening cycle (Wagner and Schwenk, 2000; Bels, 2003). The slow translation of the tongue as it is protruded forces the mouth open slowly, producing a "slow opening" (SO) phase (Fig. 10.2B) before the jaws rapidly open and then close (Fig. 10.3A: FO, C on gray profiles) during prey retraction. The inclusion of a slow open phase in strike kinematics is characteristic of all iguanians studied to date (Bels, 2003; Meyers *et al.*, 2002) and is not observed in the gape profiles of tongue prehension strikes of *Sphenodon* (Herrel *et al.*, 1995) or non-iguanians (Fig. 10.3A: capture profiles with black solid lines; Bels, 2003). Iguanians protrude the tongue during slow opening as they lunge at the prey; their tongue sticks to the prey and it is then

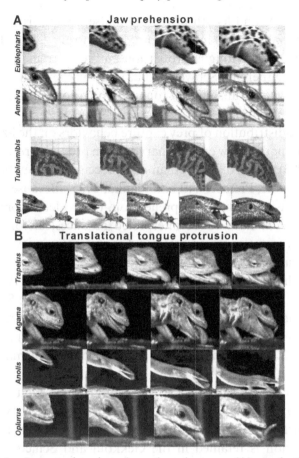

Figure 10.2. Jaw prehension in scleroglossans (A) and the unique *translational tongue protrusion* in iguanians (B). In jaw prehension (A), the tongue is clearly retracted early in the gape cycle as the mouth rapidly opens (the FO phase). In translational tongue protrusion (B), the tongue is slowly protruded, pushing the jaws open (during what is called the slow opening (SO) phase, Fig. 10.3, gray profiles) up to the point of prey contact as the animal lunges forward (up to penultimate frames for each species). *Translational tongue protrusion* in the Iguania is defined by a novel muscular mechanism that translates the tongue along a central hyobranchial rod, which produces the slow opening phase as the tongue protrudes well beyond the jaws. In translational tongue protrusion (B) the tongue retracts during fast opening (last frame in each sequence) and closing.

retracted into the mouth during fast opening and closing (Fig. 10.2B). This is in contrast to tongue prehension, where the tongue is protracted during fast opening. Even the jumping predators like *Anolis* employ this translational tongue protrusion: flying like a missile, protruded tongue first, onto the prey item (Fig. 10.2B). Within the Iguania there is an increase in the amount of tongue protrusion (and length of the SO phase, and morphological tongue translation specializations) in the Agamidae, with the Chameleontidae having developed extreme tongue protrusion and an extremely long SO phase (Smith, 1988; Wagner and Schwenk, 2000). Jaw prehension is retained in the Iguania but they use translational tongue protrusion for most prey capture (Pianka and Vitt, 2003).

Mapping the distribution of prey capture behaviors on the lizard phylogeny (Fig. 10.3B) supports a new evolutionary pattern proposed recently (Bels *et al.*, 1994; Bels, 2003) that is different than other recent interpretations (Huey and Pianka, this volume, Preface; Vitt and Pianka, 2005). Because *Sphenodon* uses both jaw prehension and tongue prehension, the evidence compels the interpretation that *both* have to be interpreted as ancestral for lepidosaurians (Fig. 10.3B). In fact, the presence of FO–C gape cycles in jaw prehension *and* tongue prehension is primitive for Tetrapoda, given that identical patterns are present for both behaviors in Amphibia (Reilly and Lauder, 1989; Nishikawa, 2000; O'Reilly, 2000), and Chelonia (Bels *et al.*, 1994, 1997; Schwenk, 2000), and in jaw prehension in Crocodilia as well (Cleuren and DeVree, 2000). Thus, jaw prehension with FO–C phases (Fig. 10.2A, Fig. 10.3A) is used by all of Reptilia (Bels, 2003) and is retained throughout the Squamata (Fig. 10.3B, solid lines).

The role of the tongue in prey capture exhibits three evolutionary states in lizards (Fig. 10.3B). First, as detailed above, the Iguania have translational tongue protrusion. Second, the primitive condition of jaw prehension and tongue prehension is retained in the Gekkota and Scincoidea. Data from gekkotans are extremely limited because of the difficulty of persuading geckos to feed under light conditions necessary for high-speed filming. Evidence for the retention of tongue prehension in feeding comes from anecdotal evidence of lingual prey capture in *Phelsuma* (Schwenk, 2000) and the widespread use of tongue prehension in frugivory (Delheusy *et al.*, 1995; Delheusy and Bels, 2000). The Gekkota commonly use jaw prehension (Delheusy *et al.*, 1995; Delheusy and Bels, 2000); however, the eublepharids appear to have lost tongue protrusion in feeding (Fig. 10.2). Interpretations of patterns within Gekkota depend on the phylogenetic placement of the Eublepharidae, which are considered derived (Cooper, this volume, Chapter 8) or primitive (Bauer, this volume, Chapter 12). However, assuming tongue prehension is common to the Gekkonidae, which are considered as ambush foragers (Bauer, this volume, Chapter 12), the eublepharids (whether a basal or a crown group) are derived in having lost tongue protrusion in prey capture. Quantitative data on tongue use in prey capture are badly needed to confirm whether gekkonids retain tongue protrusion.

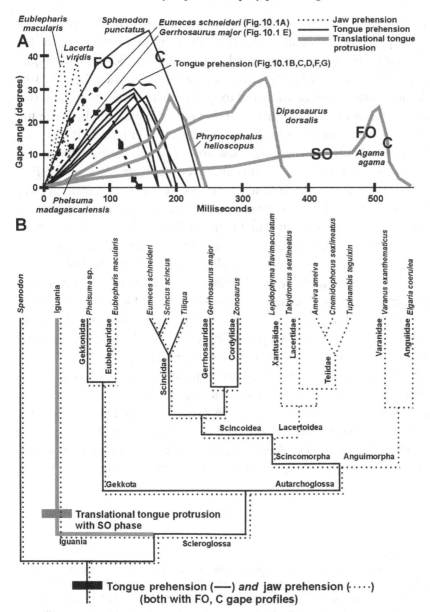

Figure 10.3. Kinematics of mouth opening during prey capture (A) and the evolution of the prey capture behavior in lizards (B). (A) Jaw prehension (dotted profiles) and tongue prehension (solid black profiles) prey capture behaviors have peaked gape cycles, indicating that there is fast opening (FO) and then closing (C) of the jaws in these behaviors. They differ in that the tongue is *retracted* during FO in jaw prehension but *protracted* during FO in tongue prehension. Iguanians use a novel behavior, translational tongue protrusion, where the tongue slides on the hyobranchium during an extended slow opening phase (SO phase: long, initially low slopes on gray profiles) before the jaws rapidly open (FO) and close (C) during the peaked portion of gray profiles. As in jaw prehension, tongue retraction after

Our interpretation that tongue prehension is retained in the Scincoidea is based on previous kinematic evidence of tongue prehension in the Gerrhosauridae (*Zonosaurus*: Urbani and Bels, 1995) and Scincidae (*Tiliqua*: Gans *et al.*, 1985; Smith *et al.*, 1999) and our additional evidence (Figs. 10.1 and 10.3A) that both jaw prehension and tongue prehension are present in two other skink species and in the Gerrhosauridae as well. We found tongue use in capture to be very common in *Eumeces schneideri* (Fig. 10.1B–D) with tongue retraction being successfully used for all small items like cricket legs (Fig. 10.1C) and tarsi (Fig. 10.1D) and even occasionally used on full-sized crickets (Fig. 10.1B). Tongue use was less frequent in *Scincus scincus* and *Gerrhosaurus major* but was regularly and successfully used to pick up smaller prey items such as legs and tarsi (Fig. 10.1F, G). Some have regarded tongue prehension in Gekkonidae and scincoideans as special circumstances and continue to maintain that tongue prehension is lost at the level of the Scleroglossa (see, for example, Vitt and Pianka, 2005). However, given that three scleroglossan families (Scincidae, Gerrhosauridae, Gekkonidae) clearly use tongue prehension *and* jaw prehension, we conclude that tongue prehension is retained in the Gekkota (except Eublepharidae) and the Scincoidea, and thus the Scleroglossa are not limited to jaw prehension.

The independent loss of tongue prehension in several scleroglossan taxa is the third pattern of tongue use in squamates (Fig. 10.3B). Available evidence shows that this has occurred once in the Gekkota and twice within the Autarchoglossa: based on the absence of tongue prehension (Fig. 10.1), the ground-dwelling eublepharid geckos, the Lacertoidea, and the Anguimorpha have each lost tongue prehension yet retain jaw prehension (Figs. 10.2A, 10.3B).

Caption for Figure 10.3 (cont.)

translational tongue protrusion begins with fast opening. Profiles are from *Eublepharis, Gerrhosaurus*, and *Eumeces* (Fig. 10.1), *Phelsuma* (Delheusy and Bels, 2000), *Sphenodon* (digitized from Gorniak *et al.*, 1982, Fig. 4), *Lacerta* (Urbani and Bels, 1995), *Dipsosaurus* and *Phrynocephalus* (Schwenk and Throckmorton, 1989), and *Agama* (Kraklau, 1991). (B). Prey capture patterns in *Sphenodon* and lizards reveal that jaw prehension (dotted lines) and tongue prehension (solid black lines) are primitive for Lepidosauria and both are retained in the Scleroglossa except for the loss of tongue prehension in the Eublepharidae, Lacertoidea and Anguimorpha. The Iguania retain jaw prehension, but have evolved their own novel capture behavior, translational tongue protrusion (gray line), characterized by lingual translation (and many associated tongue features) and the slow opening phase.

Post-capture prey processing behavior in lizards

Mean occurrence data for prey processing behaviors for each species are presented at the top of Figure 10.4. All species used transports (T), and puncture crushing (PC) behaviors and the palatal crushing (PLC) behavior was found in all but the anguimorphs. Our transport behavior (T) category did not include simple inertial transports where the head was thrust forward over the prey item with no associated movement of the tongue (Gans, 1969) (these simple inertial transports are indicated in parentheses in Fig. 10.4 and are included in TOT). Inertial transports were observed in two teiids (*Ameiva ameiva, Tupinambis teguixin*) and in the varanid *Varanus exanthematicus* (Elias *et al.*, 2000) (Fig. 10.4). All species exhibited at least some interspersibility (INT) of prey transport behaviors among prey chewing behaviors, and several groups independently increased this variable.

The correlation of prey processing behaviors among lizards was tested using pooled means for each family and the results are illustrated in Fig. 10.4. Independent contrasts analysis revealed that PC, TOT and INT were significantly correlated after the effects of phylogeny were removed (Table 10.1). This indicates that an increase in the total number of behaviors used in prey consumption is achieved by adding more PCs and INTs and that both of them increase together. The addition of six new species increased the magnitude of correlation coefficients observed by McBrayer and Reilly (2002), strengthening their conclusion that these three variables are tightly linked among lizards.

To integrate the trends in prey processing behaviors and illustrate patterns of prey processing evolution revealed by the independent contrasts analysis (based on directional trends in ancestral nodal values from GLS method of Martins and Hansen, 1997), we gap coded (Archie, 1985) the tip values for species means (for those traits significantly correlated, Table 10.1) and indicated the higher trait values in boldface and heavy lines on the phylogeny (Fig. 10.4). Gaps were PC: < 3.3, > 5.4; TOT: < 9.5, > 11.4, and INT: < 0.9, > 1.3. The evolutionary patterns indicate that the Iguania have fewer cycles of prey processing, less interspersibility, and a predominance of PLCs in reducing prey. The Gekkota are similar to the Iguania in prey processing frequencies, except in using puncture crushing over palatal crushing. The use of puncture crushing over palatal crushing seems to be a scleroglossan trait. The Lacertoidea and Varanidae use more total prey processing cycles (about double that of iguanians except *Ameiva*), extensive use of puncture crushing, and more interspersibility of chewing cycles to reduce prey. Given similar bite forces in wide and ambush foragers (McBrayer, 2004) this means it takes more

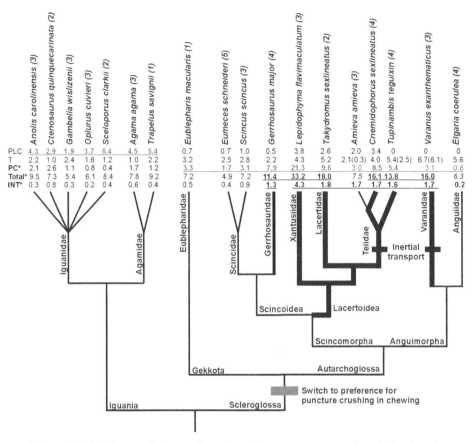

Figure 10.4. The evolution of post-capture prey processing behavior in lizards. Mean occurrence data for prey processing behaviors in lizards from high-speed video recordings of lizards feeding on prey adjusted to the same relative head length ($n = 429$, 5 feedings minimum per individual, individuals per species in parentheses). *Lepidophyma* was scored by eye. Three variables (asterisked) were found to be highly correlated after the removal of the effects of phylogeny (Table 10.1). These variables are gap-coded to show the transition to intermediate numbers of behaviors in the Gerrhosauridae and more extensive prey processing in the Lacertoidea and Varanidae (black underscore, bold lineages). A preference for puncture crushing over palatal crushing (gray bar) appeared in the Scleroglossa (transition in the predominance of PC vs. PLC, gray underscore). The Teiidae and Varanidae have independently developed inertial transport behaviors (black bars). Abbreviations: interspersion (INT) is the mean number of times a transport behavior (T, or inertial transports in parentheses) is followed by a chewing behavior (PC, puncture crushing, or PLC, palatal crushing), indicating the greater degree of prey handling involved in repositioning the prey item for chewing; TOT is the total cycles of processing per feeding bout.

Table 10.1 *Correlations of post-capture prey processing behaviors after removal of the effects of phylogeny by using independent contrasts analysis*

Total, interspersibility, and puncture crushing are significantly intercorrelated ($*p < 0.05$, df $= 5$) among 18 species representing 10 families. Species means and patterns of prey processing evolution in lizards are shown in Figure 10.4.

Variable	PLC	T	PC	TOT
Palatal crushing (PLC)	—			
Transport (T)	−0.42	—		
Puncture crushing (PC)	0.18	0.08	—	
Total (TOT)	0.22	0.13	**0.95***	—
Interspersibility (INT)	0.21	0.12	**0.94***	**0.98***

effort to process prey prior to swallowing but that the prey may be more reduced and punctured. The prey processing values within the Scincoidea are between the iguanian and autarchoglossan values. The Scincidae is essentially identical to Gekkota, and *Gerrhosaurus* exhibits prey processing more like the Lacertoidea except in having lower total numbers of cycles per prey item. *Lepidophyma* has extremely complex and extensive prey processing. We interpret these results as indicating that prey processing behavior appears to be becoming more simple in the Iguania (or staying simple, depending on how *Sphenodon* processes prey), whereas the Gekkota, Scincoidea, and Anguidae are somewhat intermediate and the Lacertoidea and Varanidae have more complex and more extensive prey processing for the same sized prey.

In summary, compared with the iguanian condition (low interspersibility and numbers of cycles, preference for palatal crushing) there are three major shifts in prey processing behaviors in lizards. First, scleroglossans appear to have shifted from palatal crushing to puncture crushing as the dominant post-capture prey processing behavior. Second, the Lacertoidea and Varanidae have independently shifted to more numerous and more complex prey processing repertoires. Third, Teiidae and Varanidae appear to have independently evolved inertial prey transport.

Patterns of evolution in feeding behavior, tongue function, and foraging mode

Although we are far from fully understanding squamate feeding function, sufficient information on tongue morphology, feeding behavior, the vomeronasal system, prey discrimination ability, and foraging mode is available to illuminate some general patterns of tongue function in lizards. Without

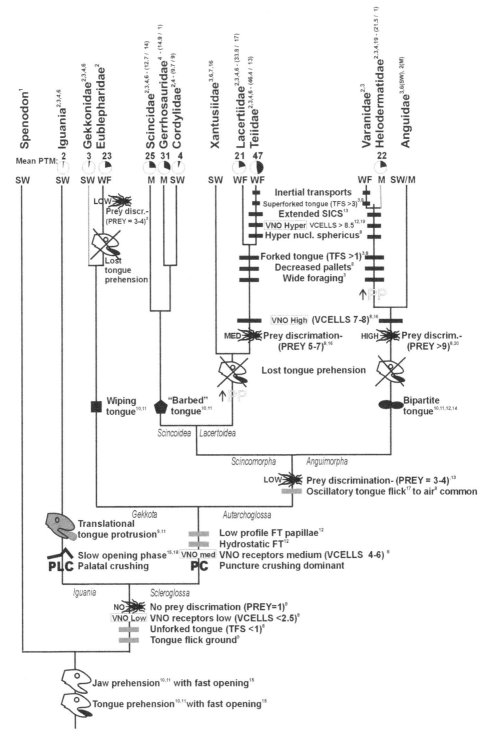

Figure 10.5. Patterns of evolution in feeding behavior and chemosensory function in lizards. From the primitive pattern of tongue and jaw prehension and visual prey recognition in *Sphenodon*, lizards have radiated into five basic

exception more data from representatives of all of the major clades would be desirable. However, in Figure 10.5 we integrate the existing data on the prey capture and processing in relation to foraging mode and specific quantitative variables available on tongue structure, prey recognition and the vomeronasal system. Feeding behavior patterns are taken from the major points of the above discussions of prey capture (Fig. 10.3) and prey processing data (Fig. 10.4). The mapping for traits on vomeronasal system morphology, prey discrimination abilities, and tongue forkedness (Table 10.2) are based on species or familial means for the VCELLS, PREY, and TFS variables from data in Cooper (1997a, Table 2) and Schwenk (1994). Other tongue character

---

Caption for Figure 10.5 (cont.)

> trophochemosensory patterns (Iguania, Gekkota, Scincoidea, Lacertoidea, Anguimorpha). The Iguania developed the unique system of translational tongue protrusion and prey processing dominated by palatal crushing (PLC) and low interspersability. The Scleroglossa developed hydrostatic foretongues and shifted to puncture crushing (PC) as the dominant chewing behavior. The Gekkota retained jaw and tongue prehension and simple prey processing but adapted the hydrostatic tongue as a wiping tongue (with the Eublepharids losing tongue prehension and gaining low prey discriminatory abilities). The Autarchoglossa developed oscillatory aerial tongue flicking for vomeronasal olfaction and moderate numbers of VNO receptor cells. The Scincoidea retain the primitive autarchoglossan conditions with a derived "barbed" tongue morphology. The Lacertoidea and Anguimorpha independently lost tongue prehension, increased prey processing behavior, and exhibit striking patterns of convergence in many aspects that facilitate chemosensory function in their most derived taxa (black bars), which have shifted to wide foraging. See text for discussion. Foraging mode assignments (sit-and-wait (SW), wide foraging (WF), or mixed (M) follow published qualitative assignments (references listed with taxa) and assignments based on familial means (indicated on taxa where available) that we calculated for percent time moving data (PTM: <10 (SW), 10–25 (M), >30 (WF)). Trait mapping (reference sources numbered) is based on feeding patterns from Fig. 10.3 and 10.4, published character states for higher taxa, published suites of characters indicating several derived tongue types, and gap-coded familial means for the following quantitative variables listed in Table 10.2: TFS (degree of tongue forking: <1 (unforked), 1–2 (forked), >3 highly forked). PREY (strength of prey discrimination measured experimentally: 1 (none detected), 3–4 (low), 5–7 (medium), >9 (high) prey discrimination. VCELLS (abundance of vomeronasal (VNO) receptor cells: <2.5 (low), 4–6 (medium), >7–8 (high), >8.5 (hyperabundant)). References and notes: 1, Walls (1981); 2, Cooper (1994b, table 5.1); 3, Schwenk (1994); 4, Perry (1999, this volume); 5, Cooper (1990); 6, Dunham *et al.* (1988); Fellers and Drost (1991); 8, Cooper (1997a); 9, Schwenk and Wagner (2001); 10, Schwenk, (1988); 11, Schwenk (2000); 12, Schwenk (1993); 13, Cooper (1994a); 14 McDowell (1972); 15, Bels (2003); 16, relevant tongue data are lacking for Xantusiidae; 17, Gove (1979); 18, Bels *et al.* (1994); 19, VNO ratio data are lacking for helodermatids, and note *Heloderma* has low prey discrimination, (PREY = 3.5).

states are specific traits or derived suites of states for specific taxa from published studies. References for each trait/state are indicated on specific traits or taxa in Figure 10.5. Character states are mapped by using parsimony on the consensus phylogeny widely used in comparative lizard studies (Estes *et al.*, 1988) based on published states for specific taxa or published family-level means with the widely supported assumption that within-family patterns are highly conserved (Schwenk, 1988, 1995; Cooper, 1994a, 1995a; Perry, 1999, this volume, Chapter 1).

Perry (1999, this volume, Chapter 1) discusses the problems and the various ways in which foraging mode has been described (from dichotomous to continuum) and used quantitative data on foraging movement timing to confirm the qualitative observations that there is family-level conservatism in foraging mode (Cooper, 1994a, 1995a,b; Schwenk, 1995). Accordingly, the assignment of foraging mode states (sit-and-wait/mixed/wide foraging) was based on familial means (Fig. 10.5) that we calculated for percent time moving (PTM) data from Perry (1999, this volume, Chapter 1) and/or are taken from published qualitative assignments of foraging mode (for taxa with no PTM data, references indicated on the taxa in Fig. 10.5). Note that phylogenetically reconstructed family nodal values are essentially identical to the raw familial means and provide the same foraging mode assignments (D. B. Miles, unpublished data). Sit-and-wait taxa (SW) are clearly described as sedentary and/or have PTM means <10%. Wide-foraging taxa (WF) are clearly described as highly active or have PTM means >30%. Following the obvious gaps in mean PTM values, we chose "mixed" (M) to describe taxa that were variable in foraging mode (both intra- and interspecifically) and/or have PTM means between 10% and 25%. We refer readers to Anderson (this volume, Chapter 15) and Perry (this volume, Chapter 1) for discussions of the versatility and ecological basis of these intermediate foraging modes. Note that mean PTM matches qualitative published assignments except for the Helodermatidae, which are qualitatively regarded as wide foraging but whose PTM data indicate they are mixed foragers. In addition, foraging mode patterns strongly support Schwenk's hypothesis (1994) that wide foraging evolved independently in the Lacertoidea and Anguimorpha.

Despite the limited number of species, the characters we use are either quantitative (Table 10.2) or unambiguous published characters for these taxa and they reveal rather striking patterns of divergence and convergence. Starting with the outgroup, *Sphenodon*, we present what these data reveal as the emergent patterns of association and evolutionary change in feeding behavior, tongue function, and foraging mode in the major lizard clades.

Table 10.2 *Familial means for the quantitative variables for the strength of prey discrimination measured experimentally* (PREY), *the degree of tongue forking* (TFS), *and the relative abundance of vomeronasal receptor cells* (VCELLS)

PREY coded as: 1, none detected; 3–4, low; 5–7, medium; >9, high prey discrimination. TFS coded as: <1, unforked; 1–2 forked; >3 highly forked. VCELLS coded as: <2.5 (low), 4–6 (medium), >7–8 (high), >8.5 (hyperabundant)). These gap codings were used for the character state mapping in Figure 10.5. Detailed descriptions of these variables are given in Cooper (1997a).

Family	Prey chemical discrimination (PREY)	Degree of lingual forking (TFS)	Vomeronasal cell abundance (VCELLS)
Phrynosomatidae (Iguania)	1.00 (none)	0.15 (unforked)	2.44 (low)
Polychrotidae (Iguania)	1.00 (none)	0.00 (unforked)	1.19 (low)
Chameleontidae (Iguania)	1.00 (none)	0.18 (unforked)	1.70 (low)
Gekkonidae	1.00 (none)	0.22 (unforked)	5.36 (medium)
Scincidae	3.65 (low)	0.28 (unforked)	4.96 (medium)
Xantusiidae		0.10 (unforked)	
Lacertidae	6.63 (medium)	1.51, *1.70* (forked)	6.98 (high)
Teiidae	5.92 (medium)	3.90, *3.26* (highly forked)	10.38 (hyperabundant)
Varanidae	15.66 (high)	7.14, *6.43* (highly forked)	8.57 (hyperabundant)
Helodermatidae	3.57 (low)	1.14, *1.08* (forked)	—
Anguidae	9.58 (high)	0.68, *0.53* (unforked)	7.69 (high)

Source: Data from Cooper (1997a, Table 2) and Schwenk (1994, Table 1, in italics).

Sphenodon: the outgroup condition

The putative outgroup taxon *Sphenodon* uses both jaw and tongue prehension (Figs. 10.3B and 10.5; Gorniak *et al.*, 1982; Herrel *et al.*, 1995; Schwenk, 2000) but whether one or the other is preferred or whether there is a size effect on tongue use is unknown. *Sphenodon* appears to be sufficiently different from the higher taxa to show that both the Iguania and Scleroglossa diverged from the putative ancestral condition based on *Sphenodon*. For example, *Sphenodon* has unique features in the transversus muscle (Smith, 1988), and in the hyoglossus and verticalis muscles of the tongue (Schwenk, 1986). Its tongue is attached ventrally nearly to the tip and the tongue is protracted tip first early in the strike, in contrast to the unattached foretongues and ventrally curling tongue protraction of other lizards (Schwenk, 1986). The tongue is broad and short with a round, uncleft apex. It has long, rather generalized, high-profile surface papillae and the entire tongue is linked to the hyobranchium. Cineradiography has shown that tongue prehension in *Sphenodon* is produced by hyobranchial protraction that flips the tongue anteroventrally over its ventral attachment to the floor of the mouth (Gorniak *et al.*, 1982). Tongue muscles probably affect the shape of the tongue tip but there is no evidence of translation of the tongue on the hyobranchial elements (Gorniak *et al.*, 1982; Schwenk, 2000). In *Sphenodon* the tongue is not used for chemosensory sampling outside the mouth and prey are recognized visually (Schwenk, 2000; Pianka and Vitt, 2003).

Evolutionary novelty and radiation in the Iguania

The translational tongue protrusion system in the Iguania (Fig. 10.5) has added a novel component to the basic system in *Sphenodon*. As in *Sphenodon*, protraction is produced by the combined actions of the hyobranchial muscles that protract the hyobranchial apparatus and the intrinsic and extrinsic tongue muscles that control the shape of the tongue. However, in the Iguania the tongue actually moves on a specialized central rod in the hyobranchial apparatus: the lingual process (Smith, 1984; Schwenk, 1986; Smith and Kier, 1989). The tongue is protruded and flips anteroventrally by the summed actions of hyobranchial protraction, translation of the tongue along the lingual process, and deformation of the tongue itself. Thus, although both use the entire tongue in prey capture, tongue protraction in the Iguania differs from *Sphenodon* in the details of tongue musculature and the addition of lingual translation (Wagner and Schwenk, 2000). Furthermore, the key innovation of lingual translation varies within the Iguania. Although the hyoid musculature and extrinsic protractor and retractor muscles of the tongue are similar for all iguanians (Bels *et al.*, 1994), the complexity of the intrinsic tongue musculature and the degree of translation of the tongue on the lingual process varies among the Iguania (Schwenk, 2000; Wagner and Schwenk, 2000). The agamids and chameleontids have developed different lingual rods and more complex sphincter-like muscles around them to effect greater protrusion distances compared with the iguanids (Schwenk and Bell, 1988; Smith, 1988; Delheusy and Bels, 1992; Bels *et al.*, 1994; Herrel *et al.*, 2001). Lingual translation is associated, phylogenetically, with the appearance of the SO phase (Fig. 10.3A) and it appears that the greater the translation, the longer the SO phase. Together these traits define translational tongue protrusion (Bels, 2003) in the Iguania as derived from tongue prehension in *Sphenodon* (Herrel *et al.*, 1995). In addition, the Iguania have long filamentous papillae on the tongue surface covered with mucous cells that function to adhere to the prey item during the strike (reviewed in Schwenk, 2000). Like the degree of lingual translation, these papillae have evolved within the Iguania into various plumate, arborate, and reticular forms serving to increase prey adhesion with the tongue (Schwenk, 2000). In sum, the Iguania (Fig. 10.5) should not be regarded as retaining the primitive condition from *Sphenodon*. Iguanians developed a unique tongue protrusion system different from tongue prehension in many aspects of the musculature, tongue surface histology, biomechanics, and kinematics of the strike. They have been extremely successful with this model "evolutionarily stable configuration" and have modified it to increase protrusion distance and prey adhesion ability (Wagner and Schwenk, 2000).

In terms of prey foraging traits, the Iguania (Fig. 10.5) are sit-and-wait foragers, have unforked tongues (TFS < 1), have low VNO receptor ratios (VCELLS < 2.5), lack the ability to discriminate prey (PREY score = 1), and they tongue-flick only to the substrate and prey, not the air (Herrel *et al.*, 1998). Thus, they have rudimentary chemical prey detection morphology and location abilities but have combined visual prey location with a sit-and-wait foraging strategy and a novel sticky, protrusile tongue.

The Scleroglossa: a second radiation in tongue function

From the condition seen in *Sphenodon*, the Scleroglossa also uniquely diverged in tongue function (Fig. 10.5). The aptly named Scleroglossa shifted to more streamlined and hydrostatic foretongues with low-profile surface papillae. Here the foretongue is no longer functionally attached to the hyobranchial apparatus, and it has a circular muscle system and a more posterior attachment of the genioglossus (Smith, 1984) that control independent hydrostatic elongation and control of foretongue movements (Schwenk, 2000). Tongue protraction involves hyobranchium/hindtongue movement and independent hydrostatic movements of the foretongue (Schwenk, 2000). It is interesting that coincident with the transition to a hydrostatic foretongue is a shift to a preference for puncture crushing over palatal crushing (Figs. 10.4, and 10.5, PC). This shift suggests that increased reliance on the jaws rather than the tongue for crushing the prey item during prey processing has occurred as a consequence of the shift to a hydrostatic foretongue. The vomeronasal system appears to increase to a medium level of receptor abundance in the Scleroglossa because the Gekkonidae (VCELLS = 5.36) and the Autarchoglossa (VCELLS = 4.96 and up) have higher scores relative to the Iguania (Fig. 10.5).

However, the increase to medium levels of VNO tissue is not clearly associated with the appearance of aerial tongue flicking in prey recognition. The shift to tongue use for prey recognition is not a scleroglossan trait but has appeared twice in higher taxa (Fig. 10.5). It appears in the Autarchoglossa because it is here that the common use of oscillatory tongue flicking to air appears (Gove, 1979; Cooper, 1997b) together with detectable levels of prey discrimination performance (PREY = 3–4). It also seems to have appeared independently in the Eublepharidae (Cooper, 1995c) where tongue flicking in prey discrimination and low levels of prey discrimination are found in contrast to the absence of prey discrimination and tongue flicking in the Gekkonidae, which is considered to be plesiomorphic for Gekkota (Cooper, 1994a, b).

The Gekkota: the tongue wipers

It has been difficult to obtain data from these secretive and nocturnal lizards but the Gekkota commonly use jaw prehension (Delheusy and Bels, 2000; Schwenk, 2000) and appear to retain tongue prehension based on reports from one species (*Phelsuma*: Schwenk, 2000) and the observation that frugivorous species commonly use the tongue to feed on flowers and fruit (Delheusy and Bels, 2000; Bauer, this volume, Chapter 12). The hydrostatic foretongue in geckos has a series of unique modifications (Schwenk, 1988: Characters 8, 15, 23, 30, 35) that show that the tongue functions primarily as an "eye wiping tongue" (Fig. 10.5, ■ : Schwenk, 2000). Geckos have a spatulate, untapered foretongue, unique peg-like tongue papillae specialized for applying secretions during eye wiping, diffusely scattered sublingual glands (rather than a single pair) all across the floor of the mouth under the free part of the tongue, uniquely subdivided hyoglossus muscles, and hypertrophied longitudinalis muscles that control wiping movements (Schwenk, 1988, 2000). In addition, the papillae in the portion of the tongue used in eye wiping in Gekkota are exceptionally smooth, lightly keratinized, non-glandular, and filled by large vascular sinuses (Schwenk, 2000). The gecko tongue is clearly highly specialized for eye licking and we propose that this is the primary function of the tongue and may actually conflict with direct aerial chemosensory and feeding functions. Their spatulate, uncleft tongues do not facilitate aerial lingual prey detection, which the Gekkonids lack (Gove, 1979; Cooper 1994a, 1995a; Cooper and Habegger, 2000; Table 10.2: PREY = 1). Dedication of the tongue to eye wiping may contribute to why the geckos have become olfactory specialists (Vitt *et al.*, 2003) with the most highly developed nasal olfactory system in lizards, which does not require aerial tongue sampling (Gabe and Saint Girons, 1976; Schwenk, 1993, 1995; Cooper, this volume, Chapter 8). Interestingly, the Gekkota have unforked tongues (Table 10.2) and do not deliver aerial samples to it via oscillatory tongue flicking, but they do have medium abundance of VNO receptors (Table 10.2). It may be that airborne molecules hitting the eyes are delivered to the vomeronasal system directly in liquids via the nasolacrimal duct, as has been shown in snakes (a tetrapod trait) (Rehorek *et al.*, 1999, 2000, Hillenius and Rehorek, 2001) or from the eyes to the vomeronasal system via eye wiping, as has been shown in frogs (Rehorek *et al.*, 1999; Hillenius and Rehorek, 2001). In fact, gecko eyes, like those of frogs, may be acting as molecule collectors with eye wiping serving more for vomerolfaction than in eye maintenance.

The aglandular low-profile papillae are not conducive to prey adhesion and the derived muscular arrangement for extreme hydrostatic expansion and curling to control of wiping movements may conflict with tongue protrusion and retraction as seen in the hypertrophied system of the Iguania. Details of the capture gape profiles and tongue use in gekkotans are badly needed to know how common tongue prehension is in Gekkota, how it compares to that of other lizards, and whether it may have actually have been lost in animal prey capture as a consequence of the conflict with eye wiping function. The Gekkota as a whole has shifted to smaller prey than the Iguania (Vitt *et al.*, 2003) which could be in part a consequence of prey capture limitations of their derived tongues. Prey processing behavior in geckos (based only on *Eublepharis*) is intermediate between the Iguania and Autarchoglossa in that they have low numbers of cycles per feeding bout, like iguanians, but exhibit more puncture crushing than palatal crushing, like autarchoglossans, (Fig. 10.4). This suggests that the number of prey processing cycles per prey has not been affected by their wiping tongue system but that their tongue's flatness has affected their shift to puncture crushing in chewing. In sum, strong morphological evidence and available functional data suggest that generalized geckos have devoted the primary hydrostatic functions of the foretongue to eye wiping while relying on vision and their advanced olfactory sense to locate prey and plant food.

The ground geckos, the Eublepharidae, are unique among the Gekkota in having lost tongue prehension and gaining the ability to discriminate airborne prey chemicals with olfaction, though apparently not with aerial tongue sampling (Cooper, 1994a, 1995a; Cooper and Habegger, 2000). They have also switched to nocturnal wide foraging. Interestingly, the Xantusiidae, like the eublepharids, also have a spatulate, untapered foretongue, unique peg-like tongue papillae specialized for applying secretions, eye wiping behavior, similar "low" prey discrimination skills, and have lost tongue prehension, but they are diurnal, sit-and-wait foragers (oscillatory aerial tongue flicking appears to be absent; pers. obs.).

The Autarchoglossa: three patterns of tongue structure and function

The hydrostatic foretongue in the Autarchoglossa is clearly associated with the oscillatory tongue flicking to air behavior (used as the delivery mechanism for aerial vomerolfaction) and they have gained the ability to discriminate prey (Fig. 10.5). All autarchoglossans examined to date (all taxa indicated except for Xantusiidae) are able to discriminate prey chemicals with aerial tongue flicking vomerolfaction. However, only a low level of prey discrimination (PREY = 3–4) appears at the level of Autarchoglossa (Fig. 10.5). VNO receptor abundance (which might be expected to have a correlated increase) had already changed to a medium level (VCELLS to 4–6) at the level of the Scleroglossa (Fig. 10.5). Thus, medium levels of VNO receptor abundance appear to have been present (perhaps as exaptations) before the appearance of aerial tongue flicking in prey discrimination.

Although the transition to aerial vomerolfactory tongue function appears to be a synapomorphy for the Autarchoglossa, patterns in prey discrimination level, VNO receptor abundance, tongue forkedness, tongue structure, tongue use in prey capture and prey processing, and several other traits do not change at this level, and clearly illustrate that there are multiple patterns of evolution within the Autarchoglossa (Fig. 10.5). One pattern involves the retention of basal autarchoglossan condition (in the Scincoidea). The other two patterns are revealed by striking patterns of convergence in tongue function, morphology, and behavior within the Lacertoidea and Anguimorpha (Fig. 10.5). In these two clades, tongue prehension is lost and the terminal taxa have independently evolved derived tongues and hypertrophied vomeronasal systems dedicated to chemosensory function.

The Scincoidea: radiation in intermediate foraging

The Scincoidea have retained essentially all of the primitive characteristics of the Autarchoglossa or Scleroglossa (Fig. 10.5). They retain both jaw and tongue prehension (Fig. 10.1, 10.3), prefer puncture crushing in prey processing (Fig. 10.4), and have unforked tongues (TFS > 1), low levels of prey discrimination (PREY = 3–4), and medium levels of VNO receptor abundance (VCELLS = 4–6). The Scincoidea are derived, however, in sharing unique features in both the hind- and foretongue (a suite of characters we term the "barbed tongue" (Fig. 10.5,⬟, following Schwenk, 1988). They have broad hindtongues compared with other autarchoglossans (Schwenk, 1995). This enlarged hindtongue may improve prey handling because of the tongue's increased surface area and adhesive abilities (Schwenk, 2000) or be used in different behaviors (Herrel *et al.*, 1999). Their relatively larger hindtongues may explain why they have intermediate prey processing behavior. In addition, the Scincoidea have three derived foretongue characters (Schwenk, 1988, 2000). First, they have irregular trailing edges on lingual scales. Second, scincoidean tongues are unique in having bundles of the genioglossus lateralis muscle (usually a hindtongue muscle) extending into the foretongue (Schwenk, 2000). This condition appears to be a novel addition to the foretongue and may aid in prey retraction with a hydrostatic foretongue. Furthermore, this unique elaboration of the foretongue muscles causes the foretongue to be significantly wider than the underlying hyoglossus muscles, again potentially aiding in prey handling. Third, they have a derived tongue tip morphology that Schwenk (1988) termed "barbed". A barbed or arrowhead-shaped tongue tip is formed by lateral expansion of the epithelium underlying the tongue tip so that it is visible in the dorsal view. We propose that the wider and more muscular foretongue with increased retractile ability could be adaptations to facilitate tongue prehension with the scincoidean hydrostatic tongue.

As such, the Scincoidea retain primitive features for the Autarchoglossa, using the hydrostatic foretongue for low levels of prey recognition with medium levels of VNO receptor abundance without sacrificing the ability to use the tongue in prey capture and aerial tongue flicking. However, their tongues appear to have developed muscular and shape adaptations to facilitate tongue prehension with their hydrostatic foretongues (e.g. barbs). They are, in a sense, intermediate between the iguanian condition and other autarchoglossans. In addition, scincoideans have intermediate values for prey processing behaviors: they retain the iguanian pattern of lower numbers of cycles per feeding bout and less interspersion, but have the autarchoglossan tendency for puncture crushing instead of palatal crushing (Fig. 10.4). Among the Scincoidea, the Gerrhosauridae appear to shift to relatively greater chemosensory function because they use tongue prehension less frequently and have more prey processing cycles and more interspersibility than skinks (Fig. 10.4). Not surprisingly, the Scincoidea also use prey types intermediate (in volume and mobility) between those of ambush and wide-foraging taxa (Vitt and Pianka, this volume, Chapter 5).

Lacertoidea and Anguimorpha: true wide foraging has evolved twice

Within the Lacertoidea and the Anguimorpha, patterns of character states show that extremely derived vomerolfactory systems have independently evolved after tongue prehension was lost (Fig. 10.5). The Lacertoidea retain a generalized autarchoglossan tongue (not "barbed" and not "bipartite"), have lost tongue prehension, and have shifted to increased cycles of prey processing during chewing (Fig. 10.5). This basic configuration is retained in Xantusiidae.

High VNO receptor abundance (VCELLS 7+) and medium prey recognition (PREY 5–7) are found in the Lacertidae and Teiidae. (Data are lacking for Xantusiidae, but they are clearly primitive in every other trait as well, except that they independently gained eye wiping.)

The Anguimorpha also lost tongue prehension but are derived in having a radically modified suite of characters defining the "bipartite" (or diploglossan) tongue (McDowell, 1972; Schwenk, 1988, 1993, 2000). The bipartite tongue (Fig. 10.5: ●●) is unique in that all elongation and shortening of the tongue is localized within the foretongue, which has also lost its glands (Schwenk, 2000), and has its own muscle, the geniomyoideus (McDowell, 1972). The tongue of the anguimorphs is visually, histologically, and functionally separated into a foretongue devoted to chemosensory function and a hindtongue retaining prey processing function (Schwenk, 2000). The advent of the bipartite tongue appears to be coincident with the shift to high prey discrimination (PREY > 9) and high VNO receptor abundance (VCELLS = 7+); however, the Anguidae appear to have these characteristics without having forked tongues (Table 10.2) or any further modifications of the tongue or chemosensory system (Fig. 10.5).

In the crown taxa of the Lacertoidea (the Lacertidae and Teiidae) and Anguimorpha (the Varanidae and Helodermatidae) slightly different but clearly convergent patterns of chemosensory evolution are evident in many characteristics (Fig. 10.5: heavy black bars). Two tongue features change in these crown groups. Forked tongues (TFS > 1) appear (Schwenk, 1994) and there is a sharp decrease in the presence of ventral tongue pallets (ventral surfaces at the tip of the tongue that contact substrates during ground tongue flicking) (McDowell, 1972). From here the Teiidae and (Varanidae + Helodermatidae) converge further in developing hypertrophied vomeronasal integration centers in the brain (the nucleus sphericus), hypertrophied VNO receptor abundance (VCELLS > 8.5), and long-distance and long-duration strike-induced chemosensory searching behavior (SICS). The Teiidae and Varanidae are further convergent in having highly forked tongues (TFS 3.9+) and the presence of inertial prey transport. In addition, they have independently evolved similar jaw muscle force biomechanics on independently evolved "pointed" skulls (McBrayer and Corbin, this volume, Chapter 9).

In sum, from a basic system in which the hyobranchium flipped an attached tongue to capture prey recognized visually, prey capture in lizards diverged in two fundamentally different ways. The Iguania adopted lingual translation and an associated slow opening phase with highly sticky tongue morphology to develop tongue protrusion, which has radiated in the Iguanidae, Agamidae, and Chameleontidae. The stout tongue is used as a major component of chewing, as evidenced by the preponderance of palatal crushing in the Iguania. With their use of palatal crushing, iguanians are able to consume prey items (to the point of swallowing) with only about 9 or fewer prey processing cycles. The ancestor of the scleroglossans probably had a hydrostatic foretongue and switched to puncture crushing as the primary chewing behavior. This led to the several ways in which extant scleroglossans deal with the needs of prey recognition and consumption. The Gekkota modified the hydrostatic tongue for wiping and licking, increased the sensitivity of the olfactory system, and retained jaw prehension and tongue prehension (Delheusy and Bels, 2000). The ancestral Autarchoglossa co-opted the hydrostatic foretongue for low levels of vomerolfactory prey discrimination; however, the chemosensory system radiated in three ways. The Scincoidea remained somewhat generalized, taking advantage of some aspects of vomerolfaction with tongue flicking with their barbed tongues without giving up the utility of the tongue in prey capture. Many aspects of tongue morphology, feeding behavior, and chemosensory behavior and ability make scincoideans appear intermediate between the typical conditions seen in most iguanians and other autarchoglossans. Both the Lacertoidea and Anguimorpha have lost tongue prehension and convergently developed specialized tongues and advanced vomerosensory morphology and performance in their terminal taxa. These specializations appear to have affected prey processing efficiency based on increases in the amount of processing needed to consume prey (Fig. 10.5: PP). Both the Lacertoidea and Anguimorpha have relatively generalized outgroups (Xantusiidae and Anguidae) that help illustrate patterns of convergence and specialization in chemosensory systems that peak in the Teiidae and Varanidae.

Feeding behavior, chemosensory adaptation, and foraging mode

The Iguania and Gekkota are considered to be sit-and-wait foragers (Perry, this volume, Chapter 1; Bauer, this volume, Chapter 12). However, these taxa differ remarkably in terms of feeding and chemosensory function. The diurnal Iguania identify prey visually and have evolved a unique translational tongue protrusion system with its derivations of the tongue, skull (McBrayer and Corbin, this volume, Chapter 9), and gape cycle (SO phase), a large reliance on tongue use in reducing the prey (their preference for palatal crushing), and ability to reduce and swallow prey with little interspersibility and relatively few cycles of prey processing (Fig. 10.4). The Iguania should not be thought of as retaining the primitive condition because their skull and tongue function are considerably different from those of the outgroup *Sphenodon*. The Iguania represents a unique radiation in the details of tongue design based on their key innovation of lingual translation (Wagner and Schwenk, 2000). This is the

most common of the feeding strategies in lizards based on the fact that about
one third of all lizard species use it.

The Gekkota (except for the Eublepharids) are the other ambush foragers but
they have retained the ambush foraging mode after the transition to a more
hydrostatic tongue. The Gekkota are still a black box behaviorally (more data for
non-eublepharid feeding are badly needed) but they are believed to retain both
tongue and jaw prehension (Schwenk, 2000). They have no evidence of aerial
tongue flicking in vomerochemoreception (instead they have hypertrophied
olfactory systems) and their unique muscle specializations and papillae show
that their hydrostatic tongues are highly modified for the eye wiping function.
Their switch to nocturnality has been hypothesized to be driven by competition
for food with competitively superior, wide-foraging autarchoglossans (Vitt *et al.*,
2003). The lack of aerial chemosensory tongue flicking in the Gekkota (although
the eublepharids can discriminate prey) reveals one of the little-considered patterns in scleroglossan evolution (Fig. 10.5): that the hydrostatic tongue evolved in
the Scleroglossa before it was adapted to aerial oscillatory tongue flicking in the
Autarchoglossa. However, the Gekkota may be using eye wiping in vomeronasal
function. Our prey processing data reveal that the appearance of the hydrostatic
tongue is coincident with the switch to puncture crushing. This pattern indicates a
decrement in the ability of the derived scleroglossan tongue to crush prey against
the palate. However, the Gekkota have been shown to have switched to smaller
prey (Vitt *et al.*, 2003) which may not require as much reduction. The Gekkota
with their nocturnal ambush strategy have also been very successful with about
one quarter of all lizards using this strategy; however, much remains to be learned
about this radiation in terms of both foraging biology and feeding behavior
(Bauer, this volume, Chapter 12).

Is the wide-foraging strategy correlated with
chemosensory specialization?

It has been widely held, with a few exceptions (Schwenk, 1994), that prey
foraging behavior shifted from an ambush or sit-and-wait (SW) mode in
Sphenodon, the Iguania, and Gekkota to an active or wide-foraging (WF)
mode at the level of the Autarchoglossa. However, Perry (1999, this volume,
Chapter 1; Fig. 10.5; Miles *et al.*, this volume, Chapter 2) argues that there is
not a single discrete shift to wide foraging at the level of the autarchoglossans.
Foraging time and movement data show that SW and/or mixed (M, familial
PTM means between 10–30) foraging modes describe the Scincoidea and the
basal taxa within the Lacertoidea and Anguimorpha, and that the distribution
of WF (PTM > 30) corroborates Schwenk's (1994: Fig. 10.3) proposal that

wide foraging evolved independently in the ancestor of Lacertidae and Teiidae, and in the ancestor of the Varanidae and Helodermatidae. This independent appearance of wide foraging in the crown autarchoglossans is strikingly congruent with the patterns of convergence in chemosensory, feeding, and tongue characters in these groups (Fig. 10.5: black bars).

The recognition that the Scincoidea have a mixed foraging mode and clearly use tongue prehension (Figs. 10.1, 10.5) illustrates the intermediate nature of this group and shifts our attention to the concerted patterns of evolution of wide foraging and chemosensory specialization in the Lacertoidea and Anguimorpha. The Scincoidea retain unforked tongues and low prey discrimination abilities, are intermediate in prey processing cycles and interspersibility, and have developed their own derived suite of "barbed" tongue features but use it in aerial oscillatory tongue flicking. They are also intermediate in prey use (Vitt and Pianka, 2005; this volume, Chapter 5). The intermediate condition of the Scincoidea reveals three important new implications for our understanding of lizard evolution. First, the adaptation of the hydrostatic foretongue for aerial chemosensory sampling does not require the loss of tongue use in prehension; this jack-of-all-trades approach remains as a very successful strategy, occurring in 25% of autarchoglossan lizards (the skinks). Second, the loss of tongue prehension has occurred twice *within* the Autarchoglossa, once within the Lacertoidea and independently within the Anguimorpha. Third, because of multiple occurrences, it strengthens the hypothesis that the loss of tongue prehension is the key innovation associated with the evolution of chemosensory specialization. In both autarchoglossan cases the loss of tongue use in prey capture is associated with subsequent evolution of extremely derived states in most aspects of lizard processing behavior, tongue and vomeronasal morphology, and chemosensory abilities in terminal taxa. The degree of convergence in these two groups is striking; however, there are clear differences that support the interpretation that they are independent transitions. The Lacertoidea have the "whole tongue" hydrostatic function like that of the Scincoidea (Schwenk, 2000) whereas the Anguimorpha have the derived bipartite tongue, and the details of character transition are different in the two groups. For example, the Lacertoidea never progress beyond medium prey discrimination ability (PREY 5–7) but ultimately attain the highest VNO receptor abundance in the teiids (10.38 vs. 8.57 in Varanidae).

The loss of tongue prehension also appears to be related to the decrease in, or loss of, tongue use for prey processing and a significant increase in number of processing cycles and repositioning movements used to consume prey. The iguanians prefer palatal crushing and use fewer than 10 cycles per prey item, whereas the Lacertoidea and Anguimorpha prefer puncture crushing and have clearly moved to using more (13–33) cycles of processing per prey item

(McBrayer and Reilly, 2002). The shift to puncture crushing and more processing could be a consequence of losing the use of the tongue in processing (i.e. it takes more cycles to prepare prey for swallowing without the use of the tongue). Alternatively, it may indicate (as in mammals) that the prey are actually being better reduced and punctured prior to swallowing to increase digestive efficiency, which has been proposed to be related to higher metabolic needs of wide foraging (McBrayer and Reilly, 2002; Brown and Nagy, this volume, Chapter 4; Reilly *et al.*, 2001). The evolution of true wide foraging is clearly associated with changes in chemosensory function in lizards.

Conclusions and future directions

The evolutionary patterns discussed here have been optimized on the Estes *et al.* (1988) phylogeny. An alternative phylogeny (Lee, 1998) has little effect on our major conclusions. It does not affect our conclusions about the Iguania, Gekkota, or Anguimorpha. The Lee phylogeny scatters the Lacertoidea, putting the Scincidae and Cordylidae (which includes the Gerrhosauridae) between the Lacertiformes and Anguimorpha. This actually strengthens support for the independent loss of tongue prehension in the latter taxa.

In terms of the general patterns, there are two different radiations of ambush predators, one mixed forager radiation, and three different radiations of wide foragers in lizards. The Iguania and Gekkota are fundamentally different in tongue function in prey capture but they have similar prey processing behaviors: gekkotans differ in preferring puncture crushing which is likely to be correlated with their possession of a hydrostatic tongue. The geckos show that the development of the hydrostatic tongue was not necessarily related to foraging mode. The Scincoidea are intermediate, because they retain primitive autarchoglossan features and tongue prehension but have adopted a mixed foraging mode. Their retention of tongue prehension and low levels of prey discrimination may limit them to the mixed foraging strategy. In fact, one family in the clade, the Cordylidae, employs sit-and-wait foraging and lacks prey discrimination (Cooper and Van Wyck, 1994; Whiting, this volume, Chapter 13).

The switch to chemosensory tongue function within scleroglossan lizards is certainly linked to their wide-foraging strategy and remains a central element of the sit-and-wait (ambush) vs. wide-foraging paradigm and the dominance of active foragers in lizard communities (Vitt *et al.*, 2003). However, our analysis reveals and amplifies the complexity within the Scleroglossa and we argue that analyses that attempt to correlate trait evolution with foraging mode should keep in mind that foraging modes are not purely dichotomous at the level of the Autarchoglossa and that the Scincoidea exhibit more generalized intermediate trait patterns than previously thought before their tongue use in prey capture and relatively intermediate chemosensory function was recognized. This more complex evolutionary pattern (different radiations within Iguania, Gekkota, Scincoidea, and within the Lacertoidea, and Anguimorpha) may actually be more interesting and explanatory in revealing the nature of foraging mode evolution when data are re-evaluated with the knowledge that there are five groups for comparison rather than two (i.e. iguanians and scleroglossans). In addition, this interpretation does not require the "reacquistion" of jaw prehension, which is clearly primitive for, and retained in, Lepidosauria. The hypothesis that the loss of tongue prehension was a key innovation that frees the foretongue for more dedicated prey chemosensory function has even stronger support because it has occurred, not once, but with each of the three transitions to wide foraging (in the Eublepharidae, and within the Lacertoidea and Anguimorpha).

We need to re-evaluate past results and emphasize the need for data on key taxa to evaluate the differences in these five radiations and, perhaps more importantly, to better probe the patterns within each radiation. Certainly the novel and most successful system of the Iguania deserves more respect as a novel radiation, whether it is primitive or not. The addition of translational tongue protrusion (and the SO phase) to jaw prehension is the key innovation in this evolutionarily stable configuration (Wagner and Schwenk, 2000) of the Iguania. It appears that the Gekkota (except eublepharids) lack animal prey discrimination, so we must determine what they are doing kinematically during tongue feeding and wiping and the contribution of nasolacrimal or eye wiping delivery of airborne signals to the vomeronasal system. We have shown that tongue prehension has not been simply lost in the Scincoidea. However, since there are over 1000 species in this clade, we need much more kinematic data to begin to understand the variation of tongue use in feeding and chemosensory function that may exist in this very speciose group. Furthermore, we need more data to test whether other correlates of the cordylid shift back to the SW strategy exist. Gape kinematics during the various tongue behaviors and jaw prehension are needed for many more species across the Lepidosauria. There at least ten families of lizards about which we know very little in terms of morphology, function, behavior, and unfortunately their placements on the phylogeny. It is clear, however, that there are multiple ways to be an ambush or wide-foraging lizard.

Acknowledgements

We thank Anthony Herrel, Vincent Bels, Donald Miles, Bill Cooper, Laurie Vitt, Eric Pianka, Kevin de Queiroz, Scott Moody, Peter Larson, Eric McElroy, Tom White, and two anonymous reviewers for insights, discussion and comments on the manuscript. We thank Tom White, Roger Anderson, and Haakon Kalkvik for animal loans. The behavioral analysis was completed with the assistance of Kristen Hickey, Andrew Clifford, Paul Bedocs, Jason Elias, Andrew Parchman, Lori Gill, Bethany Suchowiecki, Mike Mullins, Stephanie Poduska, Jill Radnov, Chris Akers, Anu Gupta, and Jillian Pokelsek. This work was supported by a John Houk Memorial Research Award and a Claude Kantner Fellowship (to LDM) and undergraduate research fellowships from the Voinovich Center for Leadership and Public Affairs, the Summer Undergraduate Research Program of the College of Medicine, and several Honors Tutorial College Apprenticeships. The research was funded by grants from the Ohio University Research Committee, the Ohio University Research Challenge program, and the National Science Foundation (IBN 0080158).

References

Archie, J. W. (1985). Methods for coding variable morphological features for numerical taxonomic analysis. *Syst. Zool.* **34**, 326–45.

Bels, V. L. (2003) Evaluating the complexity of the trophic system in Reptilia. In *Vertebrate Biomechanics and Evolution*, ed. V. L. Bels, J. P Gasc and A. Casinos, pp. 185–202. Oxford: Bios Scientific Publishers.

Bels, V. L. and Goosse, V. (1990). Comparative kinematic analysis of prey capture in *Anolis carolinensis* (Iguania) and *Oplurus cuvieri* (Oplurinae). *Belg. J. Zool.* **122**, 223–34.

Bels, V. L, Chardon, M. and Kardong, K. (1994). Biomechanics of the hyolingual system in Squamata. In *Biomechanics of Feeding in Vertebrates*, ed. V. L. Bels, M. Chardon and P. Vandewalle (*Adv. Comp. Environ. Physiol.*, vol. 18.), pp. 197–240. Berlin: Springer-Verlag.

Bels, V. L., Davenport, J. and Renous, S. (1997). Kinematic analysis of the feeding behavior in the box turtle *Terrepene carolina* (Reptilia: Emydidae). *J. Exp. Biol.* **277**, 198–212.

Cleuren, J. and De Vree, F. (2000). Feeding in Crocodilia. In *Feeding*, ed. K. Schwenk, pp. 337–58. New York: Academic Press.

Cooper, W. E. (1990). Prey odor discrimination by anguid lizards. *Herpetologica* **46**, 183–90.

Cooper, W. E. (1994a). Chemical discrimination by tongue-flicking in lizards: a review with hypotheses on its origin and its ecological and phylogenetic relationships. *J. Chem. Ecol.* **20**, 439–87.

Cooper, W. E. (1994b). Prey chemical discrimination, foraging mode, and phylogeny. In *Lizard Ecology*, ed. L. J. Vitt and E. R. Pianka, pp. 95–116. Princeton, NJ: Princeton University Press.

Cooper, W. E. (1995a). Foraging mode, prey chemical discrimination, and phylogeny in lizards. *Anim. Behav.* **50**, 973–85.

Cooper, W. E. (1995b) Evolution and function of lingual shape in lizards, with emphasis on elongation, extensibility, and chemical sampling. *J. Chem. Ecol.* **21**, 477–505.

Cooper, W. E. (1995c). Prey chemical discrimination and foraging mode in gekkonid lizards. *Herpetol. Monogr.* **9**, 120–9.

Cooper, W. E. (1996). Variation and evolution of forked tongues in squamate reptiles. *Herp. Nat. Hist.* **4**, 135–50.

Cooper, W. E. (1997a). Correlated evolution of prey chemical discrimination with foraging, lingual morphology and vomeronasal chemoreceptor abundance in lizards. *Behav. Ecol. Sociobiol.* **41**, 257–65.

Cooper, W. E. (1997b). Independent evolution of squamate olfaction and vomerolfaction and correlated evolution of vomerolfaction and lingual structure. *Amph.-Rept.* **18**, 85–105.

Cooper, W. E. and Van Wyck, J. H. (1994). Absence of prey chemical discrimination by tongue flicking in an ambush-foraging lizard having actively foraging ancestors. *Ethology* **97**, 317–28.

Cooper, W. E. and Habegger, J. J. (2000). Lingual and biting responses to food chemicals by some eublepharid and gekkonid geckos. *J. Herpetol.* **34**, 360–8.

Delheusy, V. G. and Bels, V. L. (1992). Kinematics of feeding behavior in *Oplurus cuvieri* (Reptilia: Iguanidae). *J. Exp. Biol.* **170**, 155–86.

Delheusy, V. G. and Bels, V. L. (2000). Kinematics of feeding of *Phelsuma madagascariensis* (Reptilia: Gekkonidae). *J. Exp. Biol.* **202**, 3715–30.

Delheusy, V., Brillet, C. and Bels, V. L. (1995). Etude cinématique de la prise de nourriture che *Eublepharis macularis* (Reptilia: Gekkota) et comparison au sien des geckos. *Amph.-Rept.* **15**, 185–201.

Delheusy, V., Toubeau, G. and Bels, V. L. (1994). Tongue structure and function in *Oplurus cuvieri* (Reptilia: Iguanidae). *Anat. Rec.* **238**, 263–76.

Dunham, A. E., Miles, D. B. and Reznick, D. N. (1988). Life history patterns in squamate reptiles. In *Biology of the Reptilia*, vol. 16, ed. C. Gans and Ray Huey, pp. 441–519 New York: A. R. Liss.

Elias, J. A., McBrayer, L. D. and Reilly, S. M. (2000). Prey transport kinematics in *Tupinambis teguixin* and *Varanus exanthematicus*: conservation of feeding behavior in 'chemosensory-tongued' lizards. *J. Exp. Biol.* **203**, 791–801.

Estes, R., de Queiroz, K. and Gauthier, J. (1988). Phylogenetic relationships within Squamata. In *Phylogenetic Relationships of the Lizard Families: Essays Commemorating Charles L. Camp*, ed R. Estes and G. Pregill, pp. 119–281. Stanford, CA: Stanford University Press.

Fellers, G. M. and Drost, C. A. (1991). Ecology of the island night lizard, *Xantusia riversiana*, on Santa Barbara Island, California. *Herp. Monogr.* **5**, 28–78.

Felsenstein, J. (1985). Phylogenies and the comparative method. *Am. Nat.* **125**, 1–15.

Gabe, M. and Saint Girons, H. (1976). Contribution à la morphologie comparée des fosses nasales et des leurs annexes chez les lépidosoriens. *Mem. Natl. Mus. Nat. Hist. Paris* A**98**, 1–87.

Gans, C. (1969). Comments on inertial feeding. *Copeia* **1969**, 855–7.

Gans, C., De Vree, F. and Carrier, D. (1985). Usage pattern of the complex masticatory muscles in the shingleback lizard, *Trachydosaurus rugosus*: a model for muscle placement. *Am. J. Anat.* **173**, 219–40.

Gorniak, G. C., Rosenberg, H. I. and Gans, C. (1982). Mastication in the tuatara, *Sphenodon punctatus* (Reptilia: Rhynchocephalia): structure and activity of the motor system. *J. Morphol.* **171**, 321–53.

Goosse, V. and Bels, V. L. (1992). Kinematic and functional analysis of feeding behavior in *Lacerta viridia* (Reptilia: Lacertidae). *Zool. Jb. Anat.* **122**, 187–202.

Gove, D. (1979). A comparative study of snake and lizard tongue flicking with an evolutionary hypothesis. *Z. Tierpsychol.* **51**, 58–76.

Herrel, A., Cleuren, J. and De Vree, F. (1995). Prey capture in the lizard *Agama stellio*. *J. Morphol.* **224**, 313–29.

Herrel, A., Meyers, J. M., Aerts, P. and Nishikawa, K. C. (2001). Functional implications of supercontracting muscle in the chameleon tongue retractors. *J. Exp. Biol.* **204**, 3621–37.

Herrel, A., Timmermans, J.-P. and De Vree. F. (1998). Tongue flicking in agamid lizards: morphology, kinematics, and muscle activity patterns. *Anat. Rec.* **252**, 102–16.

Herrel, A., Verstappen, M. and De Vree, F. (1999). Modulatory complexity of the feeding repertoire in scincid lizards. *J. Comp. Physiol.* A**184**, 501–18.

Hillenius, W. J. and Rehorek, S. J. (2001). Eyes to nose: the Harderian gland as part of the vomeronasal system. *J. Morphol.* **248**, 240.

Kraklau, D. M. (1991). Kinematics of prey capture and chewing in the lizard *Agama agama*. *J. Morphol.* **210**, 195–212.

Lee, M. S. Y. (1998). Convergent evolution and character correlation in burrowing reptiles: towards a resolution of squamate relationships. *Biol. J. Linn. Soc.* **65**, 369–453.

Martins, E. P. and Hansen, T. F. (1997). Phylogenies and the comparative method: a general approach to incorporating phylogenetic information into analysis of interspecific data. *Am. Nat.* **149**, 646–67.

McBrayer, L. D. (2004). The relationship between skull morphology, biting performance and foraging mode in Kalahari lacertid lizards. *Zool. J. Linn. Soc.* **140**, 403–16.

McBrayer, L. D. and Reilly, S. M. (2002). Prey processing in lizards: behavioral variation in sit-and-wait and widely foraging taxa. *Can. J. Zool.* **80**, 882–92.

McDowell, S. B. (1972). The evolution of the tongue of snakes, and its bearing on snake origins. In *Evolutionary Ecology*, vol. 6. ed. T. Dobzhansky, M. K. Hecht and W. C. Steere, pp. 191–273. New York: Appleton-Century-Crofts.

Meyers, J. J., Herrel. A. and Birch, J. (2002). Scaling of morphology, bite force, and feeding kinematics in an iguanian and a scleroglossan lizard. In *Topics in Functional and Ecological Vertebrate Morphology*, ed. P. Aerts, K. D'Aout, A. Herrel and R. Van Damme, pp. 47–62. New York: Shaker Publishing.

Nishikawa, K. C. (2000) Feeding in frogs. In *Feeding*, ed. K. Schwenk, pp. 117–47. New York: Academic Press.

O'Reilly, J. C. (2000). Feeding in caecilians. In *Feeding*, ed. K. Schwenk, pp. 149–66. New York: Academic Press.

Perry, G. (1999). The evolution of search modes: ecological versus phylogenetic perspectives. *Am. Nat.* **153**, 98–109.

Pianka, E. R. and Vitt, L. J. (2003). *Lizards: Windows to the Evolution of Diversity*. Berkeley, CA: University of California Press.

Rehorek, S. J., Hillenius, W. J., Quan, W. and Halpern, M. (1999). Keeping an eye on the nose: the Harderian gland is part of the VN system. *Am. Zool.* **39**, 96A.

Rehorek, S. J., Hillenius, W. J., Quan, W. and Halpern, M. (2000). Passage of Harderian gland secretions to the vomeronasal organ of *Thamnophis sirtalis* (Serpentes: Colubridae). *Can. J. Zool.* **78**, 1284–8.

Reilly, S. M. and G. V. Lauder. (1989). Kinetics of tongue projection in *Ambystoma tigrinum*: quantitative kinematics, muscle function, evolutionary hypotheses. *J. Morphol.* **199**, 223–43.

Reilly, S. M., McBrayer, L. D. and White, T. D. (2001). Prey processing in amniotes: biomechanical and behavioral patterns of food reduction. *Comp. Biochem. Physiol.* A**128**, 397–415.

Schwenk, K. (1986). Morphology of the tongue in tuatara, *Sphenodon punctatus* (Reptilia: Lepidosauria), with comments on function and phylogeny. *J. Morphol.* **188**, 129–56.

Schwenk, K. (1988). Comparative morphology of the lepidosaur tongue and its relevance to squamate phylogeny. In *Phylogenetic Relationships of the Lizard Families: Essays Commemorating Charles L. Camp*, ed. R. Estes and G. Pregill, pp. 569–97. Stanford, CA: Stanford University Press.

Schwenk, K. (1993) The evolution of chemoreception in squamate reptiles: a phylogenetic approach. *Brain Behav. Evol.* **41**, 124–37.

Schwenk, K. (1994). Why snakes have forked tongues. *Science* **263**, 1573–7.

Schwenk, K. (1995). Of tongues and noses: chemoreception in lizards and snakes. *Trends Ecol. Evol.* **10**, 7–12.

Schwenk, K. (2000). Feeding in lepidosaurs. In *Feeding*, ed. K. Schwenk, pp. 175–291. New York: Academic Press.

Schwenk, K. and Bell, D. A. (1988). Chameleon-like tongue protrusion in an agamid lizard. *Experientia* **44**, 697–700.

Schwenk, K. and Throckmorton, G. S. (1989). Functional and evolutionary morphology of lingual feeding in squamate reptiles: phylogenetics and kinematics. *J. Zool. Lond.* **219**, 153–75.

Schwenk, K. and Wagner, G. P. (2001). Function and the evolution of phenotypic stability: connecting pattern to process. *Am. Zool.* **41**, 552–63.

Smith, K. K. (1984). The use of the tongue and hyoid apparatus during feeding in lizards (*Ctenosaura similis* and *Tupinambis nigropunctatus*). *J. Zool. Lond.* **202**, 115–43.

Smith, K. K. (1988). Form and function of the tongue in agamid lizards with comments on its significance. *J. Morphol.* **196**, 157–71.

Smith, K. K. and Kier, W. M. (1989). Trunks, tongues, and tentacles: moving with skeletons of muscle. *Am. Sci.* **77**, 29–35.

Smith, T. L., Kardong, K. V. and Bels, V. L. (1999). Prey capture behavior in the blue-tongued skink, *Tiliqua scincoides*. *J. Herpetol.* **33**, 362–9.

Urbani, J.-M. and Bels, V. L. (1995). Feeding behaviour in two scleroglossan lizards: *Lacerta viridis* (Lacertidae) and *Zonosaurus laticaudatus* (Cordylidae). *J. Zool. Lond.* **236**, 265–90.

Vitt, L. J. and Pianka, E. R. (2005). Deep history impacts present day ecology and biodiversity. *Proc. Nat. Acad Sci. USA* **102**, 7877–81.

Vitt, L. J., Pianka, E. R., Cooper, W. E. and Schwenk, K. (2003). History and the global ecology of squamate reptiles. *Am. Nat.* **162**, 44–60.

Wagner, G. P. and Schwenk, K. (2000). Evolutionary stable configurations: functional integration and the evolution of phenotypic stability. *Evol. Biol.* **31**, 155–217.

Walls, G. Y. (1981). Feeding ecology of the tuatara (*Sphenodon punctatus*) on Stephens Island, Cook Strait. *New Zealand J. Ecol.* **4**, 89–97.

11

The meaning and consequences of foraging mode in snakes

STEVEN J. BEAUPRE AND CHAD E. MONTGOMERY

Department of Biological Sciences, University of Arkansas

Introduction

We examine current knowledge of the foraging modes of snakes with particular reference to the "syndrome hypothesis" (McLaughlin, 1989) as described for other organisms. Foraging modes of snakes are often flexible, with some species exhibiting both ambush and active tactics (Hailey and Davies, 1986; Duvall *et al.*, 1990; Greene, 1997; see Table 11.1). In this chapter we focus on inter-specific comparisons and broad-scale patterns among snakes; intraspecific variation in foraging mode is considered elsewhere (Shine and Wall, this volume, Chapter 6). In general, we find some evidence that snakes exhibit variation that is broadly consistent with the syndrome hypothesis. However, physiological and morphological data are scant, and phylogenetic relationships are poorly known. In addition, we question the criteria for defining foraging modes among snakes, and the dichotomous descriptive classification of snakes into "ambush" or "active" modes. We present evidence from bioenergetic simulation that suggests these dichotomous classes may be adaptive peaks; however, intermediate trait values are certainly attainable under permissive circumstances.

How are snakes different from lizards?

All snakes are carnivorous, gape-limited predators that swallow food whole. Therefore, snakes must supply energy to a relatively large body mass by ingesting potentially large prey through a relatively small mouth (Greene, 1997). The squamate reptile skull has evolved greatly from a relatively rigid (non-kinetic) form typical of lizards to a highly modified and highly kinetic form typical of the advanced snakes (alethinophidians, Fig. 11.1) (Greene, 1997; Cundall and Greene, 2000). The scolecophidians (Fig. 11.1) are the sister

Lizard Ecology: The Evolutionary Consequences of Foraging Mode, ed. S. M. Reilly, L. D. McBrayer and D. B. Miles. Published by Cambridge University Press. © Cambridge University Press 2007.

Table 11.1 *A list of North American snakes that utilize flexible foraging strategies, which apparently mix both active searching for prey and ambush tactics*

Information was obtained from recent taxonomic accounts of North American species, or from anecdotal accounts in other literature. Note that little is known about the extent to which flexible foraging tactics are used.

Family/species	Literature source
Colubridae	
Elaphe obsoleta	Greene, 1997
Elaphe quadrivirgata	Greene, 1997
Elaphe vulpina	Ernst and Barbour, 1989
Lampropeltis calligaster	Ernst and Barbour, 1989
Nerodia erythrogaster	Ernst and Barbour, 1989; Gibbons and Dorcas, 2004
Nerodia harteri	Gibbons and Dorcas, 2004
Nerodia rhombifera	Ernst and Barbour, 1989; Gibbons and Dorcas, 2004
Nerodia sipedon	Gibbons and Dorcas, 2004
Nerodia taxispilota	Gibbons and Dorcas, 2004
Thamnophis atratus	Rossman *et al.*, 1996; Lind and Welsh, 1994
Thamnophis cyrtopsis	Rossman *et al.*, 1996; Greene, 1997
Elapidae	
Pelamis platurus	Ernst, 1992
Viperidae	
Agkistrodon contortrix	Ernst, 1992; Greene, 1997
Agkistrodon piscivorus	Ernst, 1992; Hill, 2004
Crotalus adamanteus	Ernst, 1992
Crotalus atrox	Ernst, 1992
Crotalus horridus	Ernst, 1992 (evidence circumstantial)
Crotalus mitchelli	Ernst, 1992
Crotalus molossus	Ernst, 1992
Crotalus pricei	Ernst, 1992
Crotalus scutulatus	Ernst, 1992
Crotalus tigris	Ernst, 1992
Crotalus viridis	Ernst, 1992; Duvall *et al.*, 1990
Crotalus willardi	Ernst, 1992
Sistrurus miliarius	Ernst, 1992

group to Alethinophidians and most likely the oldest extant lineage of snakes (Greene, 1983). Scolecophidians are typically fossorial and secretive, feeding on small invertebrates (Greene, 1983, 1997; Cundall and Greene, 2000). The skull of the Scolecophidia is compact, with short mandibles and limited maxillary and mandibular movement (Cundall and Greene, 2000). The relatively non-kinetic skull and short mandibles of the scolecophidians results in a small gape relative to macrostomates (Cundall and Greene, 2000) (Fig. 11.1). In a progression from

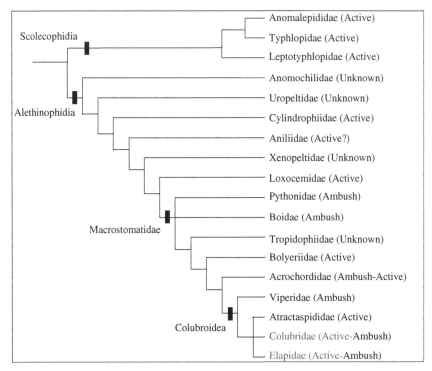

Figure 11.1. Hypothesized phylogenetic relationships of the familial relationships of snakes (redrawn from Cundall and Greene, 2000). Origin in an active foraging varanoïd ancestor is assumed. Predominantly ambush foraging lineages are indicated in gray print. Information on foraging mode is scarce or non-existent for many families. Our assignment of foraging modes to families is based on a combination of literature information (Greene, 1997; Zug *et al.*, 2001), and on consultation with experienced individuals. We note that Acrochordidae in particular have been described as active searchers for sleeping fish (Houston and Shine, 1993) and as ambush foragers (H. B. Lillywhite, pers. comm.). Reversals and foraging mode flexibility are discussed in more detail in the text, and indicated in Table 11.1, respectively. For those taxa where both active and ambush are indicated, we list the predominant mode first, and list the alternative second because of known exceptions (Table 11.1, and see text).

the basal alethinophidians to the Macrostomata, there is an increase in the kinesis of the skull, and an increase in mandible length, allowing increased gape size (Cundall and Greene, 2000). The Colubroidea (Fig. 11.1) exhibit highly kinetic skulls, and include all venomous taxa (Cundall and Greene, 2000). An evolutionary trend of increasing skull kinesis and maxillary liberation is coincident with the impressive radiation within Colubroidea and the diverse feeding adaptations found therein (Cundall and Greene, 2000).

Along with increased gape, allowing for larger prey to be ingested, snakes have evolved more diverse modes of prey subjugation than those found among lizards. Because of the vulnerability of the head during a predatory attack on relatively large prey, more derived snakes probably assume greater risk when subduing larger, potentially dangerous prey (Cundall and Beaupre, 2001; Kardong and Smith, 2002). Snakes typically show three broad forms of prey subjugation: grasping and ingesting live, constriction, and envenomation (Greene, 1997; Cundall and Greene, 2000). Overpowering small prey and consuming it live is exhibited by many colubroid snakes; however, the prey is typically invertebrates, fish or small amphibians that have limited capacity for physical defense. For example, the turtle-headed sea snake (*Emydocephalus annulatus*) eats large numbers of small fish eggs (Shine *et al.*, 2004) and garter snakes are known for grasping and ingesting small frogs, salamanders, and their larvae alive (Rossman *et al.*, 1996). Constriction, tightening of coils around the prey resulting in respiratory and/or circulatory failure, probably arose early among alethinophidians and has been variously retained among advanced lineages (Greene, 1997). The use of envenomation in prey subjugation has evolved several times in the varanoid line: among helodermatid lizards, in the Viperidae, the Elapidae, and probably more than once in the Colubridae. Indeed, recent work isolating toxins from Asian ratsnakes (*Coelognathus radiatus*) suggests the evolution of venoms early in the radiation of the Colubroidea, implying secondary losses in several extant lineages (Fry *et al.*, 2003). Whether the system of delivery is a passive introduction of the Duvernoy's gland secretions via enlarged maxillary teeth, or active injection of venom gland secretions through hollow fangs, envenomation is a highly derived prey subjugation strategy (Cundall and Greene, 2000). Prey that are envenomated and released must be relocated in the environment, a source of selection that is probably responsible for remarkable behavioral and chemosensory adaptations that are also unique to venomous snakes (Chiszar *et al.*, 1992; Kardong and Smith, 2002).

Time–energy allocation and foraging mode

In any environment, there are a limited number of life history alternatives that will allow an organism to survive and reproduce. The number of feasible life history alternatives is limited by environmental and physiological constraints acting on three fundamental allocation decisions of life history: (i) allocation of available time to competing behaviors, (ii) allocation of available mass and energy to competing functions (maintenance, activity, growth, reproduction and storage) and (iii) the packaging of energy and mass allocated to reproduction into individual offspring (Dunham *et al.*, 1989). Animals that differ in foraging mode also tend to differ in bioenergetic variables such as standard metabolic rate, field metabolic rate, and cost of digestion. Because both the rate at which energy and mass can be extracted from the environment, and the rate at which it is allocated to competing functions, can be expected to exert a major influence on optimal time–energy allocation, it seems reasonable that foraging mode will strongly affect the evolution of life history. As the rate of energy assimilation and cost of foraging vary with foraging mode, the allocation strategies associated with the optimal life history must also vary. Suites of traits adapted for a particular forging mode and rate of mass–energy intake are unlikely to be equally adaptive across a wide range of prey types and availability distributions.

The syndrome hypothesis

The foraging mode of an organism generally falls along a continuous spectrum between ambush foraging (sit-and-wait) and active foraging (widely foraging). However, only vague definitions for these foraging modes have been proposed. For example, ambush foragers have been defined as animals that allow potential prey to enter attack range, whereas active foragers move into attack range of potential prey (McLaughlin, 1989; Perry, 1999). In studies of lizards, foraging mode has been quantified by using mean number of movements made per minute, and percent of time spent moving (Huey and Pianka, 1981; Perry, 1999). Use of these movement data to quantify foraging mode implicitly assumes that movements relate to foraging rather than to other functions (e.g., thermoregulation, mate search, migration). Alternatively, Cooper *et al.* (1999) proposed the use of the proportion of attacks on prey discovered while lizard predators are moving in relation to total attacks made (*PAM*). Use of *PAM* carries the advantage of considering only movements specific to foraging. However, because snakes are generally secretive, and feed infrequently relative to lizards, these techniques are difficult to apply. Furthermore, ambush foraging snakes may expend considerable effort searching for high-quality ambush sites (Duvall *et al.*, 1990); such movements are rarely considered relevant when assessing foraging mode (although Greene [1992] described such animals as "mobile ambushers"). In part because of difficulties in defining foraging mode, few studies of snakes have attempted to independently characterize foraging mode when making comparisons of traits related to foraging mode among species.

Foraging mode may affect and be affected by ecological, behavioral, thermal, reproductive, and bioenergetic aspects of the life history of an organism. Commonly considered factors affecting the strategy of active vs. sit-and-wait foraging are habitat type, prey availability, type of prey, and predation

Table 11.2 *Suite of characteristics associated with each extreme of the foraging mode continuum* Taken together, these trait associations have been termed the "syndrome hypothesis" (McLaughlin, 1989).

	Ambush	Active	Reference
Feeding rate	low	high	Hailey and Davies, 1986; Secor, 1995; Anderson and Karasov, 1981; Nagy et al., 1984
Foraging time	long	short	Secor, 1995; Merker and Nagy, 1984
Prey type	active	active and ambush	Huey and Pianka, 1981; Naulleau and Bonnet, 1995; Secor, 1995
Diet diversity	low	high	Naulleau and Bonnet, 1995; Secor, 1995
Home range size	low	high	Secor, 1995; Naulleau and Bonnet, 1995
Movement rate/distance	low	high	Hailey and Davies, 1986; Naulleau and Bonnet, 1995; Secor, 1995; Perry, 1999
Standard metabolic rate	low	high	Fig. 11.2
Field metabolic rate	low	high	Secor and Nagy, 1994; Nagy et al., 1984; Anderson and Karasov, 1981; Bennett and Gorman, 1979; Fig. 11.3
Water flux rate	low	high	Secor, 1995
Specific dynamic action	high	low	Secor and Diamond, 2000; Secor, 2001
Endurance	lower	higher	Ruben, 1976
Thermal performance Breadth	high	low	Secor, 1995
Growth rate and maturation	slow	fast	Webb et al., 2003
Relative clutch mass	high	low	Vitt and Congdon, 1978; Huey and Pianka, 1981; Vitt and Price, 1982
Morphology	stout	slim	Vitt and Congdon, 1978; Huey and Pianka, 1981; Secor, 1995; Figs. 11.4 and 11.5
Predation risk	low	high	Secor, 1995; Bonnet et al., 1999; Webb et al., 2003
Predator avoidance strategy	crypsis	speed	Huey and Pianka, 1981; Secor, 1995

risk associated with moving, among others (Norberg, 1977; Janetos, 1982). Previous studies (Table 11.2) suggest that widely foraging animals have more encounters with prey, a more generalist diet, higher field metabolic rates, greater water flux, lower relative clutch size, and more streamlined morphology than ambush foragers (Vitt and Congdon, 1978; Huey and Pianka, 1981; Secor and Nagy, 1994; Secor, 1995) and typically forage for a shorter period (Merker and Nagy, 1984; Nagy *et al.*, 1984). The correlation of a particular suite of characteristics with the active foraging mode and the opposite characteristics with the ambush foraging mode has been termed the "syndrome hypothesis" (McLaughlin, 1989) (Table 11.2). Here, we examine available data for snakes, and where possible assess support for the syndrome hypothesis.

Data quality and availability

We reviewed publications concerning foraging mode in snakes and examined data on morphological and physiological correlates of foraging mode. We were impressed by several shortcomings of the available data.

1. Snake researchers lack a clear definition of foraging mode that is independent of the syndrome hypothesis. Without clear and objective (repeatable) definitions of foraging modes, researchers resort to examination of syndrome-related variables to categorize organisms. At best, this practice risks circularity, and renders tests of the syndrome hypothesis suspect.
2. There is a lack of hard data with which to assess foraging mode in snakes. Usually, assignment of foraging mode is made by assertion after brief observation, after consultation with an expert, or by familial/phylogenetic association. Lack of data, coupled with ambiguous criteria for assigning foraging mode, leads to subjectivity, which will no doubt interfere with clarity in hypothesis testing.
3. Poor or incomplete data for the "syndrome." In many cases, only a single syndrome-related variable (e.g. standard metabolic rate, *SMR*, or length–mass relationship) has been measured, hampering our ability to assess the syndrome hypothesis as a whole.
4. Troubles with the adjustment of physiological variables for body size in comparative studies. Often, physiological variables (e.g. *SMR*; field metabolic rate, *FMR*) have been adjusted for body mass by generating a mass-specific ratio (i.e. kJ g^{-1}). Mass-specific ratios correct for body mass only when the relationship between mass and response is isometric (exponential slope $= 1$, intercept $= 0$), a condition rarely, if ever, met in physiological data, which typically scale allometrically (Packard and Boardman, 1988, 1999; Hayes, 2001). Rampant use of mass-specific ratios poses a major problem for comparative studies of animals that differ in

body size. Such comparisons are best accomplished by using allometric scaling relations ($Y = aX^b$), which minimize error (Packard and Boardman, 1988, 1999; Hayes, 2001).

5. The phylogenetic relationships of advanced snakes, especially the Colubroidea, are uncertain (Cundall and Greene, 2000). Most specifically, the family Colubridae may be paraphyletic, consisting of numerous lineages of uncertain placement (Cadle, 1994; Vidal and Hedges, 2002). Phylogenetic research among major groups is active, including New World Crotalinae (Parkinson, 1999; Schuett *et al.*, 2002), Elapidae (Keogh, 1998), and Colubroidea (Vidal and Hedges, 2002). However, a complete understanding of major lineages within families remains elusive and information regarding branch lengths commonly required for phylogenetic correction is available for only some groups.

6. Among snakes, there are apparently few clearly documented reversals of foraging mode within families, which hampers independent tests of correlated evolution of foraging mode and "syndrome" responses. Furthermore, many snakes defy classification into dichotomous foraging mode because of their natural flexibility in foraging tactics. Some species that belong to families that exhibit predominantly one foraging mode have been described as using alternative foraging tactics (Table 11.1) (Lind and Welsh, 1994; Fitzgerald *et al.*, 2004; Shine and Wall, this volume, Chapter 6). These snakes cannot be classified unambiguously, because they utilize both strategies, and the extent to which alternative strategies are used remains unknown.

Our discussion of the syndrome hypothesis is necessarily constrained, given the above shortcomings of available data. Below, we discuss in turn the phylogenetic, ecological, physiological, and morphological implications of foraging mode in snakes. We review relevant studies and summarize existing data in an attempt to represent the current state of understanding of foraging mode and its consequences in snakes.

Phylogenetic implications

Shifts in foraging mode within individuals have been observed in the field or induced experimentally in representatives of a number of lineages including scolopendrid centipedes, aquatic invertebrates, spiders, insects, fish, turtles, amphibians, birds, and mammals (Ehlinger, 1989; Helfman, 1990; Fausch *et al.*, 1997). However, in squamate reptiles, foraging mode is thought to be phylogenetically stable and relatively conserved within families (McLaughlin, 1989; Perry, 1999). Nevertheless, species within several squamate lineages exhibit foraging modes atypical of their families. Lineages containing reversals include but are not limited to the scincid genus *Mabuya* (Cooper and Whiting,

2000), the family Lacertidae (Huey *et al.*, 1984), geckos (Huey and Pianka, 1981; see Perry, 1999 for other lizard examples), the Colubridae (Hailey and Davies, 1986; Daltry *et al.*, 2001), and Australian elapid snakes (Shine, 1980; Webb *et al.*, 2003). A shift from active to ambush foraging is thought to be more likely than a shift from ambush to active foraging owing to the physiological constraints (e.g. cost of locomotion, endurance, aerobic capacity) operating on heavy-bodied ambush foragers (Huey and Pianka, 1981). Assuming that snakes evolved from active-foraging varanoid lizards, it is likely that ambush foraging has evolved several times, most notably among Boidae and Viperidae (Fig. 11.1).

Within the Serpentes the stability of predominant foraging mode within families is perhaps more highly conserved than in lizards. Again, some exceptions (reversals) exist. Within the family Elapidae, the death adder (*Acanthophis* sp.) is convergent on the body form and foraging mode of terrestrial viperids (Shine, 1980), and the broad-headed snake (*Hoplocephalus bungaroides*) is an ambush forager that exhibits some evidence of directional evolution in life history traits towards those more typical of pit-vipers (Webb *et al.*, 2003). Furthermore, within the family Colubridae there are three well-studied reversals. Adult *Natrix maura* are ambush foragers on fish (Hailey and Davies, 1986) which is atypical of the genus *Natrix*. The Antiguan racer (*Alsophis antiguae*) ambushes small lizards from the concealment of leaf litter (Daltry *et al.*, 2001). The Oregon gartersnake (*Thamnophis atratus hydrophilus*) exhibits an ontogenetic shift in foraging mode with neonates using ambush tactics, adults using active foraging, and juveniles exhibiting both modes (Lind and Welsh, 1994). Unfortunately, few syndrome-related data are available for these reversals, making detailed analysis of trait evolution impracticable for these cases. Owing to the highly conserved nature of foraging mode within taxonomic groups, and the lack of specific knowledge regarding foraging mode and phylogeny, it is difficult to find circumstances that allow comparisons of behavior, physiology, and morphology between alternative foraging modes in sympatric, closely related snakes.

Ecological implications

The syndrome hypothesis includes a number of ecological characters that are associated with foraging mode (Table 11.2). Active foragers show increased movement rates relative to ambush foragers (as indicated by the use of movements per minute (*MPM*) to quantify foraging mode in lizards). The actively foraging coachwhip, *Masticophis flagellum*, moved more than twice as frequently and greater than twice as far as the sympatric ambush foraging sidewinder, *Crotalus cerastes* (Secor, 1995). The activity range of *M. flagellum* was also greater than twice that of *C. cerastes*, although snake size was not accounted for (Secor, 1995). In a comparison of two closely related natricine snakes, the active foraging *Natrix natrix* was active more often than the ambush foraging *N. maura* (Hailey and Davies, 1986). Likewise, the ambush foraging terrestrial asp viper (Viperidae: *Vipera aspis*) and the active foraging semi-arboreal Aesculapian snake (Colubridae: *Elaphe longissima*) differ in activity level and home range size, consistent with expectations based on foraging mode (Naulleau and Bonnett, 1995).

Diet and feeding rates

Active foragers are thought to have a more diverse diet (Secor, 1995) and to encounter prey and feed more frequently (Anderson and Karasov, 1981; Andrews, 1984; Nagy *et al.*, 1984) than ambush foragers. The difference in diet diversity is partly due to active foragers encountering both sedentary and active prey, whereas ambush foragers will encounter predominantly active prey (Huey and Pianka, 1981; McLaughlin, 1989). For example, in a study of predators on active and ambush lizards, *Bitis caudalis* (an ambush forager) fed on an active foraging lizard more often than would be expected by chance (Huey and Pianka, 1981). In addition, *Masticophis flagellum* fed twice as often, on a more diverse diet (nestling birds, snakes, lizards and mammals), as the ambush foraging *Crotalus cerastes* (mammals and lizards) (Secor, 1995). The ambush foraging *Natrix maura* fed less frequently on more active prey than the sympatric, con-generic, active foraging *Natrix natrix* (Hailey and Davies, 1986). The active foraging *Elaphe longissima* consumed birds, bird eggs, and small mammals, whereas the ambush foraging *Vipera aspis* consumed only small mammals (Naulleau and Bonnet, 1995). A potential relationship between foraging mode and prey size is complicated by changes in prey geometry (Cundall and Greene, 2000). For example, pit-vipers (primarily ambush foragers) take prey that may exceed 160% of their body mass (Greene, 1992). Actively foraging coral snakes (*Micrurus fulvius*) may take meals of nearly equal proportion (137%) by mass (Greene, 1997); however, large meals in ophiophagous and saurophagous coral snakes are facilitated by the cylindrical body form of snakes and lizards. In addition, some active foragers (e.g. turtle-headed sea snakes, *Emydocephalus annulatus*; Shine *et al.*, 2004) feed on large numbers of small prey, which may result in relatively large total meal sizes. However, there is no direct evidence for prey size selectivity between ambush foraging and active foraging snakes (Downes, 2002).

Predation risk

Ambush foragers rely on crypsis and a sedentary lifestyle to avoid predation; active foragers are more visible to predators, since they show greater activity, and must flee from potential predators (Huey and Pianka, 1981). Therefore, the active foraging lifestyle carries a greater risk of predation (Huey and Pianka, 1981; Webb *et al.*, 2003). In a study relating activity to road mortality in four snake species, active foraging snakes were killed more often than ambush foragers (Bonnet *et al.*, 1999). Juvenile survival was much greater in the ambush foraging *Hoplocephalus bungaroides* when compared with the active foraging *Rhinoplocephalus nigrescens* (Webb *et al.*, 2003). However, adult survival was similar in these species, possibly because of increased risk during extensive summer movements by ambush foraging *H. bungaroides* (Webb *et al.*, 2003). Active foraging *Masticophis flagellum* was depredated twice as often as ambush foraging *Crotalus cerastes* over a three-year period (Secor, 1995).

Relative clutch mass

The syndrome hypothesis suggests that active foragers should have a lower relative clutch mass (*RCM*) than ambush foragers (Vitt and Congdon, 1978; Vitt and Price, 1982). Lower *RCM* in active foragers may result, in part, from lower maternal body volume available for developing offspring, owing to the decreased mass – *SVL* relation. In addition, increased *RCM* results in greater total mass being transported by the mother during pregnancy or gravidity, which would result in a greater energetic cost of locomotion. Cost of transport for greater body mass may hinder escape from potential predators and therefore increase predation risk to the mother (Vitt and Congdon, 1978); this probably influences active foragers far more than sedentary ambush foragers, which rely heavily on crypsis. Unfortunately, *RCM* is computed by generating a ratio of clutch or litter mass to maternal mass. Such ratios assume isometric scaling between ratio components, a condition that is rarely met when considering body size. As previously mentioned, the use of simple ratios to represent allometric relations and to statistically compare groups that differ in body size has been criticized (Packard and Boardman, 1988, 1999). Such comparisons are subject to estimation errors and spurious correlations. In addition, *RCM* is sensitive to resource environment because clutch or litter size are sensitive to resource environment, and at least some low-energy specialists (e.g. rattlesnakes) frequently produce litters that are well below their capacity (S. Beaupre, personal observation). Finally, a comparison of RCM between foraging modes would be partly confounded by reproductive mode (oviparity, viviparity). The majority of Viperidae are viviparous, whereas most Elapidae and Colubridae are oviparous. Because of problems with the statistical comparison of *RCM* among groups that differ in size, the variable nature of *RCM*, and the complexity of dealing with confounding factors such as reproductive mode, we did not pursue comparisons of *RCM* between active and ambush foraging snakes. Such an analysis is beyond the scope of the current chapter.

Physiological implications

Fundamental constraints dictated by the laws of thermodynamics set bounds on the evolution of life history (Dunham *et al.*, 1989). Survival, growth and reproduction require energy intake (Bryant, 1988); thus, it is reasonable that aspects of digestive and metabolic physiology should be finely tuned to optimize net energy gain in response to feeding rates and food volumes associated with different foraging modes. Such tuning may not be in direct response to foraging mode *per se*, but rather to a suite of variables that may or may not covary with foraging mode such as feeding frequency (Secor and Diamond, 2000), prey size (Forsman and Lindell, 1993) and prey abundance (Forsman and Lindell, 1997). For example, three large ambush foraging radiations of snakes (boas, pythons, and pit-vipers) have also been described as "low-energy" systems because of comparatively low metabolic rates (Chappell and Ellis, 1987; Beaupre 1993; Beaupre and Duvall, 1998; Beaupre and Zaidan, 2001), relative to active foraging colubrids and lizards. Major bioenergetic features that appear to be sensitive to foraging mode (and related variables) include standard metabolic rates, field metabolic rates, and the magnitude of the cost of digestion (specific dynamic action: *SDA*). We consider each of these in turn.

Metabolism

Energy allocation to standard or resting metabolic rate can account for a sizeable portion of the total annual energy budget in snakes: 20%–30% in *Crotalus lepidus* depending on source population (Beaupre, 1996), approximately 30% in *Crotalus cerastes*, and 23% in *Masticophis flagellum* (Secor and Nagy, 1994). Therefore, reductions in *SMR* can result in significant energy savings when accumulated over the course of several months.

Metabolic rate is affected by abiotic (temperature, time of day, season) and biotic factors (body size, reproductive condition, digestive status) and therefore comparisons among taxa must be made at common values of these factors. Although the allometry between body size and metabolic rate has been established in reptiles (Andrews and Pough, 1985) including rattlesnakes (Beaupre, 1993; Beaupre and Duvall, 1998; Beaupre and Zaidan, 2001), colubrids (Peterson et al., 1998), and boids (Chappell and Ellis, 1987), a large proportion of researchers still report metabolic rates in mass-specific form (e.g. kJ $g^{-1}h^{-1}$ or ml O_2 g^{-1} h^{-1}). As mentioned above, use of ratios that assume isometry with variables that are allometrically related will induce error in comparative studies (Packard and Boardman, 1988, 1999). The preponderance of mass-specific metabolic rates in the literature poses a problem for the comparison of standard and resting metabolic rates between snakes that differ in foraging mode and body size.

Nevertheless, the metabolism of several snakes has been examined in comparisons of the costs and benefits of foraging mode. For example, the ambush foraging vipers *Vipera berus* and *Cerastes cerastes* exhibited significantly lower resting metabolic rate (*RMR*) than the actively foraging colubrid *Malpolon moilensis* (Al-Sadoon, 1991). The actively foraging colubrid *Natrix natrix* had greater aerobic scope at high temperatures than ambush foraging *N. maura* (Hailey and Davies, 1986). The actively foraging coachwhip (Colubridae: *Masticophis flagellum*) had significantly greater *RMR* and field metabolic rate (*FMR*) in comparison with the sit-and-wait foraging sidewinder (Viperidae: *Crotalus cerastes*) (Secor and Nagy, 1994; Secor, 1995). Likewise, in a comparison of metabolic rate between the ambush foraging prairie rattlesnake (Viperidae: *Crotalus viridis*), the active foraging racer (Colubridae: *Coluber constrictor*) and coachwhip (Colubridae: *Masticophis flagellum*), and the ambush foraging rosy boa (Boidae: *Lichanura roseofusca*), no differences were found in *SMR*; however, the active foragers had higher rates of metabolism during physical activity (Ruben, 1976). In general, resting and standard metabolic rates are thought to be greater in active foragers relative to ambush foragers.

To further evaluate the relationship between foraging mode and metabolic rate, we collated literature values of *SMR* for both active and ambush foraging snakes. We limited our analysis to only those studies where data were available to estimate metabolic rate at 30 °C, and where either an allometric scaling equation or a mass-specific rate (along with an average mass of the animals studied) was reported (Data: Appendix 11.1). We estimated whole animal metabolic rate for an animal of average body mass for the species by using an allometric relation, or for the average sample body mass of animals measured when given a mass-specific rate. Using the average mass of the sample is the best approach when the *SMR* is reported on a mass-specific basis because, even though the allometric relation is not known, the whole animal estimate should be accurate at the average sample mass. When available, and clearly defined by the investigators, we used estimated *SMR*; otherwise we used the lowest

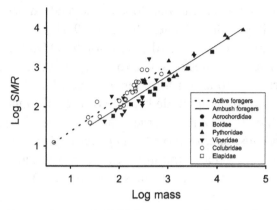

Figure 11.2. Comparisons of *SMR* between ambush and active foraging snakes. Standard metabolic rates were obtained from literature accounts and converted to a mean value for each represented species as a function of mean species or sample body size (see text for details).

metabolic rates measured and took these to be a reasonable estimate of *SMR*. All *SMR* values were converted to common units (kJ h^{-1}) by using conversion factors supplied by Gessaman and Nagy (1988). We used ANCOVA (Fit Model Platform, JMP ver 5.1, 2003, SAS Institute, Cary, NC) to assess the effects of body mass and foraging mode on *SMR* (model: Log*SMR* = LogMass, Mode, LogMass*Mode) on 56 observations of *SMR* from the literature.

As expected, body mass significantly affected *SMR* (df = 1, $F = 213.6$, $p < 0.000\ 1$; Fig. 11.2) as did foraging mode (df = 1, $F = 10.7$, $p < 0.001\ 9$; Fig. 11.2). Furthermore, the slope of the relation between body mass and SMR was similar between active and ambush foragers (df = 1, $F = 0.19$, $p < 0.662\ 3$; Fig. 11.2). The above model explained 87.2% of the variation in the dataset. At face value, this analysis supports the notion that active foraging snakes have higher *SMR* than ambush foraging snakes. However, we note that an appropriate phylogenetic correction should first be applied. We did not pursue phylogenetic correction in this analysis because foraging mode and family are completely confounded (i.e. nearly all Colubridae in the sample are active, all Viperidae are ambush, all Boidae are ambush) and the phylogenetic relationships of these groups are poorly known. To illustrate the confounding of family and foraging mode, we ran a second analysis with the model: Log*SMR* = LogMass, Family, LogMass*Family. The elapids ($n = 2$) and our single acrochordid were dropped from the analysis because of lack of replication at the family level, resulting in four remaining families for comparison (Boidae, Pythonidae, Colubridae, and Viperidae). In this case again, body mass significantly affected *SMR* (df = 1, $F = 203.1$, $p < 0.000\ 1$) as did family (df $= 1$, $F = 4.1$, $p < 0.0121$), and the slope of the relation between body mass and SMR was similar among families (df $= 3$, $F = 0.63$, $p < 0.5991$). Body mass and family also explained 87.2% of the variation in SMR. Based upon current data, standard metabolic rates of active foraging snakes appear to be significantly elevated relative to ambush foraging snakes, but this effect cannot be cleanly attributed to foraging mode because of the confounding influence of family membership. However, we urge caution in the interpretation of our SMR analysis, because species most likely differ in their allometric relations between SMR and body mass, and little scaling information was available to us (because of pervasive reporting of mass-specific rates).

Field metabolic rate

Field metabolic rates (FMR) using the doubly labeled water method have been infrequently measured in snakes. However, there are several instances of measurement within Colubridae (*Masticophis flagellum*, Secor and Nagy, 1994; *Coluber constrictor*, Plummer and Congdon, 1996; *Thamnophis sirtalis*, Peterson *et al.*, 1998), and several within Viperidae (*Crotalus atrox*, S. Beaupre *et al.*, in prep; *Crotalus cerastes*, Secor and Nagy, 1994; *Crotalus horridus*, S. Beaupre, in prep; *Crotalus lepidus*, Beaupre, 1996). We present these values as a combination of regression lines (when scaling relations were available) and as whole-animal responses vs. mean body mass of the sample when only mass-specific rates were available (Fig. 11.3). Clearly, active foragers in the sample (*C. constrictor*, *M. flagellum*, and *T. sirtalis*) have elevated FMR relative to ambush foragers (*C. atrox*, *C. cerastes*, *C. horridus*, and *C. lepidus*). However, in this case there are only two families represented, and again, foraging mode is completely confounded with family membership. Furthermore, *Masticophis* and *Coluber* are thought to be closely related, and all of the vipers represented are from a single genus. Nevertheless, variation in FMR in relation to foraging mode in snakes measured to date is consistent with the syndrome hypothesis.

Specific dynamic action

Specific dynamic action (SDA) is defined as the cost of digestion and includes energy expenditures associated with intestinal up-regulation and growth, active transport, peristalsis, intermediary protein metabolism, and synthesis associated with the processing of a meal. In a comparison of ambush foraging *Crotalus cerastes* and active foraging *Masticophis flagellum*, Secor and Nagy (1994) and Secor *et al.* (1994) noted that the SDA of the ambush foraging

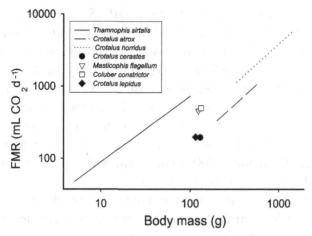

Figure 11.3. Field metabolic rates (*FMR*) in the form of allometric relations or point estimates for small samples where allometric relations were not available. Ambush foraging snakes tend to exhibit lower *FMR* than active foragers; however, comparisons are completely confounded by family membership (see text for details).

rattlesnake was much greater than that of the widely foraging coachwhip. The factorial increase in metabolic rate due to *SDA* ranged as much as 6.6-fold in *C. cerastes* (Secor *et al.*, 1994) and only 3.7-fold in *M. flagellum* (Secor and Nagy, 1994). Secor and Diamond (2000) have since developed the hypothesis that feeding frequency drives the evolution of gut function, such that infrequent feeding (typical of ambush foragers) selects for down-regulation of gut function during fasting periods to reduce whole-animal metabolic costs (with concurrent high cost of *SDA* due to added need for up-regulation upon feeding). Conversely, frequent feeding (typical of active foragers) selects for constant readiness of the gut to process food, resulting in higher whole-animal metabolic rates and lower costs of *SDA*. Additional data, including phylogenetically corrected comparative analyses (Secor and Diamond, 2000; Secor, 2001; Zaidan and Beaupre, 2003) support the contention that the magnitude of *SDA* is related to feeding frequency (hence, indirectly to foraging mode), and that infrequent feeders examined had *SDA* that was approximately 1.8 times greater than the *SDA* of frequent feeders (Secor and Diamond, 2000).

Morphological implications

Morphological variation also appears to be related to foraging mode in snakes. Active foragers typically have a more streamlined body form, generally expressed as a lower mass to snout–vent length (*SVL*) relation, and a higher tail length to snout–vent length relation compared with ambush foragers (Vitt and Congdon, 1978). Foraging mode, specifically movement rates, may constrain body form. Active foragers typically exhibit greater movement rates (Perry, 1999) and higher metabolic cost of searching for prey relative to ambush foragers (Nagy *et al.*, 1984). The energetic cost of locomotion increases with an increase in the mass being moved, all other things being equal. The observation that active foragers have less mass at a given body length (Vitt and Congdon, 1978) is consistent with the prediction that active foragers should reduce total cost of transport.

Mass–length relations can be used to compare among diverse taxa that differ in foraging mode. We searched the snake literature for mass–length allometric equations or data from which such allometric equations could be generated. We found ample data for snakes in three families: Colubridae (Kaufman and Gibbons, 1975; A. T. Holycross, unpublished data, 2003), Elapidae (Greer and Shine, 2000), and Viperidae (Kaufman and Gibbons, 1975; A. T. Holycross, unpublished data, 2002; J. Hobert, unpublished data, 2002; S. J. Beaupre, unpublished data, 2002; S. J. Beaupre and C. E. Montgomery, unpublished data, 2003; J. G. Hill, unpublished data, 2003; F. Zaidan, unpublished data, 2002). For each species, we obtained or generated a scaling relationship of the form: $Mass = aSVL^b$ where mass is in grams, *SVL* is snout–vent length (cm), and *a* and *b* are fitted constants, hereafter referred to as intercept and exponent, respectively. In general, larger *a* values correspond to fatter animals per unit length. Larger *b* values imply a faster increase in mass with increasing length, but this would most reliably correspond to fatter animals per unit length if *a* values were similar between taxa, or larger in the fatter taxon. If the syndrome hypothesis is correct, we predicted that ambush foraging snakes (Viperidae) could have greater intercept and/or exponent values, indicating greater body mass at a given length or faster increase in mass with increases in length. To test this hypothesis, we plotted each species in two-dimensional space represented by values of *a* and *b*, and examined the result for structure. We noted non-linearity and log-transformed both variables. Mass–length relations group with family; ambush foraging Viperidae exhibit greater values of log *a* at a given value of log *b* (Fig. 11.4) and greater values of log *b* at a given value of log *a*, both of which suggest larger mass at length among the Viperidae. To examine this pattern for statistical significance, we conducted ANCOVA (model log*a* = log*b*, family, log*b**family). We found significant effects ($r^2 = 0.954$) of log *b* (df = 1, $F = 369.1$, $p < 0.000\ 1$) and family (df = 2, $F = 14.8$, $p < 0.000\ 1$) on log *a*. Furthermore, the log *b*×family interaction was non-significant (df = 2, $F = 0.97$, $p < 0.389\ 8$), indicating parallel slopes

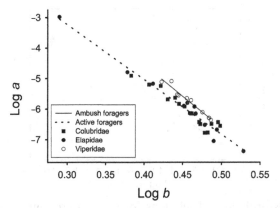

Figure 11.4. Morphological comparison among families. Allometric relations of *SVL* and body mass of the families Colubridae, Elapidae, and Viperidae (see text for data sources). The pattern of a decrease in mass to *SVL* relation in active foragers is apparent, suggesting relatively thinner body form. However, comparisons are again confounded by familial membership.

among families. The Viperidae were significantly different from Colubridae and Elapidae (by Tukey's HSD; $p < 0.05$). A family effect, with ambush foraging Viperidae exhibiting greater mass at length, is consistent with the syndrome hypothesis. Because foraging mode and family were confounded in this analysis, the same amount of variance ($r^2 = 0.955$) was explained when mode was substituted for family.

If morphology evolves in response to foraging mode, then it is reasonable to expect that snakes within a predominantly ambush foraging family (e.g. Viperidae) that exhibit more active foraging might also exhibit evolution towards more slender body form. For example, within North American crotalines the copperhead, *Agkistrodon contortrix*, has a significantly different relation between body mass and *SVL* when compared with most other crotalines (Fig. 11.5). Raw data (*SVL* and body mass) were available for ten species (*Agkistrodon piscivorus*, F. Zaidan, unpublished data, 2002; *Crotalus atrox*, S. J. Beaupre, unpublished data, 2002; A. T. Holycross, unpublished data, 2002; *Crotalus horridus*, S. J. Beaupre, unpublished data, 2002; *Crotalus lepidus*, S. J. Beaupre, unpublished data, 2002; A. T. Holycross, unpublished data, 2002; *Crotalus molossus*, S. J. Beaupre, unpublished data, 2002; A. T. Holycross, unpublished data, 2002; *Crotalus scutulatus*, A. T. Holycross, unpublished data, 2002; *Crotalus viridis*, A. T. Holycross, unpublished data, 2002; *Crotalus willardi*, A. T. Holycross, unpublished data, 2002; *Sistrurus catenatus*, A. T. Holycross, unpublished data, 2002; J. Hobert, unpublished data, 2002; and *Agkistrodon contortrix*, C. E. Montgomery and S. J. Beaupre,

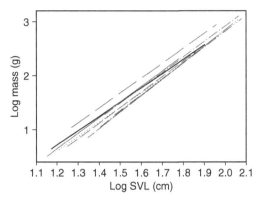

Figure 11.5. Morphological relations for selected crotaline snakes (genera *Agkistrodon, Sistrurus*, and *Crotalus*; see text for details). Log *SVL* versus log body mass are plotted for larger North American crotalines. All species (Table 11.3) except for *Agkistrodon contortrix* are plotted in gray. *A. contortrix* appears unique in slope and intercept.

unpublished data, 2003; J. G. Hill, unpublished data, 2003). We conducted a slope heterogeneity test (**PROC GLM**, SAS Institute, 1985) with *posthoc* pairwise comparisons of species to determine which groups were responsible for slope heterogeneity. Experiment-wise Type I error was held to 0.05 by employing the Dunn–Šidák method (Sokal and Rolhf, 1995) for multiple comparisons (corrected pairwise $\alpha' = 0.001$). The copperhead (*Agkistrodon contortrix*) was significantly different (with lower b value) from most other North American pit-vipers, many of which were not significantly different from each other (Table 11.3).

Copperheads have been implicated as relatively active foragers (Greene, 1997) and have also been observed actively tracking prey in a three-dimensional habitat (Beaupre and Roberts, 2001). The lower exponent value of the copperhead suggests slower increase in mass with increases in length, or more slender body form, consistent with the syndrome hypothesis. The copperhead also exhibits a slightly higher a value in comparison with other crotalines, suggesting that it is somewhat stockier at birth. Interestingly, the juvenile copperhead is a known ambush forager that uses its bright yellow tail for caudal luring (Fitch, 1961; Carpenter and Gillingham, 1990). Both luring behavior and yellow tail are lost in the adult form (Carpenter and Gillingham, 1990). Thus, the ontogenetic shift in body form of the copperhead from stocky as a neonate (relative to other crotalines) to more slender as an adult may be mirrored by an ontogenetic shift in foraging mode from ambush caudal luring to greater reliance on active foraging.

Table 11.3 *Pairwise comparisons of slope relating log SVL to log body mass in selected North American crotalines*

Significant differences (at pair wise comparison rate of 0.001, see text) are indicated by "*". Non-significant comparisons are denoted by "N". Note *A. contortrix* exhibits the greatest number of significant differences from other crotalines.

Species	Crotalus willardi	Crotalus viridis	Crotalus horridus	Crotalus scutulatus	Crotalus molossus	Crotalus lepidus	Crotalus atrox	Sistrurus catenatus	Agkistrodon piscivorus	Agkistrodon contortrix
A. contortrix	*	*	*	*	*	*	*	N	N	—
A. piscivorus	N	*	N	*	N	N	*	N	—	
S. catenatus	N	*	N	*	N	N	*	—		
C. atrox	N	N	*	N	N	N	—			
C. lepidus	N	N	N	N	N	—				
C. molossus	N	N	N	N	—					
C. scutulatus	N	N	*	—						
C. horridus	N	*	—							
C. viridis	N	—								
C. willardi	—									

Bioenergetic modeling

It is tempting to force organisms into the dichotomous extremes of active or ambush foraging. However, it may be more reasonable to think of foraging mode as a flexible continuum where, under the right circumstances, organisms could successfully exhibit phenotypes intermediate between the extremes represented by the dichotomy (Perry, 1999). It is reasonable then to ask: what are the fitness consequences of intermediate phenotypes, and what is the relative disadvantage of adopting one strategy in an environment that best supports the other? Such questions are difficult, at best, to answer in the field. However, bioenergetic simulations may provide some insight.

The first author has developed a bioenergetic simulation of growth and reproduction based on data from ambush foraging rattlesnakes (Beaupre, 2002). The simulation represents interactions between environment (temperature, food availability), behavioral decisions (to forage or engage in reproductive activities), and physiological processes (digestion, SDA, metabolism, and growth). Each day of an individual's life is simulated, from birth until death or simulation end. Because the interactions among multiple bioenergetic variables can be represented simultaneously, the simulation provides an opportunity to investigate consequences of foraging mode and associated syndrome patterns on fitness responses such as growth, while holding other variables constant.

The reader is referred to the original model description for details on structure and specific subroutine contents (Beaupre, 2002). Here, we present a baseline simulation of an ambush forager and then modify program structure to represent an active forager. Prior to simulation for this chapter, we updated functions for the cost of digestion (SDA) to reflect new data available for *Crotalus horridus* (SDA (ml CO_2) = 12.158 $SW^{0.16}PW^{1.03}$, where SW is snake wet mass, and PW is prey wet mass), (Zaidan and Beaupre, 2003). The model takes as input an initial file of state variables (sex, age, SVL, wet mass, and stored energy) that represent a cohort of 30 neonate animals. Simulated individuals then forage and grow in a user-defined food environment represented by mean foraging success (MFS; the probability of capturing food on any given day in which foraging occurs), and a maximum size for available food items (in this case, 200 g wet mass). Figure 11.6 shows the mean 10-year growth trajectory ($n = 30$ replicates) of a typical ambush forager under conditions of 0.05 MFS (5% chance of capturing food on any given day). Note that the simulation produces a slow sigmoidal growth curve and a body size asymptote that represents an energetic balance between size-dependent maintenance (including activity) and mean energy intake.

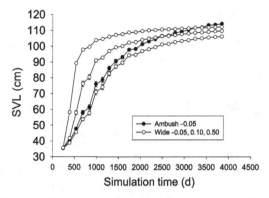

Figure 11.6. Simulated growth trajectories for ambush and active foragers. Mean foraging success (*MFS*) is defined as the probability of encountering prey on any day that foraging occurs. Note at similar *MFS*, ambush foragers grow more rapidly, and across most *MFS*, ambush foragers attain larger asymptotic body size (see text for details on simulation conditions and discussion).

We modified the simulation to represent an active forager by doubling *RMR* (which, because of sustained metabolic scope assumptions, resulted in a doubling of *FMR*) and by reducing total *SDA* by half. Whereas these factorial changes in energetic variables are broad and simplistic, they are not far removed from typical differences observed between active and ambush foragers (Secor and Nagy, 1994; Secor and Diamond, 2000). We then simulated an active forager under food availability conditions that were identical to those of the ambush forager (*MFS* = 0.05, maximum food size 200 g), and noted both decreased growth rate and smaller asymptotic body size relative to the ambush forager (Fig. 11.6). Furthermore, arbitrary increases in mean foraging success (*MFS* = 0.10, 0.50) for the active forager resulted in faster growth at small sizes, but had little effect on asymptotic body size (Fig. 11.6). Apparently, at small sizes, increased metabolism can easily be compensated for by increases in food availability. However, at large size, a doubling of metabolism effectively reduces asymptotic size even under conditions of abundant food. We believe this occurs because, even with abundant food, simulated snakes reach their maximum capacity to process (digestion takes time), forcing a lower equilibrium point for the trade-off between size-specific maintenance and net energy intake. Multiple simulations were conducted to explore the relation between physiological differences associated with foraging mode, food availability, and asymptotic body size (Fig. 11.7). Simulated ambush foragers attain larger asymptotic size than active foragers at any given food availability (Fig. 11.7) because their lower whole-body metabolic costs yield greater net energy for production. We note that in nature some of the largest

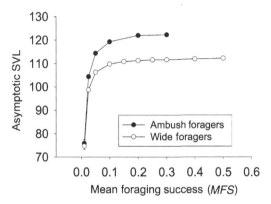

Figure 11.7. Simulated asymptotic body size for ambush and active foragers across a broad range of food availability and encounter rates (*MFS*). Under comparable *MFS*, ambush foraging snakes achieve greater asymptotic body size.

snakes can be found among ambush foraging lineages (i.e. Boidae, Pythonidae, Viperidae), although other factors such as the metabolic cost of transport clearly play an important role.

Our simulations are simplistic, especially in the sense that we have held all factors except *RMR*, *FMR*, *SDA*, and *MFS* equal. Nevertheless, the simulation results suggest that higher metabolic rates of active foragers must be compensated for by increased food capture rates or increased processing efficiency. The simulations have implications for the evolution of intermediate phenotypes between the ambush and active conditions. Clearly, departures from the tuned suite of characteristics for ambush foraging can only be supported under ecologically permissive circumstances (i.e. where food is predictable or abundant). Likewise, it seems unlikely that active foraging animals with high metabolic expenditures could easily switch to a more ambush foraging lifestyle without risking starvation. Simulation results support the perspective that, although foraging modes most likely range on a continuum between active and ambush, some points on the continuum (combinations of traits) are probably more difficult to achieve than others.

Closing remarks

In general, to the extent that available data allow responsible comparisons, snakes appear to follow the syndrome hypothesis in its coarsest form. However, during the preparation of this chapter, we became keenly aware of several problems with the uncritical application of the syndrome hypothesis to snakes. First, and perhaps foremost, a satisfactory and objective definition of foraging mode in snakes is lacking. We were tempted to apply the simplistic definition that an active forager moves itself into striking range of prey, whereas an ambush forager waits until the prey enters striking range (McLaughlin, 1989; Perry, 1999); however, this dichotomy seems inadequate to account for dramatic differences in behavior, physiology and morphology associated with foraging mode in snakes (i.e. cost of movement, frequency of feeding). Likewise, definitions that quantify movement rates in relation to foraging are problematic in snakes, which are generally secretive and feed infrequently. Even ambush foraging snakes may move considerable distances in search of productive ambush sites (Duvall *et al.*, 1990). Greene (1997, pp. 62–6) eloquently discusses multiple exceptions and caveats among snakes that obviate the use of a simple dichotomous classification (ambush vs. active foragers). Furthermore, careful reading of detailed species accounts in taxonomic literature reveals remarkable flexibility in many snakes that shift foraging tactics during ontogeny, and in response to changes in food availability (Table 11.1). The plethora of snakes with apparently flexible foraging tactics defies their assignment into a simplistic dichotomy. Nevertheless, as documented by Fig. 11.2, 11.3, and 11.4, the foraging mode dichotomy appears to hold some predictive power relative to syndrome variables.

Second, the quality of data for comparison among snakes must be improved in several ways. The phylogenetic relationships of the Colubroidea need to be resolved into well-supported, defensible monophyletic groups. Researchers that measure syndrome-related variables should consider how the data may be used in comparative studies. Specifically, ratio data, such as *RCM* and mass-specific physiological rates, should be replaced with carefully estimated and well-sampled allometric relations. Finally, efforts should be made to incorporate data from taxa of more ambiguous foraging strategy (such as those in Table 11.1) and to study known reversals in greater detail.

We also perceive an unfortunate parallel between the syndrome hypothesis and "*r*- and *K*-selection" theory. Both approaches present dichotomous categories and list characteristics of organisms that guide their classification. However, most organisms do not fit cleanly into either category, rendering both classification schemes as non-mechanistic general models that explain no special cases (Dunham and Beaupre, 1998). Perhaps a more productive way to think about the consequences of foraging mode is to consider a multivariate suite of co-adapted traits (Arnold, 1983; Lande and Arnold, 1983). Ambush and active foraging may be local optima where more intermediate trait combinations are allowed only under permissive circumstances, such as predictable or abundant food. An exclusive focus on foraging mode as a causative factor driving behavior, physiology, and morphology is probably myopic. Foraging mode is correlated with other forcing factors, such as feeding frequency, which may more directly influence the evolution of morphology and physiology (e.g. the regulation of gut function) (Secor, 2001) and selection for low-energy lifestyles, of which some snakes (e.g. rattlesnakes) may be extreme examples (Beaupre and Zaidan, 2001).

Finally, other large patterns in snake morphology may be related to foraging mode, including the evolution of caudal luring, macrophagy, and the evolution of crypsis. For example, ontogenetic shifts in morphology of the copperhead (see above) may be related to an ontogenetic shift from reliance on ambush caudal luring by neonates to more active foraging in the adult. Likewise, some extreme examples of macrophagy occur among ambush foraging snakes (Greene, 1992; Shine and Wall, this volume, Chapter 6). The evolution of macrophagy, and associated increases in viable prey size, is one possible solution to decreased energy input in infrequently feeding ambush foragers. Color pattern is also certain to shift in response to foraging mode. Widely foraging snakes with active lifestyles would benefit from coloration that enhances the use of "flicker fusion" to escape predation (Shine and Madsen, 1994; Lindell and Forsman, 1996), whereas ambush foragers would benefit from cryptic coloration that conceals them from predators and prey alike. The evolution of larger-scale patterns among snakes should enjoy greater attention as knowledge of phylogenetics and natural history increases within the group.

Acknowledgments

We thank A. T. Holycross, J. Hobert, J. G. Hill, and F. Zaidan for allowing us to use their unpublished data in various analyses. The manuscript was significantly improved by comments from Rick Shine and two anonymous referees. This research was supported by funds from the Arkansas Audubon Trust, the David Causey Grant-in-Aid, the Professor Delbert Swartz Endowed Fellowship, and NSF-EPSCoR (EPS-9871906) Graduate Fellowship to CEM, and the University of Arkansas Research Incentive Fund, the Arkansas Science and Technology Authority (grant 97-B-06), and the National Science Foundation (IBN-9728470, and IBN-0130633) to SJB.

Appendix 11.1 *Data used in SMR comparisons*
See text for details.

Species	n	Mode	Temp	b	a (J/h)	Mass (g)	SMR (V_{CO_2} or V_{O_2}/h)	SMR (J/h)	Log mass	Log SMR	Source	Reference
ACROCHORDIDAE												
Acrochordus arafurae	4	W	30	—	—	1048	24.7	485.3	3.02	2.69	table 1	Bedford and Christian, 1998
BOIDAE												
Acrantophis dumerili	7	A	30	0.62	7.57	2550	48.2	949.4	3.41	2.98	table 1	Chappell and Ellis, 1987
Boa constrictor	35	A	30	0.74	3.18	7825	126.1	2482.5	3.89	3.40	table 1	Chappell and Ellis, 1987
Candoia carinatus	14	A	30	0.71	2.41	525	50.5	205.7	2.72	2.31	table 1	Chappell and Ellis, 1987
Corallus caninus	7	A	30	0.88	1.12	550	14.6	287.6	2.74	2.46	table 1	Chappell and Ellis, 1987
Corallus enhydris	7	A	30	0.77	2.13	800	18.8	369.2	2.90	2.57	table 1	Chappell and Ellis, 1987
Epicrates cenchria	11	A	30	0.57	7.69	425	12.3	242.2	2.63	2.38	table 1	Chappell and Ellis, 1987
Eryx colubrinus	9	A	30	0.76	2.21	75	3.0	59.4	1.88	1.77	table 1	Chappell and Ellis, 1987
Lichanura roseofusca	6	A	30	—	—	314	22.0	432.6	2.50	2.64	fig. 1	Ruben, 1976
Lichanura trivirgata	12	A	30	0.66	3.79	175	5.9	116.2	2.24	2.07	table 1	Chappell and Ellis, 1987

Appendix 11.1 (cont.)

Species	n	Mode	Temp	b	a (J/h)	Mass (g)	SMR (V_{CO_2} or V_{O_2}/h)	SMR (J/h)	Log mass	Log SMR	Source	Reference
COLUBRIDAE												
Boiga irregularis	5	W	30	—	—	133	5.4	149.4	2.13	2.17	table 1 (low 0–8)	Anderson et al., 2003
Coluber constrictor	5	W	30	—	—	223	21.9	430.1	2.35	2.63	table 1	Secor and Diamond, 2000
Coluber ravergieri	5	W	30	—	—	136	7.9	155.5	2.13	2.19	fig. 2	Dmi'el, 1972
Diadophis punctatus	10	W	30	1.02	2.62	5	0.6	12.4	0.66	1.09	table 2	Buikema and Armitage, 1969
Helicops modestus	34	W	30	0.59	9.60	100	7.2	142.1	2.00	2.15	table 1	Abe and Mendes, 1980
Lampropeltis getula	5	W	30	—	—	188	11.8	233.1	2.27	2.37	table 1	Secor and Diamond, 2000
Liophis miliaris	29	W	30	0.80	5.06	200	18.0	354.3	2.30	2.55	table 1	Abe and Mendes, 1980
Malpolon moilensis	10	W	30	—	—	145	11.5	225.4	2.16	2.35	table 1 (adults)	Al-Sadoon, 1991
Masticophis flagellum	11	W	30	—	—	124	5.5	108.7	2.09	2.04	p. 1606	Secor and Nagy, 1994
Natrix maura	7	A	30	—	—	25	2.0	39.4	1.40	1.60	table 2	Hailey and Davies, 1986
Natrix n. helvetica	5	W	30	—	—	100	13.0	255.2	2.00	2.41	table 2	Hailey and Davies, 1986
Natrix n. persa	7	W	30	—	—	40	2.9	56.3	1.60	1.75	table 2	Hailey and Davies, 1986
Nerodia rhombifera	20	W	30	—	—	250	21.4	420.7	2.40	2.62	p. 442	Jacobson and Whitford, 1970
Nerodia sipedon	10	W	30	—	—	292		873.6	2.47	2.94	table 4	Blem and Blem, 1990
Nerodia taxispilota	10	W	30	—	—	372		861.6	2.57	2.94	table 4	Blem and Blem, 1990
Pituophis melanoleucus	9	W	30	0.69	4.94	431	11.5	315.0	2.63	2.50	p. 453	Zaidan and Beaupre, 2003

Spalerosophis diademae	4	W	30	—	—	218	12.7	249.3	2.34	2.40	fig. 2	Dmi'el, 1972
Thamnophis proximus	20	W	30	—	—	35	6.8	133.4	1.54	2.13	p. 442	Jacobson and Whitford, 1970
Thamnophis sirtalis	72	W	30	0.71	5.76	23	2.7	53.6	1.36	1.73	table 2	Peterson *et al.*, 1998
ELAPIDAE												
Acanthophis praelongus	3	A	30	—	—	106	5.0	99.2	2.02	2.00	table 1	Bedford and Christian, 1998
Pseudonaja nuchalis	3	W	30	—	—	214	14.4	283.4	2.33	2.45	table 1	Bedford and Christian, 1998
PYTHONIDAE												
Antaresia childreni	8	A	30	—	—	332	24.2	477.0	2.52	2.68	table 1	Bedford and Christian, 1998
Antaresia stimsoni	5	A	30	—	—	350	20.2	397.9	2.54	2.60	table 1	Bedford and Christian, 1998
Aspidites melanocephalus	3	A	30	—	—	1028	69.8	1473.6	3.01	3.17	table 1	Bedford and Christian, 1998
Liasis fuscus	4	A	30	—	—	1307	29.3	577.4	3.12	2.76	table 1	Bedford and Christian, 1998
Liasis olivaceus	5	A	30	—	—	3323	111.4	2192.7	3.52	3.34	table 1	Bedford and Christian, 1998
Morelia spilota	8	A	30	0.84	2.24	1050	37.9	746.0	3.02	2.87	table 1	Chappell and Ellis, 1987
Python curtis	8	A	30	0.86	1.09	2375	45.8	900.7	3.38	2.96	table 1	Chappell and Ellis, 1987
Python molurus	20	A	30	0.71	5.21	33950	455.0	8954.6	4.53	3.95	table 1	Chappell and Ellis, 1987
Python regius	53	A	30	—	—	1523	31.9	627.1	3.18	2.80	fig. 2A	Ellis and Chappell, 1987
Python reticulatus	15	A	30	0.75	4.93	14325	334.5	6582.8	4.16	3.82	table 1	Chappell and Ellis, 1987
Python sebae	10	A	30	0.76	3.68	16150	286.8	5644.3	4.21	3.75	table 1	Chappell and Ellis, 1987

Appendix 11.1 (cont.)

Species	n	Mode	Temp	b	a (J/h)	Mass (g)	SMR (V_{CO_2} or V_{O_2}/h)	SMR (J/h)	Log mass	Log SMR	Source	Reference
VIPERIDAE												
Agkistrodon piscivorus	23	A	30	0.71	2.82	250	5.2	142.1	2.40	2.15	table 4, AR, EP	Zaidan, 2002
Bothrops moojeni	51	A	30	0.69	14.96	410	84.7	1655.8	2.61	3.22	table 2 (adults)	Cruz-Neto and Abe, 1994
Cerastes cerastes	10	A	30	—	—	121	8.6	169.1	2.08	2.23	table 1 (adults)	Al-Sadoon, 1991
Crotalus adamanteus	5	A	30	0.93	1.06	3373	102.8	2023.1	3.53	3.31	p. 148	Dorcas *et al.*, 2004
Crotalus atrox	41	S	30	0.66	4.17	300	9.1	179.8	2.48	2.26	—	S. Beaupre *et al.*, unpubl. data
Crotalus cerastes	16	A	30	—	—	129	4.0	78.7	2.11	1.90	p. 1606	Secor and Nagy, 1994
Crotalus durrisus	15	A	30	—	—	50	2.2	43.3	1.70	1.64	p. 231	Cruz-Neto *et al.*, 1999
Crotalus horridus	36	A	30	0.78	2.00	300	6.1	168.4	2.48	2.23	table 4	Beaupre and Zaidan, 2001
Crotalus lepidus	16	A	30	0.60	2.87	115	2.5	69.0	2.06	1.84	table 3 F4, GVH	Beaupre, 1993
Crotalus molossus	16	A	30	0.60	2.87	300	4.5	122.6	2.48	2.09	table 3 F4, GVH	Beaupre, 1993
Crotalus viridis	6	A	30	—	—	301	24.1	473.9	2.48	2.68	fig. 1	Ruben, 1976
Macrovipera palaestinae	8	A	30	—	—	581	24.4	480.2	2.76	2.68	fig. 2	Dmi'el, 1972
Vipera aspis	8	A	32	—	—	85	4.3	83.6	1.93	1.92	table 3	Ladyman *et al.*, 2003
Vipera berus	10	A	30	—	—	70	9.5	186.0	1.85	2.27	table 1	Al-Sadoon, 1991

References

Abe, A. S. and Mendes, E. G. (1980). Effect of body size and temperature on oxygen uptake in the water snakes *Helicops modestus* and *Liophis miliaris* (Colubridae). *Comp. Biochem. Physiol.* **65A**, 367–70.

Al-Sadoon, M. D. (1991). Metabolic rate-temperature curves of the horned viper, *Cerastes cerastes gasperetti*, the moila snake, *Malpolon moilensis*, and the adder *Vipera berus*. *Comp. Biochem. Physiol.* **99A**, 119–22.

Anderson, N. L., Hetherington, T. E. and Williams, J. B. (2003). Validation of the doubly labeled water method under low and high humidity to estimate metabolic rate and water flux in a tropical snake (*Boiga irregularis*). *J. Appl. Physiol.* **95**, 184–91.

Anderson, R. A. and Karasov, W. H. (1981). Contrasts in energy intake and expenditure in sit-and-wait and widely foraging lizards. *Oecologia* **49**, 67–72.

Andrews, R. M. (1984). Energetics of sit-and-wait and widely-foraging lizard predators. In *Vertebrate Ecology and Systematics: A Tribute to Henry S. Fitch*, ed. R. A. Seigel, L. E. Hunt, J. L. Knight, L. Malaret, and N. L. Zuschlag, pp. 137–45. Lawrence, KS: University of Kansas, Museum of Natural History, Special Publication No.10.

Andrews, R. M. and Pough, F. H. (1985). Metabolism of squamate reptiles: allometric and ecological relationships. *Physiol. Zool.* **58**, 214–31.

Arnold, S. J. (1983). Morphology, performance and fitness. *Am. Zool.* **23**, 347–61.

Beaupre, S. J. (1993). An ecological study of oxygen consumption in the mottled rock rattlesnake, *Crotalus lepidus lepidus*, and the black-tailed rattlesnake, *Crotalus molossus molossus*, from two populations. *Physiol. Zool.* **66**, 437–54.

Beaupre, S. J. (1996). Field metabolic rate, water flux, and energy budgets of mottled rock rattlesnakes, *Crotalus lepidus*, from two populations. *Copeia* **1996**, 319–29.

Beaupre, S. J. (2002). Modeling time-energy allocation in vipers: individual responses to environmental variation and implications for populations. In *Biology of Vipers*, ed. G. Schuett, M. Hoggren, M. E. Douglas and H. W. Greene, pp. 463–81. Eagle Mountain UT: Eagle Mountain Publishing LC.

Beaupre, S. J. and Duvall, D. (1998). Variation in oxygen consumption of the western diamondback rattlesnake, *Crotalus atrox*: implications for sexual size dimorphism. *J. Comp. Physiol.* **B168**, 497–506.

Beaupre, S. J. and Roberts, K. (2001). *Agkistrodon contortrix contortrix* (southern copperhead), chemotaxis, diet and arboreality. *Herp. Rev.* **32**, 44–5.

Beaupre, S. J. and Zaidan, F. III. (2001). Scaling of CO_2 production in the timber rattlesnake (*Crotalus horridus*), with comments on cost of growth in neonates and comparative patterns. *Physiol. Biochem. Zool.* **74**, 757–68.

Bedford, G. S. and Christian, K. A. (1998). Standard metabolic rate and preferred body temperatures in some Australian pythons. *Aust. J. Zool.* **46**, 317–28.

Bennett, A. F. and Gorman, G. C. (1979). Population density and energetics of lizards on a tropical island. *Oecologia* **42**, 339–58.

Blem, C. R. and Blem, K. L. (1990). Metabolic acclimation in three species of sympatric, semi-aquatic snakes. *Comp. Biochem. Physiol.* **97A**, 259–64.

Bonnet, X., Naulleau, G. and Shine, R. (1999). The dangers of leaving home: dispersal and mortality in snakes. *Biol. Cons.* **89**, 39–50.

Bryant, D. M. (1988). Determination of respiration rates of free-living animals by the double-labelling technique. In *Towards a More Exact Ecology*, ed. P. J. Grubb and J. B. Whittaker, pp. 85–112. Oxford: Blackwell Scientific Publications.

Buikema, A. L. Jr. and Armitage, K. B. (1969). The effect of temperature on the metabolism of the prairie ringneck snake, *Diadophis punctatus arnyi* Kennicott. *Herpetologica* **25**, 194–206.

Cadle, J. E. (1994). The colubrid radiation in Africa (Serpentes: Colubridae): phylogenetic relationships and evolutionary patterns based on immunological evidence. *Zool. J. Linn. Soc.* **110**, 103–40.

Carpenter, C. C. and Gillingham, J. C. (1990). Ritualized behavior in *Agkistrodon* and allied genera. In *Snakes of the* Agkistrodon *Complex*, ed. H. K. Gloyd and R. Conant. *Soc. Study Amphib. Rept. Contr. Herpetol.* **6**, 523–31.

Chappell, M. A. and Ellis. T. M. (1987). Resting metabolic rates in boid snakes: allometric relationships and temperature effects. *J. Comp. Physiol.* **157B**, 227–35.

Chiszar, D., Lee, R. K. K., Radcliffe, C. W. and Smith, H. M. (1992). Searching behaviors by rattlesnakes following predatory strikes. In *Biology of the Pitvipers*, ed. J. A. Campbell and E. D. Brodie, pp. 369–82. Tyler, TX: Selva.

Cooper, W. E. Jr. and Whiting, M. J. (2000). Ambush and active foraging modes both occur in the scincid genus *Mabuya*. *Copeia* **2000**, 112–18.

Cooper, W. E. Jr., Whiting, M. J., Van Wyk, J. H. and Mouton, P. le F. N. (1999). Movement and attack-based indices of foraging mode and ambush foraging in some gekkonid and agamine lizards from southern Africa. *Amph.-Rept.* **20**, 391–9.

Cruz-Neto, A. P. and Abe, A. S. (1994). Ontogenetic variation of oxygen uptake in the pitviper *Bothrops moojeni* (Serpentes, Viperidae). *Comp. Biochem. Physiol.* **108A**, 549–54.

Cruz-Neto, A. P., Andrade, D. V. and Abe, A. S. (1999). Energetic cost of predation: aerobic metabolism during prey ingestion by juvenile rattlesnakes, *Crotalus durissus*. *J. Herpetol.* **33**, 229–34.

Cundall, D. and Beaupre, S. J. (2001). Field records of predatory strike kinematics in timber rattlesnakes, *Crotalus horridus*. *Amph.-Rept.* **22**, 492–8.

Cundall, D. and Greene, H. W. (2000). Feeding in snakes. In *Feeding*, ed. K. Schwenk, pp. 293–333 San Diego: Academic Press.

Daltry, J. C., Bloxam, Q., Cooper, G. *et al.* (2001). Five years of conserving the "world's rarest snake", the Antiguan racer, *Alsophis antiguae*. *Oryx* **35**, 119–27.

Dmi'el, R. (1972). Effect of activity and temperature on metabolism and water loss in snakes. *Am. J. Physiol.* **223**, 510–16.

Dorcas, M. E., Hopkins, W. A. and Roe, J. H. (2004). Effects of body mass and temperature on standard metabolic rate in the eastern diamondback rattlesnake (*Crotalus adamanteus*). *Copeia* **2004**, 145–51.

Downes, S. J. (2002). Size-dependent predation by snakes: selective foraging or differential prey vulnerability? *Behav. Ecol.* **13**, 551–60.

Dunham, A. E. and Beaupre, S. J. (1998). Ecological experiments: scale, phenomenology, mechanism and the illusion of generality. In *Experimental Ecology: Issues and Perspectives*, ed. J. Bernardo and W. Resetarits, pp. 27–49 New York: Oxford University Press.

Dunham, A. E., Grant, B. W. and Overall, K. L. (1989). Interfaces between biophysical and physiological ecology and the population ecology of terrestrial vertebrate ectotherms. *Physiol. Zool.* **62**, 335–55.

Duvall, D., Goode, M. J., Hayes, W. K., Leonhardt, J. K. and Brown, D. G. (1990). Prairie rattlesnake vernal migration: field experimental analysis and survival value. *Nat. Geog. Res.* **6**, 457–69.

Ehlinger, T. J. (1989). Foraging mode switches in the golden shiner (*Notemigonus crysoleucas*). *Can. J. Fish. Aquat. Sci.* **46**, 1250–4.

Ellis, T. M. and Chappell, M. A. (1987). Metabolism, temperature relations, maternal behavior, and reproductive energetics in the ball python (*Python regius*). *J. Comp. Physiol.* **157B**, 393–402.

Ernst, C. H. (1992). *Venomous Reptiles of North America*. Washington, DC: Smithsonian Institution Press.

Ernst, C. H. and Barbour, R. W. (1989). *Snakes of Eastern North America*. Fairfax, VA: George Mason University Press.

Fausch, K. D., Nakano, S. and Kitano, S. (1997). Experimentally induced foraging mode shift by sympatric charrs in a Japanese mountain stream. *Behav. Ecol.* **8**, 414–20.

Fitch, H. S. (1961). Autecology of the Copperhead. *Univ. Kansas Pub. Mus. Nat. Hist.* **13**, 87–288.

Fitzgerald, M., Shine, R. and Lemckert, F. (2004). Life history attributes of the threatened Australian snake (Stephen's banded snake *Hoplocephalus stephensii*, Elapidae). *Biol. Cons.* **119**, 121–8.

Forsman, A. and Lindell, L. E. (1993). The advantage of a big head: swallowing performance in adders, *Vipera berus*. *Funct. Ecol.* **7**, 183–9.

Forsman, A. and Lindell, L. E. (1997). Responses of a predator to variation in prey abundance: survival and emmigration of adders in relation to vole density. *Can. J. Zool.* **75**, 1099–108.

Fry, B. G., Lumsden, N. G., Wüster, W. *et al.* (2003). Isolation of a neurotoxin (α-colubritoxin) from a nonvenomous colubrid: evidence for early origin of venom in snakes. *J. Molec. Evol.* **57**, 446–52.

Gessaman, J. A. and Nagy, K. A. (1988). Energy metabolism: errors in gas-exchange conversion factors. *Physiol. Zool.* **61**, 507–13.

Gibbons, J. W. and Dorcas, M. E. (2004). *North American Watersnakes, A Natural History*. Norman, OK: University of Oklahoma Press.

Greene, H. W. (1983). Dietary correlates of the origin and radiation of snakes. *Am. Zool.* **23**, 431–41.

Greene, H. W. (1992). The ecological and behavioral context for pitviper evolution. In *Biology of the Pitvipers*, ed. J. A. Campbell and E. D. Brodie, pp. 107–18. Tyler, TX: Selva.

Greene, H. W. (1997). *Snakes, the Evolution of Mystery in Nature*. Berkeley, CA: University of California Press.

Greer, A. E. and Shine. R. (2000). Relationship between mass and length in Australian elapid snakes. *Mem. Queensl. Mus.* **45**, 375–80.

Hailey, A. and Davies, P. M. C. (1986). Lifestyle, latitude and activity metabolism of natricine snakes. *J. Zool.* **209**, 461–76.

Hayes, J. P. (2001). Mass-specific and whole-animal metabolism are not the same concept. *Physiol. Biochem. Zool.* **74**, 147–50.

Helfman, G. S. (1990). Mode selection and mode switching in foraging animals. *Adv. Stud. Behav.* **19**, 249–98.

Hill, J. G. III (2004). Natural history of the western cottonmouth (*Agkistrodon piscivorus leucostoma*) from an upland lotic population in the Ozark Mountains of northwest Arkansas. Ph.D. dissertation, University of Arkansas, Fayetteville, AR.

Houston, D. and Shine, R. (1993). Sexual dimorphism and niche divergence: feeding habits of the Arafura filesnake. *J. Anim. Ecol.* **62**, 737–48.

Huey, R. B. and Pianka, E. R. (1981). Ecological consequences of foraging mode. *Ecology* **62**, 991–9.

Huey, R. B., Bennett, A. F., John-Alder, H. and Nagy, K. A. (1984). Locomotor capacity and foraging behavior of Kalahari lacertid lizards. *Anim. Behav.* **32**, 41–50.

Jacobson, E. R. and Whitford, W. G. (1970). The effect of acclimation on physiological responses to temperature in the snakes *Thamnophis proximus* and *Natrix rhombifera. Comp. Biochem. Physiol.* **35**, 439–49.

Janetos, A. C. (1982). Active foragers vs. sit-and-wait predators: a simple model. *J. Theoret. Biol.* **95**, 381–5.

Kardong, K. V. and Smith, T. L. (2002). Proximate factors involved in rattlesnake predatory behavior: a review. In *Biology of the Vipers*, ed. G. W. Schuett, M. Hoggren, M. E. Douglas and H. W. Greene, pp. 253–66. Eagle Mountain, UT: Eagle Mountain Publishing.

Kaufman G. A. and Gibbons, J. W. (1975). Weight-length relationships in thirteen species of snakes in the southeastern United States. *Herpetologica* **31**, 31–7.

Keogh, J. S. (1998). Molecular phylogeny of elapid snakes and a consideration of their biogeographic history. *Biol. J. Linn. Soc.* **63**, 77–203.

Ladyman, M., Bonnet, X., Lourdais, O., Bradshaw, D. and Naulleau, G. (2003). Gestation, thermoregulation, and metabolism in a viviparous snake, *Vipera aspis*: evidence for fecundity-independent costs. *Physiol. Biochem. Zool.* **76**, 497–510.

Lande, R., and Arnold, S. J. (1983). The measurement of selection on correlated characters. *Evolution* **37**, 1210–26.

Lind, A. J. and Welsh, H. H. Jr. (1994). Ontogenetic changes in foraging behavior and habitat use by the Oregon garter snake, *Thamnophis atratus hydrophilus. Anim. Behav.* **48**, 1261–73.

Lindell, L. E. and Forsman, A. (1996). Sexual dichromatism in snakes: support for the flicker fusion hypothesis. *Can. J. Zool.* **74**, 2254–6.

McLaughlin, R. L. (1989). Search modes of birds and lizards: evidence for alternative movement patterns. *Am. Nat.* **133**, 654–70.

Merker, G. P. and Nagy, K. A. (1984). Energy utilization by free-ranging *Sceloporus virgatus* lizards. *Ecology* **65**, 575–81.

Nagy, K. A., Huey, R. B. and Bennett, A. F. (1984). Field energetics and foraging mode of Kalahari lacertid lizards. *Ecology* **65**, 588–96.

Naulleau, G. and Bonnet, X. (1995). Reproductive ecology, body fat reserves and foraging mode in females of two contrasted snake species: *Vipera aspis* (terrestrial, viviparous) and *Elaphe longissima* (semi-arboreal, oviparous). *Amph.-Rept.* **16**, 37–46.

Norberg, R. A. (1977). An ecological theory on foraging time and energetics and choice of optimal searching method. *J. Anim. Ecol.* **46**, 511–29.

Packard, G. C. and Boardman, T. J. (1988). The misuse of ratios, indices and percentages in ecophysiological research. *Physiol. Zool.* **61**, 1–9.

Packard, G. C. and Boardman, T. J. (1999). The use of percentages and size-specific indices to normalize physiological data for variation in body size: wasted time, wasted effort? *Comp. Biochem. Physiol.* **A122**, 37–44.

Parkinson, C. L. (1999). Molecular systematics and biogeographical history of pitvipers as determined by mitochondrial ribosomal DNA sequences. *Copeia* **1999**, 576–86.

Perry, G. (1999). The evolution of search modes: ecological versus phylogenetic perspectives. *Am. Nat.* **153**, 98–109.

Peterson, C. C., Walton, B. M. and Bennett, A. F. (1998). Intrapopulation variation in ecological energetics of the garter snake, *Thamnophis sirtalis*, with analysis of the precision of doubly labeled water measurements. *Physiol. Zool.* **71**, 333–49.

Plummer, M. V. and Congdon, J. D. (1996). Rates of metabolism and water flux in free-ranging racers, *Coluber constrictor*. *Copeia* **1996**, 8–14.

Rossman, D. A., Ford, N. B. and Seigel, R. A. (1996). *The Garter Snakes*. Norman, OK: University of Oklahoma Press.

Ruben, J. A. (1976). Aerobic and anaerobic metabolism during activity in snakes. *J. Comp. Physiol.* **B109**, 147–57.

SAS Institute, Inc. (1985). *SAS User's Guide: Statistics*. Version 5 Edition. Cary, NC: SAS Institute.

Schuett, G. W., Hoggren, M., Douglas, M. E. and Greene, H. W. (eds.) (2002). *Biology of the Vipers*. Eagle Mountain, UT: Eagle Mountain Publishing.

Secor, S. M. (1995). Ecological aspects of foraging mode for the snakes *Crotalus cerastes* and *Masticophis flagellum*. *Herpetol. Monogr.* **9**, 169–86.

Secor, S. M. (2001). Regulation of digestive performance: a proposed adaptive response. *Comp. Biochem. Physiol.* **128A**, 565–77.

Secor, S. M. and Diamond, J. M. (2000). Evolution of regulatory responses in feeding in snakes. *Physiol. Biochem. Zool.* **73**, 123–41.

Secor, S. M. and Nagy, K. A. (1994). Bioenergetic correlates of foraging mode for the snakes *Crotalus cerastes* and *Masticophis flagellum*. *Ecology* **75**, 1600–14.

Secor, S. M., Stein, E. D. and Diamond, J. M. (1994). Rapid up-regulation of snake intestine in response to feeding: a new model of intestinal adaptation. *Am. J. Physiol.* **266**, G695–G705.

Shine, R. (1980). Ecology of the Australian death adder *Acanthophis* antarcticus (Elapidae): evidence for convergence with the Viperidae. *Herpetologica* **36**, 281–9.

Shine, R. and Madsen, T. (1994). Sexual dichromatism in snakes of the genus *Vipera*: a review and new evolutionary hypothesis. *J. Herpetol.* **28**, 114–17.

Shine, R., Bonnet, X., Elphick, M. J. and Barrott, E. G. (2004). A novel foraging mode in snakes: browsing by the sea snake *Emydocephalus annulatus* (Serpentes, Hydrophiidae). *Funct. Ecol.* **18**, 16–24.

Sokal, R. R. and Rohlf, F. J. (1995). *Biometry*, 3rd edn. New York: Freeman.

Vidal, N. and Hedges, S. B. (2002). Higher-level relationships of caenophidian snakes inferred from four nuclear and mitochondrial genes. *C. R. Biologies* **325**, 987–95.

Vitt, L. J. and Congdon, J. D. (1978). Body shape, reproductive effort, and relative clutch mass in lizards: resolution of a paradox. *Am. Nat.* **112**, 595–608.

Vitt, L. J. and Price, H. J. (1982). Ecological and evolutionary determinants of relative clutch mass in lizards. *Herpetologica* **38**, 237–55.

Webb, J. K., Brook, B. W. and Shine, R. (2003). Does foraging mode influence life history traits? A comparative study of growth, maturation and survival of two species of sympatric snakes from south-eastern Australia. *Aust. Ecol.* **28**, 601–10.

Zaidan, III, F. (2002). Geographic physiological variation and northern range limits in the cottonmouth (*Agkistrodon piscivorus leucostoma*). Ph.D. dissertation, University of Arkansas, Fayetteville, AR.

Zaidan, F. III and Beaupre, S. J. (2003). Effects of body mass, meal size, fast length, and temperature on specific dynamic action in the timber rattlesnake (*Crotalus horridus*). *Physiol. Biochem. Zool.* **74**, 447–58.

Zug, G. R., Vitt, L. J. and Caldwell, J. P. (2001). *Herpetology: An Introductory Biology of Amphibians and Reptiles*. San Diego, CA: Academic Press.

II

Environmental influences on foraging mode

12

The foraging biology of the Gekkota: life in the middle

AARON M. BAUER

Department of Biology, Villanova University

Introduction

Although the basic dichotomy in lizard foraging modes established by Pianka (1966) has become a paradigm supported by decades of subsequent research, not all lizards can be neatly characterized as either sit-and-wait predators or wide foragers. Although some (perhaps even most) lizard species may indeed fit relatively conveniently into one of these categories, the recognition that foraging mode can change in different habitats (Polynova and Lobachev, 1981; Ananjeva and Tsellarius, 1986), seasonally (Pietruszka, 1986; Nemes, 2002), with changing food abundance (Dunham, 1983; Durtsche, 1995), with ontogeny (Huey and Pianka, 1981), or even with different prey types (Greeff and Whiting, 2000) has shattered any expectation that it can be assumed to be an invariant characteristic of a species.

However, intraspecific variability in foraging mode does not preclude the existence of higher-order patterns of foraging mode distribution across lizards more broadly. Both Pietruszka (1986) and McLaughlin (1989) found, based on movement data, that the dichotomy between sit-and-wait and wide-foraging strategies was real. Perry *et al.* (1990), however, argued that the distribution is not bimodal and that intermediate modes, as defined by movement patterns, do exist. Indeed, Pianka (1971, 1974) argued that ambush and wide-foraging strategies were the ends of a continuum and Magnusson *et al.* (1985) recognized that species characterized by intermittent movements were not appropriately characterized by either of the recognized strategies. The terms "cruise foraging" (Regal, 1978, 1983) and "saltatory searching" (O'Brien *et al.*, 1989) have been proposed for such lizards that move, stop, scan for prey, then move again. Although these terms distinguish a particular behavioral pattern clearly evidenced by some lizards, neither has been widely accepted, either because the foraging mode they describe is uncommon (which seems unlikely; Moermond,

Lizard Ecology: The Evolutionary Consequences of Foraging Mode, ed. S. M. Reilly, L. D. McBrayer and D. B. Miles. Published by Cambridge University Press. © Cambridge University Press 2007.

1979), or because the seductiveness of the existing binary paradigm causes authors to downplay intermediate or alternative characterizations of foraging mode.

A strong phylogenetic component to foraging mode has been recognized (Huey and Pianka, 1981; Cooper, 1994a; Perry, 1999, this volume, Chapter 1) and the majority of lizard families have been categorized as being either sit-and-wait or wide foragers. Such generalizations, on the whole, appear warranted. Iguanian families have been regarded as ambushers, whereas autarchoglossans are chiefly wide foragers. However, there are exceptions to this pattern at all hierarchical levels. For example, although phrynosomatids are generally sit-and-wait predators, species of *Phrynosoma* move little and slowly, but may actively "trapline" ants by visiting a series of nests in succession (Huey and Pianka, 1981). Among lacertids and scincids most species are typical autarchoglossan wide foragers, but some species are more appropriately identified as ambushers (Huey and Pianka, 1981; Perry *et al.*, 1990; Castanzo and Bauer, 1998; Cooper, 2000b; Cooper and Whiting, 2000); at the familial level, cordylids buck the autarchoglossan trend entirely, with most species thus far examined exhibiting sit-and-wait tactics (Cooper *et al.*, 1997; Cooper and Steele, 1999, Whiting, this volume, Chapter 13).

Unlike iguanians and autarchoglossans, the Gekkota (geckos and flap-footed lizards), a large and largely nocturnal lineage, has played a relatively small role in the establishment and reinforcement of the sit-and-wait versus active foraging paradigm. Indeed, neither Pietruszka (1986) nor McLaughlin (1989), whose studies supported the basic dichotomy in lizard foraging modes, included geckos in their analyses.

Regal (1978) suggested that geckos were essentially sit-and-wait foragers, a view supported by many subsequent authors (e.g. Vitt and Congdon, 1978; Vitt and Price, 1982; Pianka, 1986; Dunham *et al.*, 1988). Cooper (1994a) reviewed the data then available and, based on foraging mode determinations culled from the literature, identified 52 non-eublepharid gecko species in 18 genera to be sit-and-wait predators and another 8 species in 6 genera to be mixed or active predators. The only two eublepharid species considered were deemed wide foragers, whereas the sole pygopodid was a sit-and-wait forager. Based on these data Cooper (1995a, 1996a; Cooper *et al.*, 1999) still characterized most geckos as ambush foragers, but noted that active or mixed foraging, especially in terrestrial nocturnal geckos, was not inconsequential. Likewise, Schwenk (2000) interpreted geckos as primitively ambush foragers with some representatives that approached an active foraging style. Werner and colleagues (Werner *et al.*, 1997a,b; Werner, 1998) have taken this view further and have provided evidence that many gecko species in fact exhibit intermediate or fluctuating foraging modes, thus calling into question the traditional view of these lizards as ambush predators and highlighting that no real consensus has been reached with respect to their dominant foraging mode (Werner *et al.*, 1997a).

The Gekkota

The Gekkota comprises some 1100+ species (Table 12.1) in 106 genera that have most recently been placed into the families Gekkonidae, Eublepharidae, and Pygopodidae (Kluge, 1987) or Gekkonidae, Eublepharidae, Diplodactylidae, and Pygopodidae (Bauer, 2002). Recent analysis of partial sequences of the c-*mos* nuclear gene (Han *et al.*, 2004) suggests the existence of five or six major clades within the Gekkota that correspond largely, but not entirely, to currently recognized higher-order groupings (Fig. 12.1). These major clades, Eublepharidae (6 genera, 28 species), Carphodactylidae (5 genera, 24 species), Diplodactylidae (11 genera, 110 species), Pygopodidae (7 genera, 36 species), Gekkoninae (72 genera, 768 species), and Sphaerodactylinae (5 genera, 142 species), constitute the basic phylogenetic units of discussion in this chapter. A formal classification can be found in Han *et al.* (2004). The term geckos, as used herein, refers to the paraphyletic group of gekkotans that excludes the reduced-limb pygopodids. Gecko and gekkotan are not used interchangeably in this chapter, as the majority of the gekkotan foraging literature deals only with the fully limbed taxa and many of the comments in this chapter apply to geckos alone.

The monophyly of gekkotans as a whole is strongly supported by both morphological and molecular datasets (Kluge, 1987; Estes *et al.*, 1988; Saint *et al.*, 1998; Harris *et al.*, 1999, 2001). Modern phylogenetic hypotheses have supported the position of the Gekkota as the sister group of the Autarchoglossa (see, for example, Estes *et al.*, 1988; but see Macey *et al.*, 1997; Lee, 1998), but in much of the ecological and behavioral literature geckos have been regarded as an "intermediate" group. This intermediacy is largely the legacy of the influential work on lizard relationships by Camp (1923). Although he recognized that the phylogenetic affinities of the Gekkota were with the Autarchoglossa, Camp's classificatory scheme emphasized symplesiomorphic features and grouped geckos with iguanians in the Ascalabota, a group of visually oriented taxa with broad tongues used chiefly for feeding rather than chemoreception.

As reviewed by Schwenk (1993b, 1994), Camp's (1923) classification ultimately became associated with a presumed dichotomy in the development of the chemical senses (weakly developed in the Ascalabota vs. well developed in the Autarchoglossa), resulting in the interpretation of gekkotans as chiefly

Table 12.1 *Foraging mode (FM) in the Gekkota*

Genera are grouped by major clade within the Gekkota (see Fig. 12.1). No. spec. is the number of described species in each genus (also summed by major clade). No. FM is the number of species for which foraging mode has been reported in the literature. Distribution lists the principal areas of distribution for each genus. Habitats are broadly categorized as: A, arboreal; F, fossorial; S, saxicolous; or T, terrestrial. Act, activity period: D, diurnal; N, nocturnal; in the case of D/N or N/D, the predominant activity period appears first. Species lists the species for which FM has been recorded. An asterisk following the species name indicates that the designation of FM was based, at least in part, on quantitative movement pattern data. FM is the reported foraging mode: CF, cruise foraging; H, herbivorous (frugivory and/or nectivory); SW, sit-and-wait; WF, wide foraging; superscripts correspond to those in the Sources column in cases in which more than one FM has been reported for a given species; congeners sharing the same FM have been grouped together. Attribution of FM to a particular source does not necessarily imply that the source excluded the possibility of alternative FMs. Herbivory is reported at the generic level only.

The list of taxa is derived from Rösler (2000), Kluge (2001) and Henkel and Schmidt (2003) but incorporates my own corrections, updates, and generic allocations. Species totals also include undescribed species in instances in which I have examined specimens or seen descriptions in draft.

Abbreviations for distribution: Afr, Africa; Aus, Australia; CAm, Central America; EAfr, East Africa; EqAfr, Equatorial Africa; EAs, East Asia; ESAm, Eastern South America; IAus, Indo-Australian Archipelago; Ind, India and Sri Lanka; IndO, islands of the tropical Indian Ocean; Mad, Madagascar; NAfr, North Africa; NAm, North America; NAtl, islands of North Atlantic Ocean (Macaronesia); NCal, New Caledonia; NGui, New Guinea; NSAm, Northern South America; PacO, islands of Pacific Ocean; SAfr, Southern Africa; SEur, Southern Europe; Sey, Seychelles; SNAm, Southern North America; SSAm, Southern South America; SWAs, Southwest Asia; WAs, Western Asia; WI, West Indies.

Genus	No. spec.	No. FM	Distribution	Habitat	Act.	Species	FM	Sources
Eublepharidae								
Aeluroscalabotes	1	1	EAs, IAus	T, A	N	*felinus*	SW	8
Coleonyx	8	2	NAm, CAm	T	N	*brevis*	SW[a],WF[b]	11[b],29[b],33[a]
						variegatus*	SW[a],WF[b], SW/WF[c]	7[b],8,20[c],33[a], 37[a],38[a]
Eublepharis	5	1	SWAs	T	N	*macularius*	WF	33
Goniurosaurus	10	2	EAs	T	N	*kuroiwae orientalis**	WF	49
						luii	WF	50
Hemitheconyx	2	0	Afr	T	N			

	T	N						
Holodactylus	2	0	EAfr					
Total	28	6						
Carphodactylidae								
Carphodactylus	1	0	Aus	T, A	N			
Nephrurus	11	2	Aus	T	N	*laevissimus, levis*	SW	38
Orraya	1	0	Aus	S	N			
Phyllurus	7	1	Aus	S, A	N	*platurus*	SW	12
Saltuarius	4	0	Aus	A, S	N			
Total	24	3						
Diplodactylidae								
Bavayia	25	0	NCal	A	N			
Crenadactylus	1	0	Aus	T	N			
Diplodactylus	25	5	Aus	T/A	N	*damaeus*	WF	16,17,19
						conspicillatus, pulcher, tessellatus	SW	16,17,19,38
						stenodactylus	SW[a],WF[b]	38[a],48[b]
Eurydactylodes	3	0	NCal	A	N			
Hoplodactylus	9	0	New Zealand	A, S, T	N		H	
Naultinus	8	0	New Zealand	A	D		H	
Oedura	14	4	Aus	A/S	N	*ocellata*	SW[a],WF[b]	13[a],47[b]
						reticulata, tryoni	SW	14
						marmorata	WF	48
Pseudothecadactylus	3	0	Aus	A/S	N			
Rhacodactylus	6	0	NCal	A	N		H	
Rhynchoedura	1	1	Aus	T	N	*ornata*	SW[a],WF[b]	38[a],48[b]
Strophurus	15	3	Aus	T, A	N	*ciliaris, elderi, strophurus*	SW	38
						ciliaris aberrans	WF	48
Total	110	13						

Table 12.1 (*cont.*)

Genus	No. spec.	No. FM	Distribution	Habitat	Act.	Species	FM	Sources
Pygopodidae								
Aprasia	12	8	Aus	F	N	*haroldi, inaurita, parapulchella, pseudopulchella, pulchella, repens, smithi, striolata*	WF	44
Delma	17	3	Aus	T, F	N/D	*fraseri, inornata, nasuta*	WF	22
Lialis	2	1	Aus, NGui	T	N	*burtonis*	SW[a], WF[b]	5[b],6,21[a],22[a]
Ophidiocephalus	1	0	Aus	F	N			
Paradelma	1	0	Aus	T	N			
Pletholax	1	0	Aus	F	N/D			
Pygopus	2	2	Aus	T	N/D	*lepidopodus*	WF	22
						nigriceps	SW	26
Total	36	14						
Gekkoninae								
Afroedura	9	0	SAfr	S, A	N			
Afrogecko	3	1	SAfr	S, T	N	*porphyreus*	SW	10
Agamura	4	0	SWAs	T, S	N			
Ailuronyx	3	0	Sey	A	N			
Alsophylax	7	0	SWAs	T	N			
Aristelliger	7	0	WI	A	N			
Asaccus	7	0	SWAs	S	N			
Asiocolotes	2	0	SWAs	T, S	N			
Blaesodactylus	3	0	Mad	A	N			
Bogertia	1	0	ESAm	A	N			
Briba	1	0	ESAm	A	N			
Bunopus	3	1	SWAs	T	N	*tuberculatus*	SW/WF	2
Calodactylodes	2	0	Ind	S	D/N			

Genus			Distribution	Microhabitat	Activity	Species	Locomotion	Reference
Carinatogecko	2	0	SWAs	T, A	N	*angulifer*	SW	38
Chondrodactylus	1	1	SAfr	T	N	*marmoratus*	SW	33
Christinus	3	1	Aus	A	N		SW	46
Cnemaspis	46	1	EAs, EqAfr	A, S	D/N	*kendalli**	SW	38
Colopus	1	1	SAfr	T	N	*wahlbergi*	SW	
Cosymbotus	2	0	EAs, IAus	A	N			
Crossobamon	2	1	SWAs	T	N	*eversmanni*	SW[a], WF[b]	1[a], 29[b]
Cryptactites	1	0	SAfr	T	N			
Cyrtodactylus	66	0	EAs, IAus	A, S	N			
Cyrtopodion	7	0	WAs	T, S	N			
Dixonius	3	0	EAs	T	N			
Ebenavia	2	0	IndO	A, S	N			
Euleptes	1	0	SEur, NAfr	S, T	N			
Geckoella	6	0	Ind	T	N			
Geckolepis	6	0	Mad	A	N			
Gehyra	33	4	EAs, IAus, PacO, Aus	A, S	N	*mutilata**, *oceanica**, *variegata*	SW SW[a], WF[b]	45 13[a],14[a],17[a],18[a],19[a], 38[a],47[b],48[b]
Gekko	28	3	EAs, IAus	A, S	N	*punctata*, *gecko**, *japonicus**, *hokouensis**	WF SW WF/SW	48 30,33,47 47
Goggia	8	0	SAfr	S, T	N			
Gymnodactylus	3	2	ESAm	T, S	N	*darwinii*, *geckoides*	SW	
Haemodracon	2	0	Socotra Is.	A, S, T	N			
Hemidactylus	82	4	cosmopolitan	A	N	*frenatus**, *mabouia**, *palaichthus*, *turcicus**	SW	31,34,35,36 23,25,31,33,34,35, 36,40,45
Hemiphyllodactylus	5	0	EAs, IAus, PacO	A	N		SW[a], WF[b]	
Heteronotia	3	1	Aus	T	N	*binoei*		13[a],16[b],18[b],19[b],38[a], 47[b],48[b]

Table 12.1 (*cont.*)

Genus	No. spec.	No. FM	Distribution	Habitat	Act.	Species	FM	Sources
Homonota	10	0	SSAm	T	N			
Homopholis	3	0	EAfr, SAfr	A	N			
Lepidodactylus	29	1	EAs, IAus, PacO	A	N	*lugubris**	SW (WF in one)	33,38,45
Luperosaurus	9	0	EAs	A	N			
Lygodactylus	59	3	Afr, Mad, ESAm	A, S	D	*klugei*	SW	14,34,35,36
						picturatus, somalicus	SW[a], WF[b]	14[a],47[b]
Matoatoa	2	0	Mad	A	N			
Mediodactylus	6	1	SEur, SWAs	T, S	N	*kotschyi*	SW	32
Microgecko	3	0	SWAs	T, S	N			
Microscalabotes	1	0	Mad	A	D			
Nactus	9	2	Masc, Aus, PacO	T, A, S	N	*coindemirensis, serpeninsula*	SW	4
Narudasia	1	0	SAfr	T, S	D/N	*bibronii, capensis, turneri*	SW	10,38
Pachydactylus	45	3	SAfr	S, T, A	N			
Palmatogecko	2	0	SAfr	T	N			
Paragehyra	2	0	Mad	S	N			
Paroedura	15	0	Mad	T, S, A	N			
Perochirus	3	0	PacO	A	N			
Phelsuma	41	1	IndO, Afr	A, S	D	*andamanensis*	H, SW	27
Phyllodactylus	42	0	SAm, CAm, NAm, WI	T, S	N			
Phyllopezus	2	1	SAm	S	N	*pollicaris*	SW	14,34,35,36
Pristurus	21	5	SWAs, NAfr	S, T	D/N	*carteri, gallagheri, minimus, rupestris*	SW	2,3

Genus			Distribution			Species		
Pseudogekko	4	0	Philippines	A	N	*cellerimus*	CF	3
Ptenopus	3	1	SAfr	T	N	*garrulus*	SW	38
Ptychozoon	6	0	EAs, IAus	A	N			
Ptyodactylus	6	1	SWAs, NAfr	S	D/N	*guttatus*	WF	24
Quedenfeldtia	2	1	NAfr	S	D	*trachyblepharus*	SW	3
Rhoptropus	8	3	SAfr	S, T	D	*afer*, barnardi* boultoni**	SW	10
Saurodactylus	2	0	NAfr		N/D			
Stenodactylus	11	5	SWAs, NAfr	T	N	*arabicus, khobarensis doriae, slevini, leptocosymbotes*	SW intermed.	2 2
Tarentola	20	1	SEur, NAfr, WI, NAtl	S, A	N	*mauritanica*	SW	15
Temuidactylus	17	0	WAs	T, S	N			
Teratolepis	2	0	WAs, Ind	T	N			
Teratoscincus	5	3	SWAs	T	N	*roborowskii*, scincus przewalskii*	SW WF	1,29,47 28
Thecadactylus	1	1	CAm, SAm, WI	A	N	*rapicauda*	SW	39,40,41
Tropiocolotes	6	0	NAfr, SWAs	T, S	N			
Urocotyledon	4	0	EqAfr, Sey	A	N			
Uroplatus	10	0	Mad	A	N			
Total	768	54						
Sphaerodactylinae								
Coleodactylus	6	2	NSAm	T, A	D	*amazonicus, septentrionalis*	SW	40
Gonatodes	17	3	CAm, WI, NSAm	A, S, T	D	*concinnatus, hasemani, humeralis* "some species"	SW	33,39,40,42,43
Lepidoblepharis	18	1	CAm, NSAm	T	D/N	*xanthostigma*	WF SW	3 41

Table 12.1 (cont.)

Genus	No. spec.	No. FM	Distribution	Habitat	Act.	Species	FM	Sources
Pseudogonatodes	7	1	NSAm	T	D/N	*guianensis*	SW	39
Sphaerodactylus	94	1	SNAm, CAm, NSAm, WI	T, A	D	*millipunctatus*	SW	41
Total	142	8						
Gekkota total	**1108**	**98**						

Literature sources: 1, Ananjeva and Tsellarius (1986); 2, Arnold (1984); 3, Arnold (1993); 4, Arnold and Jones, (1994); 5, Bradshaw *et al.* (1980); 6, Cooper (1994a); 7, Cooper (1994b); 8, Cooper (1995c); 10, Cooper *et al.* (1999); 11, Dial (1978a); 12, Doughty and Shine (1995); 13, Dunham and Miles (1985); 14, Dunham *et al.* (1988); 15, Gil *et al.* (1994a); 16, Henle (1990a); 17, Henle (1990b); 18, Henle (1990c); 19, Henle (1991); 20, Kingsbury (1989); 21, Murray *et al.*, (1991); 22, Patchell and Shine (1986); 23, Paulissen and Buchanan (1991); 24, Perry and Brandeis (1992); 25, Perry (1999); 26, Pianka (1986); 27, Ratnam (1993); 28, Semenov and Borkin (1992); 29, Shenbrot *et al.* (1991); 30, Stanner *et al.* (1998); 31, Teixeira (2001); 32, Valakos and Vlachopanos (1989); 33, Van Damme and Vanhooydonck (2001); 34, Vitt (1983); 35, Vitt (1986); 36, Vitt (1990); 37, Vitt and Congdon (1978); 38, Vitt and Price (1982); 39, Vitt and Zani (1996); 40, Vitt and Zani (1998a); 41, Vitt and Zani (1998b); 42, Vitt *et al.* (1997); 43, Vitt *et al.* (1997); 43, Vitt *et al.* (2000); 44, Webb and Shine (1994); 45, Werner (1998); 46, Werner and Chou (2002); 47, Werner *et al.* (1997a); 48, Werner *et al.* (1997b); 49, Werner *et al.* (2004); 50, Cooper and Habegger (2000). *Source for generic records of herbivory*: Cooper and Vitt (2002).

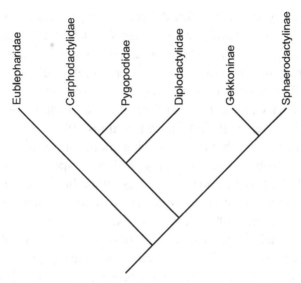

Figure 12.1. Pattern of relationship among major gekkotan clades as discussed in this chapter. Based on c-*mos* nuclear DNA sequence data as analyzed by Han *et al.* (2004).

visual predators. This view was widely promulgated and remained standard well into the 1990s (Underwood, 1951, 1971; Bogert and Martin del Campo, 1956; Evans, 1967; Simon, 1983; Halpern, 1992). Subsequently, the correlation between predominant sensory modality and foraging mode (Cooper, this volume, Chapter 8) led to the general assumption that geckos functioned chiefly as sit-and-wait predators. More recently the increased appreciation of phylogenetics by non-systematists has lead to the recognition that a better understanding of the gekkotan condition might be crucial to reconstructing the evolution of foraging mode across lizards as a whole (Schwenk, 1993b, 1994; Cooper, 1994a,b, 1995a,b,c, 1996a, 1997b).

The purpose of this chapter is to evaluate the available data to determine what generalities may be drawn about foraging in the Gekkota and to assess the extent to which the gekkotan condition can inform the discussion of the shift from the iguanian sit-and-wait strategy to the autarchoglossan wide foraging mode.

Foraging mode in the Gekkota

Sources and limitations of the data

Why have geckos played a peripheral role in the study of lizard foraging biology? One reason is the difficulty in obtaining data from them. Widely accepted criteria for the categorization of lizard foraging mode require the availability of quantitative data on lizard movement patterns (Huey and Pianka, 1981; Perry, 1999, this volume, Chapter 1; Miles *et al.*, this volume, Chapter 2). This is typically reported in the form of moves per minute (MPM) and/or percent time moving (PTM). Both measures require observation of unrestrained and undisturbed animals in the field, and are thus difficult to obtain for small nocturnal species (Henle, 1990a). Indeed, Stamps (1977) was unable to draw meaningful conclusions about the foraging mode of geckos because of this limitation. Such problems do not apply to diurnal species (see, for example, Cooper *et al.*, 1999; Werner and Chou, 2002) and perhaps not to anthropophilic nocturnal species that are often active around artificial lighting (see, for example, Paulissen and Buchanan, 1991; Petren and Case, 1996), but these species make up a relatively small percentage of gekkotans (Table 12.1). Alternate, and perhaps more appropriate, foraging metrics are equally problematic. Cooper *et al.* (1999) proposed PAM (proportion of attacks on prey discovered while lizards are moving) as a more biologically relevant indicator of foraging mode, but were only able to obtain data for a single species of diurnal gecko.

In general, workers have not specified how their field observations on nocturnal geckos were obtained, although the assumption must be made that some artificial light source was used. It seems likely that any such observations are suspect at best, as visible light is very likely to disrupt normal behavior. This limits reliable observations to those using more specialized techniques. For example, Werner *et al.* (1997a, 2004) and Werner (1998) used a flashlight covered by red cellophane to observe geckos and found that the animals were not disturbed. Night vision devices using natural light amplification with or without infrared illumination may also be used to observe gekkotans with minimal disruption (pers. obs.). However, these provide a limited field of vision and may be impracticable for smaller species or those in complex three-dimensional habitats.

As a consequence of the difficulty in obtaining quantitative data on move-ment patterns, most statements regarding foraging mode in nocturnal gekko-tans (as well as some diurnal forms) are based on opportunistic observations of actual predation attempts or unquantified impressions obtained through extended field experience (see, for example, Paulissen and Buchanan, 1991; Arnold, 1984, 1993; Werner *et al.*, 1997b). Alternatively, foraging mode has been inferred from proxy data, chiefly dietary. Indeed, Werner *et al.* (1997a) showed that many key summary references to gecko foraging modes (see, for example, Vitt and Price, 1982; Dunham and Miles, 1985; Dunham *et al.*, 1988) that cited other sources as authorities for such designations ultimately traced to papers that did not actually refer to foraging mode *per se*, but rather provided dietary data (Bustard, 1968a,b, 1970, 1971; Pianka and Pianka, 1976; Pianka and Huey, 1978). This is problematic if the hypothesized correlations between diet and foraging mode (see, for example, Huey and Pianka, 1981; Vitt and Pianka, this volume, Chapter 5) are spurious (see below).

I have summarized gekkotan foraging data from the literature in Table 12.1. Although not complete by any means, I have cited most of the examples from the ecological literature for which foraging mode of gekkotans has been explicitly stated or described. Some additional citations from species descriptions or other sources are also included (see, for example, Ratnam, 1993; Arnold and Jones, 1994). I have made no attempt to independently interpret dietary or other data in the context of foraging mode, nor have I made judgments as to the reliability of the interpretations of others (but see below). I have, however, marked with an asterisk (*) those species for which foraging mode has been determined, at least in part, on the basis of quantitative movement pattern data. Below I present brief qualitative discussions of each of several of the biological bases for foraging patterns in gekkotans and conclude with a summary of the available data relating to foraging mode in geckos and pygopodids.

Movement patterns

Perry (1999) reviewed foraging behavior indices and reported that data from only eight species in four genera of gecko were available. Werner (1998), Cooper *et al.* (1999), Werner and Chou (2002), and Werner *et al.* (2004) increased this number to 17 species. Based on criteria of MPM and PTM geckos as a group appear to be largely sedentary, supporting their traditional interpretation as sit-and-wait foragers. Traditionally MPM and PTM have been based on short observation periods of as little as one minute (see, for example, Huey and Pianka, 1981). Magnusson *et al.* (1985), however, suggested that such periods were too short to reveal meaningful foraging patterns in taxa that tended to move in intermittent bouts. Werner *et al.* (1997a, 2004) and Werner (1998), believing that gecko movements were probably underestimated by short sampling periods, used focal animal observations of approximately 30 min in studies of a number of gecko taxa. Although extended observations confirmed that most species observed were sit-and-wait predators (Table 12.1), the eublepharid *Goniurosaurus kuroiwae orientalis* behaved chiefly as a wide forager (Werner *et al.*, 2004) and at least one gekkonid species, *Gekko hokouensis*, exhibited the MPM of an ambush forager, but the PTM of a wide forager (Werner *et al.*, 1997a). Extended observations also revealed a high degree of individual variation in movement patterns across many species (Stanner *et al.*, 1998; Werner *et al.*, 1997a, 2004). Werner (1998) even identified a single individual of a sit-and-wait species, *Lepidodactylus lugubris*, that, based on movement patterns, was a wide forager.

Diet

There are two widely accepted dietary correlates of foraging mode (Huey and Pianka, 1981; Perry and Pianka, 1997; Zug *et al.*, 2001): sit-and-wait predators will tend to eat larger, more active prey and will encounter prey less frequently, whereas wide foragers will tend to eat sedentary or spatially unpredictable prey and will have higher prey encounter rates. Prey encounter rates are, in practice, determined from stomach and gut content data and many gekkotans do appear to have high incidences of empty stomachs. Although this might be indicative of sit-and-wait foraging, Huey *et al.* (2001) reported that nocturnal lizards (of which geckos were a majority) more frequently have empty stomachs than do diurnal lizards, regardless of foraging mode.

The dietary basis for foraging mode categorization is most frequently derived from data on prey size and type. In some instances the inference is fairly unambiguous. For example, Doughty and Shine (1995) regarded *Phyllurus platurus* as chiefly an ambush predator because it eats mainly large, mobile prey items. Avery (1981), however, found that although the diet of *Hemidactylus brookii* contained a large proportion of large prey items, many of these were represented by sedentary forms, such as lepidopteran larvae. The interpretation of diet is further confounded by the fact that putatively mobile prey items may be captured when either active or inactive (Huey and Pianka, 1983). Bustard (1968a,b, 1971) reported on the diets of *Oedura ocellata*, *Heteronotia binoei*, and *Gehyra variegata* and noted that diurnal grasshoppers were probably consumed by geckos as they foraged at night when the insects were inactive. The fact that large, mobile prey were consumed was taken by Vitt and Price (1982), Dunham and Miles (1985), and Dunham *et al.* (1988) as evidence of sit-and-wait foraging, but Werner *et al.* (1997a) saw Bustard's remarks as evidence of wide foraging on sedentary prey.

Dietary evidence from a diversity of gekkotans, including *Coleonyx brevis*, *Ptyodactylus guttatus*, *Homonota gaudichaudi*, and *Tarentola mauritanica*, reveals that geckos eat a diversity of arthropods that are differentially active in different microhabitats (Dial, 1978a; Marquet *et al.*, 1990; Perry and Brandeis, 1992; Gil *et al.*, 1994a). Such evidence can also bear on the question of foraging mode but, in the absence of detailed behavioral observations, is not conclusive; the ingestion of prey from multiple microhabitats could infer either wide-foraging behavior (Dial, 1978a; Perry and Brandeis, 1992), or a sit-and-wait strategy using multiple ambush sites (Gil *et al.*, 1994a; Marquet *et al.*, 1990).

The interpretation of social insects in the context of foraging behavior raises special problems. Termites, being spatially and temporally unpredictable in surface activity, should predominate in the diets of wide foragers (Huey and Pianka, 1981). None the less, they are certainly capitalized upon by a diversity of geckos (Pianka and Pianka, 1976; Pianka and Huey, 1978; Bauer *et al.*, 1990; Petren and Case, 1996), including species that have been characterized as sit-and-wait foragers (see Table 12.1). In the burrowing gecko *Ptenopus garrulous* the consumption of termites during swarming has been seen as a response to increased prey availability (Huey and Pianka, 1981), and thus evidence of plasticity in the foraging mode of an otherwise ambush predator. Alternatively, even if only sporadically exploited, swarming termites can constitute a tremendous energetic windfall for *Ptenopus* (Bauer *et al.*, 1990), and the ability of the geckos to employ an active foraging strategy may be more appropriately interpreted as evidence of a fluctuating foraging mode (Werner *et al.*, 1997a,b; Werner, 1998).

The case of swarming insects aside, the location of widely dispersed termite nests or other centers of activity would appear to constitute wide foraging. However, once such a resource is located, the predator essentially takes up an ambush strategy. This strategy is similar to the traplining used by ant-feeding *Prynosoma* (Huey and Pianka, 1981) and blurs the distinction between the extremes of foraging mode. The problem is illustrated by Greer's (1967) account of ant predation by *Lygodactylus*, which has subsequently been interpreted as evidence of both sit-and-wait (Dunham *et al.*, 1988) and wide foraging (Werner *et al.*, 1997a).

Cooper (1994b) considered herbivory as a third foraging mode used by lizards. The consumption of plant material may be regarded as analogous to that of social insects: the "prey" must be located by active foraging, but once located lizards may remain relatively stationary for long periods (Eifler, 1995). Herbivory, or at least specialized forms thereof (frugivory, nectivory), has been reported in at least four genera of geckos (Table 12.1), principally in species occurring on islands (Eifler, 1995; Cooper and Vitt, 2002; Olesen and Valido, 2003). Although few other correlates of foraging mode have been investigated in herbivorous geckos, Cooper (2000a, 2002) found that omnivorous *Rhacodactylus* exhibited lingually mediated food chemical discrimination for both plant and animal foods.

Chemoreception

Sit-and-wait foragers have been regarded as chiefly visual predators, whereas wide foragers have been linked to the use of chemosensory cues (Cooper, 1994a,

1995a,c, 1996a,b; Schwenk, 2000; Cooper, this volume, Chapter 8). Extensive experimental evidence has demonstrated that well-developed vomerolfactory abilities are often correlated with active foraging (Cooper, 1994a,b, 1997b). Dial's (1978a) evidence from both dietary information and chemoreceptive ability in *Coleonyx* spp. has served as the basis of the categorization (by some authors; see Table 12.1) of these eublepharids as wide foragers. Although early work did not always distinguish between olfaction and vomerolfaction, it has subsequently been demonstrated that the latter is probably the chief chemical sense of most scleroglossan lizards (Cooper, 1994a,b; Vitt *et al.*, 2003).

The potential functional significance of olfaction in geckos was, however, demonstrated by Chou *et al.* (1988), who noted decreased ability to capture flies by geckos with olfactory deafferentation. This was confirmed on morphological grounds by Schwenk (1993a) and Dial and Schwenk (1996), who identified geckos as olfactory specialists. Characterizing lizard chemosensory development on the basis of the ratio of sensory cells to supporting cells, Cooper (1996a) found that gekkotans as a whole were intermediate in vomerolfactory development and related lingual features (Cooper, 1995b, 1996b, 1997a) between iguanians and autarchoglossans, but confirmed that gekkotans, exclusive of pygopodids, exhibited the highest development of olfaction among squamates. Within gekkotans Cooper (1996a) identified an increase in vomerolfactory potential among eublepharids. The few species studied seem able to detect lingually mediated prey chemicals (Cooper, 1995a, 1996a), as well as pheromones (Mason and Gutzke, 1990; Brillet, 1990, 1991; Cooper and Steele, 1997), and both tongue flicking and labial licking are used by these lizards (Cooper *et al.*, 1996b; Cooper, 1998). Eublepharids thus exhibit the predicted chemosensory skills associated with wide foraging.

The significance of gekkonid olfactory development for foraging is unclear, as it has generally been assumed that vomerolfactory abilities were a requirement for wide foraging. Despite their moderate development of the vomerolfactory system, at least *Gekko gecko* shows post-strike elevation in tongue flicking rate, suggesting some ability to detect prey chemically. However, *G. gecko* lacks strike-induced chemosensory searching, as is typical of ambush predators (Cooper *et al.*, 1996a). Both animal and plant prey chemical detection are demonstrated by the omnivorous diplodactylid *Rhacodactylus* (Cooper, 2000a, 2002). Whether this capability is more generalized among gekkotans and has escaped detection owing to limited sampling (Cooper, 1996a) remains open to question. *Phelsuma madagascariensis*, a diurnal gekkonid that is also omnivorous, exhibits well-developed vomerolfactory structures (Cooper, 1996a), suggesting that lingually mediated prey detection may be present in at least some other herbivorous species of gecko. Alternatively, there is evidence that some geckos may prefer particular fruit colors (Lord and Marshall, 2001). If this is true, or if more generally frugivorous and nectivorous herbivorous geckos use olfaction and/or vision to locate plant food resources, geckos may prove an exception to the general rule that herbivory evolves from wide foragers that use vomerolfaction to locate prey (Cooper and Vitt, 2002).

Constraints on foraging correlates in Gekkota

In addition to diet and sensory modalities, foraging mode has been correlated with a variety of morphological, physiological, and life history correlates (this volume). Most of these correlates are remarkable not for what they can tell us about gecko foraging modes, but for what they cannot.

Morphology and performance

The relations between morphology, performance, and foraging mode are not especially well supported (Garland and Losos, 1994; Van Damme and Vanhooydonck, 2001). None the less, foraging mode has been correlated with a variety of aspects of lizard morphology. For example, sit-and-wait foragers generally have shorter, stouter bodies than do wide foragers (Vitt and Congdon, 1978; Anderson and Karasov, 1988). Other relations, including those with trophic morphology, limb length, and tail length (Vitt and Congdon, 1978; Vitt and Price, 1982; Vitt, 1983) have also been proposed. Geckos in general clearly conform to the stout-bodied model, but this is not necessarily indicative of ambush foraging. Body form is largely phylogenetically constrained by an overall paedomorphic pattern (Rieppel, 1984, 1994; Bauer, 1986) in most gekkotans. Although geckos vary widely in many aspects, as a group they possess relatively short, stout bodies, short tails, and large heads (pygopodids are the obvious exception, but are themselves also constrained to a serpentiform *Bauplan* with only tiny hindlimb flaps). Whatever the origins, adaptive or neutral, of this paedomorphic syndrome, it is essentially fixed in geckos. As a result, in any broad analyses including geckos as well as iguanians and autarchoglossans, geckos will tend to group with the stouter-bodied iguanians.

Reproduction

High relative clutch mass (RCM) is typically associated with ambush predation (Vitt and Congdon, 1978; Vitt and Price, 1982; Anderson and Karasov,

1988). This relation is relatively robust across lizards as a whole and is related to a decrease in locomotor performance incurred by high RCM (Garland and Losos, 1994), but geckos have been interpreted as having anomalously low RCMs for sit-and-wait predators (Vitt and Price, 1982; Vitt, 1986). Werner (1989; Werner *et al.* 1997a) regarded gecko RCMs to be intermediate between classically sit-and-wait and wide-foraging lizards and used this as evidence against the interpretation of geckos as strict ambushers. Within gekkotans few comparisons of species with differing foraging modes have been made, but Henle (1990a,c, 1991) reported lower (although not significantly so) RCMs in the wide-foraging Australian species *Heteronotia binoei* and *Diplodactylus damaeus* than in the sympatric sit-and-wait forager *Diplodactylus tessellatus*.

Within gekkotans clutch size is essentially fixed (Fitch, 1970), with two eggs or young in nearly all species, and some smaller species producing a single egg clutch and a few species occasionally producing three eggs at a time. Clutch mass may thus increase only through variation in egg size (Doughty, 1996, 1997), which itself may be limited by constraints of pelvic and cloacal anatomy. Thus, as noted by Perry (1999), a phylogenetic constraint on the clade as a whole may produce false correlations between foraging mode and reproductive capacity. Such reproductive constraints may also carry over to the seasonality of breeding. For example, Vitt (1983, 1986, 1990) characterized that in the Brazilian caatinga sit-and-wait foraging lizards reproduce seasonally. The exceptions to this were geckos, possibly because of constraints on clutch size and thus relative clutch mass.

Metabolism and endurance

Another set of constraints applies specifically to nocturnal geckos. It is hypothesized that active foragers should exhibit higher metabolic rates and higher endurance than ambushers in association with their increased locomotor demands. Empirically this is true, although not significantly so (Andrews and Pough, 1985).

For nocturnal geckos, opportunities for heat gain during an activity period are limited (Bustard, 1967; Angilletta *et al.*, 1999). Gecko preferred temperatures are lower than for diurnal iguanians and scincids, but still higher than the temperatures they actually experience in the field (Huey and Slatkin, 1976; Dial, 1978b; Gil *et al.*, 1994b; Angilletta and Werner, 1998). As a consequence, nocturnal geckos are generally active as early as possible in the evening, when ambient temperatures are still relatively high. Nocturnality dictates that many geckos, unlike diurnal lizards, are typically active at temperatures that are significantly suboptimal with respect to endurance, sprint speed and other functions (Huey *et al.*, 1989; Losos, 1990; Autumn *et al.*, 1994; Autumn and DeNardo, 1995). Measurement of these parameters at biologically relevant temperatures yields low values for nocturnal geckos, which may, in turn, be interpreted as confirmation of an ambush foraging mode. For example, Garland (1994, 1999) found that the geckos *Hemidactylus* and *Coleonyx* had lower endurance than many other lizards and that this was correlated with percentage time moving (and hence foraging mode).

Andrews and Pough (1985) found that geckos have lower metabolic rates than other lizards; Peterson (1990) found lower than expected metabolic rate even in diurnal *Rhoptropus* and presented evidence to suggest that geckos do have low metabolic rates as a group. Arad (1995), however, found higher than expected rates in three species of *Ptyodactylus* and suggested that the allometry of metabolic rate within geckos relative to other groups might need to be recalculated.

Autumn *et al.* (1994), however, found that although performance variables, such as sprint speed, are submaximal at active body temperatures in the terrestrial Asian gecko *Teratoscincus przewalskii*, metabolic cost of locomotion is reduced. This increase in locomotor performance capacity at low temperatures provides the potential for nocturnal lizards to be active foragers. Nocturnal representatives of the Gekkonidae, Eublepharidae, Carphodactylidae, and Diplodactylidae all show a reduction in the minimum cost of locomotion and a concomitant increase in maximum aerobic speed relative to the iguanian *Phrynosoma douglassi* (Autumn *et al.*, 1997, 1999). This advantage is lost in secondarily diurnal gecko taxa (*Phelsuma* and *Rhoptropus*), which can and do function at preferred body temperatures (Autumn, 1999).

Other correlates

Most other correlates of foraging mode have not been characterized for gekkotans, or have been noted only for exemplar species. For example, *Gekko gecko* appears to have a lower percentage of fast-twitch glycolytic fibers, generally indicative of species capable of short bursts of high speed, in locomotor muscles than do most other lizards (Mirwald and Perry, 1991; Bonine *et al.*, 2001). However, there exists no basis for comparison between this sit-and-wait forager and gekkotans employing mixed or wide-foraging strategies.

Gekko gecko has also been the subject of functional morphological studies of the gekkotan feeding apparatus. Andrews and Bertram (1997) demonstrated that the force and mechanical work to bite prey items was both size- and prey-type-dependent and that prey handling behavior also varied with these parameters. The study suggested that the energetics of the adductor musculature might play a role in determining feeding behavior in geckos, but this hypothesis has not subsequently been tested, and no comparative data are available for other gekkotans. Herrel *et al.* (1999, 2000) investigated cranial kinesis in *G. gecko* and another sit-and-wait predator, *Phelsuma madagascariensis*.The amphikinesis exhibited by geckos in general is probably both a function of the evolution of nocturnality and conditions imposed by large eye size, and a determining factor in gecko prey size and type handling capacity.

The evolution of foraging mode in gekkotans

Despite the deceptively large number of taxa for which foraging mode assessments have been made (Table 12.1), it is not possible to characterize gekkotans as a whole as either sit-and-wait or wide foraging. There are several reasons for this. First, gekkotans as a group account for approximately 25% of lizard species (Bauer, 2002), yet foraging mode has been assessed in any fashion for only 9% of species in 44% of genera (Tables 12.1, 12.2). Only among the sphaerodactylines have representatives of all genera been considered, but in this clade only 6% of species have been studied, none with respect to movement pattern data. Sampling is especially problematic in the Gekkoninae, where only 39% of genera and 7% of species have been associated with a particular foraging mode. This group includes not only the bulk of gekkotan species, but also has by far the greatest geographic range and the most ecological diversity. Only pygopodids can be said to have been adequately sampled, but these are the most aberrant of gekkotans and the least relevant for deriving generalities about foraging biology. Independent of the quality of the available data, assessment of foraging mode within the Gekkota has been so limited that generalizations are inappropriate and potentially misleading.

Second, until recently, a large number, if not the majority, of lizard ecologists have explicitly or implicitly accepted the dichotomy offered by the prevailing foraging paradigm. As a consequence, when assessing the foraging mode of geckos, they have considered themselves presented with only two choices, and although geckos may not exactly fit the bill for classic sit-and-wait predators, most are clearly not wide foragers, so have tended to be placed in the least discrepant category. This has tended to result in "either/or" categorization and only a few authors (e.g. Arnold, 1984, 1993; Werner *et al.*, 1997a, 2004) have considered mixed foraging modes or intermediate modes, such as cruise foraging, as possibilities to describe gekkotan foraging. This problem of

Table 12.2 *Summary of gekkotan foraging modes by phylogenetic group and diel activity category*

The mixed foraging category includes species reported as cruise or intermediate foragers, as well as all those for which both SW and WF have been reported. Percentages may not sum to 100, owing to rounding.

Group	# Genera (% gekkotan gen.)	Genera w/data (% genera in grp)	# Species (% all gekkotan spp.)	Species w/data (% species in grp)	SW Only (% species with data)	WF Only	Mixed
Eublepharidae	6 (6%)	4 (67%)	28 (3%)	6 (21%)	1 (17%)	3 (50%)	2 (33%)
Carphodactylidae	5 (5%)	2 (40%)	24 (2%)	3 (12%)	3 (100%)	0 (0%)	0 (0%)
Diplodactylidae	11 (10%)	4 (36%)	110 (10%)	13 (12%)	7 (58%)	2 (17%)	4 (33%)
Pygopodidae	7 (7%)	4 (57%)	36 (3%)	14 (39%)	1 (7%)	12 (86%)	1 (7%)
Gekkoninae	72 (68%)	28 (39%)	768 (69%)	54 (7%)	40 (74%)	3 (6%)	11 (20%)
Sphaerodactylinae	5 (5%)	5 (100%)	142 (13%)	8 (6%)	8 (100%)	0 (0%)	0 (0%)
Nocturnal genera	86 (81%)	33 (38%)	759 (69%)	70 (9%)	40 (57%)	15 (21%)	15 (21%)
Nocturnal/diurnal	11 (10%)	7 (64%)	123 (11%)	14 (11%)	8 (57%)	5 (36%)	1 (7%)
Diurnal genera	9 (8%)	7 (78%)	226 (20%)	14 (6%)	12 (86%)	0 (0%)	2 (14%)
All gekkotans	106	47 (44%)	1108	98 (9%)	60 (61%)	20 (20%)	18 (19%)

pigeonholing has been exacerbated by the fact that for much of the twentieth century all gekkotans, exclusive of pygopods, were placed in a single family, obscuring the evolutionary diversity of the group and further encouraging generalizations.

Finally, the determinations of foraging mode that have been made (Table 12.1), derived chiefly from dietary studies, are highly subjective. They mostly presuppose the nature of the relationship between foraging and prey size, type, and frequency and can, therefore, serve only to reinforce the paradigm. Other correlates of foraging mode have been little studied among geckos and pygopods, but most are strongly constrained, either by the paedomorphic syndrome expressed by geckos, by limitations of fixed clutch size, or by the physiological limits imposed by nocturnality.

Data from movement patterns would appear to be the most objective and robust source of information on gekkotan foraging. Only for 17 species, or about 1.5%, of gekkotans are such data available (Table 12.1), and no movement data exist for representatives of four of the six major gekkotan lineages. By themselves, the movement-based data suggest only that most gekkonids ($n = 15$) are sit-and-wait predators and that the single eublepharid is at least partly a wide forager. The more extensive data set of foraging categorizations based on a diversity of criteria (Table 12.1) reveals similar patterns for these two clades. Wide foraging has been recorded (alone or along with sit-and-wait predation) in five of six eublepharids studied. Gekkonines appear to be largely ambush predators, but mixed foraging has been recorded for 20% of species considered, and three species have been noted only in the context of wide foraging (Table 12.2).

Both the carphodactylids and sphaerodactylines appear, on the basis of small samples, to be exclusively sit-and-wait foragers. Diplodactylids, with 40% wide foragers or mixed foragers (Table 12.2) would seem, as a group, to be more diverse than most groups, but it should be noted that this is largely reflective of the identification by Werner *et al.* (1997b) of wide-foraging "tendencies" among certain species, based on opportunistic observations. Finally, pygopodids appear to be the only truly wide-foraging members of the Gekkota (Table 12.2).

Unfortunately, I have little faith in these generalizations, certainly not enough to consider any quantitative analysis justified. However, taken in conjunction with other information, as outlined above, and the recognition that the paradigm itself may have biased earlier interpretations, at least a qualitative summary of gekkotan foraging modes is possible. To be sure, there are geckos that are classical ambush predators. Werner and Chou (2002) demonstrated that *Cnemaspis kendallii* is an extreme sit-and-wait predator, moving hardly at all and preying exclusively on arthropods that pass within a narrow radius of its perch. However, opportunistic observations of geckos in the field reveal that many species move extensively in search of food. Werner *et al.* (1997b) reported on observations of a diversity of Australian geckos (*Strophurs ciliaris aberrans, Diplodactlyus stenodactylus, Gehyra punctata, G. variegata, Heteronotia binoei, Oedura marmorata,* and *Rhynchoedura ornata*), most of which had been considered as sit-and-wait foragers, active in areas far from their normal retreats and engaged in what can best be interpreted as foraging behavior. To this list can be added many other species. In southern Africa, for example, the saxicolous species *Pachydactylus bibronii, P. turneri, P. fitzsimonsi, P. bicolor,* and *P. scutatus,* amongst others, may regularly be encountered moving on roads at night, tens of meters from the nearest suitable rock outcrop, feeding on arthropods (Bauer *et al.,* 1990; pers. obs.). Likewise, in New Caledonia, the arboreal *Rhacodactylus auriculatus* may be found moving across roads, sometimes in large numbers (Bauer and Vindum, 1990), ostensibly foraging.

Werner *et al.* (1997a, 2004) and Werner (1998) have demonstrated that extending the focal observation period for geckos to 30 min reveals significant variation in movement patterns and, in some cases, suggests wide-foraging behavior that would have been missed by observations of shorter duration. Werner *et al.* (1997a) considered geckos as a group to be mixed strategists, using a combination of sit-and-wait and wide-foraging tactics. It may also be argued that such a mixed strategy is inherently used by all of those gekkotans that feed either on social insects or on plant products (see above).

My own (admittedly subjective) interpretation of casual observations of approximately 250 species of gekkotans in 47 genera in all major clades is that many diurnal species are principally sit-and-wait strategists, but that a majority of nocturnal species either remain near their diurnal retreats (presumably employing sit-and-wait tactics) or move widely, essentially using a type of cruise foraging. These geckos stop periodically and appear responsive to both visual and auditory cues. Pauses may be brief or may last for up to an hour or more. From my observations there is a continuum in this behavior from slow, but nearly continuous active foraging punctuated by brief pauses, to what amounts to the serial exploitation of several ambush sites, separated by distance. Although the slow, nearly continuous movement may best be regarded as cruise or saltatory foraging (Regal, 1978, 1983; O'Brien *et al.,* 1989), I propose "serial ambushers" as an appropriately descriptive term for geckos using the more punctuated form of foraging behavior.

Within the Gekkota, the phylogenetically basal eublepharids do tend to exhibit more wide-foraging behavior, on average, than do representatives of other clades (Kingsbury, 1989; Werner *et al.,* 2004). This is complemented by their relatively well developed vomeronasal apparatus and prey chemical detection capabilities (Cooper, 1995a, 1998; Cooper *et al.*, 1996b; Cooper and Habegger, 2000). Sphaerodactylines are perhaps the most sedentary predators. This may have some phylogenetic basis, or it may reflect the diurnality of most species in the group and a greater reliance on visual cues. However, even this apparent uniformity among species in the group may be artificial, as it seems likely that many of the tiny forms in this clade cruise forage or are otherwise active within the leaf litter. Pygopodids, not surprisingly, given their aberrant morphologies for the Gekkota, have perhaps the most atypical feeding biology. Unlike most geckos, which are generalist arthropod feeders, many pygopods are dietary specialists (Patchell and Shine, 1986; Webb and Shine, 1994) and employ specialized wide-foraging methods to locate prey. Two of the remaining lineages, the Gekkoninae and the Diplodactylidae, each include taxa that employ mixed foraging strategies (Table 12.2). This is likely to be the case for carphodactylids as well, particularly *Nephrurus* spp., but requires further study. Within the highly diverse Gekkoninae relationships remain too poorly known to determine whether any further phylogenetic trends exist, but it appears that diurnal and/or arboreal species move less, on average, when foraging than do other species (Tables 12.1, 12.2; pers. obs.). This partly supports Cooper's (1995a, 1996a; Cooper *et al.*, 1999) contention that active or mixed foraging was most likely in terrestrial nocturnal geckos rather than among arboreal or saxicolous taxa.

Unfortunately the distribution of foraging modes within the Gekkota (and the ambiguity in assigning modes to particular clades) precludes an unambiguous reconstruction of the ancestral foraging mode for gekkotans as a whole (Werner *et al.*, 2004). I propose, however, that the ancestral gekkotan was a mixed strategist. In the common ancestor of all non-eublepharids (Fig. 12.1) there was perhaps a decrease in vomerolfaction and an increase in olfactory abilities that was concomitant with a shift away from active foraging (Schwenk, 1993a,b, 1994; Cooper, 1996a; Vitt *et al.*, 2003). In most of the descendant lineages a majority of species are either sit-and-wait predators to a large extent, or use the slow cruise foraging or serial ambushing strategy described above. The two exceptions are pygopodids, which employ chiefly active foraging techniques in association with their highly specialized diets (Patchell and Shine, 1986; Webb and Shine, 1994), and sphaerodactylines, many of which have adopted an exclusively sedentary foraging style. Similar extremes occur among certain groups of diplodactylids, gekkonines, and carphodactylids, although each of these lineages show a great deal of intraclade variation that is apparently related to habitat and diel activity (Cooper, 1995a, 1996a; Cooper *et al.*, 1999; Vitt *et al.*, 2003; Table 12.2) as well as phylogeny.

The reinterpretation of the Gekkota as a clade characterized generally by mixed foraging strategy has implications for the evolution of foraging mode in lizards as a whole. It suggests that at least some of the characteristics of wide foraging evolved at the level of the Scleroglossa, perhaps chiefly in association with chemosensory developments. Within scleroglossans, wide foraging probably evolved as a *predominant* foraging mode at least three times, once in the Autarchoglossa (with subsequent secondary loss in some clades, such as the Cordylidae), once in the Pygopodidae, and once in the Eublepharidae. However, the extent to which these manifestations of wide foraging can be regarded as homologous to one another is unclear. Because of broader constraints associated with paedomorphosis, nocturnality, and fixed clutch size (which themselves may be evolutionarily linked), geckos do not express the full range of foraging correlates that are typical of autarchoglossans. It is perhaps best to regard as homologous only the shared potential to employ chemoreception as a foraging tool. As such, gekkotans may be viewed, not as an intermediate between iguanians and autarchoglossans, but rather as a group that, starting with the same chemosensory potential as autarchoglossans, has adopted alternative evolutionary pathways with respect to foraging. Whereas autarchoglossans have largely "abandoned" visual predation, and with it sit-and-wait foraging, gekkotans have retained both, but to varying degrees in different lineages have exploited other sensory modalities to converge on more active foraging tactics (from serial ambushing to true wide foraging). Beyond this generality, however, little more can be definitively concluded. Existing data are sufficient to demonstrate that geckos do not fit neatly into a dichotomous paradigm of lizard foraging modes, but they are inadequate to suggest a robust replacement. More and better movement pattern data (derived from extended observations and made by using night vision, infrared, and radiotelemetric techniques) as well as far more representative lineage sampling across major gecko groups are minimal prerequisites for the resolution of questions regarding the characterization and evolution of foraging mode within the Gekkota.

References

Ananjeva, N. B. and Tsellarius, A. Yu. (1986). On the factors determining desert lizards' diet. In *Studies in Herpetology*, ed. Z. Roček, pp. 445–8. Prague: Charles University and Societas Europaea Herpetologica.

Anderson, R. A. and Karasov, W. H. (1988). Energetics of the lizard *Cnemidophorus tigris* and life history consequences of food-acquisition mode. *Ecol. Monogr.* **58**, 79–110.

Andrews, C. and Bertram, J. E. A. (1997). Mechanical work as a determinant of prey-handling behavior in the tokay gecko (*Gekko gecko*). *Physiol. Zool.* **70**, 193–201.

Andrews, R. M. and Pough, F. H. (1985). Metabolism of squamate reptiles: allometric and ecological relationships. *Physiol. Zool.* **58**, 214–31.

Angilletta, M. J. Jr. and Werner, Y. L. (1998). Australian geckos do not display diel variation in thermoregulatory behavior. *Copeia* **1998**, 736–42.

Angilletta, M. J. Jr., Montgomery, L. G. and Werner, Y. L. (1999). Temperature preference in geckos: diel variation in juveniles and adults. *Herpetologica* **55**, 212–22.

Arad, Z. (1995). Physiological responses to increasing ambient temperature in three ecologically different, congeneric lizards (Gekkoninae: *Ptyodactylus*). *Comp. Biochem. Physiol.* **112A**, 305–11.

Arnold, E. N. (1984). Ecology of lowland lizards in the eastern United Arab Emirates. *J. Zool. Lond.* **204**, 329–54.

Arnold, E. N. (1993). Historical changes in the ecology and behaviour of semaphore geckos (*Pristurus*, Gekkonidae) and their relatives. *J. Zool. Lond.* **229**, 353–84.

Arnold, E. N. and Jones, C. G. (1994). The night geckos of the genus *Nactus* in the Mascarene Islands with a description of the distinctive population on Round Island. *Dodo* **30**, 119–31.

Autumn, K. (1999). Secondarily diurnal geckos return to cost of locomotion typical of diurnal lizards. *Physiol. Biochem. Zool.* **72**, 339–51.

Autumn, K. and DeNardo, D. F. (1995). Behavioral thermoregulation increases growth rate in a nocturnal lizard. *J. Herpetol.* **29**, 157–62.

Autumn, K., Farley, C. T., Enshwiller, M. and Full, R. J. (1997). Low cost of locomotion in the banded gecko: a test of the nocturnality hypothesis. *Physiol. Biochem. Zool.* **70**, 660–9.

Autumn, K., Jindrich, D., DeNardo, D. and Mueller, R. (1999). Locomotor performance at low temperature and the evolution of nocturnality in geckos. *Evolution* **53**, 580–99.

Autumn, K., Weinstein, R. B. and Full, R. J. (1994). Low cost of locomotion increases performance at low temperature in a nocturnal lizard. *Physiol. Zool.* **67**, 238–62.

Avery, R. A. (1981). Feeding ecology of the nocturnal gecko *Hemidactylus brookii* in Ghana. *Amph.-Rept.* **1**, 269–76.

Bauer, A. M. (1986). Systematics, biogeography and evolutionary morphology of the Carphodactylini (Reptilia: Gekkonidae). Ph.D. dissertation, University of California, Berkeley.

Bauer, A. M. (2002). Lizards. In *Encyclopedia of Amphibians and Reptiles*, ed. T. Halliday and K. Adler, pp. 138–75. Abingdon, UK: Andromeda Oxford Ltd.

Bauer, A. M. and Vindum, J. V. (1990). A checklist and key to the herpetofauna of New Caledonia, with remarks on biogeography. *Proc. Calif. Acad. Sci.* **47**, 17–45.

Bauer, A. M., Russell, A. P. and Edgar, B. D. (1990 [1989]). Utilization of the termite *Hodotermes mossambicus* (Hagen) by gekkonid lizards near Keetmanshoop, South West Africa. *S. Afr. J. Zool.* **24**, 239–43.

Bogert, C. M. and Martin del Campo, R. (1956). The Gila monster and its allies. The relationships, habits, and behavior of the lizards of the family Helodermatidae. *Bull. Amer. Mus. Nat. Hist.* **109**, 1–238.

Bonine, K. E., Gleeson, T. T. and Garland, T., Jr. (2001). Comparative analysis of fiber-type composition in the iliofibularis muscle of phrynosomatid lizards (Squamata). *J. Morphol.* **250**, 265–80.

Bradshaw, S. D., Gans, C. and Saint Girons, H. (1980). Behavioral thermoregulation in a pygopodid lizard. *Copeia* **1980**, 738–43.

Brillet, C. (1990). Rôle des informations olfactives et visuelles dans la discrimination du sexe chez deux espèces de geckos nocturnes: *Eublepharis macularius* et *Paroedura pictus*. *Biol. Behav.* **15**, 1–22.

Brillet, C. (1991). Analyse comparative de la structure du comportement sexuel chez deux espèces de geckos nocturnes: *Eublepharis macularius* et *Paroedura pictus* (Sauria, Gekkonidae). *Behaviour* **117**, 117–43.

Bustard, H. R. (1967). Activity cycle and thermoregulation in the Australian gecko *Gehyra variegata*. *Copeia* **1967**, 753–8.

Bustard, H. R. (1968a). The ecology of the Australian gecko *Gehyra variegata* in northern New South Wales. *J. Zool. Lond.* **154**, 113–38.

Bustard, H. R. (1968b). The ecology of the Australian gecko *Heteronotia binoei* in northern New South Wales. *J. Zool. Lond.* **156**, 483–97.

Bustard, H. R. (1970). *Australian Lizards*. Brisbane: William Collins (Australia) Ltd.

Bustard, H. R. (1971). A population study of the eyed gecko, *Oedura ocellata* Boulenger, in northern New South Wales, Australia. *Copeia* **1971**, 658–69.

Camp. C. (1923). Classification of the lizards. *Bull. Amer. Mus. Nat. Hist.* **48**, 289–481.

Castanzo, R. A. and Bauer, A. M. (1998). Comparative aspects of the ecology of *Mabuya acutilabris* (Squamata: Scincidae), a lacertid-like skink from arid south western Africa. *J. Afr. Zool.* **112**, 109–22.

Chou, L. M., Leong, C. F. and Choo, B. L. (1988). The role of optic, auditory and olfactory senses in prey hunting by two species of geckos. *J. Herpetol.* **22**, 349–51.

Cooper, W. E. Jr. (1994a). Chemical discrimination by tongue-flicking in lizards: a review with hypotheses on its origin and its ecological and phylogenetic relationships. *J. Chem. Ecol.* **20**, 439–87.

Cooper, W. E. Jr. (1994b). Prey chemical discrimination, foraging mode, and phylogeny. In *Lizard Ecology: Historical and Experimental Perspectives*, ed. L. J. Vitt and E. R. Pianka, pp. 95–116. Princeton, NJ: Princeton University Press.

Cooper, W. E. Jr. (1995a). Prey chemical discrimination and foraging mode in gekkonoid lizards. *Herpetol. Monogr.* **9**, 120–9.

Cooper, W. E. Jr. (1995b). Evolution and function of lingual shape in lizards, with emphasis on elongation, extensibility, and chemical sampling. *J. Chem. Ecol.* **21**, 477–505.

Cooper, W. E. Jr. (1995c). Foraging mode, prey chemical discrimination, and phylogeny in lizards. *Anim. Behav.* **50**, 973–85.

Cooper, W. E. Jr. (1996a). Preliminary reconstructions of nasal chemosensory evolution in Squamata. *Amph.-Rept.* **17**, 395–415.

Cooper, W. E. Jr. (1996b). Variation and evolution of forked tongues in squamate reptiles. *Herpetol. Nat. Hist.* **4**, 135–50.

Cooper, W. E. Jr. (1997a). Independent evolution of squamate olfaction and vomerolfaction and correlated evolution of vomerolfaction and lingual structure. *Amph.-Rept.* **18**, 85–105.

Cooper, W. E. Jr. (1997b). Correlated evolution of prey chemical discrimination with foraging, lingual morphology and vomeronasal chemoreceptor abundance in lizards. *Behav. Ecol. Sociobiol.* **41**, 257–65.

Cooper, W. E. Jr. (1998). Prey chemical discrimination indicated by tongue-flicking in the eublepharid gecko *Coleonyx variegatus*. *J. Exp. Zool.* **281**, 21–5.

Cooper, W. E. Jr. (2000a). Correspondence between diet and food chemical discriminations by omnivorous geckos (*Rhacodactylus*). *J. Chem. Ecol.* **26**, 755–63.

Cooper, W. E. Jr. (200b). An adaptive difference in the relationship between foraging mode and responses to prey chemicals in two congeneric scincid lizards. *Ethology* **106**, 193–206.

Cooper, W. E. Jr. (2002). Convergent evolution of plant chemical discrimination by omnivorous and herbivorous scleroglossan lizards. *J. Zool. Lond.* **257**, 53–66.

Cooper, W. E. Jr. and Habegger, J. J. (2000). Lingual and biting response to food chemicals by some eublepharid and gekkonid geckos. *J. Herpetol.* **34**, 360–8.

Cooper, W. E. Jr. and Steele, L. J. (1997). Pheromonal discrimination of sex by male and female leopard geckos (*Eublepharis macularius*). *J. Chem. Ecol.* **23**, 2967–77.

Cooper, W. E. Jr. and Steele, L. J. (1999). Lingually mediated discriminations among prey chemicals and control stimuli in cordyliform lizards: presence in a gerrhosaurid and absence in two cordylids. *Herpetologica* **55**, 361–8.

Cooper, W. E. Jr. and Vitt, L. J. (2002). Distribution, extent, and evolution of plant consumption by lizards. *J. Zool. Lond.* **257**, 487–517.

Cooper, W. E. Jr. and Whiting, M. J. (2000). Ambush and active foraging modes both occur in the scincid genus *Mabuya*. *Copeia* **2000**, 112–18.

Cooper, W. E. Jr., DePerno, C. S. and Steele, L. J. (1996a). Do lingual behaviors and locomotion by two gekkotan lizards after experimental loss of bitten prey indicate chemosensory search. *Amph.-Rept.* **17**, 217–31.

Cooper, W. E. Jr., DePerno, C. S. and Steele, L. J. (1996b). Effects of movement and eating on chemosensory tongue-flicking and on labial-licking in the leopard gecko (*Eublepharis macularius*). *Chemoecology* **7**, 179–83.

Cooper, W. E. Jr., Whiting, M. J. and Van Wyk, J. H. (1997). Foraging modes of cordyliform lizards. *S. Afr. J. Zool.* **32**, 9–13.

Cooper, W. E. Jr., Whiting, M. J., Van Wyk, J. H. and Mouton, P. leF. N. (1999). Movement- and attack-based indices of foraging mode and ambush foraging in some gekkonid and agamine lizards from southern Africa. *Amph.-Rept.* **20**, 391–9.

Dial, B. E. (1978a). Aspects of the behavioral ecology of two Chihuahuan desert geckos (Reptilia, Lacertilia, Gekkonidae). *J. Herpetol.* **12**, 209–16.

Dial, B. E. (1978b). The thermal ecology of two sympatric nocturnal *Coleonyx* (Lacertilia: Gekkonidae). *Herpetologica* **34**, 194–201.

Dial, B. E. and Schwenk, K. (1996). Olfaction and predator detection in *Coleonyx brevis* (Squamata: Eublepharidae), with comments on the functional significance of buccal pulsing in geckos. *J. Exp. Zool.* **276**, 415–24.

Doughty, P. (1996). Allometry of reproduction in two species of gekkonid lizards (*Gehyra*): effects of body size miniaturization on clutch and egg sizes. *J. Zool. Lond.* **240**, 703–15.

Doughty, P. (1997). The effects of "fixed" clutch sizes on lizard life-histories: reproduction in the Australian velvet gecko, *Oedura lesueurii*. *J. Herpetol.* **31**, 266–72.

Doughty, P. and Shine, R. (1995). Life in two dimensions: natural history of the southern leaf-tailed gecko, *Phyllurus platurus*. *Herpetologica* **51**, 193–201.

Dunham, A. E. (1983). Realized niche overlap, resource abundance, and intensity of interspecific competition. In *Lizard Ecology, Studies of a Model Organism*, ed. R. B. Huey, E. R. Pianka and T. W. Schoener, pp. 261–80. Cambridge, MA: Harvard University Press.

Dunham, A. E. and Miles, D. B. (1985). Patterns of covariation in life history traits of squamate reptiles: the effects of size and phylogeny reconsidered. *Am. Nat.* **126**, 231–57.

Dunham, A. E., Miles, D. B. and Reznick, D. N. (1988). Life history patterns in squamate reptiles. In *Biology of the Reptilia*, vol. 16, ed. C. Gans and R. B. Huey, pp. 441–522. New York: Alan R. Liss.

Durtsche, R. D. (1995). Foraging ecology of the fringe-toed lizard, *Uma inornata*, during periods of high and low food abundance. *Copeia* **1995**, 915–26.

Eifler, D. A. (1995). Patterns of plant visitation by nectar-feeding lizards. *Oecologia* **101**, 228–33.

Estes, R., de Queiroz, K. and Gauther, J. (1988). Phylogenetic relationships within Squamata. In *Phylogenetic Relationships of the Lizard Families*, ed. R. Estes and G. Pregill, pp. 119–281. Stanford, CA: Stanford University Press.

Evans, L. T. (1967). Introduction. In *Lizard Ecology: A Symposium*, ed. W. W. Milstead, pp. 83–6. Columbia, MO: University of Missouri Press.

Fitch, H. S. (1970). Reproductive cycles in lizards and snakes. *Misc. Publ. Univ. Kansas Nat. Hist. Mus.* **52**, 1–247.

Garland, T. Jr. (1994). Phylogenetic analyses of lizard endurance capacity in relation to body size and body temperature. In *Lizard Ecology: Historical and Experimental Perspectives*, ed. L. J. Vitt and E. R. Pianka, pp. 237–59. Princeton, NJ: Princeton University Press.

Garland, T. Jr. (1999). Laboratory endurance capacity predicts variation in field locomotor behaviour among lizard species. *Anim. Behav.* **58**, 77–83.

Garland, T. Jr. and Losos, J. B. (1994). Ecological morphology of locomotor performance in squamate reptiles. In *Ecological Morphology: Integrative Organismal Biology*, ed. P. C. Wainwright and S. M. Reilly, pp. 240–302. Chicago, IL: University of Chicago Press.

Gil, M. J., Guerrero, F. and Pérez-Mellado, V. (1994a). Seasonal variation in diet composition and prey selection in the Mediterranean gecko *Tarentola mauritanica*. *Israel J. Zool.* **40**, 61–74.

Gil, M. J., Guerrero, F. and Pérez-Mellado, V. (1994b). Diel variation in preferred body temperatures of the Moorish gecko *Tarentola mauritanica* during summer. *Herpetol. J.* **4**, 56–9.

Greeff, J. M. and Whiting, M. J. (2000). Foraging-mode plasticity in the lizard *Platysaurus broadleyi*. *Herpetologica* **56**, 402–7.

Greer, A. E. (1967). The ecology and behavior of two sympatric *Lygodactylus* geckos. *Breviora* **268**, 1–19.

Halpern, M. (1992). Nasal chemical senses in reptiles: structure and function. In *Biology of the Reptilia*, vol. 18, ed. C. Gans and D. Crews, pp. 423–523. Chicago, IL: University of Chicago Press

Han, D., Zhou, K. and Bauer, A. M. (2004). Phylogenetic relationships among the higher taxonomic categories of gekkotan lizards inferred from C-*mos* nuclear DNA sequences. *Biol. J. Linn. Soc.* **83**, 353–68.

Harris, D. J., Marshall, J. C. and Crandall, K. A. (2001). Squamate relationships based on C-*mos* nuclear DNA sequences: increased taxon sampling improves bootstrap support. *Amph.-Rept.* **22**, 235–42.

Harris, D. J., Sinclair, E. A., Mercader, N. L., Marshall, J. C. and Crandall, K. A. (1999). Squamate relationships based on C-*mos* nuclear DNA sequences. *Herpetol. J.* **9**, 147–51.

Henkel, F. W. and Schmidt, W. (2003). *Professional Breeders Series: Geckos.* Frankfurt-am-Main: Edition Chimaira.

Henle, K. (1990a). Population ecology and life history of three terrestrial geckos in arid Australia. *Copeia* **1990**, 759–81.

Henle, K. (1990b). Population ecology and life history of the arboreal gecko *Gehyra variegata* in arid Australia. *Herpetol. Monogr.* **4**, 30–60.

Henle, K. (1990c). *Life-history-Evolution von Echsen.* In *Evolutionsprozesse im Tierreich*, ed. B. Streit, pp. 181–99. Basel: Birkhäuser.

Henle, K. (1991). Life history patterns in lizards of the arid and semiarid zone of Australia. *Oecologia* **88**, 347–58.

Herrel, A., Aerts, P. and De Vree, F. (2000). Cranial kinesis in geckoes: functional implications. *J. Exp. Biol.* **203**, 1415–23.

Herrel, A., De Vree, F., Delheusy, V. and Gans, C. (1999). Cranial kinesis in gekkonid lizards. *J. Exp. Biol.* **202**, 3687–98.

Huey, R. B. and Pianka, E. R. (1981). Ecological consequences of foraging mode. *Ecology* **62**, 991–9.

Huey, R. B. and Pianka, E. R. (1983). Temporal separation of activity and interspecific dietary overlap. In *Lizard Ecology, Studies of a Model Organism*, ed. R. B. Huey, E. R. Pianka and T. W. Schoener, pp. 281–90. Cambridge, MA: Harvard University Press.

Huey, R. B. and Slatkin, M. (1976). Cost and benefits of lizard thermoregulation. *Quart. Rev. Biol.* **51**, 363–84.

Huey, R. B., Niewiarowski, P. H., Kaufmann, J. and Herron, J. C. (1989). Thermal biology of nocturnal ectotherms: is sprint performance of geckos maximal at low body temperatures? *Physiol. Zool.* **62**, 488–504.

Huey, R. B., Pianka, E. R., and Vitt. L. J. (2001). How often do lizards "run on empty"? *Ecology* **82**, 1–7.

Kingsbury, B. A. (1989). Factors influencing activity in *Coleonyx variegatus*. *J. Herpetol.* **23**, 399–404.

Kluge, A. G. (1987). Cladistic relationships in the Gekkonoidea (Squamata, Sauria). *Misc. Publ. Mus. Zool. Univ. Michigan* **173**, i–iv, 1–54.

Kluge, A. G. (2001). Gekkotan lizard taxonomy. *Hamadryad* **26**, 1–209.

Lee, M. S. Y. (1998). Convergent evolution and character correlation in burrowing reptiles: towards a resolution of squamate relationships. *Biol. J. Linn. Soc.* **65**, 369–453.

Lord, J. M. and Marshall, J. (2001). Correlations between growth form, habitat, and fruit colour in the New Zealand flora, with reference to frugivory by lizards. *New Zealand J. Bot.* **39**, 567–76.

Losos, J. B. (1990). Thermal sensitivity of sprinting and clinging performance in the tokay gecko (*Gekko gecko*). *Asiatic Herpetol. Res.* **3**, 54–9.

Macey, J. R., Larson, A., Ananjeva, N. B. and Papenfuss, T. J. (1997). Replication slippage may cause parallel evolution in the secondary structures of mitochondrial transfer RNAs. *Mol. Biol. Evol.* **14**, 30–9.

Magnusson, W. E., Junqueira de Paiva, L., Moreira da Rocha, R. *et al.* (1985). The correlates of foraging mode in a community of Brazilian lizards. *Herpetologica* **41**, 324–32.

Marquet, P. A., Bozinović, F., Medel, R. G., Werner, Y. L. and Jaksić, F. M. (1990). Ecology of *Garthia gaudichaudi*, a gecko endemic to the semiarid region of Chile. *J. Herpetol.* **24**, 431–4.

Mason, R. T. and Gutzke, W. H. N. (1990). Sex recognition in leopard gecko, *Eublepharis macularius* (Sauria: Gekkonidae): possible mediation by skin-derived semiochemicals. *J. Chem. Ecol.* **16**, 27–36.

McLaughlin, R. L. (1989). Search modes of birds and lizards: evidence for alternative movement patterns. *Amer. Nat.* **133**, 654–70.

Mirwald, M. and Perry, S. F. (1991). Muscle fiber types in ventilatory and locomotor muscles of the tokay, *Gekko gecko*: a histochemical study. *Comp. Biochem. Physiol.* **98A**, 407–11.

Moermond, T. C. (1979). The influence of habitat structure on *Anolis* foraging behavior. *Behaviour* **70**, 147–67.

Murray, B. A., Bradshaw, S. D. and Edward, D. H. (1991). Feeding behavior and the occurrence of caudal luring in Burton's pygopodid *Lialis burtoni* (Sauria: Pygopodidae). *Copeia* **1991**, 509–16.

Nemes, S. (2002). Foraging mode of the sand lizard, *Lacerta agilis*, at the beginning of its yearly activity period. *Russ. J. Herpetol.* **9**, 57–62.

O'Brien, W. J., Evans, B. I. and Browman, H. I. (1989). Flexible search tactics and efficient foraging in salutatory searching animals. *Oecologia* **80**, 100–10.

Olesen, J. M. and Valido, A. (2003). Lizards as pollinators and seed dispersers: an island phenomenon. *Trends Ecol. Evol.* **18**, 177–80.

Patchell, F. C. and Shine, R. (1986). Food habits and reproductive biology of the Australian legless lizards (Pygopodidae). *Copeia* **1986**, 30–9.

Paulissen, M. A. and Buchanan, T. M. (1991). Observations on the natural history of the Mediterranean gecko, *Hemidactylus turcicus* (Sauria: Gekkonidae) in northwestern Arkansas. *Proc. Arkansas Acad. Sci.* **45**, 81–3.

Perry, G. (1999). The evolution of search modes: ecological versus phylogenetic perspectives. *Amer. Nat.* **153**, 98–109.

Perry, G. and Brandeis, M. (1992). Variation in stomach contents of the gecko *Ptyodactylus hasselquistii guttatus* in relation to sex, age, season and locality. *Amph.-Rept.* **13**, 275–82.

Perry, G. and Pianka, E. R. (1997). Animal foraging: past, present and future. *Trends Ecol. Evol.* **12**, 360–4.

Perry, G., Lampl, I., Lerner, A. *et al.* (1990). Foraging mode in lacertid lizards: variation and correlates. *Amph.-Rept.* **11**, 373–84.

Peterson, C. C. (1990). Paradoxically low metabolic rate of the diurnal gecko *Rhoptropus afer*. *Copeia* **1990**, 233–7.

Petren, K. and Case, T. J. (1996). An experimental demonstration of exploitation competition in an ongoing invasion. *Ecology* **77**, 118–32

Pianka, E. R. (1966). Convexity, desert lizards, and spatial heterogeneity. *Ecology* **47**, 1055–9.

Pianka, E. R. (1971). Lizard species density in the Kalahari Desert. *Ecology* **52**, 1024–9.

Pianka, E. R. (1974). *Evolutionary Ecology*. New York: Harper and Row.

Pianka, E. R. (1986). *Ecology and Natural History of Desert Lizards*. Princeton, NJ: Princeton University Press.

Pianka, E. R. and Huey, R. B. (1978). Comparative ecology, resource utilization and niche segregation among gekkonid lizards in the southern Kalahari. *Copeia* **1978**, 691–701.

Pianka, E. R. and Pianka, H. D. (1976). Comparative ecology of twelve species of nocturnal lizards (Gekkonidae) in the Western Australian Desert. *Copeia* **1976**, 125–42.

Pietruszka, R. D. (1986). Search tactics of desert lizards: how polarized are they? *Anim. Behav.* **34**, 1742–58.

Polynova, G. V. and Lobachev, V. S. (1981). Territorial relationships in *Phrynocephalus mystaceus* [in Russian]. *Zool. Zh.* **60**, 1649–57.

Ratnam, J. (1993). Status and natural history of the Andaman day gecko, *Phelsuma andamanensis*. *Dactylus* **2**(2), 59–66.

Regal, P. J. (1978). Behavioral differences between reptiles and mammals: an analysis of activity and mental capabilities. In *Behavior and Neurology of Lizards, an Interdisciplinary Colloquium*, ed. N. Greenberg and P. D. MacLean, pp. 183–202. Rockville, MD: U. S. Department of Health, Education, and Welfare.

Regal, P. J. (1983). The adaptive zone and behavior of lizards. In *Lizard Ecology, Studies of a Model Organism*, ed. R. B. Huey, E. R. Pianka and T. W. Schoener, pp. 105–18. Cambridge, MA: Harvard University Press.

Rieppel, O. (1984). The structure of the skull and jaw adductor musculature in the Gekkota, with comments on the phylogenetic relationships of the Xantusiidae (Reptilia: Lacertilia). *Zool. J. Linn. Soc.* **82**, 291–318.

Rieppel, O. (1994). Studies on skeleton formation in reptiles. Patterns of ossification in the limb skeleton of *Gehyra oceanica* (Lesson) and *Lepidodactylus lugubris* (Dumeril & Bibron). *Ann. Sci. Nat., Zool.* **13**(15), 83–92.

Rösler, H. (2000). Kommentierte Liste der rezent, subrezent und fossil bekannten Geckotaxa (Reptilia: Gekkonomorpha). *Gekkota* **2**, 28–153.

Saint, K. M., Austin, C. C., Donnellan, S. C. and Hutchinson, M. N. (1998). C-*mos*, a nuclear marker useful for squamate phylogenetic analysis. *Mol. Phylogenet. Evol.* **10**, 259–63.

Schwenk, K. (1993a). Are geckos olfactory specialists? *J. Zool. Lond.* **229**, 289–302.

Schwenk, K. (1993b). The evolution of chemoreception in squamate reptiles: a phylogenetic approach. *Brain Behav. Evol.* **41**, 124–37.

Schwenk, K. (1994). Comparative biology and the importance of cladistic classification: a case study from the sensory biology of squamate reptiles. *Biol. J. Linn. Soc.* **52**, 69–82.

Schwenk, K. (2000). Feeding in lepidosaurs. In *Feeding*, ed. K. Schwenk, pp. 175–291. San Diego, CA: Academic Press.

Semenov, D. V. and Borkin, L. J. (1992). On the ecology of Przewalski's gecko (*Teratoscincus przewalskii*) in the Transaltai Gobi, Mongolia. *Asiatic Herpetol. Res.* **4**, 99–112.

Shenbrot, G. I., Rogovin, K. A. and Surov, A. V. (1991). Comparative analysis of spatial organization of desert lizard communities in Middle Asia and Mexico. *Oikos* **61**, 157–68.

Simon, C. A. (1983). A review of lizard chemoreception. In *Lizard Ecology, Studies of a Model Organism*, ed. R. B. Huey, E. R. Pianka and T. W. Schoener, pp. 119–33. Cambridge, MA: Harvard University Press.

Stamps, J. A. (1977). Social behavior and spacing patterns in lizards. In *Biology of the Reptilia*, vol. 7, ed. C. Gans and D. W. Tinkle, pp. 265–334. London: Academic Press.

Stanner, M., Thirakhupt, K., Werner, N. and Werner Y. L. (1998). Observations and comments on the tokay in Thailand and China as predator and as prey (Reptilia: Sauria: Gekkonidae: *Gekko gecko*). *Dactylus* **3**(2), 69–84.

Teixeira, R. L. (2001). Comunidade de lagartos da restinga de Guriri, São Mateus-Es sudeste do Brasil. *Atlantica* **23**, 77–84.

Underwood, G. (1951). Reptilian retinas. *Nature* **167**, 183.

Underwood, G. (1971). A modern appreciation of Camp's "Classification of the lizards." In *Classification of the Lizards*, C. L. Camp, pp. vii–xvii. Lawrence, KS: Society for the Study of Amphibians and Reptiles.

Valakos, E. and Vlachopanos, A. (1989). Note on the ecology of *Cyrtodactylus kotschyi* (Reptilia – Gekkonidae) in an insular ecosystem of the Aegean. *Biol. Gallo-hellenica* **15**, 179–84.

Van Damme, R. and Vanhooydonck, B. (2001). Origins of interspecific variation in lizard sprint capacity. *Funct. Ecol.* **15**, 186–202.

Vitt, L. J. (1983). Tail loss in lizards: the significance of foraging and predator escape modes. *Herpetologica* **39**, 151–62.

Vitt, L. J. (1986). Reproductive tactics of sympatric gekkonid lizards with a comment on the evolutionary and ecological consequences of invariant clutch size. *Copeia* **1986**, 773–86.

Vitt, L. J. (1990). The influence of foraging mode and phylogeny on seasonality of tropical lizard reproduction. *Pap. Avulsos Zool.* **37**, 107–23.

Vitt, L. J. and Congdon, J. D. (1978). Body shape, reproductive effort, and relative clutch mass in lizards: resolution of a paradox. *Amer. Nat.* **112**, 595–608.

Vitt, L. J. and Price, H. J. (1982). Ecological and evolutionary determinants of relative clutch mass in lizards. *Herpetologica* **38**, 237–55.

Vitt, L. J. and Zani, P. A. (1996). Organization of a taxonomically diverse lizard assemblage in Amazonian Ecuador. *Can. J. Zool.* **74**, 1313–35.

Vitt, L. J. and Zani, P. A. (1998a). Ecological relationships among sympatric lizards in a transitional forest in the northern Amazon of *Brazil. J. Trop. Ecol.* **14**, 63–86.

Vitt, L. J. and Zani, P. A. (1998b). Prey use among sympatric lizard species in lowland rain forest of Nicaragua. *J. Trop. Ecol.* **14**, 537–59.

Vitt, L. J., Pianka, E. R., Cooper, W. E. Jr. and Schwenk, K. (2003). History and the global ecology of squamate reptiles. *Amer. Nat.* **162**, 44–60.

Vitt, L. J., Souza, R. A., Sartorius, S. S., Avila-Pires, T. C. S. and Espósito, M. C. (2000). Comparative ecology of sympatric *Gonatodes* (Squamata: Gekkonidae) in the western Amazon of Brazil. *Copeia* **2000**, 83–95.

Vitt, L. J., Zani, P. A. and Monteiro de Barros, A. A. (1997). Ecological variation among populations of the gekkonid lizard *Gonatodes humeralis* in the Amazon Basin. *Copeia* **1997**, 32–43.

Webb, J. K. and Shine, R. (1994). Feeding habits and reproductive biology of Australian pygopodid lizards of the genus *Aprasia. Copeia* **1994**, 390–8.

Werner, Y. L. (1989). Egg size and egg shape in Near-Eastern gekkonid lizards. *Israel J. Zool.* **35**, 199–213.

Werner, Y. L. (1998). Preliminary observations on foraging mode in a community of house geckos on Tahiti and a comment on competition. *Trop. Ecol.* **39**, 89–96.

Werner, Y. L. and Chou, L. M. (2002). Observations on the ecology and foraging mode of the arrhythmic equatorial gecko *Cnemaspis kendallii* in Singapore (Sauria: Gekkoninae). *Raffles Bull. Zool.* **50**, 185–96.

Werner, Y. L., Bouskila, A., Davies, S. J. J. F. and Werner, N. (1997a). Observations and comments on active foraging in geckos. *Russ. J. Herpetol.* **4**, 34–9.

Werner, Y. L., Okada, S., Ota, H., Perry, G. and Tokunaga, S. (1997b). Varied and fluctuating foraging modes in nocturnal lizards of the family Gekkonidae. *Asiatic Herpetol. Res.* **7**, 153–65.

Werner, Y. L., Takahashi, H., Yasukawa, Y. and Ota, H. (2004). The varied foraging mode of the subtropical eublepharid gecko *Goniurosaurus kuroiwae orientalis*. *J. Nat. Hist.* **38**, 119–34.

Zug, G. R., Vitt, L. J., Caldwell, J. P. (2001). *Herpetology: An Introductory Biology of Amphibians and Reptiles*, 2nd edn. San Diego, CA: Academic Press.

13

Foraging mode in the African cordylids and plasticity of foraging behavior in *Platysaurus broadleyi*

MARTIN J. WHITING

School of Animal, Plant and Environmental Sciences, University of the Witwatersrand

Introduction

Understanding the evolution of key life history traits frequently involves searching for broad-scale patterns among diverse organisms (see, for example, Vitt *et al.*, 2003). When consistent patterns emerge, particularly among members of multiple clades, our understanding of how suites of traits co-evolve is improved. In the process, new hypotheses are generated, allowing further testing of patterns and the mechanisms that generate them. The simplified goal of this book is to disentangle the evolution of foraging mode within a major vertebrate lineage: lizards. In particular, we wish to understand the relationship between foraging mode and a suite of co-evolved characters. Fundamental to lizard foraging theory is the tenet that foraging mode has greatly constrained (or shaped) the evolution of certain traits, including aspects of morphology and physiology. For example, ambush foragers are often tank-like and have a high relative clutch mass, relatively slow sprint speed and lower metabolic rate than active foragers, which are slender, relatively fast, and with lower relative clutch mass (Huey and Pianka, 1981; see references in Greeff and Whiting, 2000). More recently, there is accumulating evidence that although foraging mode has had a profound influence on aspects of lizard biology, foraging mode is deeply rooted in history (phylogeny) and only one of many traits explaining current lizard diversity (Perry, 1999; Vitt *et al.*, 2003). Lizard foraging mode is also remarkably stable within entire clades (e.g. families) of lizards (Cooper, 1994, 1995). Just as identifying broad-scale patterns is important for understanding the evolution of foraging mode in lizards, identifying deviations from these patterns may be equally, if not more, informative. What factors allow an organism to alter its foraging mode and foraging behavior to exploit another resource, and when is it beneficial to do this? I tackle this question by reviewing foraging mode for one clade of

Lizard Ecology: The Evolutionary Consequences of Foraging Mode, ed. S. M. Reilly, L. D. McBrayer and D. B. Miles. Published by Cambridge University Press. © Cambridge University Press 2007.

lizards, the African Cordylidae. Specifically, I focus most of my review on one particular species (*Platysaurus broadleyi*) that exhibits remarkable foraging behavior and plasticity of foraging mode within a clade that exhibits relatively inflexible foraging mode.

Systematic placement of Cordylidae within the Cordyliformes

The Cordyliformes are a monophyletic clade of scincomorph lizards consisting of the Old World families Cordylidae (Africa) and Gerrhosauridae (Africa and Madagascar) (Odierna *et al.*, 2002). There is some disagreement on whether cordylids and gerrhosaurids are separate families, or subfamilies, within the Cordylidae (Lang, 1991; Mouton and Van Wyk, 1997; Zug *et al.*, 2001). However, there is little dispute that as a clade the Cordyliformes are mono-phyletic (see Odierna *et al.*, 2002). I follow Lang (1991) and Odierna *et al.* (2002) in considering gerrhosaurids as a separate family, and focus this review on Cordylidae *sensu stricto*. Furthermore, until recently, the Cordylidae consisted of the genera *Cordylus*, *Pseudocordylus*, *Chamaesaura*, and *Platysaurus*. A recent molecular phylogeny for all four genera demonstrated a monophyletic grouping of *Cordylus + Pseudocordylus + Chamaesaura*, placing them within *Cordylus* with *Platysaurus* as the sister taxon (Frost *et al.*, 2001). Since Frost *et al.* (2001), some papers have followed this convention. However, the use of *Cordylus sensu lato* has created nomenclatural instability; additional systematic work on the Cordylidae may result in further name change at the level of genus (P. Mouton, pers. comm.). For this reason, and because the majority of pub-lished work on cordylid foraging behavior follows the original convention, my review will focus on the traditional four genera within Cordylidae (*Cordylus*, *Pseudocordylus*, *Chamaesaura*, and *Platysaurus*) (Fig. 13.1).

Natural history of cordylids

Cordylus (girdled lizards; Fig. 13.1A) currently consists of 31 species (Frost *et al.*, 2001) and ranges from the southern Cape coast of South Africa as far north as southern Ethiopia (Odierna *et al.*, 2002). The primary radiation appears to have been in southern Africa, where the majority of species occur. Most taxa are rupicolous and occur either on rocky mountains or small rock outcrops (Mouton and Van Wyk, 1997). One species (*C. giganteus*) is colonial and lives in self-excavated burrows in Highveld grasslands (Van Wyk, 2000). Several other species are also terrestrial (e.g. *C. macropholis*, *C. tasmani*) or semi-arboreal, using dead trees or logs (e.g. *C. tropidosternum*). Most species are colonial and all are primarily insectivorous, although some taxa will eat

Figure 13.1. Representative cordylids. (A) *Cordylus cataphractus*; (B) *Pseudo-cordylus microlepidotus*; (C) *Chamaesaura anguina*; (D) *Platysaurus broadleyi* (male left, female right). (A–C) ©le Fras Mouton, (D) ©Martin Whiting.

small quantities of plant material (see, for example, Mouton *et al.*, 2000). *Cordylus* have osteoderms, which are thought to provide a degree of armament useful against predatory attack (Mouton and Van Wyk, 1997). Interestingly, *Cordylus* vary in their degree of armature such that more heavily armoured species have short legs, slow sprint speed, and are more likely to seek refuge in a crevice (Losos *et al.*, 2002).

Pseudocordylus (crag lizards; Fig. 13.1B) currently consists of seven species, all of which are found in several mountain ranges in the southern African subregion (Branch, 1998). All *Pseudocordylus* are rupicolous and insectivorous. Unlike most *Cordylus*, *Pseudocordylus* males are often conspicuously colored. Compared with *Cordylus*, *Pseudocordylus* are also less heavily spined (an index of armature), have longer hindlimbs, and sprint faster (Losos *et al.*, 2002).

Chamaesaura (grass lizards; Fig. 13.1C) currently consists of three species distributed from the coastal Cape of South Africa as far north as East Africa (Rwanda, Burundi, S. Kenya) (Lang, 1991). Among cordylids, *Chamaesaura* are unique because of their vestigial limbs and serpentiform morphology and locomotion. Furthermore, unlike other cordylids, which are largely rupicolous, *Chamaesaura* occur in grasslands, generally on mountain slopes and plateaus. *Chamaesaura* are insectivorous and possess long tails, which can be autotomized (Branch, 1998).

Platysaurus (flat lizards; Fig. 13.1D) currently consists of 24 taxa (15 species + subspecies) distributed from South Africa to southern Tanzania. They are so named because of their extreme dorso-ventral flattening, enabling them to fit into tight crevices (Scott *et al.*, 2004). All *Platysaurus* are rupicolous and largely insectivorous, although they will eat plant matter (small fruits, seeds, flower petals, young leaves) when available (Broadley, 1978; Whiting and Greeff, 1997). Currently, only the Augrabies flat lizard (*P. broadleyi*; formerly *P. capensis*) from a single population (Augrabies Falls National Park; hereafter Augrabies) has been studied in any detail in the field. *Platysaurus broadleyi* occurs predominantly along the granite banks of the Orange River beginning just east of the southern border between Namibia and South Africa and extending for a distance of several hundred kilometres to the west of Augrabies (Branch and Whiting, 1997; Scott *et al.*, 2004). The greatest densities occur at Augrabies. Male *P. broadleyi* are elaborately colored (Fig. 13.1D; Whiting *et al.*, 2003) while females are drab and retain the stripes they have as juveniles. Males are also larger than females, are aggressive and territorial, and may adopt alternative reproductive tactics (Whiting, 1999; Whiting *et al.*, 2003; Whiting *et al.*, 2006).

Foraging mode in Cordylidae

The current approach to determining lizard foraging mode is to quantify movement related to foraging behavior and the percent time spent moving while foraging (Pianka, 1966; Huey and Pianka, 1981; Perry, 1999; Chapter 1). Lizards that spend less than 10% of their time moving are considered ambush or sit-and-wait foragers; those that move for greater than 10% of their active time are considered active foragers. Although these criteria are arbitrary, this convention has been useful for categorising foraging mode and allowing broad-scale comparative analyses for a number of species (see, for example, Cooper, 1995; Perry, 1999). The problem arises when species do not fall neatly into either of the two modes. As such, foraging modes are no longer accepted as bimodal and are now thought to represent a continuum (see, for example, Pietruszka, 1986; Perry, 1999). Alternatively, more than one mode may exist (Cooper, 2005). For example, saltatory foraging has been suggested for when an animal makes many short moves punctuated by long pauses (pause–travel; Cooper, 2005). Alternatively, other authors consider saltatory foraging to consist of infrequent, but long, movements (Eifler and Eifler, 1999). Nevertheless, Cooper (2005) suggests that clusters of foraging mode still remain useful for seeking correlates with other variables. Given that many of the cordylids studied to date fall neatly into a traditional sit-and-wait mode of foraging (Cooper *et al.*, 1997; Mouton *et al.*, 2000; du Toit *et al.*, 2002), I use the 10% cut-off as a criterion for identifying foraging mode, for the purposes of this review.

Foraging mode has been quantified for nine cordylids (five *Cordylus*, two *Pseudocordylus*, one *Chamaesaura*, and one *Platysaurus*). Data therefore currently exist for the entire family. *Cordylus* is the best studied genus; reasonable data exist for four of the five taxa presented here (Fig. 13.2A, B). (*Cordylus imkae* foraging mode is unfortunately preliminary, being inferred from only a single data point (Cooper *et al.*, 1997) during which the lizard remained immobile.) *Cordylus* studied to date use elevated rock platforms, from where they make short foraging trips to capture insects (Cooper *et al.*, 1997). Data on movement rate (moves per minute) and activity (percent time moving) for the remaining four *Cordylus* strongly suggests that members of this genus are extreme ambush foragers (MPM range 0–0.23; PTM range 0–2.2; Fig. 13.2A, B).

The two species of *Pseudocordylus* that have been studied conform to an ambush foraging mode (Fig. 13.2A, B), but interestingly, during foraging bouts *P. capensis* may make longer movements (sometimes > 20 s) than is typical of ambush foragers (Cooper *et al.*, 1997). More data are needed on *Pseudocordylus* to properly evaluate its foraging mode and any degree of foraging mode plasticity.

It was originally believed that *Chamaesaura*, in part because of its serpentiform morphology (Fig. 13.1C) and speed, might be an active forager (Branch, 1998). However, foraging mode was quantified for *Chamaesaura anguina* in seminatural outdoor enclosures, where they behaved like typical ambushers (du Toit *et al.*, 2002) (Fig. 13.2).

In summary, the paradigm of a dichotomous foraging mode works well for the cordylids. When feeding on insects, all species adequately meet the criteria of a sit-and-wait foraging mode.

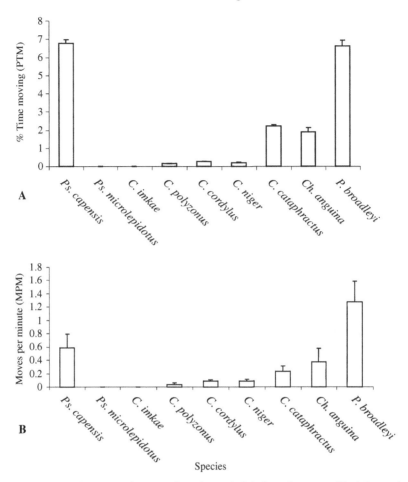

Figure 13.2. (A) Foraging mode of cordylid lizards quantified by using percent time moving (PTM). (B) Foraging mode of cordylid lizards quantified by using moves per minute (MPM). Data are from Cooper *et al.* (1997), Mouton *et al.* (2000), and du Toit *et al.*, (2002). *Cordylus* are all classical ambush foragers; *Pseudocordylus capensis* and *Platysaurus broadleyi* (Cooper *et al.*, 1997) had higher values, but were still within the range for ambush foraging.

The special case of *Platysaurus broadleyi*: plasticity of foraging mode

To find a fig: use of heterospecific cues

Platysaurus broadleyi is a facultative omnivore that feeds on Namaqua figs when they are available. Namaqua fig trees are pollinated by fig wasps and fruit asynchronously as a result (Berg and Wiebes, 1992). As such, figs are unpredictable in time and space. Ripe figs are energetically rich and constitute a valuable resource (Greeff and Whiting, 1999). Therefore, fruiting fig trees represent a high pay-off if discovered by lizards. Because of the unpredictable timing of fruiting, systematically or randomly searching for figs by travelling to trees that are dispersed in the landscape would be too energetically costly for lizards. In addition, lizards may be more vulnerable to predation. Instead, *P. broadleyi* use heterospecific cues to locate fruiting trees. When figs are ripe, large aggregations of birds are attracted to fig trees to feed on the fruit. During foraging, birds constantly move about the tree, flying short distances to locate new figs. This makes for almost continuous movement in the tree. *Platysaurus broadleyi* has a 'sensory bias' for movement and is attracted to fluttering birds. This was demonstrated by Whiting and Greeff (1997) experimentally, by showing that lizards are attracted to the movement of caged birds on bare rock, away from the context of the fig tree. More importantly, *P. broadleyi* were more likely to approach a non-fruiting fig tree containing caged birds than a fig tree with an empty cage. As such, lizards take advantage of high bird activity in fruiting trees as a cue to locate ripe figs, presumably as a result of this sensory bias for movement (Whiting and Greeff, 1997, 1999). Whether lizards make a cognitive connection between high bird activity and a fruiting fig tree is debatable but also immaterial: the pay-off is the same. Once fruiting trees are discovered, lizards return daily to feed on figs until this resource is spent.

Facultative frugivory: how to deal with a fig, fig stealing, and influence on foraging mode

Platysaurus broadleyi will either feed on fallen figs or climb into trees and remove ripe figs still attached to the tree (pers. obs.). During this time, lizard aggregations under these trees can number in the hundreds (pers. obs.; Greeff and Whiting, 1999). Lizards are in competition for this relatively limited resource with birds, rock hyraxes, and monkeys (Greeff and Whiting, 1999). Intraspecific competition for figs is also intense. Lizards attempt fig stealing (33% of individuals) if a nearby conspecific discovers a fig; up to five individuals have been recorded to approach a focal individual in possession of a fig (Whiting and Greeff, 1997). This behavior is largely unsuccessful (10.3% success rate), but forces lizards to move away from the source of competition before consuming any figs they may find. Furthermore, the distance lizards move to consume figs is a function of their perceived risk of interference competition: the presence of more conspecifics results in lizards moving further away (Whiting and Greeff, 1997). The biggest perpetrators of interference competition through fig stealing were adult females, although adult males also attempted this behavior (Whiting and Greeff, 1997).

Table 13.1 *Summarized foraging data on moves per minute (MPM) and percent time moving (PTM) for* Platysaurus broadleyi *from Augrabies Falls National Park by sex and cohort*

Data are expanded from Greeff and Whiting (2000) for lizards observed at an insect-rich site, an insect-poor site, and a fig site. Included for comparison are data from Cooper *et al.* (1997). Mean values are reported ± 1 SE.

Site	Sex/age	N	MPM	SE	Range	PTM	SE	Range
Insect-poor	males	21	0.75	0.14	0–2.7	5.1	1.11	0–17.67
	females	16	0.87	0.25	0–3.85	2.54	0.63	0–7.86
	adults	37	0.8	0.13	0–3.85	3.99	0.71	0–17.67
	juveniles	24	1.54	0.18	0.1–3.46	9.55	1.43	0.17–24.31
	all lizards	61	1.09	0.12	0–3.85	6.18	0.78	0–24.31
Insect-rich [a]	males	22	0.72	0.11	0–1.95	4.35	0.6	0.01–11.58
	females	20	1.76	0.49	0.3–9	5.14	0.54	0.5–9.9
	adults	42	1.22	0.25	0–9	4.73	0.4	0.01–11.58
All insects	males	43	0.74	0.09	0–2.7	4.72	0.62	0–17.67
	females	36	1.36	0.3	0–9	3.98	0.46	0–9.9
	adults	79	1.02	0.15	0–9	4.38	0.4	0–17.67
	all lizards	103	1.14	0.12	0–9	5.59	0.5	0–24.31
Fig site	males	13	1.51	0.18	0.4–2.4	13.6	2.33	1–30.13
	females	19	2.08	0.16	0.6–3.59	19.2	2.15	7–42.0
	adults	32	1.85	0.13	0.4–3.59	16.93	1.64	1–42.0
	juveniles	17	1.61	0.15	0.7–3	9.93	1.77	2.5–28.8
	all lizards	49	1.77	0.1	0.4–3.59	14.5	1.32	1–42.0
Cooper *et al.*, (1997)	adults	22	1.27	0.32	0–6.82	6.62	1.58	0–24.17

[a] Insufficient data for juveniles.

The very nature of a fig requires a different foraging behavior for a normally insectivorous lizard. Compared with their primary prey, which is aerial (black flies, *Simulium* spp.), figs are stationary and dispersed in a localized area in or under the tree, often in a substrate containing debris. Lizards therefore adjust their foraging behavior to suit this resource. *Platysaurus broadleyi* shift to a more active foraging mode, during which time they make more movements and cover more ground while actively searching for ripe figs (Fig. 13.3). Adult lizards performed significantly more moves per minute and spent longer moving while foraging for figs compared to foraging for insects (Greeff and Whiting, 2000) (Table 13.1; Figs. 13.3 and 13.4). Juvenile lizards, on the other hand, did not shift their foraging mode significantly in the context of fig compared with insect foraging (Fig. 13.3). Values for mean PTM during foraging for figs were in the range of those reported for some active foragers (see, for example, Perry, 1999) (Fig. 13.3A). Similarly, *P. broadleyi* made frequent movements in search of figs (Fig. 13.3B). Female lizards had the highest mean PTM (19.2%); one individual

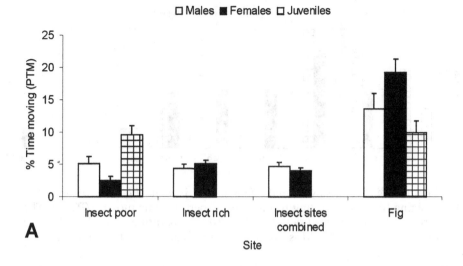

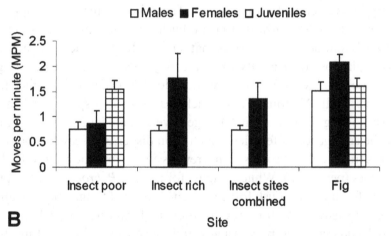

Figure 13.3. Foraging mode of *Platysaurus broadleyi*, quantified by using (A) percent time moving (PTM) and (B) moves per minute (MPM) for two insect sites of variable density and a fig site. Adults of both sexes switched to active foraging when searching for figs, but ambushed insects in similar ways at both insect sites. Juveniles had similar values for PTM and MPM at both the insect-poor site and the fig site (there were insufficient data for the insect rich site); these values bordered on active foraging.

spent 42% of observation time actively foraging for figs (Table 13.1). Evaluating the range of values for PTM, in addition to simply using mean PTM, was informative. These data show a number of individuals with high values of PTM: 65% ($n = 32/49$) were above the 10% cut-off for active foragers (Fig. 13.4).

While foraging for figs, lizards tongue-flicked the substrate (pers. obs.) and either nudged or tongue-flicked figs before eating them (Whiting and Greeff,

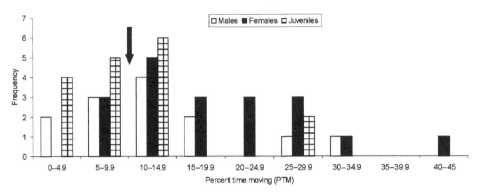

Figure 13.4. Frequency of male, female and juvenile *Platysaurus broadleyi* foraging for figs according to intervals of percent time moving (PTM). The arrow indicates the traditional cut-off (10%) separating ambush and active foragers (Perry, 1999). Most individuals (65%) are above that cut-off.

1997). This is contrary to expectation for an ambush forager. For a sit-and-wait predator, crypsis is essential for the element of surprise and may be disrupted by tongue flicking. Therefore, lizards that ambush prey do not tongue-flick; in addition, lizards that belong to lineages that have shifted from active to ambush foraging typically lose their ability to detect prey chemicals and cease tongue flicking potential food (Chapter 8). Why, then, do *P. broadleyi* tongue-flick figs?

At Augrabies Falls National Park, blackflies (*Simulium* spp.) breed in the shallow, fast-flowing sections of the Orange River, where large plumes of male flies await the emergence of females, which soon disperse in search of a blood meal. Lizards travel defined routes from crevices away from the river to feed on these fly plumes (pers. obs.). While foraging for insects, *P. broadleyi* are typical ambushers. A feature of ambush foragers is an absence of tongue flicking and a dependence on visual cues for prey location and capture (see Chapter 8). *Platysaurus broadleyi* locate prey visually and lunge forward to capture them in the air, sometimes performing a complete 360° aerial flip (pers. obs.). No tongue flicking is used during insect foraging. In contrast, *P. broadleyi* readily sample figs by using tongue flicking. While visual cues are used for initial assessment of fig quality, *P. broadleyi* tongue-flick or nudge figs prior to consumption (Whiting and Greeff, 1997). These observations suggest that *P. broadleyi* uses tongue flicking either to evaluate fig quality (possibly hydration) or to detect fig chemicals through vomerolfaction.

Food chemical discrimination

Whiting and Cooper (2003) tested the hypothesis that *P. broadleyi* can discriminate fig chemicals. Free-ranging *Platysaurus broadleyi* were presented with fig and fly extract on laminated cardboard tiles; the number of tongue-flicks directed at these stimuli, and the time spent at these stimuli, were compared with an odorless control. Lizards directed more tongue-flicks at the fig-labeled stimulus, spent more time at the fig-labeled tile, and attempted to eat this tile significantly more often than the insect or control stimulus. These results are surprising because previously tested omnivores within scleroglossan lizards have the ability to discriminate both insect and plant chemicals (Chapter 8). However, one alternative explanation is that *P. broadleyi* respond to fig cues by using gustation, rather than vomerolfaction (Whiting and Cooper, 2003). *Platysaurus broadleyi* were also observed licking the fig-labeled tile, a behavior seen in other lizards capable of gustation (Eifler, 1995; Cooper *et al.*, 2002). Furthermore, lizards will lick discarded soft fruit, such as over-ripe banana, that would not normally occur in their environment. Nevertheless, this behavior may be an adaptive response that allows *P. broadleyi* to assess the suitability of figs for consumption.

Age-specific foraging behavior

Like adults, juvenile *P. broadleyi* at Augrabies also feed predominantly on blackflies (Greeff and Whiting, 2000). Unlike adults, juvenile *P. broadleyi* do not eat entire figs, but tend to eat fig seeds and fragments of figs, presumably because of their smaller gape size (Whiting and Greeff, 1997). When feeding on fig fragments, juveniles used head shaking and 'pressing' (using the substrate as leverage) to obtain manageable pieces for ingestion (Whiting and Greeff, 1997). Adults of both sexes also engaged in head shaking and 'pressing' during fig handling, and always used head shaking in combination with 'pressing', but not *vice versa*. Adult males had the shortest fig handling time and both males and females ate entire figs, but unlike juveniles, they never ate fig seeds. When foraging for figs, adults generally ignored insects, whereas juveniles were equally likely to attempt insect prey captures. In addition to blackflies, juveniles were also observed eating small ants, something adults never did (Greeff and Whiting, 2000). While adults shifted foraging mode in response to resource type, juveniles did not, and had similar values for PTM and MPM for insect and fig foraging (Table 13.1). However, juveniles spent more time moving while foraging for insects than adults, and hovered around the 10% cut-off for an active foraging mode (Table 13.1) (Greeff and Whiting, 2000).

Thus, *P. broadleyi* show age-specific differences in foraging behavior partly explained by differences in morphology (body and gape size). Optimal foraging theory predicts that animals ignore lower-quality prey items if the pay-off does not warrant the time and effort expended in capture and prey handling

(Shafir and Roughgarden, 1998). Given the energetic rewards of a ripe fig, the expectation is that lizards would focus their efforts on finding figs when in the presence of a fruiting tree. Why *P. broadleyi* ignore insects while foraging for figs as adults, but not as juveniles, is an intriguing question that may be related to different energetic pay-offs for different lizard–prey body size ratios. It may be that for juvenile lizards there is less of a difference between an insect and a fragment of fruit. Alternatively, age-specific differences in foraging behavior may be explained by learning and/or copying of foraging behavior. We tend to think of lizard foraging behavior as very 'hard wired' and often ignore cognitive processes that may influence a lizard's foraging decisions (but see Day *et al.*, 1999). *Platysaurus broadleyi* appear to be acutely aware of conspecific behavior during foraging. In lizard species that aggregate, copying and learning could be very important processes shaping foraging behavior and even foraging mode. The influence of cognition on foraging behavior and foraging mode, could be a very rewarding, but challenging, future avenue of research.

Risk-sensitive foraging

While searching for figs in debris, lizards appear to engage in risk-sensitive foraging by concentrating their effort along the rock–debris interface. Figs were no more abundant in this area than closer to the tree base. However, *P. broadleyi* are rock specialists and only occasionally cross sandy areas between rock. By staying close to rock, they presumably reduce the risk of predation on a suboptimal substrate.

Influence of insect prey density on foraging mode

The spatial availability, density, and type of prey can influence predator space use and activity in several ways. For example, male lions either hunt for large prey (buffalo) in male coalitions in savanna woodlands where buffalo are more abundant, or associate with females (and scavenge more) in open savanna, during which time they feed on smaller prey (e.g. zebra and wildebeest) (Funston *et al.*, 1998). In lizards, these patterns are less dramatic, but may nevertheless result in a shift of the core activity area of a home range and greater spatial overlap among neighbors (Eifler, 1996). What is less well understood is how lizards alter their foraging behavior and foraging mode under variable food availability. The only cordylid to be studied under variable insect prey availability is *P. broadleyi*. Extreme concentrations of blackflies occur in fast-flowing sections of the Orange River at Augrabies and quickly

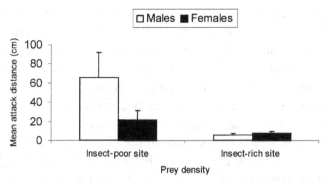

Figure 13.5. Mean (±1 SE) attack distance for adult *Platysaurus broadleyi* at an insect-rich and an insect-poor site. Attack distance was significantly greater for adults attempting to capture blackflies at the insect poor site.

taper off as a function of distance from the river. The highest aggregations of these lizards occur in the region of these plumes (pers. obs.). If insects are abundant, optimal foraging theory predicts that lizards should be able to expend less energy foraging by only selecting insects in close range if food quality is uniform (see, for example, Shafir and Roughgarden, 1998). If insects are abundant, optimal foraging theory predicts that lizards should be more selective and focus on profitable prey (see, for example, Stephens and Krebs, 1986; Shafir and Roughgarden, 1998). Alternatively, if prey quality is uniform, lizards can expend less energy foraging by only selecting insects in close range. This scenario is easily tested in *P. broadleyi* because blackflies are relatively invariant in size. Greeff and Whiting (2000) tested this hypothesis by measuring attack distance in an area of high and low insect density. *Platysaurus broadleyi* had shorter mean attack distances at the insect-rich area (Fig. 13.5). However, *P. broadleyi* did not adjust its foraging mode in any significant way in response to insect density. Lizards performed similar numbers of movements and spent similar amounts of time moving (Table 13.1; Fig. 13.3). What sets *P. broadleyi* apart from other cordylids and many other ambush foragers of insect prey is their high frequency of short movements (Cooper *et al.*, 1997; Greeff and Whiting, 2000). These movements are at least partly dictated by their insect prey. To catch enough small flies to meet their energy budget, *P. broadleyi* are expected to make many, short foraging movements.

Cooper *et al.* (1997) quantified foraging mode in *P. broadleyi* (*Platysaurus capensis* in their paper) at Augrabies. Mean PTM was comfortably within the range of typical ambush foragers (6.6%) (Fig. 13.2A), but they noted that *P. broadleyi* moved frequently during foraging (mean MPM: 1.27) (Fig. 13.2B). However, these movements were brief, resulting in relatively low PTM. They also found wide variation among individuals in both PTM (range: 0–24.17) and

MPM (range 0–6.82), with some individuals exhibiting values comfortably in the range of active foraging (e.g. maximum PTM = 24.2%). Greeff and Whiting (2000) reported measures of foraging mode for males, females, and juveniles from the same population, but under high and low prey availability. I have re-examined this data and present a slight expansion here (Table 13.1). I constructed frequency histograms to graphically capture variation in PTM at the insect sites and to allow visual comparison with the fig site. Frequency distribution of PTM values depicts classic ambush foraging in lizards that were feeding on insects, but active foraging in lizards that were feeding on figs (Fig. 13.6A). I also plotted the frequency distribution for MPM for adults at the two insect

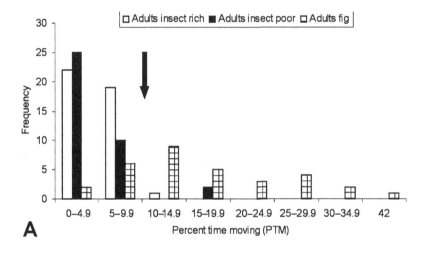

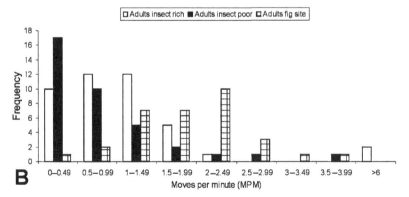

Figure 13.6. Frequency of adult male and female *Platysaurus broadleyi* by (A) intervals of percent time moving (PTM) and (B) intervals of moves per minute (MPM) for the insect-rich site, the insect-poor site, and the fig site. Values for the insect sites cluster around those predicted for an ambush forager; values for the fig site conform to active foraging and are extreme in some cases (e.g. 42% PTM).

sites for comparison with the fig site. Again, MPM values were similar for the two insect sites but significantly lower than for lizards foraging for figs (see Greeff and Whiting (2000) for statistical analysis) (Fig. 13.6B).

How do these patterns relate to phylogeny?

Scincidae (skinks) are basal to the Cordyliformes and considered their sister group (Lang, 1991; Odierna *et al.*, 2002) (Fig. 13.7). As a clade, active foraging in skinks is stable, with only a few exceptions (see, for example, Cooper and Whiting, 2000). Furthermore, gerrhosaurids (the sister group to Cordylidae), based on data for two taxa and parsimony, are considered active foragers (Cooper *et al.*, 1997). Therefore, the most parsimonious scenario for the evolution of foraging mode in Cordylidae is that active foraging was the ancestral state and that an ambush foraging mode has evolved once, independently, within Cordylidae (Cooper, 1995; Cooper *et al.*, 1997). Interestingly, among squamates this is a rare case of foraging mode reversal from active to ambushing within a clade. Ambush foraging is basal among squamates, characterizing the major clades Iguania and Gekkota (with some exceptions), which are basal to the Autarchoglossa, where active foraging first appears (Cooper, 1994). Among cordylids, all genera are clear-cut ambush foragers (Cooper *et al.*, 1997; Mouton

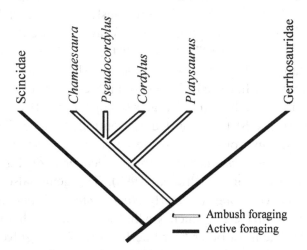

Figure 13.7. Hypothesis for the evolution of foraging mode in the Cordylidae and sister taxa (skinks + gerrhosaurids) following Cooper *et al.* (1997) and more recent data (e.g. du Toit *et al.*, 2002). Phylogenetic information follows Frost *et al.* (2001) and Odierna *et al.* (2002). Data on foraging mode in Gerrhosauridae are limited. Preliminary observations and parsimony suggest that gerrhosaurids are active foragers and therefore that the ancestor of the Cordylidae was an active forager. Branch lengths are not to scale.

et al., 2000; du Toit *et al.*, 2002) and therefore exhibit strong foraging mode stability within their clade. Of the *Platysaurus* species, *P. broadleyi* is the only member of the genus for which foraging mode has been quantified and ambushes insects (albeit with high variation among individuals) but actively searches for figs. So why does *P. broadleyi* exhibit plasticity of foraging mode and how is *P. broadleyi* different from other cordylids?

Covariation of morphology, antipredator behavior, and foraging mode in cordylids

Cordylus are typically tank-like and have short legs; several species are heavily armoured (Mouton and Van Wyk, 1997; Branch, 1998; Losos *et al.*, 2002). Morphology for 24 species, mostly of *Cordylus*, but also several of *Pseudocordylus* and *Platysaurus*, were examined by Losos *et al.* (2002). A subset of these were quantified for sprint speed and antipredator (escape) behavior. Losos *et al.* (2002) were specifically interested in quantifying the trade-off between sprint speed and degree of armature. Degree of armature was inversely related to distance fled from an approaching predator, such that more heavily armoured species took refuge following a short flight distance relative to less armoured species that sometimes did not take refuge. Heavily armoured species are all extreme ambush foragers (Cooper *et al.*, 1997; Mouton *et al.*, 2000), typifying the previously documented link between co-evolved traits such as morphology and foraging mode (Chapter 11, this volume; Vitt and Congdon, 1978; Losos, 1990).

The gerrhosaurids, the sister taxon to the cordylids, contain an interesting mix of dorsoventrally flattened, elongated, and 'cylindrical' forms. They are putatively considered to be active foragers (Cooper *et al.*, 1997). *Gerrhosaurus validus* is a large (max. SVL 285 mm), dorsoventrally flattened lizard that, like *P. broadleyi*, is omnivorous and takes refuge in rock crevices. *Gerrhosaurus flavigularis* and *G. nigrolineatus* are elongate, slender-bodied lizards that are terrestrial and insectivorous (Branch, 1998). The genus also contains the highly specialized *G. skoogi* (formerly *Angolosaurus skoogi*) (Lamb *et al.*, 2003) from the Namib Desert, which is cylindrical, psammophilous and largely herbivorous. Unlike most other lizard herbivores, *G. skoogi* feeds mostly on grasses and seeds (Pietruszka *et al.*, 1986). Considering the variation in habitat use (savannah, rock, sand), morphology (flattened, elongate, cylindrical) and diet (seeds and grasses, insects, fruits and leaves), gerrhosaurids offer enormous potential to test the foraging mode 'paradigm' and to test for additional plasticity of both foraging mode and feeding behavior. Rather than adding taxa that fit neatly into either extreme of foraging mode, we need to search for exceptions to the 'paradigm,' particularly among closely related taxa exhibiting divergent traits and which have divergent diets.

Cordylus and *Pseudocordylus* meet all the expectations of an organism with a specialized morphology and concordant evolution of associated traits linked to foraging mode, but what of *Chamaesaura* and *Platysaurus*? *Chamaesaura* are slender, and serpentiform; they have vestigial limbs that may be used for balance while stationary (Branch, 1998) but not in any significant way during locomotion. Given their slender morphology and adept locomotion in thick grass (features of an active forager) and phylogeny (predicting ambushing), their foraging mode was equivocal (Cooper *et al.*, 1997), until recently determined as sit-and-wait (du Toit *et al.*, 2002). Phylogeny has therefore played a major role in maintaining a relatively invariant foraging mode among the cordylid clade, even in a taxon (*Chamaesaura*) experiencing major morphological specialization.

Platysaurus are a separate clade within Cordylidae (Fig. 13.7) which lack protective spines and armour (Losos *et al.*, 2002); although they use rock crevices, they often only seek refuge as a last resort (*P. broadleyi*) (Whiting, 2002). Furthermore, females (and juveniles) of all species of *Platysaurus*, except one, have dorsal stripes (Broadley, 1978) that during flight may help create an optical illusion to a visual predator (Brodie, 1992) and are therefore probably adaptive. Many actively foraging lizards, for example African *Trachylepis* (formerly *Mabuya*) and many of the African lacertids, also have stripes (Branch, 1998). *Platysaurus broadleyi* have relatively high sprint speed (unpubl. data), tend to use flight to evade predators, and make many short movements during foraging. (Typical ambush foragers use crypsis to avoid detection by predators.) In comparison with *Cordylus*, they have relatively long hindlimbs (Losos *et al.*, 2002), which may facilitate more movement. As such, they do not have the suite of traits characteristic of ambush foragers. Furthermore, *Platysaurus* have extreme dorsoventral flattening, which facilitates the use of vertical rock surfaces (Losos *et al.*, 2002). That *Platysaurus* have stripes, are unconstrained by armature (cf. *Cordylus*), have relatively long hindlimbs, are relatively fast, and stay in the open, are all traits that are likely to promote active foraging.

That *P. broadleyi* is not a typical ambush forager is therefore not surprising. However, disentangling the relative effects of phylogeny and ecology, and the interplay of associated morphological and behavioral traits on foraging mode, is a challenge. *Platysaurus broadleyi* exhibits flexible foraging behavior by ambushing insects and by actively searching for figs. Furthermore, there appears to be an ontogenetic shift towards less active insect foraging as an adult, and adults exhibit high variation in foraging mode with some individuals showing values of PTM and MPM typical of active foragers. Spatial variability in insect prey is by itself insufficient to shift foraging mode, although *P. broadleyi* are more selective foragers when flies are abundant. Therefore, phylogeny has to some degree constrained *P. broadleyi* in its mode of prey capture. However, *P. broadleyi* is flexible enough to exploit another food resource (figs), which is variable in time and space, and thereby switch foraging mode to maximize intake of this resource. Broadley (1978) reported stomach contents for 13 other species of *Platysaurus*. Ten of these contained plant material, suggesting that herbivory/frugivory is widespread in the genus and deserves further attention. It may be that plasticity of foraging mode is more widespread than previously believed, particularly for species that exploit resources that are spatially and temporally variable (see, for example, Eifler and Eifler, 1999).

Conclusions and future directions

Platysaurus broadleyi is a member of a clade of lizards that are sit-and-wait foragers (Cooper, 1994, 1995). However, *P. broadleyi* vary greatly in foraging behavior depending on age, sex, or resource type. As juveniles, they spend almost 10% of their time moving during foraging for insects: right at the ambush – active foraging interface (Perry, 1999). They also make frequent movements. As adults, they spend less time moving while foraging, but still make many short movements. If figs are available, they switch to active or 'herbivorous' foraging (*sensu* Cooper, 1994) to increase the probability of finding an immobile, discrete resource. This is true of all age and sex classes of *P. broadleyi* (Whiting and Greeff, 1997).

The fitness benefits of resource switching are very difficult to measure and are unknown in *P. broadleyi*. If lizards are able to take advantage of a high-quality resource, presumably this will speed growth and fat deposition, all of which should have positive fecundity/fitness spin-offs. But whether lizards trade off the amount of time they spend foraging against reproductive behavior (courtship and territory defense), or simply maximize energy intake, is an intriguing question that would require careful study.

Biologists are always interested in exceptions to the rule, and what these mean for mechanisms explaining diversity in traits and species. With respect to cordylids, all taxa except *Platysaurus* fit into the extreme end of the sit-and-wait category of foraging mode. The *Platysaurus* system suggests two promising areas of research: (1) how widespread is foraging mode plasticity and what is the influence of a high-energy resource that is variable in time and space on foraging mode? and (2) what does having a combination of traits common to ambush and active foragers mean for foraging mode (see above)? A number of lizard species from multiple clades show omnivory and/or ontogenetic or seasonal shifts in diet (see, for example, Durtsche, 2004). We need to explicitly measure foraging mode for these species, from a range of clades. It is unlikely that shifts in foraging mode in response to prey type are rare. For example, Eifler and Eifler (1999) manipulated insect prey spatial availability (but not fruits) for the omnivorous grand skink (*Oligosoma grande*). As a result, females (but not males) switched to making fewer moves of longer duration to feed on larger insects, and ate less fruit. Plasticity of foraging mode has also been documented in non-lizard taxa such as birds and fish. For example, American robins adopt a sit-and-wait foraging mode to capture foliage insects, but actively search for terrestrial insects and fruit (although some variables measured for fruit were intermediate). Foraging mode for American robins was therefore a function of food type and distribution, in addition to habitat structure (Paszkowski, 1982). Brook charr (a fish) show divergent foraging modes in the same population because of intraspecific competition that could ultimately result in a resource polymorphism. Some individuals ambush crustaceans from the lower water column while others actively forage for insects in the upper water column. Furthermore, aggressive individuals adopted either sit-and-wait or active foraging modes, whereas non-aggressive fish had intermediate foraging modes (McLaughlin *et al.*, 1999). More studies experimentally manipulating food availability and food type are needed to test the 'paradigm' and to further explore plasticity of foraging mode as a function of a host of social and environmental variables.

The Gerrhosauridae also beg further study. The incredible range of body shapes, diets, and habitats suggest that foraging mode could be variable. A relatively invariant foraging mode within this clade would strongly support a phylogenetic influence. However, gerrhosaurids are on a continuum from insectivory to almost complete herbivory. Do they actively forage irrespective of food type, as phylogeny would suggest, or is there plasticity of foraging mode?

Finally, we need a better understanding of foraging-related cognitive processes (cognitive ecology *sensu* Dukas, 1998) that may influence foraging mode and how lizards exploit temporally and spatially variable resources (Regal, 1978). *Platysaurus broadleyi* may be a suitable candidate for such studies because they take advantage of heterospecific cues to find fruiting fig trees and have variable foraging modes. They also live in a very simple landscape, occur in large aggregations, and appear to cue in on conspecific behaviors during foraging (e.g. fig stealing) (Whiting and Greeff, 1997). Copying of conspecific behavior and cultural transmission during foraging may be more prevalent in lizards that aggregate than we have previously believed. Furthermore, the ability of animals to track changes in food availability in their environment and to respond by select-ing the most economic foraging mode is poorly understood; lizards may be suitable candidates. Investigating the cognitive ecology of foraging in *P. broadleyi* and other lizards promises to be highly rewarding.

Acknowledgements

For providing excellent photographs and for his detailed studies of cordylid foraging, I thank le Fras Mouton. My field work on flat lizards at Augrabies Falls National Park has been variously funded by the National Research Foundation in South Africa (including a grant to J. H. Van Wyk), the Transvaal Museum, and the universities of Stellenbosch and the Witwatersrand. I thank South African National Parks and the staff at Augrabies for assistance and permission to work in Augrabies. For a highly stimulating collaborative effort on foraging in *P. broadleyi*, I thank Jaco Greeff; for earlier collaborative work and field expe-rience, I thank Bill Cooper. Finally, I thank two anonymous reviewers and the editors of this volume for their constructive criticism of my contribution.

References

Berg, C. C., and Wiebes, J. T. (1992). *African Fig Trees and Fig Wasps*. Amsterdam, The Netherlands: Koninklijke Nederlandse Akademie van Wetensschappen North-Holland.

Branch, B. (1998). *Field Guide to the Snakes and Other Reptiles of Southern Africa*, 2nd edn. Cape Town: Struik Publishers.

Branch, W. R. and Whiting, M. J. (1997). A new *Platysaurus* (Squamata: Cordylidae) from the Northern Cape Province, South Africa. *Afr. J. Herpetol.* **46**, 124–36.

Broadley, D. G. (1978). A revision of the genus *Platysaurus* A. Smith (Sauria: Cordylidae). *Occ. Pap. Natl. Mus. Rhod.*, **B6**, 129–85.

Brodie, E. D. III. (1992). Correlational selection for color pattern and antipredator behavior in the garter snake *Thamnophis ordinoides*. *Evolution* **46**, 1284–98.

Cooper, W. E. Jr. (1994). Prey chemical discrimination, foraging mode, and phylogeny. In *Lizard Ecology: Historical and Experimental Perspectives*, ed. L. J. Vitt and E. R Pianka, pp. 95–116. Princeton, NJ: Princeton University Press.

Cooper, W. E. Jr. (1995). Foraging mode, prey chemical discrimination, and phylogeny in lizards. *Anim. Behav.* **50**, 973–85.

Cooper, W. E. Jr. (2005). The foraging mode controversy: both continuous variation and clustering of foraging movements occur. *J. Zool. Lond.* **267**, 179–90.

Cooper, W. E. Jr. and Whiting, M. J. (2000). Ambush and active foraging modes both occur in the scincid genus *Mabuya*. *Copeia* **2000**, 112–118.

Cooper, W. E. Jr., Perez-Mellado, V. and Vitt, L. J. (2002). Responses to major categories of food chemicals by the lizard *Podarcis lilfordi*. *J. Chem. Ecol.* **28**, 689–700.

Cooper, W. E. Jr., Whiting, M. J. and van Wyk, J. H. (1997). Foraging modes of cordyliform lizards. *S. Afr. J. Zool.* **32**, 9–13.

Day, L. B., Crews, D. and Wilczynski, W. (1999). Spatial and reversal learning in congeneric lizards with different foraging strategies. *Anim. Behav.* **57**, 393–407.

Dukas, R., ed. (1998). *Cognitive Ecology: The Evolutionary Ecology of Information Processing and Decision Making.* Chicago, IL: The University of Chicago Press.

Durtsche, R. D. 2004. Ontogenetic variation in digestion by the herbivorous lizard *Ctenosaura pectinata. Physiol. Biochem. Zool.* **77**, 459–70.

du Toit, A., Mouton, P. le F. N. M., Geertsema, H. and Flemming, A. (2002). Foraging mode of serpentiform, grass-living cordylid lizards: a case study of *Cordylus anguina. African Zool.* **37**, 141–9.

Eifler, D. A. (1995). Patterns of plant visitation by nectar-feeding lizards. *Oecologia* **101**, 228–33.

Eifler, D. A. (1996). Experimental manipulation of spacing patterns in the widely foraging lizard *Cnemidophorus uniparens. Herpetologica* **52**, 477–86.

Eifler, D. A., and Eifler, M. A. (1999). The influence of prey distribution on the foraging strategy of the lizard *Oligosoma grande* (Reptilia: Scincidae). *Behav. Ecol. Sociobiol.* **45**, 397–402.

Frost, D., Janies, D., Mouton, P. le F. N. and Titus, T. (2001). A molecular perspective on the phylogeny of the girdled lizards (Cordylidae, Squamata). *Am. Mus. Nov.* **3310**, 1–10.

Funston, P. J., Mills, M. G. L., Biggs, H. and Richardson, P. R. K. (1998). Hunting by male lions: ecological influences and socioecological implications. *Anim. Behav.* **56**, 1333–45.

Greeff, J. M. and Whiting, M. J. (1999). Dispersal of Namaqua fig seeds by the lizard *Platysaurus broadleyi* (Sauria: Cordylidae). *J. Herpetol.* **33**, 328–30.

Greeff, J. M. and Whiting, M. J. (2000). Foraging-mode plasticity in the lizard *Platysaurus broadleyi. Herpetologica* **56**, 402–7.

Huey, R. B. and Pianka, E. R. (1981). Ecological consequences of foraging mode. *Ecology* **62**, 991–9.

Lamb, T., Meeker, A. M., Bauer, A. M. and Branch, W. R. (2003). On the systematic status of the desert plated lizard (*Angolosaurus skoogi*): phylogenetic inference from DNA sequence analysis of the African Gerrhosauridae. *Biol. J. Linn. Soc.* **78**, 253–61.

Lang, M. (1991). Generic relationships within Cordyliformes (Reptilia: Squamata). *Bull. Inst. R. Sci. Nat. Belg.* **61**, 121–88.

Losos, J. B. (1990). Concordant evolution of locomotor behaviour, display rate and morphology in *Anolis* lizards. *Anim. Behav.* **39**, 879–90.

Losos, J. B., Mouton, P. le F. N., Bickel, R., Cornelius, I. and Ruddock, L. (2002). The effect of body armature on escape behaviour in cordylid lizards. *Anim. Behav.* **64**, 313–21.

McLaughlin, R. L., Ferguson, M. M. and Noakes, D. L. G. (1999). Adaptive peaks and alternative foraging tactics in brook charr: evidence of short-term divergent selection for sitting-and-waiting and actively searching. *Behav. Ecol. Sociobiol.* **45**, 386–95.

Mouton, P. le F. N. and Van Wyk, J. H. (1997). Adaptive radiation in cordyliform lizards: an overview. *Afr. J. Herpetol.* **46**, 78–88.

Mouton, P. le F. N., Geertsema, H. and Visagie, L. (2000). Foraging mode of a group-living lizard, *Cordylus cataphractus. Afr. Zool.* **35**, 1–7.

Odierna, G., Canapa, A., Andreone, F. *et al.* (2002). A phylogenetic analysis of cordyliformes (Reptilia: Squamata): comparison of molecular and karyological data. *Mol. Phylogenet. Evol.* **23**, 37–42.

Paszkowski, C. A. (1982). Vegetation, ground, and frugivorous foraging of the American robin. *Auk* **99**, 701–9.

Perry, G. (1999). Evolution of search modes: ecological versus phylogenetic perspectives. *Am. Nat.* **153**, 98–109.

Pianka, E. R. (1966). Convexity, desert lizards, and spatial heterogeneity. *Ecology* **47**, 1055–9.

Pietruszka, R. D. (1986). Search tactics of desert lizards: how polarized are they? *Anim. Behav.* **34**, 1742–58.

Pietruszka, R. D., Hanrahan, S. A., Mitchell, D. and Seely, M. K. (1986). Lizard herbivory in a sand dune environment: the diet of *Angolosaurus skoogi*. *Oecologia* **70**, 587–91.

Regal, P. J. (1978). Behavioral differences between reptiles and mammals: an analysis of activity and mental capabilities. In *Behavior and Neurobiology of Lizards*, ed. N. Greenberg and P. D. Maclean, pp. 183–202. Washington, D.C.: Department of Health, Education and Welfare.

Scott, I. A. W., Keogh, J. S. and Whiting, M. J. (2004). Shifting sands and shifty lizards: Molecular phylogeny and biogeography of African flat lizards (*Platysaurus*). *Mol. Phylogenet. Evol.* **31**, 618–29.

Shafir, S. and Roughgarden, J. (1998). Testing predictions of foraging theory for a sit-and-wait forager, *Anolis gingivinus*. *Behav. Ecol.* **9**, 74–84.

Stephens, D. W. and Krebs, J. R. (1986). *Foraging Theory*. Princeton, NJ: Princeton University Press.

Van Wyk, J. H. (2000). Seasonal variation in stomach contents and diet composition in the large girdled lizard, *Cordylus giganteus* (Reptilia: Cordylidae) in the Highveld grasslands of the northeastern Free State, South Africa. *Afr. Zool.* **35**, 9–27.

Vitt, L. J. and Congdon, J. D. (1978). Body shape, reproductive effort, and relative clutch mass in lizards: resolution of a paradox. *Am. Nat.* **112**, 595–607.

Vitt, L. J., Pianka, E. R., Cooper, W. E. Jr. and Schwenk, K. (2003). History and the global ecology of squamate reptiles. *Am. Nat.* **162**, 44–60.

Whiting, M. J. (1999). When to be neighbourly: differential agonistic responses in the lizard *Platysaurus broadleyi*. *Behav. Ecol. Sociobiol.* **46**, 210–14.

Whiting, M. J. (2002). Field experiments on intersexual differences in predation risk and escape behaviour in the lizard *Platysaurus broadleyi*. *Amph.-Rept.* **23**, 119–24.

Whiting, M. J. and Cooper, W. E. Jr. (2003). Tasty figs and tasteless flies: plant chemical discrimination but no prey chemical discrimination in the cordylid lizard *Platysaurus broadleyi*. *Acta Ethologica* **6**, 13–17.

Whiting, M. J. and Greeff, J. M. (1997). Facultative frugivory in the Cape flat lizard, *Platysaurus capensis* (Sauria: Cordylidae). *Copeia* **1997**, 811–18.

Whiting, M. J. and Greeff, J. M. (1999). Use of heterospecific cues by the lizard *Platysaurus broadleyi* for food location. *Behav. Ecol. Sociobiol.* **45**, 420–3.

Whiting, M. J., Nagy, K. A. and Bateman, P. W. (2003). Evolution and maintenance of social status signalling badges: experimental manipulations in lizards. In *Lizard Social Behavior*, ed. S. F. Fox, J. K. McCoy and T. A Baird, pp. 47–82. Baltimore, MD: Johns Hopkins University Press.

Whiting, M. J., Stuart-Fox, D. M., O'Connor, D. *et al.* (2006). Ultraviolet signals ultra-aggression in a lizard. *Anim. Behav.* **72**, 353–63.

Zug, G., Vitt, L. J. and Caldwell, J. P. (2001). *Herpetology: An Introductory Biology of Amphibians and Reptiles*, 2nd edn. San Diego, CA: Academic Press.

14

Interactions between habitat use, behavior, and the trophic niche of lacertid lizards

BIEKE VANHOOYDONCK, ANTHONY HERREL,
AND RAOUL VAN DAMME
Department of Biology, University of Antwerp

Introduction

The existence of evolutionary trade-offs prevents simultaneous optimization of different functions that require opposing biomechanical or physiological adaptations (Stearns, 1992). Consequently, trade-offs likely play an important role in niche partitioning, in that species specialized in exploiting one type of niche (e.g. microhabitat) are expected to be less proficient at exploiting others. For instance, in *Anolis* lizards, a trade-off exists between sprint speed and sure-footedness because long limbs are required to move fast, whereas short limbs aid sure-footedness (Losos and Sinervo, 1989; Losos and Irschick, 1996). Accordingly, species that predominantly move on broad surfaces (i.e. trunk–ground ecomorph) specialize for speed and have long limbs, whereas species living on narrow substrates (i.e. twig ecomorph) are specialized in slower but secure movements.

In a similar fashion, species specializing in different dietary niches may have diverged morphologically because the biomechanical demands on the feeding and/or locomotor apparatus are often not reconcilable within one phenotype. Clearly, the ability of a predator to exploit a certain prey type will depend on the functional characteristics of the prey (e.g. prey distribution, hardness, and escape response) and the performance of the feeding and locomotor system of the predator. For instance, in labrid fishes the amount of force potentially generated by the jaws trades off with the speed of jaw movement because of differences in the four-bar linkage system of the jaws and hyoid (long links aid high force outputs, but rapid movements are realized by short links) (Westneat, 1994, 1995). Consequently, species feeding mainly on hard prey will typically possess long links, whereas species feeding on evasive prey have short links, and thus have faster jaw movements (Westneat, 1994, 1995). Similarly, lizard species eating a lot of small evasive prey benefit from high jaw opening and closing

Lizard Ecology: The Evolutionary Consequences of Foraging Mode, ed. S. M. Reilly, L. D. McBrayer and D. B. Miles. Published by Cambridge University Press. © Cambridge University Press 2007.

velocities, but at the same time appear to be constrained in bite force potential (Meyers and Herrel, 2005).

In addition, trade-offs between locomotor performance traits (e.g. endurance, sprint speed) may also affect an organism's feeding ecology. For instance, species feeding mainly on clumped prey need to spend a lot of their time moving around looking for food, and thus benefit from having a high endurance capacity (Huey and Pianka, 1981; Huey et al., 1984; Garland, 1999). Since endurance and sprint speed may trade off (see, for example, Reidy et al., 2000; Vanhooydonck et al., 2001; Van Damme et al., 2002), species specialized in feeding on clumped prey may not be able to catch fast, evasive prey, thus potentially resulting in a dichotomy in foraging strategy. In general it is thought that active foraging species eat clumped prey and have a high endurance, and that sit-and-wait predators eat evasive prey and excel in sprinting capacities (Huey and Pianka, 1981; Huey et al., 1984). In addition, it has been suggested that sit-and-wait predators have high bite strengths, because evasive prey are often hard (McBrayer, 2004), suggesting that dietary niche is affected by locomotor and feeding performance simultaneously. Similarly, specialists on clumped prey benefit from high stamina, but at the same time need fast jaw movements to capture and process these small prey (Meyers and Herrel, 2005).

Lacertid lizards are an excellent study system to explore the correlates of feeding ecology and behavior because, of all lizard families, they display the greatest interspecific variation in movement patterns and rates (Cooper and Whiting, 1999; Perry, 1999). Moreover, in Kalahari lacertids, the differences in activity level appear to be correlated with differences in energy expenditure (Nagy et al., 1984), locomotor performance (speed and endurance) (Huey et al., 1984), and diet (Huey and Pianka, 1981). Additionally, lacertids exhibit precisely defined patterns of prey selection (Diaz, 1995) and vary greatly in diet (see, for example, Pollo and Pérez-Mellado, 1988; Pérez-Mellado and Corti, 1993; Herrel et al., 2001c; Verwaijen et al., 2002).

Here, we combine data on locomotor performance (i.e. sprint speed, endurance, and climbing speed) (Vanhooydonck et al, 2001; Vanhooydonck and Van Damme, 2001), bite strength (Herrel et al., 2004), and head morphology (Herrel et al., 2004) in different species of lacertid lizard to test the idea that functional trade-offs play an important role in the interspecific partitioning of the dietary niche. Moreover, we explore how functional trade-offs arising through habitat specialization and locomotion may affect diet in lacertid lizards.

We specifically address two questions. First, we investigate how the trade-off between sprint speed and endurance (Vanhooydonck et al., 2001) affects dietary niche in lacertid lizards. Since endurance capacity appears to be positively correlated with the amount of time lizards spend moving (Garland, 1999), and since species that move around frequently (i.e. widely foraging) are expected to eat more clumped and sedentary prey (see, for example, Huey and Pianka 1981), we can expect species that typically feed on sedentary prey to have great endurance capacity. However, because of the speed–endurance trade-off, specializing in stamina may hamper the same predators' ability to catch fast, evasive prey.

Second, we investigate whether a saxicolous lifestyle (i.e. occurring on vertical, rocky surfaces), typical of many lacertid lizard species, constrains their feeding performance and hence restricts their dietary niche. Lacertids occurring on vertical walls and cliffs tend to have a dorsoventrally flattened head and body (Vanhooydonck and Van Damme, 1999). In addition, a flattened head seems advantageous in climbing lacertids as they frequently hide and/or forage in crevices and cracks in between the rocks (Miles, 1994; Vanhooydonck and Van Damme, 1999). Head size (e.g. head height), however, is a determinant of bite force in lizards (Herrel *et al.*, 2001a; McBrayer, 2004), and in lacertids in particular (see, for example, Herrel *et al.*, 1999, 2001c, 2004; Verwaijen *et al.*, 2002). Because bite force and a vertical lifestyle thus seem to pose conflicting demands on the morphology of the head, we can expect a trade-off between both bite force and climbing ability (saxicolous species excel in climbing speed) (Vanhooydonck and Van Damme, 2003). Moreover, the limitations on bite force, due to a flattened head, in saxicolous species might restrict their diet to softer prey.

Methods

In the analyses, we use performance (sprint speed, endurance, climbing speed, and bite force), morphological (snout–vent length and head height), and ecological (diet) data. Sprint speed was measured on an electronic racetrack 2 m long by 0.15 m wide, lined with a cork substrate. Eight pairs of photocells, placed at 0.25 m intervals, recorded the time taken for lizards to cross each successive infrared beam. To measure climbing speed, an electronic racetrack 1 m long by 0.15 m wide was tilted to an angle of 70° and the eight pairs of photocells were positioned at 0.15 m intervals. The racetrack was lined with smooth slate to mimic a rocky surface. We measured endurance as the running time till exhaustion on a treadmill moving at a low and constant speed (0.22 m/s). Lastly, bite force was measured by using an isometric Kistler force transducer. Morphological measurements were made to the nearest 0.01 mm by using digital calipers. All performance and morphological data were published previously; we refer to these studies for a detailed description of the experimental protocol (sprint speed: Bauwens *et al.*, 1995, Vanhooydonck *et al.*, 2001; endurance: Vanhooydonck *et al.*, 2001; climbing speed: Vanhooydonck and Van Damme, 2001; bite force: Herrel *et al.*, 2004; snout–vent length: Vanhooydonck and Van Damme, 2001; Herrel *et al.*, 2004; head height: Herrel *et al.*, 2004). As for dietary information, we used published accounts that contained information on the proportion of each prey type (by number) present in the diet of the species for which morphological and performance data were available (Angelici *et al.*, 1997; Mou and Barbault, 1986; Molina-Borja, 1991; Nouira, 1983; Pérez-Mellado and Corti, 1993; Pollo and Pérez-Mellado, 1988; Roig-Fernandez, 1998; Valido and Nogales, 1994, 2003). If results from different seasons or different years were reported, we used those for which the sample size was the largest. Data for males and females were pooled, as were data obtained from both stomach and fecal pellet samples. In addition, we supplemented the literature data with new data on diet for two species, *L. oxycephala* and *P. melisellensis* (unpubl. data).

We classified each prey item into three categories according to hardness (hard, soft, intermediate) and evasiveness (evasive, sedentary, intermediate). We chose these two characteristics because they are relevant with respect to the measured performance traits, and presumably foraging strategy, of the predator. Our classification is based on the actual forces needed to crush various prey items (Herrel *et al.*, 2001c; Aguirre *et al.*, 2003, pers. obs.) and on the observed escape tactics of invertebrates (see Table 14.1). We subsequently calculated the proportion each category represented in the lizards' diet. If we obtained data from different populations of the same species, we used the average over the different populations in the analyses.

Raw data on the morphology, performance and diet for all species used in this study are given in Table 14.2 and 14.3. Prior to statistical analyses, we logarithmically ($\log_{10}$) transformed all performance traits (sprint speed, endurance, climbing speed, and bite force) and morphological measurements (SVL and head height). Proportions were arcsin transformed (see Sokal and Rohlf, 1995). We used the independent contrast approach (Felsenstein, 1985, 1988) to take into account the evolutionary relatedness of the different species. To calculate independent contrasts between two nodes, the values of the variables in question must be subtracted. The direction of subtraction is entirely arbitrary, and consequently there is ambiguity in assigning signs to contrasts. The ambiguity in assigning signs makes it necessary to force the regression of one contrast against the other through the origin (see Appendix 1 in Garland [1992] for full explanation). Independent contrasts were calculated by using PDTREE (Garland *et al.*, 1999). Information on the topology of the phylogenetic tree was drawn from various sources (Arnold, 1989, 1998; Fu, 2000;

Table 14.1 *Classification of each prey type according to hardness and evasiveness level*

The taxonomic grouping is the same as the one used in the original papers from which the diet data were taken (see text for references). Taxonomic groups are ordered alphabetically.

taxon	hardness	evasiveness
Acari	soft	sedentary
Araneae	soft	sedentary
Blattaria	soft	evasive
Chilopoda	soft	intermediate
Coleoptera	hard	intermediate
Collembola	soft	sedentary
Dermaptera	soft	intermediate
Diplopoda	soft	sedentary
Diptera	soft	evasive
Dyctioptera	soft	evasive/sedentary
Formicidae	hard	intermediate
Gastropoda	hard	sedentary
Heteroptera	soft	intermediate
Homoptera	hard	sedentary
Hymenoptera (except Formicidae)	hard	evasive
Isopoda	intermediate	sedentary
Isoptera	soft	sedentary
Larvae	soft	sedentary
Lepidoptera	soft	evasive
Lithobiomorpha	soft	intermediate
Mantodea	soft	sedentary
Mecoptera	soft	intermediate
Myriapoda	soft	sedentary/intermediate
Neuroptera	soft	intermediate
Odonata	intermediate	evasive
Oligochaeta	soft	sedentary
Opilionida	soft	intermediate
Orthoptera	intermediate	evasive
Phasmida	soft	sedentary
Pseudoscorpionida	soft	sedentary
Scorpionida	soft	sedentary
Solifugae	hard	sedentary
Stylommatophora	hard	sedentary
Thysanura	soft	sedentary
Trichoptera	soft	evasive
Vegetable matter	hard	sedentary
Vertebrate	hard	evasive

Table 14.2 *Means and standard errors of morphological and performance traits*

Means of sprint speed (sprinting, m/s), endurance (s) and climbing speed (climbing, m/s) are given per species; means of snout–vent length (SVL, mm), head height (mm) and bite force (N) are given for each sex separately. Habitat use (hab), i.e. saxicolous (sax) versus terrestrial (ter) lifestyle, is indicated per species (see Vanhooydonck and Van Damme, 2003).

species	hab	morphology SVL male	morphology SVL female	head height male	head height female	performance sprinting	endurance	climbing	bite force male	bite force female
G. galloti	ter	112.6 ± 2.6	92.1 ± 1.6	16.3 ± 0.7	11.6 ± 0.2	1.93 ± 0.13	311.38 ± 28.23	0.34 ± 0.06	107.6 ± 8.3	34.6 ± 3.2
P. algirus	ter	74.8 ± 1.1	63.2 ± 3.8	8.8 ± 0.2	6.8 ± 0.5	2.53 ± 0.15	—	—	11.5 ± 0.6	5.2 ± 1.2
P. hispanicus	ter	43.0 ± 1.0	44.2 ± 0.6	4.8 ± 0.1	4.10 ± 0.2	1.50 ± 0.35	—	—	2.3 ± 0.3	1.0 ± 0.2
L. oxycephala	sax	59.3 ± 1.1	49.7 ± 2.3	5.9 ± 0.1	4.8 ± 0.2	2.02 ± 0.09	109.6 ± 13.40	1.25 ± 0.08	3.9 ± 0.2	1.6 ± 0.2
L. bedriagae	sax	67.9 ± 2.5	63.1 ± 3.7	6.9 ± 0.3	6.2 ± 0.4	1.79 ± 0.15	437.6 ± 38.68	0.99 ± 0.08	7.2 ± 1.0	3.8 ± 0.8
L. vivipara	ter	49.7 ± 0.7	50.4 ± 1.3	5.4 ± 0.1	4.6 ± 0.1	0.87 ± 0.05	370.33 ± 105.34	—	2.1 ± 0.1	1.4 ± 0.1
P. melisellensis	ter	56.7 ± 1.5	53.7 ± 1.3	6.3 ± 0.2	5.1 ± 0.1	1.30 ± 0.12	—	—	4.0 ± 0.3	1.8 ± 0.1
P. hispanica	ter	55.4 ± 1.3	53.8 ± 0.7	5.9 ± 0.2	4.9 ± 0.1	2.03 ± 0.40	—	0.62 ± 0.06	4.3 ± 0.3	2.1 ± 0.1
P. atrata	ter	63.7 ± 0.6	58.4 ± 0.6	7.2 ± 0.2	5.4 ± 0.1	1.09 ± 0.42	—	0.19 ± 0.22	7.8 ± 0.6	2.6 ± 0.1
P. muralis	ter	56.3 ± 1.4	55.6 ± 1.4	6.3 ± 0.2	5.4 ± 0.1	2.14 ± 0.21	184.9 ± 12.39	0.29 ± 0.03	6.5 ± 0.8	3.1 ± 0.3
P. tiliguerta	ter	55.1 ± 1.2	50.1 ± 1.2	6.0 ± 0.2	4.6 ± 0.2	1.55 ± 0.13	194.79 ± 18.91	0.45 ± 0.06	3.6 ± 0.3	1.1 ± 0.1
P. sicula	ter	67.7 ± 1.4	65.3 ± 1.1	7.7 ± 0.2	6.5 ± 0.2	1.67 ± 0.10	266.07 ± 30.65	0.47 ± 0.04	6.6 ± 0.5	3.5 ± 0.3
P. lilfordi	ter	62.1 ± 0.9	60.7 ± 0.8	7.7 ± 0.3	6.5 ± 0.2	2.34 ± 0.11	—	—	10.7 ± 0.3	6.0 ± 0.4
L. bilineata	ter	90.4 ± 1.8	98.2 ± 1.2	11.8 ± 0.2	11.2 ± 0.8	2.68 ± 0.23	276.6 ± 30.18	0.28 ± 0.05	23.3 ± 0.1	14.5 ± 1.1
A. scutellatus	ter	—	—	—	—	2.80 ± 0.13	74.22 ± 7.77	0.14 ± 0.15	—	—
A. pardalis[a]	ter	59.0 ± 0.8	60.69	8.1 ± 0.2	6.69	2.62 ± 0.19	87.83 ± 4.36	0.32 ± 0.08	4.9 ± 0.3	1.88
A. erythrurus	ter	73.0 ± 1.4	—	8.2 ± 0.2	—	3.13 ± 0.14	—	—	8.4 ± 0.4	—
L. longicaudata	ter	90.4 ± 3.1	76.6 ± 3.25	9.5 ± 0.3	7.9 ± 0.2	3.34 ± 0.17	51.71 ± 4.47	1.05 ± 0.19	7.0 ± 1.2	4.6 ± 0.4

[a]Data for one female included in the analyses.

Table 14.3 *Proportion of different prey types present in the diet of the lacertid species included in this study*

Proportions per category according to evasiveness (evasive [evas], intermediate [int e], sedentary [sed]) and hardness (hard, intermediate [int h], soft).

	evasiveness			hardness		
species	evas	int e	sed	hard	int h	soft
G. galloti	0.069	0.262	0.651	0.438	0.487	0.088
P. algirus	0.145	0.345	0.502	0.548	0.040	0.403
P. hispanicus	0.147	0.294	0.534	0.429	0.032	0.519
L. oxycephala	0.009	0.801	0.176	0.835	0.035	0.127
L. vivipara	0.434	0.150	0.415	0.196	0.023	0.781
P. melisellensis	0.111	0.532	0.271	0.569	0.078	0.203
P. hispanica	0.325	0.363	0.309	0.655	0.015	0.328
P. muralis	0.256	0.347	0.387	0.524	0.050	0.415
P. tiliguerta	0.368	0.276	0.333	0.460	0.058	0.460
P. sicula	0.227	0.355	0.365	0.418	0.114	0.415
P. lilfordi	0.025	0.589	0.378	0.949	0.015	0.028
L. bilineata	0.148	0.365	0.371	0.488	0.330	0.066
A. pardalis	0.051	0.532	0.405	0.886	0.038	0.064
A. erythrurus	0.080	0.774	0.144	0.723	0.015	0.260

Harris and Arnold 1999) (Fig. 14.1). Since no information is available on divergence times, we set all branch lengths to unity.

In subsequent statistical analyses, we tested how the speed–endurance trade-off and the (potential) bite force – climbing speed trade-off affect the dietary niche of lacertid lizards. In the former case, we hypothesized that (1) sprint speed is positively correlated with the proportion of evasive prey and/or prey of intermediate evasiveness and (2) endurance is positively correlated with the proportion of sedentary prey taken; in the latter case we hypothesized that (3) bite force and climbing speed are negatively correlated, mediated by differences in head height; (4) bite force is negatively correlated with the proportion of soft prey items present, whereas climbing speed is positively correlated with the proportion of soft prey. Because head shape and bite force are known to be sexually dimorphic in lacertid lizards (see, for example, Olsson and Madsen, 1998; Herrel *et al.*, 1999, 2001c, 2004), all analyses including these traits are done for males and females separately. To test hypotheses (1), (2), and (4), we performed regressions through the origin with one of the performance traits as dependent variable and one of the diet traits as independent variable. To test the trade-off hypothesis (3) we first calculated the residuals of the contrasts in bite force, climbing speed, and head height against the contrasts in SVL. We took

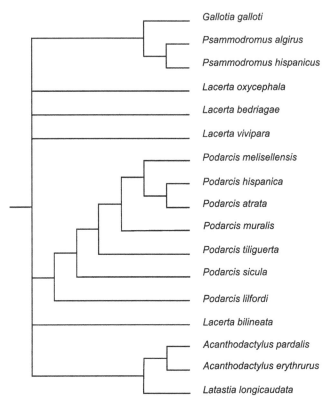

Figure 14.1. Phylogenetic relationships among the lacertid lizard species used in this study. Topology is based on both morphological and molecular data (see text for references); branch lengths are not to scale.

SVL into account in the latter analysis, because we were particularly interested in how shape (i.e. head shape) differences affect performance and whether both performance traits are related irrespective of known interspecific differences in SVL. We subsequently performed regressions through the origin of the residual contrast in bite force against the residual contrast in climbing speed, and of each of those against the residual contrast in head height.

Results and discussion

All bivariate correlations (Spearman rank) between the raw performance and diet data ($\log_{10}$ transformed) and between residual head height, climbing speed, and bite force are given in Appendixes 14.1 and 14.2, respectively. The results from regression analyses on the same variables but taking into account phylogenetic relatedness among the different species, are shown in Appendixes 14.3 and 14.4, respectively. In this chapter, however, we limit

ourselves to discussing the results related to the four specific hypotheses put
forward in the sections above.

Speed–endurance trade-offs and dietary niche

Contrary to our expectations, the contrasts in sprint speed are negatively
correlated with the contrasts in the proportion of evasive prey present in the
diet ($r = -0.54$, $F_{1, 12} = 5.02$, $p = 0.045$). However, there is a tendency for the
contrasts in the proportion of intermediate evasive prey present to be positively correlated with the contrasts in sprint speed ($r = 0.46$, $F_{1, 12} = 3.19$,
$p = 0.099$). The contrasts in endurance, on the other hand, tend to be positively
correlated with the contrasts in the proportion of sedentary prey present in the
diet ($r = 0.69$, $F_{1, 6} = 5.37$, $p = 0.060$) (Fig. 14.2).

Thus, among lacertid lizards, endurance capacity and eating sedentary prey
tend to co-evolve. This is not unexpected as lizards need to move around
frequently to encounter this type of prey, and to do so great stamina is required
(Garland, 1999). Moreover, it further corroborates the idea that sedentary and
clumped prey items are mostly eaten by widely foraging species, which in turn

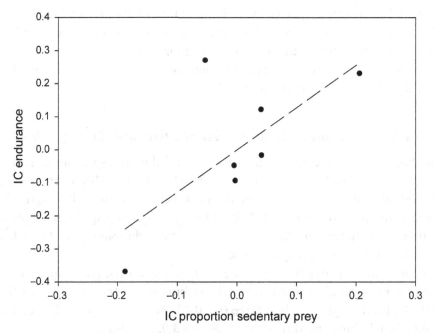

Figure 14.2. Plot of independent contrasts (IC) of endurance against the
independent contrasts of the proportion of sedentary prey present in the
samples. Among lacertids, eating sedentary prey tends to co-evolve with
endurance capacity ($r = 0.69$, $p = 0.060$).

appear to have great stamina (Huey and Pianka, 1981; McBrayer, 2004). Sprint speed, on the other hand, seems to have co-evolved with eating evasive prey but in a way opposite to that predicted: a small increase in sprint speeds results in eating more evasive prey. Clearly, among lacertids, maximal sprint speed does not seem to determine the capacity to capture evasive prey items. Sprint speed, however, does seem relevant for catching prey classified as of "intermediate evasiveness." Possibly, to catch evasive prey, predator accelera- tion capacity is more important than sprint speed *per se*. The prey items we categorized as "evasive" are prey that typically jump or fly (e.g. flies, dragon- flies, grasshoppers) (see Table 14.1). Because no pursuit is possible, at least not by lizards, these prey types can only be captured by very short, quick dashes of movement from a standstill. Prey items classified as of "intermediate evasive- ness", on the other hand, typically escape by running away (e.g. bugs, beetles, ants, centipedes) (see Table 14.1) and thus they may be captured after a prolonged, fast chase. The negative correlation between the proportion of evasive prey and sprint speed in these lizards is in itself suggestive of a trade- off between sprint speed and acceleration capacity. Unfortunately no data on acceleration capacity in lacertids are available to test this hypothesis.

In addition, sprint speed in lacertids might be selected for in other ecological contexts. For instance, species living in open microhabitats excel in sprinting ability, which also appears to be correlated with their predator avoidance strategy (Vanhooydonck and Van Damme, 2003). In turn, open microhabitats might not harbor many evasive prey so that these prey items are not available to the lizard species occurring there.

Saxicoly, sex, and trade-offs between bite force and climbing ability

In males, the residual contrasts in bite force and climbing speed are negatively correlated ($r = -0.64$, $F_{1,9} = 6.21$, $p = 0.034$) (Fig. 14.3a). Moreover, although the residual contrasts in head height are positively correlated with the residual contrasts in bite force ($r = 0.78$, $F_{1,9} = 13.81$, $p = 0.005$) (Fig. 14.3b), they are negatively correlated with the residual contrasts in climbing speed ($r = -0.84$, $F_{1,9} = 20.80$, $p = 0.001$) (Figure 14.3c).

In contrast, females exhibited none of these trade-offs. No significant cor- relations were found in bite force and climbing speed ($r = 0.12$, $F_{1,9} = 0.12$, $p = 0.74$), head height and bite force ($r = 0.40$, $F_{1,9} = 1.73$, $p = 0.22$), or head height and climbing speed ($r = -0.50$, $F_{1,9} = 3.02$, $p = 0.12$).

In male lacertid lizards, bite force and climbing ability trade-offs appear to be because of the differential effect of head height on either performance trait. Whereas males with higher heads are able to generate more bite force, males

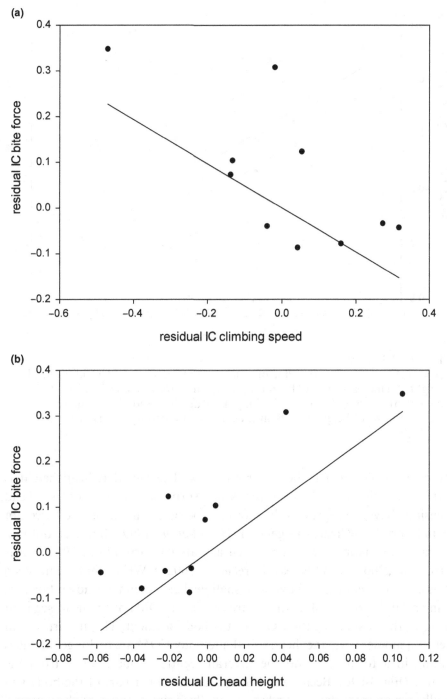

Figure 14.3. Relations between bite force, climbing speed and head height in male lacertid lizards. (a) Plot of residual contrasts in bite force against residual contrasts in climbing speed. Bite force and climbing speed trade-off

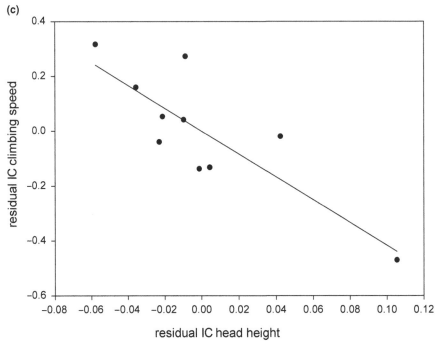

Figure 14.3. (cont.)
($r = -0.64$, $p = 0.030$). (b) Plot of residual bite force against residual head height. An increase in head height results in an increase in bite force ($r = 0.78$, $p = 0.005$). (c) Plot of residual climbing speed against residual head height. An increase in head height results in a decrease in climbing ability ($r = -0.84$, $p = 0.001$).

with lower heads seem to be faster climbers. The fact that head height is positively correlated with bite force is not surprising. Even more so, head dimensions have been previously shown to be determinant of bite force not only in lacertids, or lizards in general (Herrel *et al.*, 1999, 2001a, c), but in a wide range of taxonomic groups, such as bats (Aguirre *et al.*, 2002), birds (Herrel *et al.*, 2005), and turtles (Herrel *et al.*, 2002b). When climbing up steep inclines, on the contrary, having a shallow head could be advantageous. Having a tall head may shift the centre of mass (COM) of the head segment relative to the position of the COM of the body segment. This in turn would generate a moment arm (i.e. head COM and body COM are no longer aligned) resulting in a torque around the (overall) centre of mass away from the substrate (Fig. 14.4c). Because of this torque the upper part of the body will be pulled away from the substrate, causing the animals to topple backwards. The COM of a shallow head, on the contrary, is shifted relative to the COM of the body so that a counterclockwise torque may be generated (Fig. 14.4a).

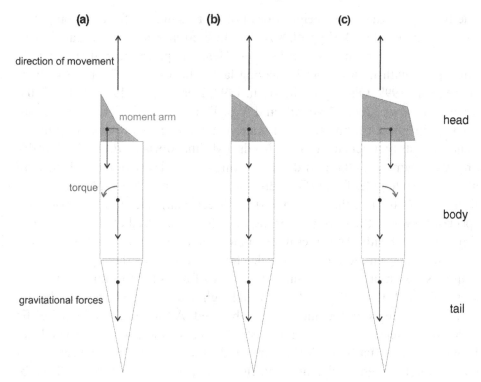

Figure 14.4. Free-body representation of a lizard moving up a vertical incline. Gravitational forces (arrows pointing down) act opposite to the direction of movement (arrows pointing up). The hypothetical COMs of the head, body and tail segment are represented by black circles. Torques are represented by arrows; moment arms by short lines. (a) Shallow heads could potentially result in a counterclockwise torque, pulling the upper part of the body towards the substrate. (b) If the COMs of the segments are in line, no torque is generated (moment arm = 0). (c) Tall heads, on the other hand, may result in a clockwise torque, pulling the upper part of the body away from the substrate.

A counterclockwise torque will pull the upper part of the body towards the substrate, thus preventing climbing animals from falling backwards (Cartmill, 1985; Pounds, 1988; Miles, 1994). Although in some lizard families saxicolous and arboreal species tend to have lower heads and/or bodies, and use a more sprawling gait to lower their stance than their terrestrial sister species (Vanhooydonck and Van Damme, 1999; Herrel *et al.*, 2001b, 2002a; Zaaf *et al.*, 2001; Bickel and Losos, 2002; Spezzano and Jayne, 2004), to our knowledge this is the first study that shows a significant correlation between head shape and climbing performance.

If we consider climbing ability and bite force in female lacertid lizards, we do not find evidence for a trade-off. The lack of a trade-off in females might be due to the fact that head height does not detrimentally affect climbing performance in females. We hypothesize that differential selection on head dimensions exists between males and females. Head shape in lacertids is sexually dimorphic, with males typically having larger heads than females (Molina-Borja *et al.*, 1997; Olsson and Madsen, 1998; Herrel *et al.*, 1999, 2001c, 2004; Verwaijen *et al.*, 2002; Gvozdik and Van Damme, 2003). The larger head, through its effect on bite force, in males is advantageous both during male–male contests (winners have larger heads) (Molina-Borja *et al.*, 1997; Gvozdik and Van Damme, 2003) and during matings, as males with larger heads are able to grasp females faster (Gvozdik and Van Damme, 2003). Male head size thus appears to be under strong sexual (directional) selection. In saxicolous species, however, the selection for high heads is countered by (natural) selection for lower heads. This does not seem to hold true for females. Possibly, the lower heads in females do not impede climbing ability as they are too low to cause a substantial torque around the centre of mass for the animal. Thus, it seems plausible that a threshold for head height may need to be reached before its negative effects on locomotion are observed. Although some interspecific variation in relative head height exists among females, none of their head heights might actually reach this threshold, thus explaining the absence of a correlation between head height and climbing speed. A plot of residual climbing speed against residual head height for both males and females (Fig. 14.5) suggests that the variation in residual head height is lower in females as compared with males and that head height detrimentally affects climbing ability only after a certain threshold has been reached (dotted line on Fig. 14.5).

That the above trade-off is indeed an ecologically relevant one is suggested by the analysis of diet: in males, the contrasts in bite force are negatively correlated with the proportion of soft prey eaten ($r = -0.62$, $F_{1, 12} = 7.51$, $p = 0.018$) (Fig. 14.6). This suggests that an increase in bite force has co-evolved with eating fewer soft prey items. Similarly, the contrasts in female bite force tend to be negatively correlated with the proportion of soft prey present in the sample ($r = -0.53$, $F_{1, 11} = 4.35$, $p = 0.061$). The contrasts in climbing speed, however, are not correlated with the contrasts in the proportion of soft prey ($r = -0.003$, $F_{1, 6} = 0.0001$, $p = 0.99$).

Our *a priori* prediction that fast climbers should eat proportionally more soft-bodied prey, and have lower bite force, is based on optimal foraging theory (McArthur and Pianka, 1966; Schoener, 1971; Arnold, 1993; Emerson *et al.*, 1994; Roughgarden, 1995). Optimal foraging theory posits that animals should balance the costs and benefits of feeding bouts by selecting those prey items that can be efficiently processed and have a high caloric content. Because an increase in bite force reduces the time it takes to subdue prey (Herrel *et al.*, 1999;

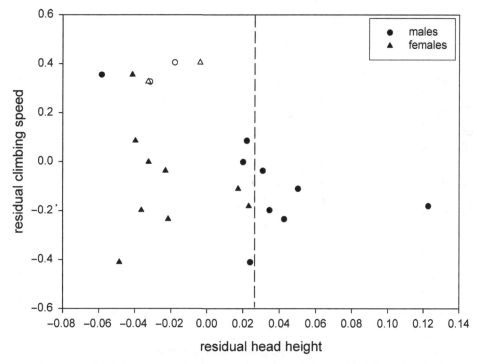

Figure 14.5. Plot of residual climbing speed against residual head height in males (circles) and females (triangles). At a residual head height of about 0.03, climbing speed drops off dramatically (dotted line), corroborating our hypothesis that head height detrimentally affects climbing speed after a certain threshold has been reached. In females, residual head height falls below this threshold. Open symbols refer to saxicolous species.

Verwaijen *et al.*, 2002), tougher prey become energetically more interesting for animals with great bite strength. This idea is corroborated by the fact that bite force and proportion of soft-bodied prey are negatively correlated among lacertids. The increased handling times of tough prey items (at least for species with decreased bite strength), however, do not seem to prevent fast climbers from eating them. Indeed, prey selection experiments in two lacertid species showed that, despite a large difference in bite capacity, both species preferred the largest of the available prey items (Herrel *et al.*, 2001c). Possibly, the cost of an increased handling time does not outweigh the energetic benefit of eating a large and/or tough prey item. In addition, prey availability may differ among micro-habitats (see also above). For instance, when comparing food availability for two sympatric species with a distinct microhabitat use, *L. oxycephala* and *P. melisellensis*, it was found that soft-bodied prey were much more common among the vegetation, and thus available to the latter species, than they were on the stone walls, home to the former species (Verwaijen *et al.*, 2002). Although it

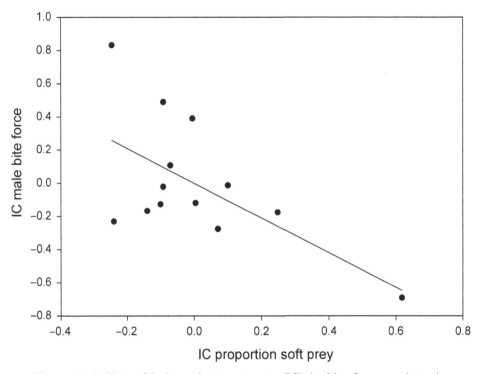

Figure 14.6. Plot of independent contrasts (IC) in bite force against the independent contrasts in proportion of soft prey present in the male samples. A decrease in proportion of soft prey eaten is paralleled by an increase in bite force ($r = -0.62$, $p = 0.018$). Similar results were found when considering females only.

is unclear at the moment how general this pattern of differential prey availability is, it may explain why typical rock climbing species eat less soft-bodied prey.

Conclusions

We found at least some indications that functional trade-offs among performance traits (speed vs. endurance and bite force vs. climbing speed) might be important in the evolution of foraging strategies and the feeding ecology of lacertid lizards. However, detailed data on rates and patterns of movement in a whole suite of species (specifically for the European clade within Lacertidae) are required to further test this hypothesis. In addition, our results clearly show the need for quantification of functionally relevant traits of different prey types (e.g. hardness, escape speed) to directly link these prey-specific performance traits to the predator's performance capacity. Lastly, data on prey availability are largely lacking. However, to be able to interpret dietary differences, for instance among species with distinct microhabitat usage, it is of prime importance to know which prey items are actually available to each predator.

References

Aguirre, L. F., Herrel, A., Van Damme, R. and Matthysen, E. (2002). Ecomorphological analysis of trophic niche partitioning in a tropical savannah bat community. *Proc. R. Soc. Lond.* **B269**, 1271–8.

Aguirre, L. F., Herrel, A., Van Damme, R. and Matthysen E. (2003). The implications of food hardness for diet in bats. *Funct. Ecol.* **17**, 201–12.

Angelici, F. M., Luiselli, L. and Rugiero, L. (1997). Food habits of the green lizard, *Lacerta bilineata*, in central Italy and a reliability test of faecal pellet analysis. *Ital. J. Zool.* **64**, 267–72.

Arnold, E. N. (1989). Towards a phylogeny and biogeography of the Lacertidae: relationships within an old-world family of lizards derived from morphology. *Bull. Brit. Mus. Nat. Hist. (Zool.)* **55**, 209–57.

Arnold, E. N. (1998). Structural niche, limb morphology and locomotion in lacertid lizards (Squamata, Lacertidae); a preliminary survey. *Bull. Brit. Mus. Nat. Hist. (Zool.)* **64**, 63–89.

Arnold, S. J. (1993). Foraging theory and prey size – predator size relations in snakes. In *Snakes: Ecology and Behaviour*, ed. R. A. Seigel and J. J. Collins, pp. 87–112. New York: McGraw-Hill.

Bauwens, D., Garland, T. Jr., Castilla, A. M. and Van Damme, R. (1995). Evolution of sprint speed in lacertid lizards: morphological, physiological, and behavioural covariation. *Evolution* **49**, 848–63.

Bickel, R. and Losos, J. B. (2002). Patterns of morphological variation and correlates of habitat use in Chameleons. *Biol. J. Linn. Soc.* **76**, 91–103.

Cartmill, M. (1985). Climbing. In *Functional Vertebrate Morphology*, ed. M. Hildebrand, D. M. Bramble, K. F. Liem and D. B. Wake, pp. 73–88. Cambridge, MA: The Belknap Press.

Cooper, W. E. Jr. and Whiting, M. J. (1999). Foraging modes in lacertid lizards from southern Africa. *Amph.-Rept.* **20**, 299–311.

Diaz, J. A. (1995). Prey selection by lacertid lizards: a short review. *Herpetol. J.* **5**, 245–51.

Emerson, S. B., Greene, H. W. and Charnov, E. L. (1994). Allometric aspects of predator-prey interactions. In *Ecological Morphology*, ed. P. C. Wainwright and S. M. Reilly, pp. 123–39. Chicago, IL: University of Chicago Press.

Felsenstein, J. (1985). Phylogenies and the comparative method. *Am. Nat.* **125**, 1–15.

Felsenstein, J. (1988). Phylogenies and quantitative characters. *Ann. Rev. Ecol. Syst.* **19**, 445–71.

Fu, J. (2000). Toward the phylogeny of the family Lacertidae – Why 4708 base pairs of mtDNA sequences cannot draw the picture. *Biol. J. Linn. Soc.* **71**, 203–17.

Garland, T. Jr. (1992). Procedures for the analysis of comparative data using phylogenetically independent contrasts. *Syst. Biol.* **41**, 18–32.

Garland, T. Jr. (1999). Laboratory endurance capacity predicts variation in field locomotor behaviour among lizard species. *Anim. Behav.* **58**, 77–83.

Garland, T. Jr., Midford, P. E. and Ives, A. R. (1999). An introduction to phylogenetically based statistical methods, with a new method for confidence intervals on ancestral states. *Am. Zool.* **39**, 374–88.

Gvozdik, L. and Van Damme, R. (2003). Evolutionary maintenance of sexual dimorphism in head size in the lizard *Zootoca vivipara*: a test of two hypotheses. *J. Zool. Lond.* **259**, 7–13.

Harris, D. J. and Arnold, E. N. (1999). Relationships of wall lizards, *Podarcis* (Reptilia: Lacertidae) based on mitochondrial DNA sequences. *Copeia* **1999**, 749–54.

Herrel, A., Spithoven, L., Van Damme, R. and De Vree, R. (1999). Sexual dimorphism of head size in *Gallotia galloti*; testing the niche divergence hypothesis by functional analyses. *Funct. Ecol.* **13**, 289–97.

Herrel, A., De Grauw, E. and Lemos-Espinal, J. A. (2001a). Head shape and bite performance in xenosaurid lizards. *J. Exp. Zool.* **290**, 101–7.

Herrel, A., Meyers, J. J. and Vanhooydonck, B. (2001b). Correlation between habitat use and body shape in a phrynosomatid lizard (*Urosaurus ornatus*): a population-level analysis. *Biol. J. Linn. Soc.* **74**, 305–14.

Herrel, A., Van Damme, R., Vanhooydonck, B. and De Vree F. (2001c). The implications of bite performance for diet in two species of lacertid lizards. *Can. J. Zool.* **79**, 662–70.

Herrel, A., Meyers, J. J. and Vanhooydonck, B. (2002a). Relations between microhabitat use and limb shape in phrynosomatid lizards. *Biol. J. Linn. Soc.* **77**, 149–63.

Herrel, A., O'Reilly, J. C. and Richmond, A. M. (2002b). Evolution of bite performance in turtles. *J. Evol. Biol.* **15**, 1083–94.

Herrel, A., Vanhooydonck, B. and Van Damme, R. (2004). Omnivory in lacertid lizards: adaptive evolution or constraint? *J. Evol. Biol.* **17**, 974–84.

Herrel, A., Podos, J., Huber, S. K. and Hendry, A. P. (2005). Bite performance and morphology in a population of Darwin's finches: implications for the evolution of beak shape. *Funct. Ecol.* **19**, 43–8.

Huey, R. B. and Pianka, E. R. (1981). Ecological consequences of foraging mode. *Ecology* **62**, 991–9.

Huey, R. B., Bennett, A. F., John-Alder, H. and Nagy, K. A. (1984). Locomotor capacity and foraging behaviour of Kalahari lacertid lizards. *Anim. Behav.* **32**, 41–50.

Losos, J. B. and Irschick, D. J. (1996). The effects of perch diameter on the escape behaviour of Anolis lizards: laboratory based predictions and field tests. *Anim. Behav.* **51**, 593–602.

Losos, J. B. and Sinervo, B. (1989). The effects of morphology and perch diameter on sprint performance of *Anolis* lizards. *J. Exp. Biol.* **245**, 23–30.

McArthur, R. H. and Pianka, E. R. (1966). On the optimal use of a patchy environment. *Am. Nat.* **100**, 603–9.

McBrayer, L. (2004). The relationship between skull morphology, biting performance and foraging mode in Kalahari lacertid lizards. *Zool. J. Linn. Soc.* **140**, 403–16.

Meyers, J. J. and Herrel, A. (2005). Prey capture kinematics of ant-eating lizards. *J. Exp. Biol.* **208**, 113–27.

Miles, D. B. (1994). Covariation between morphology and locomotory performance in Sceloporine lizards. In *Lizard Ecology: Historical and Experimental Perspectives*, ed. L. J. Vitt and E. R. Pianka, pp. 207–35. Princeton, NJ: Princeton University Press.

Molina-Borja, M. (1991). Notes on alimentary habits and spatial-temporal distribution of eating behaviour patterns in a natural population of lizards (*Gallotia galloti*). *Vieraea* **20**, 1–9.

Molina-Borja, M. Padron-Fumero, M., and Alfonso-Martin, M. T. (1997). Intrapopulation variability in morphology, coloration, and body size in two races of the lacertid lizard, *Gallotia galloti*. *J. Herpetol.* **31**, 499–507.

Mou, Y.-P. and Barbault, R. (1986). Régime alimentaire d' une population de lézard des murailles, *Podarcis muralis* (Laurent, 1768) dans le Sud-Ouest de la France. *Amph.-Rept.* **7**, 171–80.

Nagy, K. A., Huey, R. B. and Bennett, A. F. (1984). Field energetics and foraging mode of Kalahari lacertid lizards. *Ecology* **65**, 588–96.

Nouira, S. (1983). Partage des resources alimentaires entre deux Lacertidae sympatriques des iles Kerkennah (Tunisie): *Acanthodactylus pardalis* et *Eremias olivieri*. *Soc. Zool. Fr.* **1983**, 477–83.

Olsson, M. and Madsen, T. (1998). Sexual selection and sperm competition in reptiles. In *Sperm Competition and Sexual Selection*, ed. T. R. Birkhead and A. P. Moller, pp. 503–610. London: Academic Press.

Pérez-Mellado, V. and Corti, C. (1993). Dietary adaptations and herbivory in lacertid lizards of the genus *Podarcis* from western Mediterranean islands (Reptilia-Sauria). *Bonn. Zool. Beitr.* **44**, 193–220.

Perry, G. (1999). The evolution of search modes: ecological versus phylogenetic perspectives. *Am. Nat.* **153**, 98–109.

Pollo, C. J. and Pérez-Mellado, V. (1988). Trophic ecology of a taxocenosis of Mediterranean Lacertidae. *Ecolog. Mediterranea* **14**, 131–46.

Pounds, J. A. (1988). Ecomorphology, locomotion, and microhabitat structure: patterns in a tropical mainland *Anolis* community. *Ecol. Monogr.* **58**, 299–320.

Reidy, S. P., Kerr, R. and Nelson, J. A. (2000). Aerobic and anaerobic swimming performance of individual Atlantic cod. *J. Exp. Biol.* **203**, 347–57.

Roig-Fernandez, J. M. (1998). Ecologia trófica de una población pirenaica de lagartija de turbera *Zootoca vivipara* (Jacquin, 1787). Tesis de licenciatura, University of Barcelona, Spain.

Roughgarden, J. (1995). *Anolis Lizards of the Caribbean*. New York: Oxford University Press.

Schoener, T. W. (1971). Theory of feeding strategies. *Ann. Rev. Ecol. Syst.* **2**, 369–404.

Sokal, R. R. and Rohlf, F. J. (1995). *Biometry. The Principles and Practice of Statistics in Biological Research*, 3rd edn. New York: W. H. Freeman and Company.

Spezzano, L. C. and Jayne, B. C. (2004). The effects of surface diameter and incline of the hindlimb kinematics of an arboreal lizard (*Anolis sagrei*). *J. Exp. Biol.* **207**, 2115–31.

Stearns, S. C. (1992). *The Evolution of Life Histories*. New York: Oxford University Press.

Valido, A. and Nogales, M. (1994). Frugivory and seed dispersal by the lizard *Gallotia galloti* (Lacertidae) in a xeric habitat of the Canary Islands. *Oikos* **70**, 403–11.

Valido, A. and Nogales, M. (2003). Digestive ecology of two omnivorous Canarian lizard species (*Gallotia*, Lacertidae). *Amph.-Rept.* **24**, 331–44.

Van Damme, R., Wilson, R. S., Vanhooydonck, B. and Aerts, P. (2002). Performance constraints in decathletes. *Nature* **415**, 755–6.

Vanhooydonck, B. and Van Damme, R. (1999). Evolutionary relationships between body shape and habitat use in lacertid lizards. *Evol. Ecol. Res.* **1**, 785–805.

Vanhooydonck, B. and Van Damme, R. (2001). Evolutionary trade-offs in locomotor capacities in lacertid lizards: are splendid sprinters clumsy climbers? *J. Evol. Biol.* **14**, 48–54.

Vanhooydonck, B. and Van Damme, R. (2003). Relationships between locomotor performance, microhabitat use and antipredator behaviour in lacertid lizards. *Funct. Ecol.* **17**, 160–9.

Vanhooydonck, B. Van Damme, R. and Aerts, P. (2001). Speed and stamina trade-off in lacertid lizards. *Evolution* **55**, 1040–8.

Verwaijen, D., Van Damme, R. and Herrel, A. (2002). Relationships between head size, bite force, prey handling efficiency and diet in two sympatric lacertid lizards. *Funct. Ecol.* **16**, 842–50.
Westneat, M. W. (1994). Transmission forces and velocity in the feeding mechanisms of labrid fishes (Teleostei, Perciformes). *Zoomorphology* **114**, 103–18.
Westneat, M. W. (1995). Feeding, function, and phylogeny: analysis of historical biomechanics in labrid fishes using comparative methods. *Syst. Biol.* **44**, 361–83.
Zaaf A., Van Damme, R., Herrel, A. and Aerts, P. (2001). Spatio-temporal gait characteristics of level and vertical locomotion in a ground-dwelling and a climbing gecko. *J. Exp. Biol.* **204**, 1233–46.

Appendix 14.1 *Non-parametric bivariate correlations (Spearman Rank) between the performance and diet variables*

Correlations between bite force and diet are given for the sexes separately. Significant correlations (at the $p = 0.05$ level) are shown in bold.

		diet					
		evasiveness			hardness		
performance		evas	int e	sed	hard	int h	soft
sprinting	*r*	−0.39	**0.56**	−0.15	**0.60**	−0.17	0.29
	p	0.16	**0.04**	0.62	**0.02**	0.56	0.49
	n	14	**14**	14	**14**	14	14
endurance	*r*	0.52	**−0.76**	0.48	**−0.88**	0.31	0.36
	p	0.18	**0.03**	0.23	**0.004**	0.46	0.39
	n	8	**8**	8	**8**	8	8
climbing	*r*	−0.02	0.19	**−0.74**	0.07	−0.52	0.29
	p	0.96	0.65	**0.04**	0.87	0.18	0.49
	n	8	8	**8**	8	8	8
bite force, males	*r*	−0.30	0.17	0.16	0.20	0.35	**−0.60**
	p	0.19	0.55	0.58	0.48	0.22	**0.02**
	n	14	14	14	14	14	**14**
bite force, females	*r*	−0.28	0.14	0.18	0.15	0.43	**−0.60**
	p	0.36	0.64	0.57	0.62	0.14	**0.03**
	n	13	14	13	13	13	**13**

Appendix 14.2 *Non-parametric bivariate correlations (Spearman Rank) between residual head height, residual climbing speed, and residual bite force for males and females separately*

Significant correlations (at the 0.05 level) are shown in bold.

Males		residual climbing	residual bite force
residual head height	r	**−0.75**	**0.67**
	p	**0.008**	**0.02**
	n	**11**	**11**
residual climbing	r		**−0.71**
	p		**0.02**
	n		**11**
Females		residual climbing	residual bite force
residual head height	r	0.17	0.37
	p	0.61	0.26
	n	11	11
residual climbing	r		0.06
	p		0.87
	n		11

Appendix 14.3 *Regressions through the origin on the independent contrasts of the performance variables against the independent contrasts of the diet variables*

Regressions of the independent contrasts of bite force against the independent contrasts in the diet variables are given for each sex separately. Significant relationships at the $p = 0.05$ level are shown in bold; significant relations at the $p = 0.10$ level are shown in italic.

performance		evasiveness			hardness		
		evas	int e	sed	hard	int h	soft
sprinting	*r*	**−0.54**	*0.46*	−0.26	**0.58**	0.20	**−0.72**
	p	**0.05**	*0.10*	0.38	**0.03**	0.50	**0.004**
	n	**13**	*13*	13	**13**	13	**13**
endurance	*r*	*0.67*	**−0.85**	*0.69*	**−0.98**	0.53	0.54
	p	*0.07*	**0.008**	*0.06*	**<0.0001**	0.18	0.16
	n	7	**7**	7	**7**	7	7
climbing	*r*	−0.33	**0.79**	*−0.69*	0.53	−0.50	−0.003
	p	0.43	**0.02**	*0.06*	0.18	0.21	0.99
	n	7	**7**	7	7	7	7
bite force, males	*r*	−0.36	−0.001	0.30	0.08	**0.83**	**−0.62**
	p	0.21	0.99	0.30	0.79	**<0.0001**	**0.02**
	n	14	14	14	14	**14**	14
bite force, females	*r*	−0.27	−0.12	0.46	−0.02	**0.83**	*−0.53*
	p	0.37	0.69	0.11	0.96	**<0.0001**	*0.06*
	n	13	13	13	13	**13**	*13*

Appendix 14.4 *Regressions through the origin on the residual independent contrasts of head height, climbing speed and bite force for males and females separately*

Significant relations (at the $p = 0.05$ level) are shown in bold.

Males		residual climbing	residual bite force
residual head height	*r*	**−0.84**	**0.78**
	p	**0.001**	**0.005**
	n	**10**	**10**
residual climbing	*r*		**−0.64**
	p		**0.03**
	n		**10**
Females		residual climbing	residual bite force
residual head height	*r*	−0.50	0.40
	p	0.12	0.22
	n	10	10
residual climbing	*r*		0.12
	p		0.74
	n		10

15

Food acquisition modes and habitat use in lizards: questions from an integrative perspective

ROGER A. ANDERSON

Department of Biology, Western Washington University

Introduction

The four basic tasks, EPM, FAM, and habitat

One useful theoretical focus in evolutionary ecology is that an animal has four basic, autecological tasks: (1) find, acquire and utilize food; (2) avoid, evade, and deter predators; (3) cope with abiotic stresses and avoid abiotic extremes; and (4) acquire mates and reproduce. An integrative understanding would require knowing the relative influence of each of these tasks on the ecology of an individual, on the population, and on the evolution of higher taxa. In addition, it would be important to identify and understand how behavioral traits (ethotypes), physiological traits (physiotypes), and morphological traits (morphotypes) of animals are adapted to each of the four basic autecological tasks (Fig. 15.1). I refer to the sum of these traits as the EPM (the ethophysiomorph or ethophysiomorphotype). Among the four basic autecological tasks, food acquisition and utilization may be the primary, albeit sometimes indirect, cause for salient features of ethotypes, physiotypes, and morphotypes (and thus EPMs) of lizards and other animals (Anderson and Karasov, 1988).

The classic, general vertebrate mode of food acquisition as mobile, ectothermic predators on invertebrates continues to dominate in lizards (basal level in Fig. 15.1), and many features of lizards appear to be related to the basic autecological task of food acquisition (Pianka and Vitt, 2003). The set of physiological, behavioral, and morphological characteristics (the EPM) that are integrally involved in the search, detection, capture, and eating of food, may be considered an *adaptive syndrome* (Eckhardt, 1979) that I refer to as the "food acquisition mode," FAM. An *adaptive syndrome* is a coordinated set of characteristics (adaptive traits) associated with an issue of overriding importance (a core adaptation) to an organism (Eckhardt, 1979).

Lizard Ecology: The Evolutionary Consequences of Foraging Mode, ed. S. M. Reilly, L. D. McBrayer and D. B. Miles. Published by Cambridge University Press. © Cambridge University Press 2007.

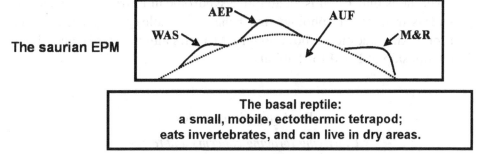

Figure 15.1. Placing the features of adaptedness of a lizard in perspective. The EPM, or ethophysiomorph, is the observable set of traits in extant taxa. The EPM is represented as an amorphous mound, because it is an integrated set of adaptations, comprising adaptedness to the four basic autecological challenges or tasks. The shape of the mound and relative sizes of the related features will depend upon phylogenetic history and the food acquisition mode under study. AUF (acquire and utilize food) is shown as the largest and primary task to which the lizard is adapted, whereas features of adaptedness directly related to each of the other three tasks are consequential, co-adapted features, being pulled or carried along as part of the EPM. The other tasks are WAS (withstand abiotic stresses), AEP (avoid and evade predators) and M&R (find mates and reproduce). See text for further discussion.

The adaptive syndrome of FAM is complex adaptive traits that dominate the EPM; hence in Figure 15.1 food acquisition is the larger and more basal area relative to the other three autecological tasks (Fig. 15.1: AUF vs. WAS, AEP, M&R). FAM is more complex than a "behavioral syndrome" in animals (Sih *et al.*, 2004) as it may incorporate sets of "evolutionarily stable configurations" (Schwenk and Wagner, 2001). The hypothesis that FAM is an adaptive syndrome has a two-part ecological assumption: the principal adaptations of food acquisition are to (1) the prey types, (2) the habitat in which those prey are sought and captured including (a) the substratum, (b) the nanohabitat (including the abiotic challenges of the medium, e.g. air, water, soil) and (c) the microhabitat (Anderson, 1993). What is meant by prey types, substratum, nanohabitat, and microhabitat will be explained below.

Ecomorphological theory assumes that the effect of an EPM's performance depends on the "habitat" in which the EPM is performing (Garland and Losos, 1994; Koehl, 1996; Irschick and Losos, 1999; Irschick and Garland, 2001). Early theoretical and empirical considerations of "foraging modes" in lizards noted (Pianka, 1966) or emphasized which types of prey were sensed, perceived, and pursued as well as how much the predator lizard moved (Regal, 1978, 1983; Anderson and Karasov, 1981; Huey and Pianka, 1981).

The hypothesis that FAM is an adaptive syndrome in lizards, then, incorporates ideas from ecomorphology and behavioral ecology. This chapter will focus on asking questions about how FAM varies among lizards, and about the relationship of FAM to "habitat."

Food acquisition modes

Clarifying foraging concepts and terms

I developed the concept of FAM (Anderson, 1986, 1993; Anderson and Karasov, 1988) rather than use the term "foraging mode," for several related reasons. First, FAM focuses attention not necessarily on the extremes of "foraging modes" (see below), but on the features of adaptedness (*sensu* Endler, 1986) related to food acquisition and use within each individual, population, species, and genus. In a macroevolutionary context, each genus (or clade) is expected to have a unique constellation of EPMs that differ from those of other genera and that is more a consequence of "food acquisition and use" than of the other four basic tasks (this issue will be revisited near the chapter's end). Moreover, if the FAM is identifiable at the genus level, then it is likely that most congeners use largely similar sets of (1) the prey types, (2) substrata, and (3) microhabitats.

One difficulty with the concept of "foraging mode" is that it has had an historical inertia of a narrow focus on patterns of movement (see, for example, Perry, 1999; this volume, Chapter 1). FAM subsumes the narrower concept that "foraging mode" has become, and from which other researchers are trying to broaden. Hence, FAM is a more explicitly inclusive phenomenon and will be used herein rather than "foraging mode" for this reason.

Another problem with the general use of the term "foraging mode" has been the terms used for one of the predation modes: "sit-and-wait ambusher" or "ambush forager," or "sit-and-wait forager." The term "sit-and-wait," which is often used in association with the term "ambusher," is redundant and should be eliminated. Moreover, the terms "sit-and-wait forager" or "*ambush forager*" can be considered oxymoronic if one uses a common word dictionary definition for *forage*: "to wander in search of food." The sedentary, visual search by an ambush predator does not fit that definition of foraging. Instead, we should reserve the term "forager" for those animals that are moving about and directly seeking food, and to use the terms "*ambush predator*" and "*ambusher*" instead of "ambush forager." Hence, using the terms *ambush predation, foraging predation*, and *herbivory* less ambiguously distinguishes among the major categories of FAM in lizards.

Herbivory and omnivory

Lizards are often cited as exemplars of contrasting "foraging modes" (= FAMs) (Huey and Pianka, 1981; Anderson and Karasov, 1981, 1988; Anderson, 1993). The three most contrasting FAMs are ambush predators, intensive foraging predators, and herbivores. Although herbivorous (folivorous) lizards are exemplified by the iguanines (see, for example, Christian *et al.*, 1984; Mautz and Nagy, 1987; Durtsche, 2000), omnivorous lizards (most eat arthropods, fruits, and flowers) can be found in most lizard families (Cooper and Vitt, 2002). Thus, individual lizards have the flexibility in their food acquisition behavior to make use of profitable feeding opportunities (Greeff and Whiting, 2000); this flexibility will be particularly important if one food type becomes rare or cannot meet energetic needs (Mautz and Nagy, 1987; Durtsche, 2000).

An ambush predator lizard moving about, ostensibly seeking an ambushing site, then finding one and stopping there, seems to be exhibiting essentially similar behavior to when an omnivore or herbivore moves and visually searches for flowers, leaves, etc., to eat, or when an omnivore simply encounters flowers or fruits and takes the opportunity to eat them (Janzen and Brodie, 1995). The apparent FAMs used by omnivorous lizards when they obtain animal prey, however, usually are either distinctly ambushing or distinctly foraging (Pianka, 1986; Cooper, 1994). Hence, the behavioral flexibility in individual lizards that adds the more easily digestible plant foods to their diet may be a small behavioral variation and not a large, sudden "switch" in behavior (and associated physiotype and morphotype), as may be the change in FAM from ambusher to intensive forager.

Characterizing food acquisition in ambushers

Ambushing lizards (many iguanians are the archetypes) (Vitt *et al.*, 2003b; Vitt and Pianka, this volume, Chapter 5) are relatively sedentary predators. Most ambushers visually detect prey that is moving or otherwise conspicuous (Huey and Pianka, 1981). It is not uncommon to see an ambusher perching at a good vantage point to visually search for prey. When prey of an ambusher walk or land within range, the type of approach by the ambusher toward prey may be pursuit-and-capture patterns known as (1) "approach–pause–strike" (Moermond, 1981), (2) "jump–strike" (Moermond 1981), (3) "stalk–strike" (Moermond, 1981; Rose, 2004), or (4) "stationary–strike" (Moermond, 1981; Wainwright *et al.*, 1991; Munger, 1984). The type of approach to capture prey, as well as the types of movement to increase the likelihood of encounter with prey, may depend not only on the characteristics of the ambusher, but also on the relative spatiotemporal patterns of availability among prey of differing food value, the evasiveness of the prey, and the substratum–nanohabitat–microhabitat (defined below) used by both predator and prey.

Occasionally the ambusher moves to a new position from which to search (Huey and Pianka, 1981). Some ambushers have a relatively long period of apparent visual search, and thus could be called long-wait ambushers (FAM type D) (Table 15.1; Fig. 15.2); an example may be *Crotaphytus collaris* (Crotaphytidae) (Cooper *et al.*, 2001), which perches on boulders and captures lizards and larger insects (rarer than smaller prey) on boulders and ground. Other ambushers have a relatively short period of apparent visual search (Table 15.1; Fig. 15.2) before they move to a new search position; thus they could be called short-wait ambushers (Anderson, 1993). Examples of short-wait ambushers may be *Tropidurus azureus* (Tropiduridae) (Ellinger *et al.*, 2001) and, at times, *Gambelia wislizenii* (Crotaphytidae) (Pietruszka, 1986; Steffen, 2002; Rose, 2004). When *G. wislizenii* is capturing grasshoppers or cicadas on the low perimeters of shrubs, it makes relatively brief stops to visually search those shrubs, then moves on to the next shrub. In contrast, if lizards are common or these large insects are rare, then *G. wislizenii* is more of a long-wait ambusher of lizards (Rose, 2004).

Problems characterizing ambushers and gleaners

The short-wait (frequently moving) and long-wait (infrequently moving) designations for these visually searching ambushers are meant to be relative, non-dichotomous, qualitative comparisons within and among individual ambushers. A long-wait ambushing lizard is reminiscent of a sallying bird (Robinson and Holmes, 1982), whereas a short-wait ambusher has some similarities to a foliage-gleaning bird, in that it has a repeated series of brief moves and brief stops-and-visual-searches for prey in a nearby small volume (Eckhardt, 1979). The ambusher, however, is ostensibly waiting for potentially evasive prey to reveal themselves by movement. In contrast, the gleaner is searching for cryptic prey that may be small and may not be moving, and hence may be easy to capture. If the principal prey are cryptic and slow-moving and the prey are seen as they move, however, then the gleaner and short-wait ambusher may have similar rates of movement, similar means of detecting prey, and similar prey capture methods (Table 15.1; Fig. 15.2).

Some biologists would refer to both gleaners and ambushers as *saltatory foragers* or *pause–travel foragers* involving "move–pause–move" foraging (O'Brien *et al.*, 1990). Several problems arise from using these pause–travel designations. One is simply identifying "non-movement," or "pausing." Another is the conflation of functions for "non-movement." Is it stopping

Table 15.1 *Basic food acquisition modes (FAM) and their ecological characteristics*

FAM type	FAM name	mesohabitats microhabitats	prey detection	prey types	search movements
A	wide, intensive forager	open prevalent smaller shrub-patches	chemoreception > vision	hidden, inactive	nearly continuous, faster between patches, slower in patches, higher speeds overall
B	short-wait ambusher	open prevalent smaller shrub patches	vision	small, mobile evaders	moderate distances and speeds between "perches;" overall high-to-moderate speeds
C	proximal forager or proximal, intensive forager	cover prevalent larger shrub-patches	vision ≥ chemoreception	hidden, inactive	nearly continuous, faster between patches, slower in patches, moderate speeds overall
D	long-wait ambusher	open prevalent near patch edges	vision	larger, mobile evaders	longer distances, faster between "perches," moderate speeds overall
E	short-wait ambusher or gleaner	vegetation prevalent densely branched	acute vision	small, low to high mobility	short distances, slow to moderate speeds between "perches"
F	secretive, proximal forager	open ≥ cover smaller patches	vision > chemoreception	larger, mobile	very short distances, slow in patches, moderate between short distances, slow to moderate
G	cryptic, long-wait ambusher	vegetation prevalent plant perimeter or visibly exposed	vision	larger, mobile evaders	very short distances, slow in patches, moderate between short distances, slow to moderate
H	cryptic, mobile ambusher	vegetation prevalent densely branched	acute vision avoiders	small, slow, short moves	very slow, nearly continuous, short pauses

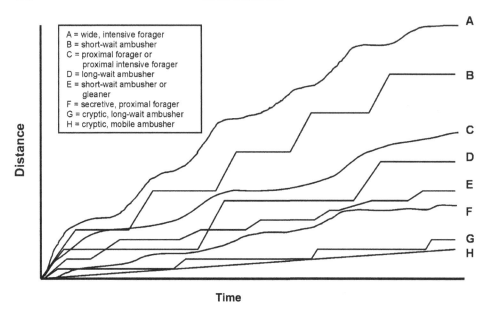

Distance

A = wide, intensive forager
B = short-wait ambusher
C = proximal forager or
      proximal intensive forager
D = long-wait ambusher
E = short-wait ambusher or
      gleaner
F = secretive, proximal forager
G = cryptic, long-wait ambusher
H = cryptic, mobile ambusher

Time

Figure 15.2. Schematic representations of patterns of movements in lizards with different food acquisition modes (FAM). Each line and corresponding letter represents a FAM. FAMs (described in Table 15.1) are defined in the context of habitat, and include methods of detecting prey and types of prey captured. Food acquisition is assumed to be the primary activity (no thermoregulation and no mate-seeking). Note that movement frequencies and distances of ambushers, B, D, E, and G are depicted as interspersed among the movement patterns of foragers A, C, F, and H. Hence, the movement patterns among lizards may be less differentiated and more complex than just a continuum. Moreover, these apparently subtle movement variations permit one to envisage a flexible shift in FAM for a lizard, albeit limited by intrinsic EPM constraints, prey availability, and the challenges of the four basic tasks in the context of habitat patterns.

to: (1) dig for prey *(pursuit)*, or (2) visually *search* for prey, or (3) look for predators? Another problem is assigning function to "moving." Is it "changing search locations (patch or perch)" or is it "searching for prey?"

Also problematic is determining when the lizard actually detected the prey item. Is it just before, or just after the lizard begins "not moving," or "moving?" Attack-based indices – based on whether the prey was pursued from a standstill or whether the lizard was on the move when it initiated attack – may help distinguish ambush predators from foraging predators (Cooper *et al.*, 1999). These indices, however, do not obviate the several aforementioned problems, particularly for cryptic and secretive lizards, which tend to move extremely slowly. Comparing the abilities of ambushers and gleaners (and other foragers) to visually identify non-moving prey (Day *et al.*, 1999) and to identify hidden prey by chemoreception (Simon, 1983; Cooper, 1994) may help reduce ambiguity of prey detection, and thereby better distinguish among FAMs in lizards.

Variation among ambushers in body form, food acquisition behavior, and habitat Although many ambushers have short, broad snouts that appear to function well for engulfing mobile insects, these lizards do vary considerably in (1) food acquisition behavior, (2) "habitat use," (3) body form, and perhaps (4) their locomotory capacities. One familiar extreme form of ambusher (probably FAM type D more often than B) (Table 15.1; Fig. 15.2) is *Callisaurus draconoides* (Phrynosomatidae), a fast, agile, ground-running insectivore that captures its mobile prey along perimeters of perennial plants and out in the open in desert scrub of the southwestern USA (Pianka and Parker, 1972; Smith *et al.*, 1987). *Callisaurus draconoides* has a moderately gracile body and long limbs (long distal elements of hindlimb) relative to intergirdle length, and is an agile bipedal sprinter (Pianka, 1986; Bonine and Garland, 1999); hence it is presumed to be an accomplished evader of predators, as would be expected of a lizard frequently seen in the open.

Arboreal and saxicolous lizards are varied in EPM and in FAM (types E, G, H in Table 15.1 and Fig. 15.2). Another well-known ambusher is *Sceloporus magister*; it is a mid-sized, robust-bodied lizard that clings and perches on desert tree trunks, large limbs, and sometimes boulders. It is likely to be FAM type E; it pursues arthropod prey on trunks and large limbs, and on the nearby ground (Vitt *et al.*, 1981). There are other common phrynosomatids in the genera *Sceloporus, Urosaurus,* and *Uta* (variations of FAM type E) that are small, agile climbers-and-perchers on trees, shrubs and rocks, and are often on ground under and very near shrubs in xeric habitats of the western USA. *Urosaurus* are more gracile than the other two genera, and are found more on branches (Miles, 1994a,b; James and M'Closkey, 2002), although there are saxicolous demes of *U. ornatus* with unique body forms (Herrel *et al.*, 2001). *Urosaurus* seem to approach the body form of the arboreal specialists.

Many lizards are arboreal specialists, such as species in the genus *Anolis* (FAM types E and G, a few may be like type H) (Table 15.1; Fig. 15.2). For example, *Anolis carolinensis* (Polychrotidae) is a cryptic yet agile climber-and-leaper with adhesive toepads. It may vary between FAM types E and G; as a user of trunk and crown microhabitats, it commonly ambushes arthropods from its perch on branches; it can leap from branch to branch and plant to plant (Spezzano and Jayne, 2004). Nocturnal arboreal geckos also have adhesive toepads; geckos are relatively fast climbing ambushers (presumably FAM types E and G). Many geckos are small and dorsoventrally flattened, coincident with the ability to rapidly run on flat vertical substrata and hide under bark and in crevices (Cooper, 1994; Pianka and Vitt, 2003). The diurnal lizard *Plica plica* (Topidurinae) of lowland tropical rainforests in Brazil is usually found clinging with recurved claws on smooth trunks; it is also dorsoventrally flattened. As is the case for geckos, the flattening may enable (1) a low center of gravity that increases climbing agility (Vitt, 1991), (2) a low profile to reduce visibility to predators, or (3) hiding under loose bark. Like geckos, it may be FAM types E or G.

In contrast, chameleons are climbing, ambush predatory lizards (FAM types G and H) (Table 15.1; Fig. 15.2); their laterally compressed bodies, moderately gracile legs, and zygodactylous feet permit grasping of branches and twigs ventromedially and maintaining balance on these narrow substrata. These ambushers are extremely slow-moving masters of crypsis (Bickel and Losos, 2002; Pianka and Vitt, 2003). Lizard species that are chameleon-like in form (laterally compressed, or with prehensile tails, or with four legs of similar length) and behavior (with slow movements) are found in *Abronia* (Anguidae), *Corucia* (Scincidae) *Anolis* (Anolidae), *Polychrus* (Polychridae), and *Enyaliodes* (Tropiduridae). *Corytophanes* and *Laemanctus* (Corytophaninae) are cryptic denizens of branch–twig nanohabitats (Pianka and Vitt, 2003). *Corytophanes* may be extremely long-wait ambushers of large, evasive arthropods (Andrews, 1979). Unresolved is which of these lizards are ambushers that move very slowly when changing ambush positions (FAM type G), and which are slow-moving foliage gleaners (FAM type H).

There are also terrestrial ambushers of open, desert scrub habitats or in the open forest floor that rely on crypsis. The ant-eating specialists in the genus *Phrynosoma* (Phrynosomatidae), for example, have body forms and pattern of movements used for food acquisition that differ considerably from syntopic *Callisaurus*. *Phrynosoma* alternates between long-wait ambushing and "trap-line ambushing" (Whitford and Bryant, 1979; Shaffer and Whitford, 1981; Munger, 1984; Pianka, 1986). Through much of their daily activity periods, these cryptic, pancake-like lizards are long-wait ambushers (FAM type G) (Table 15.1; Fig. 15.2). In contrast, these lizards may shift into "trap-line ambushing" (perhaps like FAM type B, Fig. 15.2). Trap-lining may be accomplished (1) with the aid of map-like knowledge, (2) simply by encountering the abundant, active ant colonies, or (3) by seeing the colonies from a distance and walking over to them, much as would a herbivore or omnivore seeing flowers on a shrub.

Characterizing foragers and variation among foragers

The FAM of predatory lizards that most contrasts with ambushing is variously referred to as *"active searcher," "active forager," "wide forager,"* or *"intensive forager"* (Regal, 1978; Huey and Pianka, 1981; Anderson and Karasov, 1981). I recommend using the terms "intensive forager" and "forager." If *foraging* is "wandering in search of food," then the term *intensive foraging* means "highly concentrated, thorough searching for food." Hence, an *intensive forager* wanders about in search of food, and uses several sensory modalities to permit a thorough search (Regal, 1978). At any one moment, a thorough search is probably restricted to a relatively small area and not just on a substratum, but also beneath or behind the substratum surface. The word "active," whether it is associated with "forager" or "searcher," is arguably confusing, if one considers the speeds, distances, and frequencies of movements of various ambushers and foragers as depicted in Fig. 15.2. Instead, I suggest using the terms *wide* and *proximal* as relative terms that indicate the areal extent of the search during a foraging bout. Thus, "wide forager" and "proximal forager" can be used as qualitative, comparative descriptors for the relative mobility of foragers during a foraging bout (Table 15.1; Fig. 15.2).

Intensive foragers use multiple sensory systems such as vision, olfaction, and vomerolfaction to search microhabitats for hidden sedentary prey (Cooper, 1994; Schwenk and Wagner, 2001). Intensive foraging lizards may travel widely and quickly between patches of microhabitat, but move relatively slowly within patches (one-tenth the velocity between patches), where they commonly search for prey hidden in leaf litter and near-surface soil (Anderson, 1993, 1994). Detecting hidden prey appears to require thorough, intensive searches with two or more sensory modes; hence these lizards would be called wide, intensive foragers. Archetypal examples of wide, intensive foragers (FAM type A) (Table 15.1; Fig. 15.2) are adult *Aspidoscelis tigris* (Teiidae) of western deserts in North America (Pianka, 1970).

Proximal foragers and proximal intensive foragers (variations of FAM type C) (Table 15.1; Fig. 15.2) probably search for prey in patches of microhabitat that are larger, comprise more of a mesohabitat, and are more three-dimensionally developed than the microhabitat searched by wide intensive foragers. The lizards *Eumeces inexpectatus* (Scincidae), a ground–log–trunk user, and the semi-arboreal (trunk and ground) scincid *Eumeces laticeps* are proximal foragers in the southeastern USA (Vitt and Cooper, 1986; Cooper and Vitt, 1994). An open question is how much these lizards behave like (1) flush–chasers of mobile evasive prey, and (2) foliage gleaners (i.e. detect cryptic, non-moving prey visually rather by chemoreception). Even if their prey were detected more often by vision than by chemoreception, if most of these prey were not moving or were slow-moving and did not have high evasive abilities, then these lizards still would be considered more as proximal intensive foragers rather than just proximal foragers.

Movements and foraging behaviors of *Aspidoscelis tigris* are varied (Anderson, 1986, 1993, 1994; Anderson and Vitt, 1990). I have seen *A. tigris* walking on the ground between and under plants, sometimes stop, dig, and unearth prey (Anderson, 1993). I also have seen *A. tigris* behave like a "flush–chaser," as in some birds (Robinson and Holmes, 1982); that is, some flushed prey were evasive and had to be chased, others that were slow moving were easily picked up (Anderson, 1993). Auffenberg (1994) reported similar behavior in the Bengal monitor, *Varanus bengalensis*. Less frequently, *A. tigris* climbed through shrubs and large annual plants in apparent visual search for prey (Anderson, 1993). For example, when caterpillars were common in lupines, the *A. tigris* moved from lupine to lupine, stopping to look up into them, sometimes climbing them and capturing caterpillars. These sorts of behaviors are similar to the foraging pattern of "travel, stop to visually search an array of nearby surfaces, travel," which is common in foliage-gleaning birds (Robinson and Holmes, 1982; McLaughlin, 1989), and perhaps in short-wait ambushers. Another, less common, behavior in *A. tigris* that is similar to behaviors seen in ambushers is seen when a basking *A. tigris* notices a potential prey moving nearby, then gives chase (Anderson, 1986). It seems likely that small intensive foragers with extended bouts of basking during cool periods of the activity season would have occasional, opportunistic, ambush-like encounters with prey. If the overwhelmingly primary function of "being still" is thermoregulation and not prey search, then conflating thermoregulation with a type of "foraging mode" is problematic. Nevertheless, the subtle position and posture changes of thermoregulating ambushers as they visually search for prey may be routine (Muth, 1977).

A more continuous movement pattern, akin to "cruising" instead of saltatory or intermittent locomotory search (O'Brien *et al.*, 1990; Kramer and McLaughlin, 2001), perhaps could be attributed to *A. tigris* that are moving on the ground for many minutes, in the open and along shrub perimeters, and only stopping for prey detected by chemoreception, most likely by vomerolfaction (e.g. termites, scorpions, larvae and pupae). I am excluding the intermittent momentary pauses that may be 0.2 or less (Avery and Bond, 1989), and perhaps used to scan the distance for predators (Kramer and McLaughlin, 2001). Cruising may be considered a variant of FAM type A, but with distances greater between the slow-downs that usually happen at perennials where intensive searching occurs. It may be that this FAM type A moves at a relatively constant rate until it "encounters" prey (rather than habitat "patches"), whether discovery of prey is by chemoreception, by vision, or by flushing prey (Kramer and McLauglin, 2001). This variant of FAM type A may be used by some large *Varanus* (Auffenberg, 1994) and snakes (Shine *et al.*, 2004); individuals may move at relatively constant rates among meso-habitats (perhaps more like a fast type C) until they "encounter" prey. They may ambush the prey – especially if the prey are proficient evaders – from above, below, or behind.

During the approximate middle of the morning activity period of adult *A. tigris*, putatively when temperatures are salubrious and overt thermoregulatory behaviors are not obvious, the lizards' foraging movements generally appear slow and nearly continuous. These foraging movements, similar to those of some larger-bodied snakes and *Varanus*, could be interpreted as either a fast variant of FAM type C (faster than FAM type A), a normal type C, or type A (Table 15.1; Fig. 15.2), depending on (1) how frequently the lizard scratches the litter or pokes its snout into holes, cracks, and crevices, (2) how which type of prey are discovered and how they are captured, and (3) the distances traversed. The versatility of foraging movements and the several prey detection modalities observed for *A. tigris*, during many hours of field observations (Anderson, 1986, 1993, 1994; Anderson and Vitt, 1990), and for *Varanus* (Auffenberg, 1981, 1994), generates concern at conclusions about lizard food acquisition that are derived by researchers using small observational datasets taken over short periods of time or too few locations (see also Perry, this volume, Chapter 1).

The FAM of intensive foragers, however, comprises more than the mobile search and the remarkable uses of the chemoreception system of tongue and vomeronasal organ to detect hidden prey. For example, intensive foragers commonly dig to expose prey, hence digging is a form of pursuit of prey. Other lizards dig burrows, but *Aspidoscelis tigris* is an especially rapid and proficient digger (Anderson, 1993). Another important feature of *A. tigris* as an intensive forager is the use of forceps-like jaws to precisely extricate hidden prey from soil and litter. Moreover, a gleaning-like behavior (non-moving prey are visually detected) by *A. tigris* (Anderson, 1993) and other teiid lizards, such as *Kentropyx* (perhaps FAM type C) (Fig. 15.2) on trunks and branches (Vitt and Carvalho, 1992) may sometimes be more common than the technique of physically exposing hidden prey that were detected by chemoreception. Note that, other than the non-moving prey not being as visually obvious as fruits and leaves, encountering and eating non-moving, unhidden prey by intensive foragers and gleaners (along with the frequent slow, intervening movements between prey encounters) may be behavior similar to that of omnivores and herbivores encountering and eating plants. A reasonable hypothesis that arises from the foregoing discussion is that the flexibility of foraging in intensive foragers may permit the evolution of ambush predation more readily than intensive foraging predation evolving from ambush predation. That is, the

FAM types B or D could evolve from FAM types A or C more readily than types A or C from B or D (Table 15.1; Fig. 15.2). If so, then what FAM type could be the precursor to types A or C? Perhaps FAM type F.

Intermediate, mixed, or uncertain FAMs in lizards

The FAMs of secretive lizards are difficult to categorize; they are probably FAM type F, sometimes types C, G, or perhaps rarely, type H (Table 15.1; Fig. 15.2). This current difficulty is largely because of inadequate field data on behavior (Vitt and Pianka, this volume, Chapter 5), and too few laboratory studies on prey detection and capture methods by secretive lizards. It seems likely that these secretive lizards will be adept at short, less-than-body-length lunges to capture prey that have moved in range. Thus, even the more rapid movements by these lizards behind some screen of cover (e.g. leaf litter, rock, log, or bark) would be difficult for researchers to detect.

Many secretive diurnal lizards that have snake-like body forms, such as some anguids (Fitch, 1989), pygopodids (Patchell and Shine, 1986), and amphisbaenids (Colli and Zamboni, 1999) move stealthily, many through litter and sand, some in grass (Pianka and Vitt, 2003). Stealthy movements, particularly when in a somewhat three-dimensional matrix, may enable detection of prey by any one of several modalities. That is, these behaviors permit at least four distinct methods of foraging: (1) ambushing (FAM type E or G) (Table 15.1; Fig. 15.2), with moving prey being detected visually or by sound vibrations (Hetherington, 1989); (2) proximal foraging by gleaning of sedentary prey (slow-moving or non-moving), detected visually on a surface during one of the many frequent stops between short moves (FAM type F); (3) proximal foraging (one variant of FAM type C), and flush–chasing mobile prey with modest evasion abilities; or (4) proximal intensive foraging (another variant of FAM type C), wherein hidden prey are detected by chemoreception and are unearthed.

Other lizards that may be examples of so-called "intermediate foraging modes" or "mixed foraging modes" are the ostensibly secretive lizards *Eumeces skiltonianus* (Scincidae), *Elgaria coerulea* (Anguidae), and the omnivore *Xantusia riversiana* (Xantusiidae) (Fellers and Drost, 1991; Mautz, 1993; Rutherford and Gregory, 2003, personal observations). They may behave like (1) foragers that have extremely proximal movement patterns (variant of FAM type C or type H), or (2) ambushers (FAM type G), or (3) they can flexibly employ elements of both FAMs (perhaps as an FAM type F) as do the aforementioned stealthy lizards and snakes. They are all moderately robust and short-limbed relative to intergirdle length. These lizards commonly use the cover of rocks, presumably for predator avoidance, thermoregulation, and food acquisition (Fellers and Drost, 1991; Mautz, 1993; Rutherford and Gregory, 2003). Where heavy herbaceous cover or thorny thickets are available, however, these lizards can be heard and sometimes seen moving slowly about. Although the functions of these movements are unclear, they are likely to be related to foraging, thermoregulation or mate-seeking. If these secretive lizards are relatively sedentary and stealthily approach or stalk nearby prey, then observing foraging in these lizards would be problematic. Being secretive permits hiding and foraging at the same time, but it also may necessitate thigmothermy, lower body temperatures, and thus limited exercise capacity relative to sympatric heliotherms (Mautz *et al.*, 1992). *Coleonyx*, the terrestrial, nocturnal geckos in deserts of North America, may move and pursue prey as FAM types E or G, but like the secretive diurnal lizards, *Coleonyx* may be even more variable or intermediate because they are under the "cover" of darkness, even if they may be out in the open (Dial, 1978; Kingsbury, 1989).

Consider also lizards that hide in plain sight during the day, such as the cryptic geckos, chameleons, and horned lizards. Chameleons (and chameleon-like lizards), for example, have very slow movements as they traverse small branches and twigs (some also use the open forest floor). Regardless of whether during their foraging bout, they are (1) frequently changing perch positions as an FAM type G or (2) using a nearly continuous motion as an FAM type H, then these very slow movements may be conducive to (a) seeing the slightest arthropod movement, particularly for those lizards with the advantage of turret-eyes (Butler, 2005), and (b) perhaps even recognizing motionless prey.

Discussion of FAMs would be enriched by integrating the focus on foraging *versatility* and its ecological basis (MacNally, 1994; Komers, 1997; Perry, this volume, Chapter 1) with the complementary, contrasting focus on how the combined *constraints* of prey types and habitat structure require different sets of adaptive features, including learning, for food acquisition (Robinson and Holmes, 1982; Regal, 1978, 1983). Studies of lacertids, cordylids, scincids, and gymnophthalmids that spend a significant portion of their time on topographic structures and physiognomic features in the habitat or in a matrix such as litter, soil or dense vegetation (grass) may be among the more useful taxa for study of variation in FAM within a clade, and the consequences of that variation for the EPMs. It is these secretive species and perhaps the cryptic-slow-moving species (e.g. chameleons) that may have intermediate rates of foraging movements or very slow, virtually continuous movement (hence low frequency of move–stop–move) or variable rates of foraging movements (Huey and Pianka, 1981; Vrcibradic and Rocha, 1996; Cooper and Whiting, 2000).

Discerning variation in FAMs requires more criteria to categorize food acquisition of lizards than used hitherto. Some additional useful criteria would be prey mass (includes individual or multiple prey at one location) relative to predator mass, prey mobility during detection, prey evasiveness, the pursuit methods required to capture prey, and the distances and times between prey capture episodes. Clever combinations of laboratory, mesocosm, and open field studies are also needed to document the FAMs of these secretive and cryptic lizards.

Further complications in comparing among FAMs of lizards

Regardless of the variability in food-acquisition behavior in species with putative intermediate FAMs, most lizards, not just omnivores, are opportunistic about the foods they eat. Ambushers sometimes eat relatively sedentary prey, and foragers sometimes eat relatively evasive prey (Pianka, 1966, 1986). Large terrestrial predators with proficient chemoreception, such as the larger species of *Varanus*, may be able to chemodetect the best nanohabitat or microhabitat for ambushing (Auffenberg, 1981) as well as for digging. It may be that the larger terrestrial predators with proficient, exemplary modalities of chemoreception and vision vary food acquisition behavior to enable capture of very different and large prey types as they become available (Auffenberg, 1981, 1994).

Some of the documented and assumed foregoing EPM features that are integral aspects of the FAM of intensive foragers are very different from those of ambushers and herbivores. Moreover, the unique FAM of intensive foragers may affect other features related to the other three basic tasks. For example, the mate-search polygyny mating system, the encounters of *A. tigris* with ambush predators waiting in the exact microhabitats and nanohabitats searched by *A. tigris*, and the means of avoidance and evasion of predators used by *A. tigris* may be obvious consequences of the FAM in *A. tigris* (Anderson, 1986; Anderson and Vitt, 1990; Garland, 1994). Testing whether the FAM is the principal adaptive syndrome should include, at the least, identifying the putatively consequential traits associated with the other three basic tasks and comparing the challenges within and among the four basic autecological tasks as those challenges vary spatiotemporally.

The conclusions from meta-analyses of movement on most lizards are problematic in part because a particular movement or set of movements in a lizard may be related to one or more of the other three basic autecological tasks (or avoiding human observers), not just food acquisition (Anderson, 1993; Perry, this volume, Chapter 1). The focus, heretofore, on patterns of movement to compare ambushers and foragers, specifically on the frequency of movement and the time spent moving by the lizard, is too narrow or atomized (Lauder, 1996), particularly in the sense of trying to understand the EPMs of FAM, and FAM as an adaptive syndrome. Broadening and deepening the analysis of food acquisition in lizards would require, but not be limited to, documenting (1) the sensory modalities and behavioral tactics used to detect prey, (2) when and where each type of prey was detected and pursued, and (3) how the lizard pursued, captured, and handled each prey type. That is, careful observational–descriptive field studies (natural history) (Bartholomew, 1986) and insightfully designed experimental studies of food acquisition are needed within and among taxa.

The evolutionary constraints and opportunities associated with each major clade of lizards (e.g. lingual vs. jaw prehension of prey) (Vitt *et al.*, 2003b; Vitt and Pianka, this volume, Chapter 5; Reilly and McBrayer, this volume, Chapter 10) may be more fundamental and integrated than body form and the aforementioned simple aspects of movement. Advanced chemoreceptive perception and orientation (Regal, 1978, 1983; Cooper, 1994) and the ability to visually recognize cryptic (camouflaged and non-moving) prey by form may be the sorts of characteristics that demarcate differences among lizards in FAM. Proficient chemoreception (vomerolfaction and olfaction) and perhaps some ability to visually recognize non-moving prey (and non-moving predators) should permit a lizard to find and choose among a relatively large array of prey types: eggs, sedentary or inactive larvae, pupae, nymphs, and adults of insects and other arthropods. Sedentary and inactive prey are cryptic or in refugia. Hence, it is to be expected that the spatiotemporal patterns of more peripatetic lizards that seek sedentary or hidden prey would be different from the spatiotemporal patterns of more sedentary lizards that ambush more peripatetic prey.

Framing questions and hypotheses about the assumed importance of spatiotemporal patterns for FAM

If food acquisition is one of the four basic tasks to which the animal must display adaptedness, and if FAM is the adaptive syndrome, then much of what we see as an animal's EPM is related to FAM. Hence, spatiotemporal patterns of behaviors in lizards are integrated into the necessities of food acquisition, and food acquisition behavior will affect predation risk, exposure to abiotic stress (e.g. moisture, temperature, darkness, moving in a medium), and encounter probabilities with mates, and, conversely, these three other basic tasks may constrain and modify the EPMs typical for that FAM (Fig. 15.1). It can be hypothesized that most animals are going to be "ecologically tested" for their capabilities in the context of food acquisition. Thus, if the autecological challenge of food acquisition is assumed to be at the core of the animal's adaptedness and adaptability, then some features of adaptedness for locomotion may be demonstrated to be required because of the constraints of the other three basic autecological tasks imposed on the animal during food acquisition (e.g. the benefit of encountering potential mates, the need for predator evasion, and the capacity for exercise as related to body temperature and moving through or digging through a dense medium such as soil.

Discerning the intrinsic and extrinsic constraints on each basic autecological tasks of lizards within a deme would be improved by (1) documenting or varying the spatiotemporal patterns (e.g. substrata, nanohabitat-and-medium, microhabitat, and momentary, daily, seasonally) of those lizards, (2) assigning the extent to which any of the four basic tasks is being performed in those places at those times, and (3) determining the level of challenge to perform each of the four basic tasks, alone and in combination. For example, given that wide, intensive foragers have large home ranges and encounter more habitat heterogeneity at several scales than do ambushers (or herbivores) of similar size, whereas ambushers tend to specialize in use of substratum, nanohabitat and microhabitat, then how does a lizard's FAM affect mating systems, patterns of home range use, spacing, and dispersal among mesohabitats and macrohabitats? Also, assuming small lizards have relatively great challenges to avoid and evade predators, then how does EPM, FAM and "habitat use" vary among genera and species of small-bodied lizards vs. among genera and species of large-bodied lizards?

Habitat

Habitat spatial scale and its relevance to lizards

The foregoing descriptions of lizards that vary in FAM included frequent comments and questions about where these lizards acquire their food. Organismal biologists have long recognized the fundamental importance of prey and substratum–microhabitat to animals (MacArthur and Pianka, 1966; Eckhardt, 1979; Robinson and Holmes, 1982; Anderson, 1993). Assuming that among terrestrial vertebrates lizards are small to moderate size (Pough, 1980), then most lizards should show adaptedness at the spatial levels of *nanohabitat*, *substratum*, and *microhabitat*, rather than to the *macrohabitat* level. These spatial concepts need clarification.

Habitat is a locale with an identifiable set of abiotic and biotic components, distinct from the sets of components in surrounding locales, and which is conducive to at least temporary survival of one or more individuals of a species (Block and Brennan, 1993; Litvaitis *et al.*, 1994; Hall *et al.*, 1997; Morris, 2003). Habitat has a broad, inclusive connotation, hence dispersal corridors, sites of nutrient acquisition, mate acquisition, reproduction, and refugia (momentary, daily, seasonal) from abiotic stressors (ambient temperature, solar insolation) and refugia from biotic stressors (predators, parasites, interference competitors) can all be considered habitat for an individual or set of individuals. Analyses of *habitat use* should be performed hierarchically, at different spatial scales or levels, because the influence of large scale may be very different from the influence of small scale on the spatiotemporal patterns of animals (Morris, 2003), particularly in relation to the four basic autecological tasks.

Macrohabitat is most often equated with a landscape-level feature, which for terrestrial animals may be a vegetation association (Block and Brennan, 1993) or ecosystem (Myers and Ewel, 1990). Examples would be high pine, pine flatwoods, hardwood, and marshes in Florida (Myers and Ewell, 1990) and sagebrush, shadscale, greasewood, and saltgrass associations in the Great Basin desert scrub (Franklin and Dyrness, 1988). It is expected for smaller animals without great vagility, such as lizards, that most individuals in a deme may encounter only a single or at most a few macrohabitats in their lifetimes (Hall *et al.*, 1997).

Mesohabitat is a patch of habitat that is a major, identifiable subset of the habitat, and which may form much of a small animal's home range and in which one or more of the four basic autecological tasks occurs (Fig. 15.3). One useful example of a set of *mesohabitats* that correlate with desert reptile habitat use (Steffen, 2002; Rose, 2004) are three mesohabitats within the macrohabitat of Great Basin desert scrub in southeastern Oregon: small dunes, low profile sandy patches, and shallow depressions of hardpan. Each mesohabitat comprises a different heterogeneity of plants (e.g. species richness, abundance, distribution of sizes, amount of areal cover by plants), topography, and surface substratum (Steffen, 2002; Rose, 2004). Individual lizards, however, can readily traverse among all three mesohabitats in less than one minute (Rose, 2004). Using the concept of mesohabitat helps focus attention on one of the ostensibly fundamental goals in lizard ecology: documenting spatiotemporal variation in lizard population size (the effect) and seeking the causes of that variation.

Microhabitat refers more to the specific site within a mesohabitat or macrohabitat; this site or "patch" is where at least one of the four basic autecological tasks occurs. Examples of *microhabitat* include a shrub, tree, log, boulder, or dune face. It has been proposed that lizards may select microhabitats based on

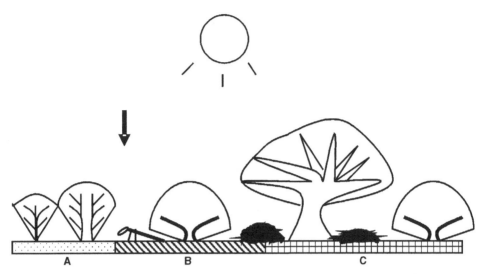

Figure 15.3. An example of overlaps and diversity in spatial scales within a mesohabitat; these overlaps challenge the researcher to determine how lizards use features of the habitat for each of the four basic tasks. An imaginary male lizard (front legs extended) is depicted below the arrow. Its body is partly shaded under shrub, partly in the open, but may be looking onto another substratum in a sunlit location under another nearby shrub. Which of the four basic tasks it is performing, or how many tasks it is accomplishing simultaneously, is unclear. Moreover, this lizard can easily move among substrata, nanohabitats, and microhabitats. Thus, the flexibility of the lizard's FAM may depend on spatiotemporal patterns of prey availability, not just the intrinsic constraints of the lizard. Details of the figure: On ground substratum A, two shrubs of the same species form a microhabitat patch; on ground substratum B is one shrub of a second species as a microhabitat; on C are two microhabitats, a tree and a shrub, but each plant provides areal cover to two ground substrata; any of these plants could also grow close enough together to form a mixed-species patch. Boulders are microhabitats; the larger lies on two substrata and is partly exposed to the sun; the other boulder is thinner, lies on a single substratum type, and, with this sunlight orientation, is shaded; both boulders are under the tree, but sunlit at some times. See text for examples of variation in nanohabitats, microhabitats, and substrata.

the kinds of prey available in them rather than capturing whatever prey are available in specific microhabitats (Vitt *et al.*, 2003a; Vitt and Pianka, this volume, Chapter 5). For example, temporal changes in microhabitat used by *A. tigris* were related to changes in arthropod prey availability in the microhabitats: under three species of dominant woody perennials in a desert scrub (Anderson, 1993, 1994). For most species of lizards, the microhabitat level may correlate well with where and when they perform the basic autecological task of food acquisition. But substrata surely must vary within and among

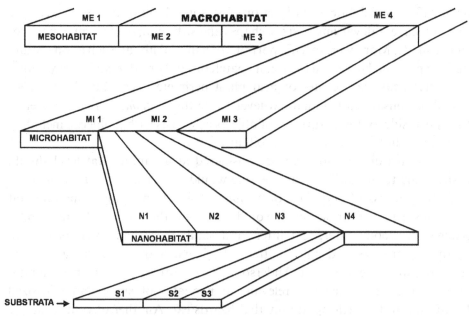

Figure 15.4. Schematic of the complex of spatial subdivisions of habitat which a lizard may face when it moves among mesohabitats in a macrohabitat. With just a single mesohabitat (ME) depicted in detail, one may infer that it would be challenging to the lizard to use and select an optimum array of spatial subdivisions. Whereas a lizard of FAM types F, G, or H may use primarily one or two substrata in one or two nanohabitats of one or two microhabitats (MI), a lizard of FAM types A or B may move among several of each spatial subdivision; these differences in spatial use may have consequences for their EPMs. Note that any of the three substrata (S1, S2, S3) shown in nanohabitat N3 also may be in the other nanohabitats, and so may be other substratum types. Similarly, any of the four types of N shown here in MI 2 (N1, N2, N3, N4) may be in other microhabitats, and so may be other nanohabitats. Any of the three types of MI depicted in ME 4 also may be in the other mesohabitats, and so may be other microhabitats. Other examples may exclude one or two subdivisions between substratum and macrohabitat. Widths of types (e.g. S1 > S2 = S3) within a subdivision can be used to represent their relative availabilities. See text for examples of these spatial subdivisions.

microhabitats (Fig. 15.4), hence more definitive understanding of food acquisition would require knowledge of how nanohabitat and substrata are used.

Nanohabitat (microhabitat element, Howes and Lougheed, 2004; or microsites, Bauwens *et al.*, 1995) can be used to designate a subset of microhabitat. *Nanohabitat* may be so small as to approximate the size of smaller animals, yet where a specific behavioral act can occur. Some of the many *nanohabitat* examples are: (1) burrow in soil, or crevice in rock or tunnel in a log, stump, snag, or tree; (2) on or under a rock or portion of log or piece of coarse woody debris; (3) in small patches of litter or loose soil; (4) on trunks or branches or twigs or leaves (all varying in sizes, shapes and stiffness); or (5) a small patch of sun or shade. Moreover, specific portions of a microhabitat, such as on ground under a perennial plant, on (6) a small patch of sand or litter, and even whether it was (7) near plant center or near plant perimeter (Kerr and Bull, 2004), would be considered nanohabitat-level. Note that *nanohabitat* could refer to the upper, side, or lower surfaces of objects as well as within objects, including also the medium (e.g., air, water, soil, litter).

How lizard distribution can be considered at a nanohabitat level should easily come to mind. For example, one would expect that thermoregulation and effectiveness of crypticity of *Sceloporus woodi* in a single microhabitat under a shrub in Florida Scrub would vary with the mosaic of nanohabitat patches of woody litter, leaf litter, and sand; these nanohabitats vary in thermal properties and color patterns both when they are dry and when they are wet (Tiebout and Anderson, 2001). Moreover, the nanohabitat level should be particularly relevant to neonates of small and mid-sized lizards and to the arthropod prey that lizards eat. Another obvious example of focusing at nanohabitat and microhabitat spatial scales would be an analysis of perch sites of rocks or branches that are available vs. those that are used. Perch sites in perennial plants differ between males and females of *Anolis frenatus* but not between sexes of *A. limifrons*; several plausible, non-exclusive explanations can be offered for these patterns (Scott *et al.*, 1976). Rocks as perch sites differ between juvenile and adult *Cophosaurus texanum* (Phrynosomatidae); juveniles perch more on larger rocks, and although the adaptedness of this difference is unclear, several plausible explanations were offered (Durtsche *et al.*, 1997).

Another example of the importance of nanohabitat may be as a refugium during the inactivity period, because the body form of lizards may display adaptedness to their refugia (or lizard distribution and abundance may be restricted by availability of refugia). For example, the dorsoventrally flat-tened bodies of (a) *Phrynosoma, Callisaurus, Cophosaurus*, and *Uma* enable proficient refuging in loose soil and litter; (b) small gekkonids such as *Hemidactylus* enable proficient refuging under bark and in crevices, (c) *Sauromalus* and *Platysaurus* enable refuging in rocky crevices, and (d) *Plica plica* may permit either retreat under bark or sleep exposed on trunks (Pianka and Vitt, 2003). Contrast the foregoing lizards with (1) the robust-bodied *Sceloporus magister*, which has nightly retreats in cracks, hollows, and tunnels in wood and crevices in rock, and (2) the cylindrical bodies of *A. tigris* and *Dipsosaurus dorsalis*, which may better enable these lizards to proficiently construct and use a burrow refugium in soil. If body form of a lizard is closely related to the type of nanohabitat (and substratum, for cryptic lizards that are inactive and exposed) used for its inactivity period, then the consequences of that body form for the lizard's activities during the activity period must be investigated.

Substratum is the precise type of surface that the animal is upon, and can vary in such features as hardness, rigidity, surface tensile strength, rugosity, curvature, and color pattern. Example categories of substrata include (1) ground (e.g. varies in texture and compaction as related to the mineral and organic components), (2) leaf and stick litter and coarse woody debris, (3) rocks, boulders, and rock-face crevices, (4) bark of tree bole, shrub stem, branch, twig, (5) leaf, (6) grass, (7) moss or lichens. Precisely where that substratum is located, however, such as upper, side, or lower surface, is stated within reference to nanohabitat and microhabitat. Assuming distribution and abundance of substrata vary within and among nanohabitats, microhabitats and larger spatial scales (Fig. 15.4), then it would seem important to know whether substrata or other features of the habitat more directly influence spatiotemporal pattern of lizards (e.g., lizards with toepads moving and clinging to undersides of structures), but this may be problematic.

Differentiating spatial scales in habitat analyses

It is not uncommon for the levels of microhabitat, nanohabitat, and substratum, such as represented in Figs. 15.3 and 15.4, to be undifferentiated in empirical papers; these combined levels have been called "substrate" (Adolph, 1990), "substrate microhabitats" (Downes, 2001), "microhabitat" (Pianka, 1986; Vitt *et al.*, 1997b, 2003a), or "habitat subsites" (James, 1994). Whether the conflation of spatial scales (see, for example, Smith *et al.*, 1987; Vitt *et al.*, 2000a,b; Herrel *et al.*, 2001; James and M'Closkey, 2002) is problematic may depend on the animals studied (Wiens, 1989), the questions asked, the habitat structures considered, the analyses performed, and the conclusions derived. For example, including (Adolph, 1990; Vitt *et al.*, 1997b; Downes, 2001) or excluding (Vitt *et al.*, 2003b) the temporally varying patches of sun and shade in the spatial analyses will depend on the importance of sun and shade on the lizards spatial distribution.

Another significant problem within each level of spatial designation is when individuals are found near edges or in transition zones. For example, if designation of being in the open is more than 0.1 m from a shrub perimeter, but the lizard is 0.5 m from the perimeter and searching the foliage perimeter for prey, then it is problematic to say that the lizard is in the open and focused on food acquisition in the open (Hager, 2001; Rose, 2004) (Fig. 15.3). And because most research on habitat use in lizards does not involve focal animal studies, the features of the habitat used for events that are brief, such as prey capture or predator evasion, may be erroneously unrecognized for their relative importance as habitat. Moreover, a single type of substratum may span several levels of habitat designation (macrohabitat, mesohabitat, microhabitat, nanohabitat) (Fig. 15.4), but its effect on the animal may be different at these different levels (Wiens, 1989), particularly within the context of each of the four basic autecological tasks. Alternatively, the same microhabitat may be found in several macrohabitats, and predator evasion may be accomplished in similar ways in those habitats (Repasky and Schluter, 1994).

Within the microhabitat and larger scales of habitat, significant variation in *habitat structure* exists. Such variables as (1) foliage height and form, (2) spatial distribution and abundance of trees, shrubs, ferns, herbs, grasses, and moss, (3) vertical and horizontal distribution of dead wood (of varying decay states) and coarse woody debris, and (4) litter depth in a habitat would contribute to measures of above-ground habitat structure that may be relevant to the animal (Morrison *et al.*, 1998). Many of these aspects of habitat structure can be differentiated (James and M'Closkey, 2002), most into nanohabitats and substrata (Fig. 15.3). If the Iguania, Gekkota, and Autarchoglossa differ markedly in use of elevated perches (Vitt *et al.*, 2003b), and if there are strong functional relationships of lizard morphotype, locomotory performance and aspects of habitat structure used by lizards (Pounds, 1988; Bonine and Garland, 1999; Buettell and Losos, 1999; Vanhooydonck *et al.*, 2000), it would seem prudent to have more precise designations of substrata, nanohabitats, and microhabitats than lizard ecologists have used heretofore. Furthermore, comparing among variants of substrata, nanohabitat, or microhabitats for their relative functionality and relative challenge for any or each of the four basic tasks performed by lizards in a deme should enable one to predict habitat quality (defined below).

Effects of spatial scale (e.g. micro-, meso-, macro-, landscape, regional) and temporal scale (hours, days, seasons, years) in *habitat use* and *habitat selection*, and the fitness consequences of habitat use and habitat selection (terms defined below) have been studied mostly in mammals (Johnson *et al.*, 2001; Morris and Davidson, 2000) and birds (Robichaud and Villard, 1999; Knick and Rotenberry, 2000), but not much of this leading research has been performed with lizards (Pianka, 1986; James, 1994; Knight and Morris, 1996; Dooley and Bowers, 1998). Early habitat use studies in lizards were largely stimulated by interest in interspecific competition (Heatwole, 1977; Heatwole and Taylor, 1985). Definitive studies of habitat use and habitat selection, however, must compare among habitats for their quality, abundance, and availability to the animal and their relative use, selection (choice) and preference by the animal (Morris, 2003). The term *habitat quality* refers to the relative benefit of that "habitat" among habitats used by that organism or population for the same basic task; the individual reproduction or population growth outcomes from use of that habitat is one way to measure habitat quality (Hall *et al.*, 1997). Although measuring *habitat availability*, the amount of habitat that is "available" to an animal, is often problematic, a suitable ersatz measure of habitat availability is *habitat abundance*, the amount or abundance of each "relevant" habitat (Litvaitis *et al.*, 1994; McClean *et al.*, 1998).

Investigating habitat use and habitat selection

Observations of *habitat use*, the spatiotemporal patterns of one or more animals in a habitat (Hall *et al.*, 1997), and *habitat distribution*, the relative abundances of individuals among habitats (Morris, 2003), are statistical measures that spatiotemporally associate the animal or animals with a particular habitat, *but use for what*? Identifying which of the four basic tasks are relevant at each time and each place would assign function to habitat use. If FAM is the principal adaptive syndrome in lizards, then it is expected that spatiotemporal patterns during the daily activity period of a lizard are associated with food acquisition. *Habitat use* is a common measure, but even a brief perusal of the microhabitat and habitat use literature in lizards (I choose not to single out these studies) reveals that spatial scales are commonly mixed in analyses, and the methods for obtaining data vary enough to question conclusions from meta-analyses should they be performed. And even more importantly, habitat use is rarely buttressed by *habitat selection* measures.

Habitat selection (*habitat choice*) connotes an uneven spatiotemporal association of the animal among "available" habitats (Morris, 2003), whereby the animal uses innate perceptions and learned decisions to choose among macrohabitats and among microhabitats to use (Hutto, 1985; Orians and Wittenberger, 1991; Stamps, 2001). Habitat selection can be established statistically by comparing the pattern of use versus the pattern of "availability" among different "habitats" (Arthur *et al.*, 1996; McClean *et al.*, 1998), assuming individuals are vagile enough to compare among habitats (Punzo and Madragon, 2002). The term *habitat preference* can be reserved for experimental tests wherein, under simplified conditions, animals have equal access to each habitat (Litvaitis *et al.*, 1994; Tiebout and Anderson, 2001; Howard *et al.*, 2003). Habitat selection is integral to the co-evolutionary interactions within and among populations of competing individuals and among those populations and their predators (Morris, 2003). Differences (including perception as well as behavioral actions) among individuals varying in age, size, gender, and EPMs in a population can be expected to be important components of these co-evolutionary interactions (Smith and Skúlason, 1996; McCairns *et al.*, 2004). For example, microhabitat and season correlated with mortality rates of juvenile *Psammodromus algirus* (Civantos and Forsman, 2000). A macrohabitat use study of *P. algirus* reached similar conclusions (Diaz and Carrascal, 1991).

The relationships of habitat selection to patterns and processes of spacing out and dispersal have not developed enough coherence in the scientific literature to be either predictive or generalized (Davis and Stamps, 2004). The studies on lizard dispersal (see, for example, Olsson *et al.*, 1997; Lena *et al.*, 1998; de Fraipont *et al.*, 2000; Clobert *et al.*, 2002) and settling (Massot *et al.*, 1994; M'Closkey *et al.*, 1998) are a very good beginning. But studies of perception, sampling, and exploration of microhabitats and habitats by lizards are rare (Zollner and Lima, 1997; Davis and Stamps, 2004). The assumptions and results of research methods among these dispersal and settling studies also vary (Doughty and Sinervo, 1994; de Fraipont *et al.*, 2000; Ronce *et al.*, 2001; Massot *et al.*, 2003), hence general conclusions cannot be made. Only one study ostensibly comparing dispersal or settling among species varying in food acquisition and phylogeny has been performed (Hokit *et al.*, 1999); the putative ambusher had limited dispersal, and an inference from limited data was that the putatively wide, intensive forager had higher rates of dispersal.

More studies are needed before meaningful meta-analyses of dispersal and settling in lizards can be performed. One goal of such a meta-analysis would be to compare vagility and dispersal among lizards (generally controlling for body size) that represent different combinations of FAM and habitat use (macro- to nano-) to enable us to know whether vagility and dispersal patterns are more a consequence of FAM or of the habitat used. One must also know where which individuals go and how those individuals fare in their respective "habitats" (Ronce *et al.*, 2001; Schlaepfer *et al.*, 2002; Sergio and Newton, 2003). Moreover, the habitat use and habitat selection by an animal in its home range may vary as a function of quality of available habitat patches (see, for example, Fig. 15.3) in the context of (1) prey availability (Mysterud and Ims, 1998; Kotler *et al.*, 2001), (2) competitor presence (Robertson, 1996; Martin and Martin, 2001), (3) refugia used for evasion or avoidance of predators (Lima and Dill, 1990; Yunger, 2004), and (4) nest site success (Rumble and Anderson, 1996; Shine *et al.*, 1997; Misenhelter and Rotenberry, 2000).

Integrating food acquisition and other activities in use of habitat

Effect of thermoregulation versus FAM on habitat use

Thermoregulation is a classic example of lizards performing one of the four basic autecological tasks: coping with abiotic stresses and avoiding abiotic extremes. For example, within microhabitats that may be valuable for one or more of the other four basic tasks, some lizards can accept extremely high temperatures and thus increase the length of their daily activity period (Grant, 1990; Grant and Dunham, 1988), and others can lower their thermal set points in cooler seasons, thereby extending their daily and seasonal activity time (Van Damme *et al.*, 1990; Christian and Bedford, 1995; Grbac and Bauwens, 2001). But accepting lower body temperatures may reduce their locomotor capacity (Van Damme *et al.*, 1989, 1990) and cognitive ability, and thus may require crypsis or secretive behavior. Thus, many lizards, at least those that do not live in warm and shady or nocturnal macrohabitats, are active only at times and locations conducive to thermoregulation (Dunham *et al.*, 1989; Bashey and Dunham, 1997; Angert *et al.*, 2002). Wide, intensive foragers in deserts and other sparsely vegetated habitats in warm climates and warm seasons may be more restricted spatiotemporally by high temperatures than are ambushers (Anderson and Karasov, 1981; Vitt *et al.*, 1993). Spatial variation in habitat use also may be related to body size: heliothermic lizards of larger body size, warmed to field active body temperatures, were able to move about further and longer in larger shady microhabitats and mesohabitats than could smaller lizards (Asplund, 1974).

The availability of subsurface refugia affects seasonal macrohabitat selection in Gila monsters (Beck and Jennings, 2003), whereas the macrohabitat affects the ambient temperatures, and thus time available for foraging, and ultimately lizard life histories in *Aspidoscelis hyperythrus* (Karasov and Anderson, 1984) and *Sceloporus merriami* (Grant and Dunham, 1990). Macrohabitat and meso-habitat (but called microhabitat) use were documented in a population of *Lacerta vivipara* where no other lizard species were present; it was hypothesized that smaller, unmeasured subdivisions of the habitat that were conducive to thermoregulation and antipredation were important for the lizard (Strijbosch, 1988). A more detailed, multiscale habitat selection study of *Lacerta lepida*, the largest European lacertid (Castilla and Bauwens, 1992), hypothesized that either (1) their point sampling technique (no focal sampling of individuals) may have biased observations toward less used subdivisions of the habitat or (2) lizards must use different "habitats" because of spatiotemporal constraints related to food availability or thermoregulation.

The consideration of spatiotemporal constraints related to thermoregulation can be broadened and integrated with other phenomena. For example, understanding the mesohabitats and microhabitats within an individual lizard's home range that are optimal for thermoregulation, both for gaining heat and avoiding overheating, are important, as these meso- and microhabitats may be unavailable to that individual because of interference competition (Adolph, 1990; Downes and Bauwens, 2002) and high predation risk (Fuentes and Cancino, 1979; Downes and Shine, 1998; Howard *et al.*, 2003). Variation in nanohabitat use may also elucidate the trade-offs between thermoregulation and the other basic tasks (Kerr and Bull, 2004).

Interspecific competition and habitat use as related to FAM

There are a number of well-known examples of ambushing lizards in the same guild using different microhabitats or nanohabitats, and coincidentally reducing the potential for interspecific competition. However, it is not clear how important the influences of interference competition or exploitative competition are for habitat selection by these lizards (Medel *et al.*, 1988; Vitt *et al.*, 2000b; James and M'Closkey, 2002).

Studies of interspecific competition of lizards on islands have commonly discovered "habitat" shifts by the behaviorally subordinate species (Jenssen *et al.*, 1988; Losos *et al.*, 1993; Petren and Case, 1996; Downes and Bauwens, 2002), but the spatial scales of habitat shifts generally are not well differentiated. Among island competitors, the gekkonids and tropidurines are ambushers, the FAMs of the island lacertids are less obvious, and some scincids may be proximal intensive foragers. Microhabitat and nanohabitat shifts associated with competition may be more likely if the competing taxa have greater use of vertical structure within a mesohabitat (James and M'Closkey, 2002) than do wide, intensive foragers, although even wide foragers climb in search of prey on occasion (Vitt and Carvalho, 1992; Anderson, 1993). Wide, intensive foraging lizards, such as teiids, tend to have large home ranges and sample numerous microhabitats during their foraging, hence spatial separation of interspecific competitors among intensive foragers is more likely to be at mesohabitat or macrohabitat levels, and include body size differences that may reduce competition among syntopic congeners (Vitt *et al.*, 2000a).

Antipredation as related to FAM and habitat use

Antipredation appears to be the primary function assumed for maximal locomotor performance in many whole-animal performance studies of lizard

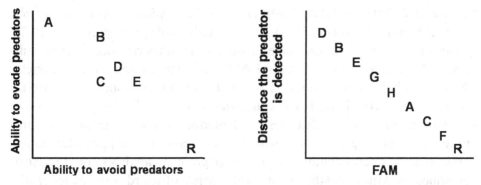

Figure 15.5. Hypothesized relationships of antipredation patterns among lizards that vary in FAM, considering also visibility in habitats each FAM type tends to occupy. Letters refer to FAM type, except R, which indicates when most lizards are in a refugium nanohabitat during their inactivity periods. See Fig. 15.2 for FAM descriptions.

locomotion (Irschick and Garland, 2001). But evading a pursuing predator usually happens in the context of other activities, and the primary activity for much of a lizard's daily activity period is food acquisition. Because lizards are common prey of larger vertebrates, antipredation in lizards is necessary regardless of the FAM. Even though within the same mesohabitats lizards with different FAMs can be captured by the same predators, the relative reliance on crypsis, vigilance, and evasion should vary among the different FAMs (Fig. 15.5). Moreover, if lizards use substrata, nanohabitats, and microhabitats that are most conducive to prey capture, then antipredation used in those locations while preoccupied with food acquisition may differ from the antipredation used on substrata and in nanohabitats and micro-habitats used by lizards in transit among the prey-occupied locations (Fig. 15.5). Thus, precise methods of the basic task of antipredation (avoiding, evading, deterring, and escaping predators) can be hypothesized to happen as a consequence of FAM.

Determining how FAM and adaptedness to substrata and habitat affect whole-animal performance

Recent studies have shown some relationship between morphotype and "habitat" (Vitt *et al.*, 1997a; Vanhooydonck and Van Damme, 1999; Herrel *et al.*, 2001) or morphotype, locomotory performance, and "habitat" (*Anolis*: Losos *et al.*, 2000; Higham *et al.*, 2001; Spezzano and Jayne, 2004; Toro *et al.*, 2004; sceloporine phrynosomatids: Miles, 1994a,b; *Niveoscincus*: Melville and Swain, 2000, 2003; lacertids: Vanhooydonck *et al.*, 2000; Vanhooydonck and

Van Damme, 2003; cordylids: Losos *et al.*, 2002; *Tropidurus*: Kohlsdorf *et al.*, 2004; gekkonids: Zaaf *et al.*, 2001). A meta-analysis of sprint speed in over 120 species of lizard from at least nine taxonomic families revealed a significant effect of "microhabitat use" but not FAM (Van Damme and Vanhooydonck, 2001). Microhabitat, however, was assigned at broad levels such as terrestrial, arboreal, and saxicolous; FAM categories were broadly assigned as well. Recall the assumption that features of adaptedness in a FAM include encountering and capturing prey on particular substrata and in particular nano-habitats and microhabitats. Thus, finer-grained analyses of the sort performed by Miles (1994b) and a greater array of lizard taxa (Miles *et al.*, this volume, Chapter 2) are needed to enable a better understanding of the effects of FAM on whole animal performance measures.

Comprehensive comparisons of locomotory performance among lizards obviously require more than just measuring maximum velocities. Other useful parameters could be acceleration, agility, length of sustained sprint, and time required to repeat near-maximal performance. Standardizing methods for deriving quantitative data on behavior in the field would also improve the ecological and evolutionary interpretations of the foregoing studies (Huey *et al.*, 1984; Anderson, 1993; Garland and Losos, 1994; Mattingly and Jayne, 2004). Studying locomotory performance in semi-natural experimental conditions that are both standardized and ecologically realistic presumably would better elucidate the relationship of "habitat" to behavior, performance, and morphology for a population of lizards (Jayne and Ellis, 1998; Losos *et al.*, 2000; Irschick and Garland, 2001; Irschick, 2003).

Consider *where* a lizard uses locomotion: (1) on substrata of a mesohabitat in transit between patches of microhabitats; and (2) within microhabitats and in or on substrata. And consider *how* a lizard uses locomotion: (a) negotiating obstacles; (b) seeking thermoregulation spots; (c) mate seeking; (d) competing for mates; (e) seeking and pursuing potential prey; and (f) evading predators. Effective judgment about which, and how, locomotion performance capacities should be tested would benefit from an extensive understanding of the lizard's autecology (Irschick *et al.*, 2005), which would include knowing, (1) when, where, and how often episodes of walking, climbing, leaping, digging, running, and other forms of locomotion occur; (2) the conditions or context of each kind of locomotory episode (such as (a) what behavioral state the lizard was in when the locomotory bout began [e.g. foraging, mate seeking, thermoregulating], (b) distances from the objective [prey, or mate, or competitor, or predator and refugium] when the bout began, (c) how far the lizard moved, (d) how fast it moved, and (e) how many direction changes were in that moving bout); and (3) spatiotemporal patterns of behavior before and after each

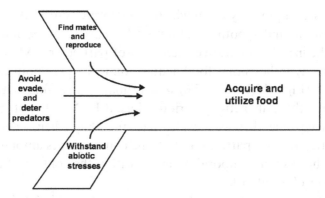

Figure 15.6. Schematic representation of the relative influences (box sizes) among the four basic tasks on evolution of lizard EPM. Food acquisition is depicted as drawing the other three tasks along, so that the EPM is largely based on the food acquisition mode and its consequences.

locomotion bout (a recovery or hiding period may be required) (Hancock and Gleeson, 2002). Field-relevant tests of locomotor performance capacity may require laboratory microcosm and field mesocosm studies wherein appropriate sets of substrata, nanohabitat, and microhabitats are used in standardized arrays of interactions with real or ersatz prey, competitors, and predator.

Clarifying relationships among the four basic tasks and when and where those tasks are performed should help one develop testable hypotheses about the evolution of FAMs. For example, when first entering a desert scrub, a proximal foraging lizard traversing the open mesohabitats and seeking hidden prey among the prey-rich shrub microhabitats would encounter challenges of two other basic tasks in the open areas: (1) evading fast-moving predators and potentially (2) withstanding high ambient temperatures. Moreover, a male moving among microhabitats in an open mesohabitat may have more difficulty encountering and guarding the widely dispersed females that are potential mates. It would seem that the evolving FAM of the lizard invading this habitat would carry (Fig. 15.1) or "pull" along the features of adaptedness to meet the challenge of the other basic tasks (Fig. 15.6); hence, a unique EPM (wide, intensive forager) would evolve.

Conclusion

In this chapter the variation in food acquisition behavior among and within ambushers and within an intensive forager was showcased. Given the lacunae in our knowledge of food acquisition behavior for the many taxa of lizards that are difficult to observe in the field, along with the problems in discerning behaviors, their functions, and precisely when which stimuli were perceived, then some of the basic questions and hypotheses about FAMs have yet to be addressed.

Even for the lizards as putative exemplars of particular FAMs, remaining untested is the hypothesis that food acquisition is the paramount basic autecological task (Figs. 15.1 and 15.6) and primary cause for spatiotemporal patterns during the daily activity period. Thus, if FAM is hypothesized to be an adaptive syndrome, then we must learn (1) the what, when, where, and how of lizard spatiotemporal patterns and (2) the consequences among lizards for variation in those spatiotemporal patterns, within the context of each of the four basic autecological tasks.

Within and among taxa of lizards, identifying the functional significance of spatial features of the environment for food acquisition and for each of the other basic autecological tasks, however, requires intensive and extensive efforts documenting (1) observational–descriptive field data on which substrata, nanohabitats, and microhabitats are used, and the consequences of those usages, and (2) whole-animal performances that are ecologically relevant to the physical challenges of the substrata, nanohabitats, and microhabitats faced by the lizards. Knowing the relationship of food acquisition in lizards to habitat selection and habitat quality will move us toward comprehensive answers to the basic ecological question: "What are the spatiotemporal patterns of individual lizards and what are the causes for these spatiotemporal patterns?"

References

Adolph, S. C. (1990). Influence of behavioral thermoregulation on microhabitat use by two *Sceloporus* lizards. *Ecology* **71**, 315–27.

Anderson, R. A. (1986). Foraging behavior, energetics of reproduction, and sexual selection in a widely-foraging lizard *Cnemidophorus tigris*. PhD. dissertation, University of California, Los Angeles.

Anderson, R. A. (1993). Analysis of foraging in a lizard, *Cnemidophorus tigris*: salient features and environmental effects. In *Biology of Whiptail Lizards (Genus Cnemidophorus)*, ed. J. W. Wright and L. J. Vitt, pp. 83–116. Norman, OK: Oklahoma Museum of Natural History.

Anderson, R. A. (1994). Functional and population responses of the lizard *Cnemidophorus tigris* to environmental fluctuations. *Am. Zool.* **34**. 409–21.

Anderson, R. A. and Karasov, W. H. (1981). Contrast in energy intake and expenditure in sit-and-wait and widely foraging lizards. *Oecologia* **49**, 67–72.

Anderson, R. A. and Karasov, W. H. (1988). Energetics of the lizard, *Cnemidophorus tigris*, and life history consequences of food acquisition mode. *Ecol. Monogr.* **58**, 79–110.

Anderson, R. A. and Vitt, L. J. (1990). Sexual selection versus alternative causes of sexual dimorphism in teiid lizards. *Oecologia* **84**, 145–57.

Andrews, R. M. (1979). The lizard *Corytophanes cristatus*: an extreme "sit-and-wait" predator. *Biotropica* **11**, 126–39.

Angert, A. L., Hutchison, D., Glossip, D. and Losos, J. B. (2002). Microhabitat use and thermal biology of the collared lizard (*Crotaphytus collaris collaris*) and the fence lizard (*Sceloporus undulatus hyacinthinus*) in Missouri Glades. *J. Herpetol.* **36**, 23–9.

Arthur, S. M., Manly, B. F. J., McDonald, L. L. and Garner, G. W. (1996). Assessing habitat selection when availability changes. *Ecology* **77**, 215–27.

Asplund, K. K. (1974). Body size and habitat utilization in whiptail lizards (*Cnemidophorus*). *Copeia* **1974**, 695–703.

Auffenberg, W. (1981). *The Behavioral Ecology of the Komodo Monitor.* Gainesville, FL: University of Florida Press.

Auffenberg, W. (1994). *The Bengal Monitor.* Gainesville, FL: University of Florida Press.

Avery, R. A. and Bond, D. J. (1989). Movement patterns of lacertid lizards: effects of temperature on speed, pauses and gait in *Lacerta vivipora*. *Amph.-Rept.* **10**, 77–84.

Bartholomew, G. A. (1986). The role of natural history in contemporary biology. *BioScience* **36**, 324–9.

Bashey, F. and Dunham, A. E. (1997). Elevational variation in the thermal constraints on and microhabitat preferences of the greater earless lizard *Cophosaurus texanus*. *Copeia* **1997**, 725–37.

Bauwens, D., Garland, T., Jr., Castilla, A. M. and Van Damme, R. (1995). Evolution of sprint speed in lacertid lizards: morphological, physiological, and behavioral covariation. *Evolution* **49**, 848–63.

Beck, D. D. and Jennings, R. D. (2003). Habitat use by gila monsters: the importance of shelters. *Herpetol. Monogr.* **17**, 111–29.

Bickel, R. and Losos, J. B. (2002). Patterns of morphological variation and correlates of habitat use in chameleons. *Biol. J. Linn. Soc.* **76**, 91–103.

Block, W. M. and Brennan, L. A. (1993). The habitat concept in ornithology: theory and applications. In *Current Ornithology*, vol. 11, ed. D. M. Power, pp. 35–91. New York: Plenum Press.

Bonine, K. E. and Garland, T. Jr. (1999). Sprint performance of phyrnosomatid lizards, measured on a high-speed treadmill, correlates with hindlimb length. *J. Zool. Lond.* **248**, 255–65.

Buettell, K. and Losos, J. B. (1999). Ecological morphology of Caribbean anoles. *Herp. Monogr.* **13**, 1–28.

Butler, M. A. (2005). Foraging mode of the chameleon, *Bradypodion pumilum*: a challenge to the sit-and-wait versus active forager paradigm? *Biol. J. Linn. Soc.* **84**, 797–808.

Castilla, A. M. and Bauwens, D. (1992). Habitat selection by the lizard *Lacerta lepida* in a Mediterranean oak forest. *Herpetol. J.* **2**, 27–30.

Christian, K. A., Tracy, C. R. and Porter, W. P. (1984). Diet, digestion, and food preferences of Galapagos land iguanas. *Herpetologica* **40**, 205–12.

Christian, K. A. and Bedford, G. S. (1995). Seasonal changes in thermoregulation by the frillneck lizard, *Chlamydosaurus kingii*, in tropical Australia. *Ecology* **76**, 124–32.

Civantos, E. and Forsman, A. (2000). Determinants of survival in juvenile *Psammodromus algirus* lizards. *Oecologia* **124**, 64–72.

Clobert, J., M. Massot, L. P. and Rossi, J. M. (2002). Condition-dependent dispersal and ontogeny of the dispersal behaviour: an experimental approach. *J. Anim. Ecol.* **71**, 253–61.

Colli, G. R. and Zamboni, D. S. (1999). Ecology of the worm-lizard *Amphisbaena alba* in the Cerrado of Central Brazil. *Copeia* **1999**, 733–42.

Cooper, W. E. Jr. (1994). Prey chemical discrimination, foraging mode, and phylogeny. In *Lizard Ecology Historical and Experimental Perspectives*, ed. L. J. Vitt and E. R. Pianka, pp. 95–116. Princeton, NJ: Princeton University Press.

Cooper, W. E. Jr. and Vitt, L. J. (2002). Distribution, extent, and evolution of plant consumption by lizards. *J. Zool. Lond.* **257**, 487–517.

Cooper, W. E. Jr. and Vitt, L. J. (1994). Tree and substrate selection in the semi-arboreal scincid lizard *Eumeces laticeps*. *Herpetol. J.* **4**, 20–3.

Cooper, W. E. Jr. and Whiting, M. J. (2000). Ambush and active foraging modes both occur in the Scincid genus *Mabuya*. *Copeia* **2000**, 112–18.

Cooper, W. E., Jr., Vitt, L. J., Caldwell, J. P., and Fox, S. F. (2001). Foraging modes of some American lizards: relationships among measurement variables and discreteness of modes. *Herpetologica* **57**, 65–76.

Cooper, W. E. Jr., Whiting, M. J., Van Wyk, J. H. and Mouton, P. F. N. (1999). Movement- and attack-based indices of foraging mode and ambush foraging in some gekkonid and agamine lizards from southern Africa. *Amph.-Rept.* **20**, 391–9.

Day, L. B., Crews, D. and Wilczynski, W. (1999). Spatial and reversal learning in congeneric lizards with different foraging strategies. *Anim. Behav.* **57**, 393–407.

Davis, J. M. and Stamps, J. A. (2004). The effect of natal experience on habitat preferences. *Trends Ecol. Evol.* **19**, 411–16.

de Fraipont, M., Clobert, J., John-Alder, H. and Meylan, S. (2000). Increased pre-natal maternal corticosterone promotes philopatry of offspring in common lizards *Lacerta vivipara*. *J. Anim. Ecol.* **69**, 404–13.

Dial, B. E. (1978). Aspects of the behavioral ecology of two Chihuahuan desert geckos (Reptilia, Lacertilia, Gekkonidae). *J. Herpetol.* **12**, 209–16.

Diaz, J. A. and Carrascal, L. M. (1991). Regional distribution of a Mediterranean lizard influence of habitat cues and prey abundance. *J. Herpetol.* **18**, 291–7.

Dooley, J. L. Jr. and Bowers, M. A. (1998). Demographic responses to habitat fragmentation: experimental tests at the landscape and patch scale. *Ecology* **79**, 969–80.

Doughty, P. and Sinervo, B. (1994). The effects of habitat, time of hatching, and body size on the dispersal of hatchling *Uta stansburiana*. *J. Herpetol.* **28**, 485–90.

Downes, S. (2001). Trading heat and food for safety: costs of predator avoidance in a lizard. *Ecology* **82**, 2870–81.

Downes, S. and Bauwens, D. (2002). An experimental demonstration of direct behavioural interference in two Mediterranean lacertid lizard species. *Anim. Behav.* **63**, 1037–46.

Downes, S. and Shine, R. (1998). Heat, safety, or solitude? Using habitat selection experiments to identify a lizard's priorities. *Anim. Behav.* **55**, 1387–96.

Dunham, A. E., Grant, B. W. and Overall, K. L. (1989). Interfaces between biophysical and physiological ecology and the population ecology of terrestrial vertebrate ectotherms. *Physiol. Zool.* **62**, 335–55.

Durtsche, R. D. (2000). Ontogenetic plasticity of food habits in the Mexican spiny-tailed iguana, *Ctenosaura pectinata*. *Oecologia* **124**, 185–95.

Durtsche, R. D., Gier, P. J., Fuller, M. M. *et al.* (1997). Ontogenic variation in the autecology of the greater earless lizard *Cophosaurus texanus*. *Ecography* **20**, 336–46.

Eckhardt, R. C. (1979). The adaptive syndromes of two guilds of insectivorous birds in the Colorado Rocky Mountains. *Ecol. Monogr.* **1979**, 129–49.

Ellinger, N., Schlatte, G., Jerome, N. and Hodl, W. (2001). Habitat use and activity patterns of the neotropical arboreal lizard *Tropidurus (Uracentron) azureus werneri* (Tropiduridae). *J. Herpetol.* **35**, 395–402.

Endler, J. A. (1986). *Natural Selection in the Wild.* Princeton, NJ: Princeton University Press.

Fellers, G. M. and Drost, C. A. (1991). Ecology of the island night lizard, *Xantusia riversiana*, on Santa Barbara Island, California. *Herpetol. Monogr.* **5**, 28–78.

Fitch, H. S. (1989). A field study of the slender glass lizard, *Ophisaurus attenuatus*, in northeastern Kansas. *Occ. Pap. Mus. Nat. Hist. Univ. Kansas* **125**, 1–50.

Franklin, J. F. and Dyrness, C. T. (1988). *Natural Vegetation of Oregon and Washington.* Corvallis, OR: Oregon State University Press.

Fuentes, E. R. and Cancino, J. (1979). Rock-ground patchiness in a simple *Liolaemus* lizard community (Reptilia, Lacertilia, Iguanidae). *J. Herpetol.* **13**, 343–50.

Garland, T. Jr. (1994). Phylogenetic analyses of lizard endurance capacity in relation to body size and body temperature. In *Lizard Ecology: Historical and Experimental Perspectives*, ed. L. J. Vitt and E. R. Pianka, pp. 237–59. Princeton, NJ: Princeton University Press.

Garland, T. Jr. and Losos, J. B. (1994). Ecological morphology of locomotor performance in squamate reptiles. In *Ecological Morphology: Integrative Organismal Biology*, ed. P. C. Wainwright and S. M. Reilly, pp. 240–302. Chicago, IL: University of Chicago Press.

Grant, B. W. (1990). Trade-offs in activity time and physiological performance for thermoregulating desert lizards, *Sceloporus merriami. Ecology* **71**, 2323–33.

Grant, B. W. and Dunham, A. E. (1988). Thermally imposed time constraints on the activity of the desert lizard *Sceloporus merriami. Ecology* **69**, 167–76.

Grant, B. W. and Dunham, A. E. (1990). Elevational covariation in environmental constraints and life histories of the desert lizard *Sceloporus merriami. Ecology* **71**, 1765–76.

Grbac, I. and Bauwens, D. (2001). Constraints on temperature regulation in two sympatric *Podarcis* lizards during autumn. *Copeia* **2001**, 178–86.

Greeff, J. M. and Whiting, M. J. (2000). Foraging-mode plasticity in the lizard *Platysaurus broadleyi. Herpetologica* **56**, 402–7.

Hager, S. B. (2001). Microhabitat use and activity patterns of *Holbrookia maculata* and *Sceloporus undulatus* at White Sands National Monument, New Mexico. *J. Herpetol.* **35**, 326–30.

Hall, L. S., Krausman, P. R. and Morrison, M. L. (1997). The habitat concept and a plea for standard terminology. *Wild. Soc. Bull.* **25**, 173–82.

Hancock, T. V. and Gleeson, T. T. (2002). Metabolic recovery in the Desert Iguana (*Dipsosaurus dorsalis*) following activities of varied intensity and duration. *Funct. Ecol.* **16**, 40–8.

Heatwole, H. (1977). Habitat selection in reptiles. In *Biology of Reptilia*, vol. 7, ed. C. Gans and D. Tinkle, pp. 137–55. New York: Academic Press.

Heatwole, H. and Taylor, J. (1985). *Ecology of Reptiles.* NSW, Australia: Surrey Beatty & Sons.

Herrel, A., Meyers, J. J. and Vanhooydonck, B. (2001). Correlations between habitat use and body shape in a phrynosomatid lizard (*Urosaurus ornatus*): a population-level analysis. *Biol. J. Linn. Soc.* **74**, 305–14.

Hetherington, T. E. (1989). Use of vibratory cues for detection of insect prey by the sandswimming lizard *Scincus scincus. Anim. Behav.* **37**, 290–7.

Higham, T. E., Davenport, M. S. and Jayne, B. C. (2001). Maneuvering in an arboreal habitat: the effects of turning angle on the locomotion of three sympatric ecomorphs of *Anolis* lizards. *J. Exp. Biol.* **204**, 4141–55.

Hokit, D. G., Stith, B. M. and Branch, L. C. (1999). Effects of landscape structure in Florida scrub: a population perspective. *Ecol. Appl.* **9**, 124–34.

Howard, R., Williamson, I. and Mather, P. (2003). Structural aspects of microhabitat selection by the skink *Lampropholis delicata. J. Herpetol.* **37**, 613–17.

Howes, B. J. and Lougheed, S. C. (2004). The importance of cover rock in northern populations of the five-lined skink *(Eumeces fasciatus). Herpetologica* **60**, 287–94.

Huey, R. B. and Pianka, E. R. (1981). Ecological consequences of foraging mode. *Ecology* **62**, 991–9.

Huey, R. B., Bennett, A. F., John-Alder, H. and Nagy, K. A. (1984). Locomotor capacity and foraging behavior of Kalahari lacertid lizards. *Anim. Behav.* **32**, 41–50.

Hutto, R. L. (1985). Habitat selection by nonbreeding, migratory land birds. In *Habitat Selection in Birds*, ed. M. L. Cody, pp. 455–76. Orlando, FL: Academic Press.

Irschick, D. J. (2003). Measuring performance in nature: implications for studies of fitness within populations. *Integr. Comp. Biol.* **43**, 396–407.

Irschick, D. J. and Garland, T. Jr. (2001). Integrating function and ecology in studies of adaptation: investigations of locomotor capacity as a model system. *Annu. Rev. Ecol. Syst.* **32**, 367–96.

Irschick, D. J. and Losos, J. B. (1999). Do lizards avoid habitats in which performance is submaximal? The relationship between sprinting capabilities and structural habitat use in Caribbean anoles. *Am. Nat.* **154**, 293–305.

Irschick, D. J., Herrell, A., Vanhooydonck, B., Huyghe, K. and Van Damme, R. (2005). Locomotor compensation creates a mismatch between laboratory and field estimates of escape speed in lizards: a cautionary tale for performance-to-fitness studies. *Evolution* **59**, 1579–87.

James, C. D. (1994). Spatial and temporal variation in structure of a diverse lizard assemblage in arid Australia. In *Lizard Ecology*, ed. L. J. Vitt and E. R. Pianka, pp. 287–317. Princeton, NJ: Princeton University Press.

James, S. E. and M'Closkey, R. T. (2002). Patterns of microhabitat use in a sympatric lizard assemblage. *Can. J. Zool.* **80**, 2226–34.

Janzen, F. J. and Brodie, E. D. III. (1995). Visually-oriented foraging in a natural population of herbivorous lizards (*Ctenosaura similis*). *J. Herpetol.* **29**, 132–6.

Jayne, B. C. and Ellis, R. V. (1998). How inclines affect the escape behaviour of a dune-dwelling lizard, *Uma scorpia. Anim. Behav.* **55**, 1115–30.

Jenssen, T. A., Marcellini, D. L. and Smith, E. P. (1988). Seasonal micro-distribution of sympatric *Anolis* lizards in Haiti. *J. Herpetol.* **22**, 226–74.

Johnson, C. J., Parker, K. L. and Heard, D. C. (2001). Foraging across a variable landscape: behavioral decisions made by woodland caribou at multiple spatial scales. *Oecologia* **127**, 590–602.

Karasov, W. H. and Anderson, R. A. (1984). Interhabitat differences in energy acquisition and expenditure in a lizard. *Ecology* **65**, 235–47.

Kerr, G. D. and Bull, C. M. (2004). Microhabitat use by the scincid lizard *Tiliqua rugosa*: exploiting natural temperature gradients beneath plant canopies. *J. Herpetol.* **38**, 436–545.

Kingsbury, B. A. (1989). Factors influencing activity in *Coleonyx variegatus. J. Herpetol.* **23**, 399–404.

Knight, T. W. and Morris, D. W. (1996). How many habitats do landscapes contain? *Ecology* **77**, 1756–64.

Knick, S. T. and Rotenberry, J. T. (2000). Ghosts of habitats past: contribution of landscape change to current habitats used by shrubland birds. *Ecology* **81**, 220–7.

Koehl, M. A. R. (1996). When does morphology matter? *Annu. Rev. Ecol. Syst.* **27**, 501–42.

Kohlsdorf, T., James, R. S., Carvalho, J. E. *et al.* (2004). Locomotor performance of closely related *Tropidurus* species: relationships with physiological parameters and ecological divergence. *J. Exp. Biol.* **207**, 1183–92.

Komers, P. E. (1997). Behavioural plasticity in variable environments. *Can. J. Zool.* **75**, 161–9.

Kotler, B. P., Brown, J. S., Oldfield, A., Thorson, J. and Cohen, D. (2001). Foraging substrate and escape substrate: patch use by three species of gerbils. *Ecology* **82**, 1781–90.

Kramer, D. L. and McLaughlin, R. L. (2001). The behavioral ecology of intermittent locomotion. *Amer. Zool.* **41**, 137–53.

Lauder, G. V. (1996). The argument from design. In *Adaptation*, ed. M. R. Rose and G. V. Lauder, pp. 55–91. San Diego, CA: Academic Press.

Lena, J. P., Clobert, J., de Fraipont, M., Lecomte, J. and Guyot, G. (1998). The relative influence of density and kinship on dispersal in the common lizard. *Behav. Ecol.* **9**, 500–7.

Lima, S. L. and Dill, L. M. (1990). Behavioral decisions made under the risk of predation: a review and prospectus. *Can. J. Zool.* **68**, 619–40.

Litvaitis, J. A., Titus, K. and Anderson, E. M. (1994). Measuring vertebrate use of terrestrial habitats and foods. In *Research and Management Techniques for Wildlife Habitats*, 5th edn., ed. T. A. Bookhout, pp. 254–74. Bethesda, MD: Wildlife Society.

Losos, J. B., Creer, D. A., Glossip, D. *et al.* (2000). Evolutionary implications of phenotypic plasticity in the hindlimb of the lizard *Anolis sagrei*. *Evolution* **54**, 301–5.

Losos, J. B., Marks, J. C. and Schoener, T. W. (1993). Habitat use and ecological interactions of an introduced and a native species of *Anolis* lizard on Grand Cayman, with a review of outcomes of anole introductions. *Oecologia* **95**, 525–32.

Losos, J. B., Mouton, P., Bickel, R., Cornelius, I. and Ruddock, L. (2002). The effect of body armature on escape behaviour in cordylid lizards. *Anim. Behav.* **64**, 313–21.

MacArthur, R. H. and Pianka, E. R. (1966). On optimal use of a patchy environment. *Am. Nat.* **100**, 603–9.

MacNally, R. C. (1994). On characterizing foraging versatility, illustrated by using birds. *Oikos* **69**, 95–106.

Martin, P. R. and Martin, T. E. (2001). Ecological and fitness consequences of species coexistence: a removal experiment with wood warblers. *Ecology* **82**, 189–206.

Massot, M., Clobert, J., Chambon, A. and Michalakis, Y. (1994). Vertebrate natal dispersal: the problem of non independence of siblings. *Oikos* **70**, 172–6.

Massot, M., Huey, R. B., Tsuji, J., and van Berkum, F. H. (2003). Genetic, prenatal, and postnatal correlates of dispersal in hatching fence lizards (*Sceloporus occidentalis*). *Behav. Ecol.* **14**, 650–5.

Mattingly, W. B. and Jayne, B. C. (2004). Resource use in arboreal habitats: structure affects locomotion of four ecomorphs of *Anolis* lizards. *Ecology* **85**, 1111–24.

Mautz, W. J. (1993). Ecology and energetics of the island night lizard, *Xantusia riversiana* on San Clemente Island. In *Recent Advances in California Islands Research, Proceedings of the Third California Islands symposium*, ed.

F. G. Hochberg, pp. 417–28. Santa Barbara, CA: Santa Barbara Natural History Museum.

Mautz, W. J. and Nagy, K. A. (1987). Ontogenetic changes in diet, field metabolic rate, and water flux in the herbivorous lizard *Dipsosaurus dorsalis*. *Physiol. Zool.* **60**, 640–58.

Mautz, W. J., Daniels, C. B. and Bennett, A. F. (1992). Thermal dependence of locomotion and aggression in a xantusiid lizard. *Herpetologica* **48**, 271–9.

M'Closkey, R. T., Hecnar, S. J., Chalcraft, D. and Cotter, J. E. (1998). Size distributions and sex ratios of colonizing lizards. *Oecologia* **116**, 501–9.

McCairns, R., Scott, J. and Fox, M. G. (2004). Habitat and home range fidelity in a trophically dimorphic pumpkinseed fish (*Lepomis gibbosus*) population. *Oecologia* **140**, 271–9.

McClean, S. A., Rumble, M. A., King, R. M. and Baker, W. L. (1998). Evaluation of resource selection methods with different definitions of availability. *J. Wildl. Manag.* **62**, 793–801.

McLaughlin, R. L. (1989). Search modes of birds and lizards: evidence for alternative movement patterns. *Am. Nat.* **133**, 654–70.

Medel, R. G., Marquet, P. A. and Jaksic, F. M. (1988). Microhabitat shifts of lizards under different contexts of sympatry: a case study with South American *Liolaemus*. *Oecologia* **76**, 567–9.

Melville, J. and Swain, R. (2000). Evolutionary relationships between morphology, performance and habitat openness in the lizard genus *Niveoscincus* (Scincidae: Lygosominae). *Biol. J. Linn. Soc.* **70**, 667–83.

Melville, J. and Swain, R. (2003). Evolutionary correlations between escape behaviour and performance ability in eight species of snow skinks (*Niveoscincus:* Lygosominae) from Tasmania. *J. Zool. Lond.* **261**, 79–89.

Miles, D. B. (1994a). Covariation between morphology and locomotory performance in sceloporine lizards. In *Lizard Ecology*, ed. L. J. Vitt and E. R. Pianka, pp. 207–35. Princeton, NJ: Princeton University Press.

Miles, D. B. (1994b). Population differentiation in locomotor performance and the potential response of a terrestrial organism to global environmental change. *Amer. Zool.* **34**, 422–36.

Misenhelter, M. D. and Rotenberry, J. T. (2000). Choices and consequences of habitat occupancy and nest site selection in sage sparrows. *Ecology* **81**, 2892–901.

Moermond, T. C. (1981). Prey-attack behavior of *Anolis* lizards. *Z. Tierpsychol.* **56**, 128–36.

Morris, D. W. (2003). Toward an ecological synthesis: a case for habitat selection. *Oecologia* **136**, 1–13.

Morris, D. W. and Davidson, D. L. (2000). Optimally foraging mice match patch use with habitat differences in fitness. *Ecology* **81**, 2061–6.

Morrison, M. L., Marcot, B. G. and Mannan, R. W. (1998). *Wildlife Habitat Relationships*, 2nd edn. Madison, WI: University of Wisconsin Press.

Munger, J. C. (1984). Optimal foraging? Patch use by horned lizards (Iguanidae: *Phrynosoma*). *Am. Nat.* **123**, 654–80.

Muth, A. (1977). Body temperatures and associated postures of the zebra-tailed lizard, *Callisaurus draconoides*. *Copeia* **1977**, 122–5.

Myers R. L. and Ewell, J. J. (1990). *Ecosystems of Florida*. Orlando, FL: University of Central Florida Press.

Mysterud, A. and Ims, R. A. (1998). Functional responses in habitat use: availability influences relative use in trade-off situations. *Ecology* **79**, 1435–41.

O'Brien, J. W., Browman, H. I. and Evans, B. I. (1990). Search strategies for foraging animals. *Am. Scient.* **78**, 152–9.

Olsson, M., Annica, G., and Tegelsstrom, H. (1997). Determinants of breeding dispersal in the sand lizard, *Lacerta agilis*, (Reptilia, Squamata). *Biol. J. Linn. Soc.* **60**, 243–56.

Orians, G. H. and Wittenberger, J. F. (1991). Spatial and temporal scales in habitat selection. *Am. Nat.* **137**, S29–S49.

Patchell, F. C. and Shine, R. (1986). Food habits and reproductive biology of the Australian legless lizard (Pygopodidae). *Copeia* **1986**, 30–9.

Perry, G. (1999). The evolution of search modes: ecological versus phylogenetic perspectives. *Am. Nat.* **153**, 98–109.

Petren, K. and Case, T. J. (1996). An experimental demonstration of exploitation competition in an ongoing invasion. *Ecology* **77**, 118–32.

Pianka, E. R. (1966). Convexity, desert lizards, and spatial heterogeneity. *Ecology* **47**, 1055–9.

Pianka, E. R. (1970). Comparative autecology of the lizard *Cnemidophorus tigris* in different parts of its geographic range. *Ecology* **51**, 703–20.

Pianka, E. R. (1986). *Ecology and Natural History of Desert Lizards*. Princeton, NJ: Princeton University Press.

Pianka, E. R. and Parker, W. S. (1972). Ecology of the iguanid lizard *Callisaurus draconoides*. *Copeia* **1972**, 493–508.

Pianka, E. R. and Vitt, L. J. (2003). *Lizards: Windows to the Evolution of Diversity*. Berkeley and Los Angeles, CA: University of California Press.

Pietruszka, R. D. (1986). Search tactics of desert lizards: how polarized are they? *Anim. Behav.* **34**, 1742–58.

Pough, F. H. (1980). The advantages of ectothermy for tetrapods. *Am. Nat.* **115**, 92–112.

Pounds, J. A. (1988). Ecomorphology, locomotion, and microhabitat structure: patterns in a tropical mainland *Anolis* community. *Ecol. Monogr.* **58**, 299–320.

Punzo, F. and Madragon, S. (2002). Spatial learning in Australian skinks of the genus *Ctenotus* (Scincidae). *Amph.-Rept.* **23**, 233–8.

Regal, P. J. (1983). The adaptive zone and behavior of lizards. In *Lizard Ecology*, ed. R. B. Huey, E. R. Pianka and T.W Schoener, pp. 105–18. Cambridge, MA: Harvard University Press.

Regal, P. J. (1978). Behavioral differences between reptiles and mammals: an analysis of activity and mental capacities. In *Behavior and Neurology of Lizards*, ed. N. Greenberg and P. D. MacLean, pp. 183–202. Rockville, MD: National Institute of Mental Health.

Repasky, R. R. and Schluter, D. (1994). Habitat distributions of wintering sparrows along an elevational gradient: tests of food, predation and microhabitat structure hypotheses. *J. Anim. Ecol.* **63**, 569–82.

Robertson, D. R. (1996). Interspecific competition controls abundance and habitat use of territorial Caribbean damselfishes. *Ecology* **77**, 885–99.

Robichaud, I. and Villard, M. (1999). Do black-throated green warblers prefer conifers? Meso- and microhabitat use in a mixedwood forest. *Condor* **101**, 262–71.

Robinson, S. K. and Holmes, R. T. (1982). Foraging behavior of forest birds: the relationships among search tactics, diet, and habitat structure. *Ecology* **63**, 1918–31.

Ronce, O., Olivieri, I., Clobert, J. and Danchin, E. (2001). Perspectives in the study of dispersal evolution. In *Dispersal*, ed. J. Clobert, E. Danchin, A. A. Dhondt and J. D. Nichols, pp. 340–57. Oxford, UK: Oxford University Press.

Rose, E. R. (2004). Foraging behavior in *Gambelia wislizenii*, the long-nosed leopard lizard, in Harney County, Oregon. Master's thesis, Western Washington University.

Rumble, M. A. and Anderson, S. H. (1996). Microhabitats of Merriam's turkeys in the Black Hills, South Dakota. *Ecol. Monogr.* **61**, 326–34.

Rutherford, P. L. and Gregory, P. T. (2003). Habitat use and movement patterns of northern alligator lizards (*Elgaria coerulea*) and western skinks (*Eumeces skiltonianus*) in Southeastern British Columbia. *J. Herpetol.* **37**, 98–106.

Schlaepfer, M. A, Runge, M. C. and Sherman, P. W. (2002). Ecological and evolutionary traps. *Trends Ecol. Evol.* **17**, 474–80.

Schwenk, K. and Wagner, G. P. (2001). Function and the evolution of phenotypic stability: connecting pattern to process. *Am. Zool.* **41**, 552–63.

Scott, N. J. Jr., Wilson, D. E., Jones, C. and Andrews, R. M. (1976). The choice of perch dimensions by lizards of the genus *Anolis* (Reptilia, Lacertilia, Iguanidae). *J. Herpetol.* **10**, 75–84.

Sergio, F. and Newton, I. (2003). Occupancy as a measure of territory quality. *J. Anim. Ecol.* **72**, 857–65.

Shaffer, D. T. and Whitford, W. G. (1981). Behavioral responses of a predator, the Round-tailed Horned lizard, *Phrynosoma modestum* and its prey, honey pot ants, *Myrmecocystus* spp. *Am. Midl. Nat.* **105**, 209–16.

Shine, R., Bonnet, X., Elphick, M. J. and Barrott, E. G. (2004). A novel foraging mode in snakes: browsing by the sea snake *Emydocephalus annulatus* (Serpentes, Hydrophiidae). *Funct. Ecol.* **18**, 16–24.

Shine, R., Elphick, M. J., and Harlow, P. S. (1997). The influence of natural incubation environments on the phenotypic traits of hatchling lizards. *Ecology* **78**, 2559–68.

Sih, A., Bell, A., Johnson, J. and Chadwick, J. (2004). Behavioral syndromes: an integrative overview. *Quart. Rev. Biol.* **79**, 241–77.

Simon, C. A. (1983). A review of lizard chemoreception. In *Lizard Ecology*, ed. R. B. Huey, E. R. Pianka and T. W. Schoener, pp. 119–33. Cambridge, MA: Harvard University Press.

Smith, D. D., Medica, P. A. and Sanborn, S. R. (1987). Ecological comparison of sympatric populations of sand lizards (*Cophosaurus texanus* and *Callisaurus draconoides*). *Great Basin Nat.* **47**, 175–85.

Smith, T. and Skúlason, S. (1996). Evolutionary significance of resource polymorphisms in fishes, amphibians, and birds. *Annu. Rev. Ecol. Syst.* **27**, 111–33.

Spezzano, L. C. Jr. and Jayne, B. C. (2004). The effects of surface diameter and incline on the hindlimb kinematics of an arboreal lizard (*Anolis sagrei*). *J. Exp. Biol.* **207**, 2115–31.

Stamps, J. A. (2001). Habitat selection by dispersers: integrating proximate and ultimate approaches. In *Dispersal*, ed. J. Clobert, E. Danchin, A. A. Dhondt and J. D. Nichols, pp. 230–42. Oxford, UK: Oxford University Press.

Steffen, J. E. (2002). The ecological correlates of habitat use for the long nose leopard lizard, *Gambelia wislizenii*, in southeast Oregon. Master's thesis, Western Washington University.

Strijbosch, H. (1988). Habitat selection of *Lacerta vivipara* in lowland environment. *Herpetol. J.* **1**, 207–10.

Tiebout, H. M. and Anderson, R. A. (2001). Mesocosm experiments on habitat choice by an endemic lizard: implications for timber management. *J. Herpetol.* **35**, 173–85.

Toro, E., Herrel, A. and Irschick, D. (2004). The evolution of jumping performance in Caribbean *Anolis* lizards: solutions to biochemical trade-offs. *Am. Nat.* **163**, 844–56.

Van Damme, R. and Vanhooydonck, B. (2001). Origins of interspecific variation in lizard sprint capacity. *Funct. Ecol.* **15**, 186–202.

Van Damme, R., Bauwens, D., and Verheyen, R. F. (1989). Effect of relative clutch mass on sprint speed in the lizard *Lacerta vivipara*. *J. Herpetol.* **23**, 459–61.

Van Damme, R., Bauwens, D. and Verheyen, R. F. (1990). Evolutionary rigidity of thermal physiology: the case of the cool temperate lizard *Lacerta vivipara*. *Oikos* **57**, 61–7.

Vanhooydonck, B. and Van Damme, R. (1999). Evolutionary relationships between body shape and habitat use in lacertid lizards. *Evol. Ecol. Res.* **1**, 785–805.

Vanhooydonck, B. and Van Damme, R. (2003). Relationships between locomotor performance, microhabitat use and anitpredator behaviour in lacertid lizards. *Funct. Ecol.* **17**, 160–9.

Vanhooydonck, B., Van Damme, R., and Aerts, P. (2000). Ecomorphological correlates of habitat partitioning in Corsican lacertid lizards. *Funct. Ecol.* **14**, 358–68.

Vitt, L. J. (1991). Ecology and life history of the scansorial lizard *Plica plica* (Iguanidae) in Amazonian Brazil. *Can. J. Zool.* **69**, 504–11.

Vitt, L. J. and Carvalho, C. M. (1992). Life in the trees: the ecology and life history of *Kentropyx striatus* (Teiidae) in the lavrado area of Roraima, Brazil. With comments on the life histories of tropical teiid lizards. *Can. J. Zool.* **70**, 1995–2006.

Vitt, L. J. and Cooper, W. E. Jr. (1986). Foraging and the diet of a diurnal predator (*Eumeces laticeps*) feeding on hidden prey. *J. Herpetol.* **20**, 408–15.

Vitt, L. J., Avila-Pires, T. C. S., Zani, P. A., Esposito, M. C. and Sartorius, S. S. (2003a). Life at the interface: ecology of *Prionodactylus oshaughnessyi* in the western Amazon and comparisons with *P. argulus* and *P. eigenmanni*. *Can. J. Zool.* **81**, 302–12.

Vitt, L. J., Caldwell, J. P., Zani, P. A. and Titus, T. A. (1997a). The role of habitat shift in the evolution of lizard morphology: evidence from tropical *Tropidurus*. *Evolution* **94**, 3828–32.

Vitt, L. J., Pianka, E. R., Cooper, W. E. Jr. and Schwenk, K. (2003b). History and global ecology of squamate reptiles. *Am. Nat.* **162**, 44–60.

Vitt, L. J., Sartorius, S. S., Avila-Pires, T. C. S., Esposito, M. C. and Miles, D. B. (2000a). Niche segregation among sympatric Amazonian teiid lizards. *Oecologia* **122**, 410–20.

Vitt, L. J., Souza, R. A., Sartorius, S. S., Avila-Pires, T. C. S. and Esposito, M. C. (2000b). Comparative ecology of sympatric *Gonatodes* (Squamata: Gekkonidae) in the western Amazon of Brazil. *Copeia* **2000**, 83–95.

Vitt, L. J., van Loben Sels, R. C. and Ohmart, R. D. (1981). Ecological relationships among arboreal desert lizards. *Ecology* **62**, 398–410.

Vitt, L. J., Zani, P. A., Caldwell, J., Araujo, P. and Magnusson, W. E. (1997b). Ecology of whiptail lizards *(Cnemidophorus)* in the Amazon region of Brazil. *Copeia* **1997**, 745–57.

Vitt, L. J., Zani, P. A., Caldwell, J. P. and Durtsche, R. D. (1993). Ecology of the whiptail lizard *Cnemidophorus deppii* on a tropical beach. *Can. J. Zool.* **71**, 2391–400.

Vrcibradic, D. and Rocha, C. F. D. (1996). Ecological differences in tropical sympatric skinks (*Mabuya macrorhyncha* and *Mabuya agilis*) in southeastern Brazil. *J. Herpetol.* **30**, 60–7.

Wainwright, P. C., Kraklau, D. M. and Bennett, A. F. (1991). Kinematics of tongue projection in *Chamaeleo oustaleti*. *J. Exp. Biol.* **159**, 109–33.

Whitford, W. G. and Bryant, M. (1979). Behavior of a predator and its prey: the horned lizard (*Phrynosoma cornutum*) and harvester ants (*Pogonomyrmex*). *Ecology* **60**, 686–94.

Wiens, J. A. (1989). Spatial scaling in ecology. *Funct. Ecol.* **3**, 385–97.

Yunger, J. A. (2004). Movement and spatial organization of small mammals following vertebrate predator exclusion. *Oecologia* **139**, 647–54.

Zaaf, A., Van Damme, R., Herrel, A. and Aerts, P. (2001). Spatiotemporal gait characteristics of level and vertical locomotion in a ground-dwelling and climbing gecko. *J. Exp. Biol.* **204**, 1233–46.

Zollner, P. A. and Lima, S. L. (1997). Landscape-level perceptual abilities in white-footed mice: perceptual range and the detection of forested habitat. *Oikos* **80**, 51–60.

16

The evolution of foraging behavior in the Galápagos marine iguana: natural and sexual selection on body size drives ecological, morphological, and behavioral specialization

MAREN N. VITOUSEK

Department of Ecology and Evolutionary Biology, Princeton University

DUSTIN R. RUBENSTEIN

Department of Neurobiology and Behavior, Cornell University

MARTIN WIKELSKI

Department of Ecology and Evolutionary Biology, Princeton University

Introduction

Each year thousands of tourists visit the Galápagos Islands and become intrigued by the unique habits of the world's only sea-going lizard, the Galápagos marine iguana (*Amblyrhynchus cristatus*), as it swims offshore and dives under the waves to feed. One of the islands' first visitors, Charles Darwin, reported fascination with watching these creatures forage, and he

opened the stomach of several, and in each case found it largely distended with minced sea-weed ... [that] grows at the bottom of the sea, at some little distance from the coast.

(Darwin, 1839)

We now know that the Galápagos marine iguana is the only terrestrial vertebrate that forages almost exclusively on macrophytic marine algae. Although marine iguanas are active foragers, their short, intense bouts of foraging activity more closely resemble the activity pattern of sit-and-wait foragers. To understand why these endemic lizards have adapted such a unique foraging strategy and how it differs from the general pattern of foraging in lizards, we must examine the social and environmental selective pressures that are unique to this species and its environment.

The Galápagos marine iguana is a model system to understand how natural and sexual selection drive morphological and behavioral adaptations. In this chapter we will show how the unique foraging strategy of the marine iguana is an adaptation resulting from the forces of both sexual selection, acting through their unique social system, and natural selection by a harsh and variable environment. We will explore a variety of environmental and physiological

Lizard Ecology: The Evolutionary Consequences of Foraging Mode, ed. S. M. Reilly, L. D. McBrayer and D. B. Miles. Published by Cambridge University Press. © Cambridge University Press 2007.

constraints (e.g. tidal cycle, the rapid loss of body heat during foraging, variation in body size) that act as overriding selective forces, and are not as salient to their terrestrial counterparts. Finally, we will examine how these constraints have resulted in the adaptation of a variety of unusual morphological characteristics that enable efficient foraging. Thus, we hope to demonstrate how contrasting social and environmental selective pressures on body size have shaped this unique foraging strategy and have resulted in the unusual morphological and behavioral characteristics that first caught Darwin's eye.

Natural and sexual selection

Marine iguanas are highly sexually dimorphic: adult males weigh approximately 70% more than adult females (Laurie and Brown, 1990a). This size dimorphism is driven primarily by sexual selection: females display a strong preference for mating with bigger males in this lekking, or arena-mating, species (Trillmich, 1983; Wikelski *et al.*, 1996, 2001, 2005; Partecke *et al.*, 2002). Male mating success is highly skewed (Mackenzie *et al.*, 1995); and a single male may be responsible for 35% of the copulations on a lek (Wikelski *et al.*, 1996, 2001). Despite sexual selection for large size, body size varies substantially between islands: maximum adult male body mass ranges from 1 kg on the island of Genovesa to 12 kg on southern Isabela Island (Laurie, 1983, 1989; Wikelski *et al.*, 1997). These differences in maximum body size are due to variability in algal productivity and sea surface temperature between islands (Wikelski and Trillmich, 1997; Wikelski and Romero, 2003; Wikelski, 2005). Large body size, however, comes at a high cost: members of the biggest size classes experience greatly increased mortality risks during periods of food shortage (Laurie and Brown, 1990b; Wikelski and Trillmich, 1997; Wikelski, 2005) owing to their higher total caloric requirement (Wikelski *et al.*, 1997). El Niño events, which result in a sharp decline in food availability, occur frequently in the Galápagos Islands, averaging once every 3.8 years (Quinn *et al.*, 1987; Laurie, 1989). Moreover, during these periods of food shortage reproductive failure may be complete (Wikelski and Trillmich, 1997). Thus, marine iguanas are subject to competing pressures of natural and sexual selection that determine optimal (and maximal) body size, which in turn shapes their unique foraging behavior.

Foraging behavior

Diet choice and foraging strategies

Although the seas around the Galápagos Islands are among the most productive in the world, marine iguanas are constrained in their foraging abilities by the cold and rough surface waters, as well as the restricted types of macrophytic marine algae upon which they can feed. Together, these constraints necessitate active foraging during a greatly restricted period that is ultimately regulated by the tidal cycle. Marine iguanas have evolved highly unique and efficient foraging strategies to overcome the temporal and spatial constraints associated with foraging on marine algae, allowing them to maximize energy intake and ultimately body size.

The preferred diet of marine iguanas consists primarily of two species of red alga *(Centroceras* sp. and *Gelidium* sp.) and one species of green alga *(Ulva sp.)*, although at least ten genera of alga are consumed regularly across the archipelago (Carpenter, 1966). The preferred algal species varies among islands (Wikelski *et al.*, 1993; Wikelski and Wrege, 2000; Rubenstein and Wikelski, 2003), although the preferred types appear to be those with the highest nutritional quality of the species present; on Santa Fe this is *Gelidium*, which has both the highest carbon and nitrogen content and C:N ratio (Rubenstein and Wikelski, 2003). Choice tests conducted on Santa Cruz Island confirmed the iguanas' preference for red algae over green algae, such as *Gelidium pusillum* var. *pacificum* and *Hypnea spinella* over *Ulva lobata* (Shepherd and Hawkes, 2005). As in their terrestrial relatives (Troyer, 1982), digestion of plants (i.e. algae) is achieved by specialized hindgut endosymbionts, shown to be capable of hydrolyzing and fermenting plant polymers that are indigestible to the host (Mackie *et al.*, 2004).

Given the strong selection for large body size and the constraints caused by variable food availability (Rubenstein and Wikelski, 2003) and environmental conditions, marine iguanas have evolved highly efficient methods of foraging to maximize energy intake. The total daily active period for marine iguanas lasts between one third and one half of the duration of their closest relative, *Conolophus pallidus* (Christian and Tracy, 1985; Wikelski and Trillmich, 1994; Drent *et al.*, 1999). Marine iguanas forage actively by grazing on intertidal and/or subtidal algae (Hobson, 1965; Carpenter, 1966; Trillmich, 1979; Buttemer and Dawson, 1993; Wikelski and Trillmich, 1994). Although marine iguanas are famed for their unique diving habits, the majority of animals forage intertidally, grazing at low tide on exposed algae. Only about 5% of each population forages subtidally by diving beneath the sea surface to feed on offshore algal beds (Wikelski and Trillmich, 1994). Body condition (body mass (g) $\times 10^6$ / SVL^3) (Laurie, 1989) does not differ between marine iguanas utilizing these two modes of obtaining energy (Wikelski and Trillmich, 1994). Subtidal and intertidal foragers may differ in the proportion of various algal species contained in the diet, but the quality of these diets is not significantly different (Wikelski *et al.*, 1993). Individuals are highly consistent in the foraging strategy used, and the choice of this strategy is dependent upon body size (Trillmich and Trillmich, 1986): most females and smaller males forage intertidally, whereas larger animals typically forage subtidally. A small group of intermediate-sized iguanas may exploit both strategies (Trillmich and Trillmich, 1986). The minimum size of diving iguanas varies widely between islands and may depend on population density, algal abundance, and the size and topography of the site (Trillmich and Trillmich, 1986; Wikelski and Trillmich, 1994). Minimum body size of subtidal foragers ranges over a five-fold difference, from 600 g on Genovesa to 3000 g on Fernandina.

Temporal and spatial constraints on foraging

Given the similar diet quality and relative energetic benefits of foraging intertidally vs. subtidally, why are there size-associated differences in foraging strategy? The shift from intertidal to subtidal foraging appears to be due to the higher total caloric requirement of large animals (despite a reduced relative energetic requirement) (Wikelski and Trillmich, 1997; Wikelski *et al.*, 1997). Intertidal foragers are both temporally and spatially constrained, whereas subtidal foragers experience reduced temporal constraints and can predominantly avoid spatial constraints. Intertidal foraging is limited to the period of low tide, when algal beds are accessible (Hobson, 1965; Trillmich and Trillmich, 1986; Buttemer and Dawson, 1993; Wikelski and Trillmich, 1994; Rubenstein and Wikelski, 2003). Environmental effects significantly impact the duration of intertidal foraging: rough seas and high low-tide levels decrease the total time devoted to foraging (Trillmich and Trillmich, 1986; Wikelski and Trillmich, 1994). The pattern of foraging is also dependent upon environmental conditions on land: on sunny days marine iguanas are able to increase body temperature more quickly following foraging, and thus shorten re-warming duration between successive bouts (Buttemer and Dawson, 1993; Wikelski and Trillmich, 1994). Foraging efficiency is constrained by body temperature: marine iguanas prefer to maintain a body temperature between 35 and 39 °C (White, 1973; Bartholomew *et al.*, 1976; Wikelski and Trillmich, 1994), much warmer than the usually cool (11–23 °C) Galápagos sea surface temperature. Voluntary body temperatures of up to 43 °C have been observed in hatchling marine iguanas before foraging. Body temperature declines linearly throughout a foraging bout, resulting in a decreasing bite rate and reduced energy intake (Fig. 16.1) (Wikelski and Trillmich, 1994). The speed of digestion also decreases at lower temperatures: gut passage time in marine iguanas has a Q_{10} of approximately 2.5 (Wikelski *et al.*, 1993).

Because of these temperature constraints, intertidally foraging marine iguanas prefer to graze during morning or midday low tides, because these

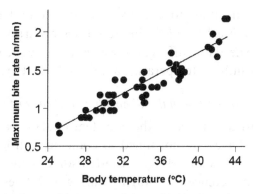

Figure 16.1. Maximum bite rate of intertidally foraging marine iguanas increases linearly with body temperature, as measured by implanted radiotelemetry tags on Santa Fe (redrawn from Wikelski and Trillmich, 1994).

periods offer both the lowest tide levels and the fastest re-warming after foraging. If low tides occur during early morning, marine iguanas are less able to achieve the warm body temperatures necessary for a successful foraging bout; if they occur in the late afternoon, animals have a difficult time re-warming sufficiently after returning from the water. When the low tide occurs after 1600 h, animals switch from foraging during the early afternoon to the subsequent morning low tide. This shift takes place approximately every 14 days (Trillmich and Trillmich, 1986; Wikelski and Hau, 1995; Rubenstein and Wikelski, 2003). Marine iguanas foraging intertidally in cold waters further maximize their body temperature by foraging slightly after low tide in the morning, and ahead of low tide in the afternoon (Trillmich and Trillmich, 1986).

The timing of intertidal foraging is also affected by algal spatial constraints. Algal blade length is greater at increased intertidal depths (Rubenstein and Wikelski, 2003). Algae grow quickly when submerged during high tide, but are rapidly grazed down by foraging iguanas. This competition causes marine iguanas on some islands to anticipate the tides, arriving at the foraging grounds prior to the time of low tide and thereby gaining a competitive advantage (Wikelski and Hau, 1995). Anticipation is greatest on islands with lower food supplies, and during the periods of poor foraging conditions associated with El Niño events. During a long-term El Niño on the island of Genovesa (which has relatively poor food availability even during 'good' years), marine iguanas that arrived at the foraging grounds earlier had higher survival rates (Wikelski and Hau, 1995). Anticipation is reduced when low tide occurs earlier in the day, and increases during afternoon low tides. Early foraging may not be efficient near the beginning of the day, when iguanas have had insufficient time to achieve warmer body temperatures. The increased anticipation exhibited during low tides occurring later in the day may result from the need for sufficient re-warming subsequent to foraging and prior to the drop in ambient temperature associated with the onset of darkness.

Ontogenetic and sexually dimorphic patterns of foraging

Subtidal foragers are not subject to the same temporal or spatial constraints as intertidal foragers. Dives are not timed to coincide with low tide (Hobson, 1965; Buttemer and Dawson, 1993), although some populations of subtidal foragers display a preference for foraging during high tides (Rubenstein and Wikelski, 2003). In general, however, the suitable foraging window is not as restrictive as it is in intertidal foraging. Subtidal foraging occurs between late morning and early afternoon (Trillmich and Trillmich, 1986); this allows re-emergence at the hottest period of the day to maximize re-warming rate (Buttemer and Dawson, 1993). Subtidally foraging animals are also able to feed during a much greater range of weather conditions. Moreover, foraging duration is more plastic in subtidal than in intertidal foraging bouts. The duration of foraging dives averages approximately three minutes, although individuals may remain voluntarily submerged for over 45 min (Hobson, 1969; Trillmich and Trillmich, 1994; M. Wikelski, personal observation). Dive duration is inversely related to algal productivity (Fig. 16.2), whereas the duration

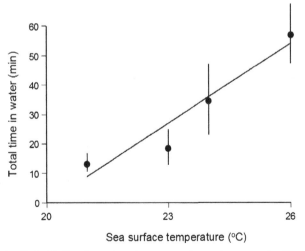

Figure 16.2. Subtidally foraging iguanas show increased dive durations during periods of high sea surface temperature and low algal productivity, and decreased dive length when sea surface temperature is low, indicating high productivity of preferred algal species (from Wikelski and Trillmich, 1994).

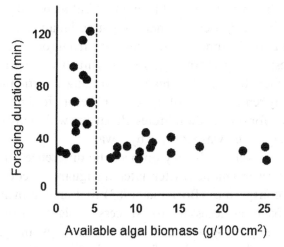

Figure 16.3. The duration of intertidal foraging bouts within a site remains constant with changing algal biomass, except below a threshold level of 5 g/100 cm², when intertidal foraging duration increases.

of intertidal foraging bouts within a site remains constant, except at very low algal biomass (Fig. 16.3).

The total time devoted to foraging varies between populations: on some islands subtidal foragers spend significantly more time feeding, whereas on others intertidal foragers devote more time to obtaining energy (Trillmich and Trillmich, 1986; Wikelski and Trillmich, 1994; Drent *et al.*, 1999). The total time spent foraging in the intertidal zone varies. Intertidal foragers show much greater variance in the duration of foraging bouts than subtidal foragers, and more frequently fast for up to several days between foraging bouts (Trillmich and Trillmich, 1986).

Despite the reduced temporal and spatial constraints associated with subtidal foraging, it is not an effective or efficient method of foraging for smaller marine iguanas; the energetic costs appear to outweigh the benefits. Larger animals are much more efficient swimmers, as both swimming speed and power generated per body undulation are directly related to body size (Bartholomew *et al.*, 1976; Vleck *et al.*, 1981). Smaller iguanas may not be able to generate enough power to overcome the buoyancy that results from diving with inflated lungs (Bartholomew *et al.*, 1976). Smaller animals also lose heat more rapidly than larger individuals (Bartholomew and Lasiewski, 1965). Because of the difference in heat loss, smaller iguanas would be forced to return to land to warm up more frequently than larger animals, or face inefficient foraging due to the temperature-associated decrease in bite rate. An increased locomotory shuttling between cold water and warm land would greatly increase the cost of subtidal foraging (Trillmich and Trillmich, 1986). Intertidal foragers, in particular small or young iguanas, remain above the water the majority of the time, and are never far from rocks on which they can warm themselves; thus, this foraging strategy is much less expensive for smaller individuals. In fact, the smallest iguanas foraging intertidally on Santa Fe island attain body temperatures of up to 43 °C before foraging, and remain in the intertidal area for only a few minutes, departing with body temperatures of 37 °C, having never been washed over by a wave.

Hatchlings and juveniles display ontogenetic differences in foraging strategy. For the first several months after hatching, iguanas predominantly forage on the feces of conspecifics (Boersma, 1983) that enable them to obtain the endosymbiotic bacteria necessary for successful digestion of algae (Troyer, 1982; Mackie *et al.*, 2004). Although older hatchlings and juveniles forage intertidally, they are restricted to foraging in the upper intertidal region. Smaller individuals are subject to the competing demands of a higher relative energetic requirement, coupled with a decreased ability to retain heat when exposed to cold sea water (Wikelski *et al.*, 1993). Running speed decreases with body temperature, and this limits the ability of smaller iguanas to elude large waves. Gripping strength, an important indicator of the ability of marine iguanas to maintain their position on rocks in the intertidal zone when buffeted by high surf, is also positively correlated with body size (Wikelski and Trillmich, 1994).

Small individuals are restricted to foraging in the upper intertidal zone where they have a reduced chance of encountering large waves that would cool them more quickly and might sweep them off of rocks. A juvenile swept into the ocean would face a much higher relative energetic cost of swimming back to shore than a large individual, and might face the additional danger of being consumed or injured by large predatory fish and marine invertebrates (Wikelski and Trillmich, 1994; M. Wikelski, personal observation; H. Snell, personal observation). Although foraging in the upper intertidal zone provides many advantages to small marine iguanas, the shorter algal blade lengths and higher density of foraging animals result in increased competition for scant resources.

Surviving times of crisis

Although large animals are generally able to obtain sufficient energy by foraging subtidally, they suffer much higher mortality during conditions of low food availability (Laurie and Brown, 1990a). Marine iguanas experience frequent but unpredictable declines in food availability during El Niño events, in which warming sea-surface temperatures (up to 32 °C) and a failure of the upwelling of cold, nutrient-rich water induces massive dieback of the preferred algal species (*Gelidium* spp., *Centroceras* spp., *Ulva* spp. and *Spermothamnium* spp.) (Laurie, 1987, 1989; Rubenstein and Wikelski, 2003). Subsequently, intertidal zones are colonized and grown over by a species of brown alga, *Giffordia mitchelliae*. Marine iguanas face difficulties digesting these brown algae, possibly because iguanas lack the hindgut endosymbionts to digest *G. mitchelliae* (Laurie, 1989; Cooper and Laurie, 1987). As a consequence, iguanas experience greatly increased mortality rates during El Niño events, due to starvation. The 1982–3 El Niño resulted in 60% mortality across the archipelago (Laurie, 1990). Mortality rates of up to 90% have been recorded on individual islands (Seymour Norte in 1997–8 and Genovesa in 1991–4) (Wikelski and Wrege, 2000; Romero and Wikelski, 2001).

During El Niños there is differential mortality among size classes, with the highest mortality rates found in the largest animals, owing to higher absolute caloric requirements (Laurie and Brown, 1990b; Wikelski and Trillmich, 1997; Wikelski *et al.*, 1997), and in the smallest size classes, which are restricted to foraging in the upper intertidal zone and thus may be subject to the strongest resource competition (Laurie and Brown, 1990a,b). The foraging efficiency of marine iguanas, measured as intake/bite (mg g$^{-0.8}$), declines with increasing size (Fig. 16.4), which may further contribute to the high mortality rates observed. An additional factor constraining small animals during El Niños may be the rising sea-surface level in the Galápagos, which can climb more than 30 cm above normal, causing the total submergence of the preferred algal species typically located in the intertidal region.

Uniquely among lizards, marine iguanas may have evolved the ability to reduce their body length during periods of food shortage to improve their

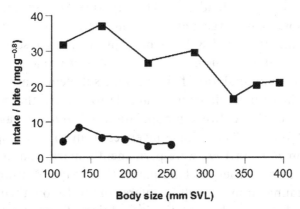

Figure 16.4. Foraging efficiency of intertidally foraging marine iguanas differs between islands (different symbols) and is negatively related to body size (redrawn from Wikelski and Romero, 2003).

survival probability. During El Niño events, marine iguanas appear to be able to decrease body length by as much as 20% (6.8 cm) within two years. Reduction of this magnitude cannot be accounted for solely by reducing cartilage and connective tissue, which together account for only *c.* 10% of body length, and suggests that these animals are resorbing bone mass (Wikelski and Thom, 2000). The amount of shrinkage is significantly associated with survival, owing to both increased foraging efficiency and decreased energy expenditure of smaller individuals. The magnitude of decrease in body length is greatest among the largest size classes, which are typically those that suffer the highest mortality during El Niño events (Wikelski and Thom, 2000).

A unique instance of dietary niche expansion

A subpopulation of marine iguanas on Seymour Norte Island found a unique way to increase maximum body size through dietary niche expansion, thereby gaining an advantage in sexual selection (Wikelski and Wrege, 2000). These iguanas exhibited the distinctive foraging behavior of supplementing their algal diet with the beach plant *Batis maritima* both before and after foraging intertidally. Although the nutrient value of *Batis* is lower than that of marine algae, and assimilation efficiency is likely to be much lower, *Batis* foraging is not subject to the same time constraints (tidal cycle and decreasing body temperature) as intertidal foraging. Through expanding dietary niche breadth *Batis* foragers were able to increase maximum body size by nearly 100% when compared with non-*Batis* eaters. The decreased dependence on algal abundance appears to have increased survival of the Seymour Norte *Batis* foragers during historic El Niño events: despite the near-total mortality of individuals in the largest size classes on many islands during El Niños, many large *Batis*-foraging males, some in excess of 2 kg, survived. During the 1997–8 El Niño, however, high sea-surface levels flooded and killed the entire population of *Batis* plants on which the Seymour Norte subpopulation fed. Without supplementation from *Batis*, the unusually large individuals were unable to survive the period of food deprivation, and the entire subpopulation of *Batis* foragers perished (Wikelski and Wrege, 2000). Some juveniles from this subpopulation that did not feed on *Batis* in 1997 were permanently marked, and in 2003 were observed ingesting a variety of land plants (Galápagos Naturalists Guides, personal communication). It will be interesting to follow their ontogeny to determine whether this subpopulation's tradition of terrestrial foraging will be successfully passed down to this generation, and whether it will enable the development of larger body size.

Morphology, behavior, and physiology

Endogenous rhythms and foraging patterns

The precise timing of foraging during the tidal cycle is vitally important to intertidally foraging marine iguanas, and may be regulated by an endogenous rhythm (Wikelski and Hau, 1995). Iguanas that have been resting out of view of the ocean for several days will arrive at the foraging grounds not at the previous time of foraging, but rather at the predicted time of low tide (or prior to low tide in populations that anticipate tidal changes). Moreover, marine iguanas experimentally housed in darkened enclosures retain foraging activity rhythms with a mean period of 24.5 h, although the period of their apparently free-running endogenous rhythm is sometimes longer or shorter than the tidal cycle. Captive animals will also shift activity levels from the evening to the morning low tide at the appropriate time during the 14-day low tide cycle. Foraging behavior does not appear to be synchronized by social cues, as initially synchronous animals housed together in darkened enclosures may develop asynchronous periodicity (Wikelski and Hau, 1995). Experimentally altering the onset of light in animals enclosed away from the shoreline does not shift their activity rhythm, indicating that foraging activity is not primarily cued by the light–dark cycle. Finally, foraging time is more closely related to the calculated time of low tide than to the actual time at which the intertidal algae beds are exposed. These results suggest that marine iguanas possess either a circatidal (c. 12.4 h) or circalunadian (c. 24.8 h) oscillatory system that enables them to predict the timing of low tide, and thus the most appropriate period for foraging (Wikelski and Hau, 1995).

Algal quality cues reproduction

Food availability also has implications for the timing of reproduction. Marine iguanas breed seasonally only once a year. Males typically establish territories two months prior to the start of reproduction (Trillmich, 1983; Partecke et al., 2002). The reproductive period varies by several months across islands, with the mean copulation date ranging from mid December on Santa Fe and Genovesa to February–March on Española and southern Isabela (Trillmich, 1979; Laurie, 1984). Within each population copulation is relatively synchronous, with all matings on a particular island occurring over a 20–25 d period (Trillmich, 1979; Partecke et al., 2002; Rubenstein and Wikelski, 2003). The timing of synchronized reproduction varies by up to several weeks each year within a population, and coincides with the peak nutritional quality of the preferred algal species (Rubenstein and Wikelski, 2003). In years when sea-surface temperatures are higher, both the median date of copulation and the peak algal quality occur earlier. This indicates that marine iguanas may use changes in algal quality, associated with seasonal but variable changes in water temperature, as a cue to initiate breeding (Rubenstein and Wikelski, 2003).

Foraging energetics

Marine iguanas have developed a variety of morphological adaptations to maximize foraging efficiency. They possess blunt heads with tricuspid teeth that are flattened laterally, enabling effective grazing on marine algae (Darwin, 1839), as well as long and sharp claws with a powerful ability to grip onto the lava (Carpenter, 1966). Their tails are flattened compared with those of other iguanids, and thereby adapted for efficient swimming (Tracy and Christian, 1985). Black coloration and circulatory heat shunts help to maintain preferred body temperature during foraging bouts in the cold sea water. Marine iguanas also possess highly developed salt glands that allow them to excrete the excess salt ingested while foraging in sea water (Dunson, 1969), and harbor endosymbiotic bacteria that allow efficient digestion of marine macrophytes. Some of these morphological adaptations contribute to a greatly decreased cost of locomotion in water compared with the cost of walking at the same speed: at 1.0 km/h, swimming has been estimated to use only one quarter of the energy expended on terrestrial locomotion (Gleeson, 1979). Marine iguanas do not, however, appear to have adapted physiologically to the low sea-surface temperatures characteristic of their foraging environment by lowering the temperature at which they can be maximally active (Bartholomew, 1966; Dawson et al., 1977; Trillmich and Trillmich, 1986). Foraging behavior is therefore highly constrained by body temperature (Fig. 16.5). Foraging accounts for approximately 8%–10% of average daily energetic expenditure, as determined by both heart rate energetics (M. Vitousek and M. Wikelski, unpublished data) and the doubly labeled water method (Gleeson, 1979; Nagy and Shoemaker, 1984; Drent et al., 1999). Energetic expenditure on foraging varies widely, however, ranging from 0 (no foraging) to over 40 kJ/d. Intertidally foraging animals may spend more than 75% of their total daytime expenditure on foraging behaviors (Fig. 16.6) (M. Vitousek, unpublished data).

Unlike many lizards, marine iguanas of different sex and size classes do not differ in diet composition, diet quality, or digestive efficiency (which, at 70%, is higher than that of most herbivorous lizards: Wikelski et al., 1993). Size

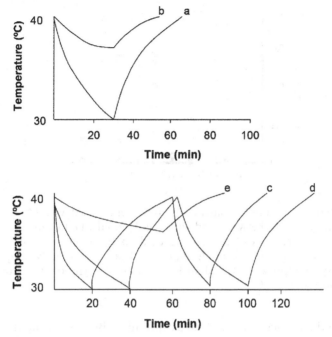

Figure 16.5. Schematic variable sawtooth model of foraging behavior of iguanas. Iguana body temperature (curved line) falls from around 40 °C while foraging and rises while basking in a sawtooth manner. Iguanas stop feeding when full or cold, whichever happens first. Curves are steeper in the cool season than in the hot season. When algal biomass is high a single feeding excursion per low tide is sufficient. Letters indicate: a, Academy Bay, Santa Cruz, in the cool season (this study); b, Tortuga Bay, Santa Cruz, in the hot season (this study) (if algal biomass were low, as is more common in the cool season, the iguana might require two or more feeding excursions); c, Santa Fe in the cool season (Wikelski and Trillmich, 1994; Wikelski et al., 1997); d, Fernandina in the cool season (Trillmich and Trillmich, 1986); e, Genovesa in the hot season (after Wikelski et al., 1997).

classes also do not differ in time devoted to foraging on a given island (Wikelski and Trillmich, 1994). Larger iguanas have a higher total intake of dry algal mass than small animals; however, the difference between food intake and energetic expenditure is much lower in large animals, indicating that these individuals are barely meeting their energetic requirements whereas small iguanas have an energetic surplus (Wikelski et al., 1997; Drent et al., 1999; Romero and Wikelski, 2003).

The average daily energetic expenditure of intertidal and subtidal foragers does not differ, although on some islands subtidal foragers spend significantly more time foraging than intertidal foragers (Drent et al., 1999). Daily energetic expenditure varies between islands; this difference probably

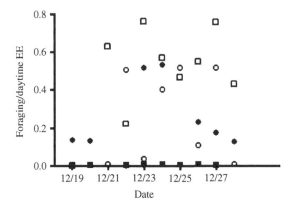

Figure 16.6. The amount of daily energetic expenditure devoted to foraging is highly variable both between and within individuals. Data show the percent of daytime energetic expenditure devoted to foraging (n = 4 Santa Fe females), as determined through implanted heart rate – body temperature. Symbol types represent individual females; closed symbols are from 1999, open symbols from 2003.

results from differences in resource availability. Both subtidal and intertidal foragers have a higher energetic expenditure on Santa Fe, an island with relatively low sea-surface temperature, rough seas, and high total algal productivity, than on neighboring Genovesa, which is characterized by warmer water, smoother seas, and a lower total algal productivity (Drent *et al.*, 1999).

Conclusion

The combination of unique social and environmental selective pressures faced by the Galápagos marine iguana have resulted in the development of a foraging style that is unique, not only among iguanids, but among all reptiles. In order to meet the energetic demands associated with large (sexually selected) body size, marine iguanas have evolved unique foraging strategies specialized for the consumption of marine macrophytic algae, which is both temporally and spatially constrained. The short, intense, and energetically costly grazing bouts of marine iguanas combine the leisurely grazing of herbivorous lizards with the active-searching foraging style of carnivorous lizards. In addition, they have evolved morphological adaptations, such as blunt heads, black coloration, and salt glands, that help them maximize energy intake during their costly foraging bouts. In summary, the Galápagos marine iguana is a model system to understand how natural and sexual selection drive morphological and behavioral adaptations.

References

Bartholomew, G. A. (1966). A field study of temperature relations in the Galapagos marine iguana. *Copeia* **1966**, 241–50.

Bartholomew, G. A., Bennett, A. F. and Dawson, W. R. (1976). Swimming, diving and lactate production of the marine iguana, *Amblyrhynchus cristatus. Copeia* **4**, 709–20.

Bartholomew, G. A. and Lasiewski, R. C. (1965). Heating and cooling rates, heart rate and simulated diving in the Galapagos marine iguana. *Comp. Biochem. Physiol.* **16**, 573–82.

Boersma, P. D. (1983). An ecological study of the Galapagos marine iguana. In *Patterns of Evolution in Galapagos Organisms*, ed. R. I. Bowman, M. Berson and A. E. Leviton, pp. 157–76. San Francisco, CA: AAAS.

Buttemer, W. A. and Dawson, W. R. (1993). Temporal pattern of foraging and microhabitat use by Galapagos marine iguanas, *Amblyrhynchus cristatus. Oecologia* **96**, 56–64.

Carpenter, C. C. (1966). The marine iguana of the Galapagos Islands, its behavior and ecology. *Proc. Cal. Acad. Sci.* **34**, 329–76.

Christian, K. A. and Tracy, C. R. (1985). Physical and biotic determinants of space utilization by the Galapagos land iguana *(Conolophus pallidus). Oecologia* **66**, 132–40.

Cooper, W. E. and Laurie, W. A. (1987). Investigation of deaths in marine iguanas *(Amblyrhynchus cristatus)* on Galapagos. *J. Comp. Pathol.* **97**, 129–36.

Darwin, C. (1839). *Journal of Researches.* Cambridge: Cambridge University Press.

Dawson, W. R., Bartholomew, G. A. and Bennett, A. F. (1977). A reappraisal of the aquatic specializations of the Galapagos marine iguana *(Amblyrhynchus cristatus). Evolution* **31**, 891–7.

Drent, J., van Marken Lichtenbelt, W. D. and Wikelski, M. (1999). Effects of foraging mode and season on the energetics of the marine iguana, *Amblyrhynchus cristatus. Funct. Ecol.* **13**, 493–9.

Dunson, W. A. (1969). Electrolyte excretion by salt gland of Galapagos marine iguanas. *Am. J. Physiol.* **216**, 995.

Gleeson, T. T. (1979). Foraging and transport costs in the Galapagos marine iguana, *Amblyrhynchus cristatus. Physiol. Zool.* **52**, 549–57.

Hobson, E. S. (1965). Observations on diving in the Galapagos marine iguana, *Amblyrhynchus cristatus* (Bell). *Copeia* **1965**, 1180–9.

Hobson, E. S. (1969). Remarks on aquatic habits of the Galapagos marine iguana, including submergence times, cleaning symbiosis, and the shark threat. *Copeia* **1969**, 401–2.

Laurie, W. A. (1983). Marine iguanas in Galapagos. *Oryx* **17**, 18–25.

Laurie, W. A. (1984). Interim report on the marine iguana situation in the aftermath of the 1982–3 El Niño. *Noticias de Galapagos* **40**, 9–11.

Laurie, W. A. (1987). Marine iguanas – living on the ocean margin. *Oceanus* **30**(2), 54–60.

Laurie, W. A. (1989). Effects of the 1982–83 El Niño-southern oscillation event on marine iguana *(Amblyrhynchus cristatus* Bell, 1825) populations on Galapagos. In *Global Ecological Consequences of the 1982–83 El Niño-Southern Oscillation*, vol. 52, ed. P. W. Glynn, pp. 361–80. New York: Elsevier.

Laurie, W. A. (1990). Population biology of marine iguanas *(Amblyrhynchus cristatus)* I. Changes in fecundity related to a population crash. *J. Anim. Ecol.* **59**, 515–28.

Laurie, W. A. and Brown, D. (1990a). Population biology of marine iguanas *(A. cristatus)* II. Changes in annual survival rates and the effects of size, sex, age and fecundity in a population crash. *J. Anim. Ecol.* **59**, 529–44.

Laurie, W. A. and Brown, D. (1990b). Population biology of marine iguanas *(Amblyrhynchus cristatus)* III. Factors affecting survival. *J. Anim. Ecol.* **59**, 545–68.

Mackenzie, A., Reynolds, J. D., Brown, V. J. and Sutherland, W. J. (1995). Variation in male mating success on leks. *Am. Nat.* **145**, 633–52.

Mackie, R. I., Rycyk, M., Ruemmler, R. L., Aminov, R. I. and Wikelski, M. (2004). Biochemical and microbiological evidence for fermentative digestion in free-living land iguanas *(Conolophus pallidus)* and marine iguanas *(Amblyrhynchus cristatus)* on the Galapagos archipelago. *Physiol. Biochem. Zool.* **77**, 127–38.

Nagy, K. A. and Shoemaker, V. H. (1984). Field energetics and food consumption of the Galapagos marine iguana, *Amblyrhynchus cristatus. Physiol. Zool.* **57**, 281–90.

Partecke, J. A., von Haenseler, A. and Wikelski, M. (2002). Territory establishment patterns support the hotshot-hypothesis in lekking marine iguanas. *Behav. Ecol. Sociobiol.* **51**, 579–87.

Quinn, W. H., Neal, V. T. and Antunez de Mayolo, S. E. (1987). El Niño occurrences over the past four and a half centuries. *J. Geophysical Res.* **C92**, 14.

Romero, L. M. and Wikelski, M. (2001). Corticosterone levels predict survival probabilities of Galapagos marine iguanas during El Niño events. *Proc. Natn. Acad. Sci. USA* **98**, 7366–70.

Rubenstein, D. R. and Wikelski, M. (2003). Seasonal changes in food quality: a proximate cue for reproductive timing in marine iguanas. *Ecology* **84**, 3013–23.

Shepherd, S. A. and Hawkes, M. W. (2005). Algal food preferences and seasonal foraging strategy of the marine iguana, *Amblyrhynchus cristatus*, on Santa Cruz, Galápagos. *Bull. Mar. Sci.* **77**, 51–72.

Tracy, C. R. and Christian, K. A. (1985). Are marine iguana tails flattened? *Brit. J. Herpetol.* **6**, 434–5.

Trillmich, K. (1979). Feeding behavior and social behavior of the marine iguana. *Noticias de Galapagos* **29**, 17–20.

Trillmich, K. (1983). The mating system of the marine iguana, *Amblyrhynchus cristatus. Z. Tierpsychol.* **63**, 141–72.

Trillmich, K. G. K. and Trillmich, F. (1986). Foraging strategies of the marine iguana, *Amblyrhynchus cristatus. Behav. Ecol. Sociobiol.* **18**, 259–66.

Troyer, K. (1982). Transfer of fermentative microbes between generations in a herbivorous lizard. *Science* **216**, 540–2.

Vleck, D., Gleeson, T. T. and Bartholomew, G. A. (1981). Oxygen consumption during swimming in Galapagos marine iguanas and its ecological correlates. *J. Comp. Physiol.* **141**, 531–6.

White, F. N. (1973). Temperature and the Galapagos marine iguana – insights into reptilian thermoregulation. *Comp. Biochem. Physiol.* **45A**, 503–13.

Wikelski, M. (2005). Evolution of body size in Galapagos marine iguanas. *Proc. R. Soc. Lond.* **B272**, 1985–93.

Wikelski, M. and Hau, M. (1995). Is there an endogenous tidal foraging rhythm in marine iguanas? *J. Biol. Rhyth.* **10**, 335–50.

Wikelski, M. and Romero, L. M. (2003). Body size, performance and fitness in Galapagos marine iguanas. *Int. Comp. Biol.* **43**, 376–86.

Wikelski, M. and Thom, C. (2000). Marine iguanas shrink to survive El Niño. *Science* **403**, 37–8.

Wikelski, M. and Trillmich, F. (1994). Foraging strategies of the Galapagos marine iguana *(Amblyrhynchus cristatus)*: adapting behavioral rules to ontogenetic size change. *Behavior* **128**, 255–79.

Wikelski, M. and Trillmich, F. (1997). Body size and sexual size dimorphism in marine iguanas fluctuate as a result of opposing natural and sexual selection: an island comparison. *Evolution* **51**, 922–36.

Wikelski, M. and Wrege, P. H. (2000). Niche expansion, body size and survival in Galapagos marine iguanas. *Oecologia* **124**, 107–15.

Wikelski, M., Carbone, C., Bednekoff, P. A., Choudhury, S. and Tebbich, S. (2001). Why is female choice not unanimous? Insights from costly mate sampling in marine iguanas. *Ethology* **107**, 623–38.

Wikelski, M., Carbone, C. and Trillmich, F. (1996). Lekking in marine iguanas: female grouping and male reproductive strategies. *Anim. Behav.* **52**, 581–96.

Wikelski, M., Carillo, V. and Trillmich, F. (1997). Energy limits to body size in a grazing reptile, the Galapagos marine iguana. *Ecology* **78**, 2204–17.

Wikelski, M., Gall, B. and Trillmich, F. (1993). Ontogenetic changes in food intake and digestion rate of the herbivorous marine iguana (*Amblyrhynchus cristatus*, Bell). *Oecologia* **94**, 373–9.

Wikelski, M., Steiger, S. S., Gall, B. and Nelson, K. N. (2005). Sex, drugs and mating role: testosterone-induced phenotype-switching in Galapagos marine iguanas. *Behav. Ecol.* **19**, 260–8.

17

The evolution of the foraging mode paradigm in lizard ecology

LANCE D. McBRAYER

Department of Biology, Georgia Southern University

DONALD B. MILES

Department of Biological Sciences, Ohio University

STEPHEN M. REILLY

Department of Biological Sciences, Ohio University

> Always question the paradigm
>
> *Carl Gans, 1986*

Sometimes a straightforward natural history observation initiates the development of a major area of research in ecology or evolutionary biology. The observation that species numbers increase with island area is one such example. Another is the description by Pianka (1966) and Schoener (1969) of two "distinct" behavioral morphs that differed in their feeding behavior forty years ago. Although other studies described the behavior (see, for example, Kennedy, 1956; Rand, 1967), it was the early publications of Pianka and Schoener that demonstrated the ecological significance of the search strategies. Ostensibly a species' movement behavior affected its foraging success and consequently was a potential mechanism for resource partitioning. Hence, understanding variation in foraging mode was a foundation for key papers in theoretical and empirical analyses of species interactions (Schoener, 1971). However, ecologists quickly realized the numerous ramifications inherent in the differences between species that ambush prey vs. those that widely search an environment for elusive or concealed prey (see, for example, Eckhardt, 1979).

In a seminal paper, Huey and Pianka (1981) formalized the foraging mode paradigm. Their study elaborated on the potential ecological consequences of variation in search behavior and presented a summary of the traits that were expected to be affected by foraging mode. Using data collected from Kalahari lizards, they corroborated several of the hypothesized differences between ambush and widely foraging lizards. One may ask why their publication was so important. First, many studies on a broad array of vertebrate and invertebrate taxa found that species were readily classified into either ambush foragers or widely foraging species. Second, foraging mode became shorthand

Lizard Ecology: The Evolutionary Consequences of Foraging Mode, ed. S. M. Reilly, L. D. McBrayer and D. B. Miles. Published by Cambridge University Press. © Cambridge University Press 2007.

for describing species attributes. Third, as McLaughlin (1989) pointed out, the foraging mode paradigm includes the opportunity to understand how foraging strategies correlate with a suite of behavioral, ecological, physiological, and morphological characters. Thus, their paper provided a framework to explain a myriad of differences between species and hence derive additional hypotheses to explain the diversification of taxa. Consequently, our thinking became focused on examining foraging mode, as a pervasive evolutionary force.

Since the publication of Huey and Pianka (1981), numerous studies related foraging modes to many aspects of lizard ecology, and because so many traits have been correlated with foraging mode, the sit-and-wait vs. active foraging (SW vs. AF) dichotomy emerged as a central paradigm in lizard ecology. Our goal at the onset of this project was to review what has been learned about the foraging biology of lizards in over three decades of research. Specifically, we wanted to know how various aspects of lizard biology covary with foraging mode and how these patterns relate to the initial dichotomous paradigm. Therefore, we asked contributors to address four questions. First, how has a given trait or suite of traits been affected by foraging mode? Second, because differences in foraging mode have been generally related to a deep split in the phylogeny of modern lizards (Perry, 1999), what is the concordance between phylogeny, foraging mode, and a given trait? Third, do traits associated with foraging mode in lizards follow a continuum or is there evidence that foraging mode patterns are clearly dichotomous? And fourth, what are the unresolved questions that need additional study in order to understand the paradigm? By unifying these contributions we believe this volume represents a review of the state of the paradigm at many levels of organismal design, function and ecology. As is evident from the chapters, there is a tremendous amount of information linked with foraging mode. However, at the same time it is also true that each chapter identified major areas of needed research.

This volume has also provided the opportunity to evaluate support for a dichotomous characterization of foraging mode. Clearly, many of the authors supported the existence of a continuum. Although many species do occupy extreme positions on a spectrum describing activity during foraging, a remarkable result emerging from the chapters in this volume is the number of species that are intermediate between ambush and active foragers. Thus, a key question is how to account for the additional variation. Should a third category or continuum be recognized? If so, how would this affect the syndrome of traits that have been intimately tied to the two foraging categories? Certainly, more support exists for the description of *relative* differences in search behavior first suggested by Pianka (1966) and reiterated by Huey and Pianka (1981). However, categorizing species into one or another mode, albeit convenient, dramatically discounts variation that may relate to other key functional traits and environmental heterogeneity. Consequently, it is prudent to use quantitative traits, moves per minute, percent time moving, or other available variables, to test hypotheses regarding the proximal and ultimate factors that structure variation in search behavior (McLaughlin, 1989).

We have striven to provide future researchers with an integrative description of the behavioral syndrome that is "foraging mode." The volume also serves as a review of the ecological consequences of behavior extended across numerous trait complexes. We know of no other example in vertebrates where an integrative behavioral variable like foraging mode has such an extensive reach into the biology and phylogenetic history of a clade. The central role of foraging mode in lizard biology has spurred investigations in a diversity of fields within ecology, morphology, evolution, and behavior and validated foraging mode as a paradigm in lizard ecology. The paradigm now involves not only the idea of foraging mode variation but the totality of the organismal traits that may or may not covary with foraging strategies.

Evaluating foraging mode variation

A long standing question within the foraging mode paradigm has been whether to treat search behavior as a continuum or a dichotomy. The first classifications of lizards into foraging modes were subjective and as described by Pianka (1973) somewhat arbitrary. The lack of objective criteria for defining the foraging mode was a key weakness early on, yet despite a dichotomous classification most authors concluded that species were arrayed along a continuum, although most taxa investigated tended to occur in the extremes of the continuum. Huey *et al.* (1984) provided some guidelines toward delineating foraging modes by suggesting the use of percent time moving (PTM) as a relative guideline for classifying species. Substantial differences were evident in the treatment of foraging mode. Some authors concluded that two categories were insufficient to capture the observed variation in foraging mode (Regal, 1978). In contrast, Magnusson *et al.* (1985) favored the treatment of foraging mode as a continuum. Most of these suggestions were developed in the absence of quantitative data. As more data were gathered, the categorization ultimately appeared to be arbitrary, with no objective criteria available for distinguishing foraging modes (McLaughlin, 1989). Two early studies (Pietruszka, 1986; McLaughlin, 1989) found partial support for the validity of a dichotomous paradigm; however, more recent studies with increasing amounts of data (Perry, 1999; this volume, Chapter 1) failed to detect bimodal distribution of search behavior.

One question we pondered was: when did the values for MPM and PTM become formalized to classify species? The use of a value of 1.0 for MPM most likely emerged from the work of McLaughlin (1989), who showed a distinct break between groups of lizard species (see figure 2 in McLaughlin [1989]). There does not appear to be a formal proposal for the use of 10% as a threshold for identifying foraging mode. Huey *et al.* (1984) focused on a range of values to place species into one or another foraging mode. They stressed the relative nature of foraging modes among coexisting taxa. Pietruszka (1986) relied on two measures of foraging behavior: movement rate (meters moved per minute) and foraging movement frequency. The latter variable could be interpreted to be similar to PTM. His study was notable for documenting considerable amounts of seasonal and individual variation in search behavior. Butler (2005) attributed the empirical cut-off of <15% for classifying species to Cooper and Whiting (1999). However, Cooper and Whiting (1999) cite Huey and Pianka (1981) and Perry's unpublished dissertation (Perry, 1995) for a 10%–15% threshold. Reilly and McBrayer (Chapter 10) distinguish a medium or mixed category in familial means between about 10 and 25 PTM, noting, as Schwenk (1994) did, that true active foraging of over *c.* 40 PTM appears independently in the families Teiidae and Varanidae. None of these studies, however, formally tested for the existence of a statistically significant gap separating foraging modes.

We feel that a valuable issue raised in the volume has been to question the biological reality of 'threshold values' to describe foraging modes. Anderson (Chapter 15) provides a thorough description of the full range and complexities of food acquisition modes in lizards. Various authors may use divergent terms to describe the foraging behavior of a species. For example, species in the teiid genus *Aspidoscelis* are often characterized as "extreme wide foragers," active foragers, or intensive foragers. Are these simply different terms for the same behavior or do they represent qualitative differences in behaviors that are not captured by using data on MPM or PTM? Anderson also alludes to similar issues in describing jump–strikes (*e.g. Sceloporus, Anolis*) and stalk–strikes (*e.g. Gambelia, Callisaurus*) in species generally thought of as sit-and-wait foragers. He also generates a testable framework of foraging mode states to encourage workers to better describe these complexities. Thus, how one can use objective measures to define a species behavior is another issue raised in the volume. How should the categories of SW or AF be objectively defined or expanded? Additional analyses are necessary to determine the subtleties of variation in feeding behavior. Is it biologically meaningful, in terms of diet breadth, or feeding success, if a species has a value of 10 for PTM, rather than 14 or 19? The use of broad categories to describe foraging behavior obscures important, yet subtle, differences in the manner in which species locate and acquire prey.

Other efforts have been made to more precisely quantify and thus objectively ascribe foraging mode states, mainly through the inclusion of more traits that characterize feeding behavior. Cooper and Whiting (1999) introduced an additional variable, percentage of attacks while moving (PAM), as a metric to describe and categorize species' foraging modes. They argued that PAM is better able to discriminate between foraging modes because for ambush foragers PAM will equal zero. They also show that PAM is highly correlated with PTM but not with MPM (Cooper and Whiting, 1999; Cooper et al., 2001). Likewise, Cooper et al. (2005) and Cooper (2005b) show that average speed, speed while moving, and average duration of movement may be correlated with PAM and PTM.

We suggest another new foraging variable, attacks while stationary (AWS). Attacks while stationary are applicable to both sit-and-wait foragers and active foragers because active foragers often dig up prey while stationary and/or capture prey while engaged in other activities (e.g. basking) although presumably at a much lower rate than while moving. The use of AWS would be complementary to other behavioral metrics. Like PAM, quantification of AWS might be tedious; long hours of observation may be required in order to obtain measurements. However, if investigators were present during the normal activity period of the focal lizards, enough data could be obtained. Although AWS, PAM, PMP, PTM, and other variables may be suitable to quantitatively characterize foraging modes, all of these variables may result in more confusion if analyses are not done to distinguish which variables are most appropriate and where statistically significant gaps exist such that foraging modes may be separated. Interestingly, each of the new variables tends to have a unimodal distribution, which provides further support for the use of a continuum model to describe foraging behavior.

Regardless of whether or not foraging mode is a two-state or multi-state character, major advances were made in lizard biology by authors using a dichotomous classification. This volume is a testament to many of these advances. For example, Vitt (1983) demonstrated significant covariation between tail morphology and autotomy, which was consistent with predictions based on expected differences among species that exploited different foraging modes. The relationship between chemoreception and active or ambush foraging modes was suggested long ago (Evans, 1961), and yet it is only over the past 15 years that Cooper and colleagues have shown this trait to be a major evolutionary force in the diversification of lizards (see, for example, Cooper, 1994a,b, 2005a). Finally, life history variation has been shown to covary with foraging mode (Dunham *et al.*, 1988; Webb *et al.*, 2003).

However, a key problem in correlative works related to foraging mode in lizards became how to determine the foraging mode of an organism. Although foraging mode was known to influence a myriad of traits, efforts to obtain data that would provide an index into a species feeding behavior have lagged behind other studies. Many questions have become increasingly vexing as we learn more about the intricacies of lizard ecology as it pertains to their foraging. For example, what are the best variables to categorize species? Could we derive thresholds to put species in one foraging group or another? People have discussed thresholds for some time, but what really constitutes a sit-and-wait forager vs. an active forager: is there a real continuum of foraging modes? And finally, does the SW–AF paradigm mirror the *r–K* selection dichotomy in life history theory? Characterizing species as either "r-" or "K-"selected focused much life history theory during the 1970s and 1980s. As additional data were collected, intermediate species were found not to exactly fit the predefined categories, and alternative hypotheses began to come into favor. Ultimately, the *r–K* continuum was supplanted with other theoretical constructs, yet like SW and AF it was instrumental in guiding our efforts to obtain a handle on the problem.

The foraging mode paradigm: new insights and the need for more data

The chapters included in this book provide the most recent summaries of the consequences and correlates of foraging mode, and meet our goal of reviewing the state of the paradigm at many levels of analysis. Many themes have emerged from these chapters and set the stage for rapid progress across all disciplines. More than anything else, all authors agree that more data are needed on more species. For example, Vitt and Pianka (Chapter 5) point out that it has taken each of them a lifetime to collect all of their dietary data. Yet this Herculean effort still only represents information on *c.* 4.25% of all lizard species. Likewise, Bauer (Chapter 12) calls attention to the fact that our understanding of gecko foraging is rudimentary in that a tiny fraction of gekkonid species have been studied. Gekkotans, which are assumed to be SW, make up 25% of all lizard species. Depending on how they relate phylogenetically to the SW Iguania and the various more widely foraging taxa, they have critical implications on inferences of foraging mode and character evolution. Future work must be focused on this critical polarizing clade. Finally, we hold a feeble understanding of foraging modes in the "other half" of squamates: snakes. Shine and Wall (Chapter 6) and Beaupre and Montgomery (Chapter 11) not only draw our attention to this, but also make salient points as to *whether* the very idea of foraging mode as we think of it in lizards is applicable to snakes.

The need for more data is also apparent within areas that have made considerable progress and valuable insight. Perry (Chapter 1) highlights several important methodological revisions in how we measure and assign foraging mode states. He also highlights interesting variation in particular groups (*Sceloporus*), as do Miles *et al.* for polychrotines (Chapter 2). Bonine (Chapter 3) and Brown and Nagy (Chapter 4) demonstrate that several physiological parameters (e.g. metabolic rates) generally evolve as predicted, although broad comparative physiological data are scarce. Herrel (Chapter 7) points out that, in the Iguania, much is left to learn about the multiple evolutionary transitions to herbivory and AF from various SW ancestors. In scleroglossans, however, the evolution of omnivory appears to be a first step towards becoming an actively foraging herbivore. Finally, support for the relationship between lingual morphology, chemoreception, and the evolution of foraging mode remains strong (Cooper, Chapter 8). New information on feeding kinematics and behavior (Reilly and McBrayer, Chapter 10) shows that the use of the tongue in prey capture was independently lost in the Anguimorpha and Varanoidea in concert with the appearance of many other convergent traits associated with an increasingly derived chemosensory role of the tongue in the teiids and varanids. Moreover, McBrayer and Corbin (Chapter 9) add the independent evolution of different designs of long, narrow skulls to this pattern. The impacts of new data are important and illustrate the way in which the paradigm continues to foster innovative research with a strong influence in the study of lizard biology.

The second half of the volume addresses plasticity in foraging mode, which remains a severely under-explored topic. Plasticity in foraging mode can arise through ontogenetic, seasonal, annual, or even daily sources of variation, as is the case of nocturnality in geckos (Bauer, Chapter 12). Shine and Wall (Chapter 6) point out that plasticity in foraging mode and diet is the rule in snakes. Their chapter evaluated several mechanistic hypotheses. One major factor dictating dietary shifts is the pronounced change in body size over ontogeny. Small snakes may be constrained to be active foragers, whereas larger snakes may adopt a sedentary, ambush mode of feeding. Seasonal and annual variation in food supplies may also induce shifts in foraging mode. Huey and Pianka (1981) provided one example; Whiting presents another case study in the cordylid species *Platysaurus broadleyi* (Chapter 13). Interestingly Pietruszka (1986) pointed to the fact that a common North American lizard

(*Gambelia wislizenii*) may shift foraging modes seasonally, yet it has not been until just recently that data were brought to bear on this (and it looks as though they do) (Rose, 2004). Habitat structure and heterogeneity have cogent effects on foraging mode plasticity. The lacertids are one of the most variable groups in terms of foraging mode; Vanhooydonck *et al.* (Chapter 14) examine patterns between habitat use, feeding and locomotor behaviour, and the trophic niche of this clade. This chapter identifies functional trade-offs among feeding and climbing performance traits as they relate to the evolution of foraging strategies, diet, and habitat use. Anderson (Chapter 15) provides a detailed essay on foraging plasticity and how we should think about it in terms of levels and patterns of habitat structure. He posits that to understand food acquisition mode we must know the what, when, where, and how of lizard spatiotemporal use patterns within the context basic autecological tasks (feeding, predator evasion, coping with abiotic enviroment, and reproduction). Finally, Vitousek *et al.* (Chapter 16) examine the most unreptilian strategy of marine herbivory in marine iguanas, where strong sexual selection and cold temperature has led to unique morphological adaptations for their unique herbivorous foraging mode.

The promise of the foraging mode paradigm

We believe that the contributions in this volume will foster and focus a revitalization of work investigating the contribution of foraging mode to understanding the diversification of taxa. We encourage other investigations to avoid diminishing the diversity of feeding behaviors or foraging mode by using dichotomous and broad categories. Such a dichotomous approach constrains the observed variation to fit into the extremes. To be sure, there are examples in the natural world of extreme sit-and-wait or actively foraging species, and these taxa also tend to conform to the predicted ecological and behavioral correlates. However, one wonders whether this too might not be an artifact of history: the teiid (really AF) and iguanid (really SW) lizards of North America have figured strongly in the historical establishment (whether purposefully or by accident) of a dichotomous view in lizards. As additional information on PTM and MPM accumulate from additional, detailed behavioral work, it is likely that more species will fall between strict SW and AF extremes. In short, we might imagine that the categories of SW and AF will ultimately acquire fuzzy boundaries, which will force the revision of the paradigm.

The advantage of the foraging mode paradigm is the clear delimitation of associated predicted ecological and evolutionary consequences. The expanding database already favors at least a trichotomy of categories, and more data will ultimately lead to a quantitative description of foraging mode. This situation again is analogous to the revision of *r–K* theory with the introduction of additional factors that can generate more complex variation in life history patterns, e.g. the consideration of how environmental variation would modify survival and reproduction and the explicit inclusion of juvenile and adult vital rates in life history models. Another consideration is how will the paradigm change as additional categories are modified, added or abandoned. Perry *et al.* (1990) demonstrated a significant negative correlation between relative clutch mass (RCM) and moves per minute in lacertid lizards. As more data are collected on the search behavior of lizards in other clades, more exceptions to the rule are discovered. For example, although chamaeleons were long believed to be SW foragers, at least one species (*Bradypodion pumilium*) has now been described as an active forager (Butler, 2005). In addition, Cooper (2005b) portrayed species in a two-dimensional space described by PTM and MPM, yet failed to demonstrate unambiguous clustering of all species into active or sit-and-wait foraging. These conclusions further support the lack of evidence for bimodality in movement data (see, for example, Perry, 1999; this volume, Chapter 1). Indeed, the boundaries separating SW from AF are becoming increasingly blurred. Although it has long been argued that foraging modes diverged early in lizard evolution, the expansion of information on and taxonomic patterns of MPM and PTM in a greater range of species supports the presence of a continuum, alternative interpretations of many organismal trait patterns, and the implications of new contrasting lizard phylogenies clearly show that a dichotomous view of the foraging mode paradigm does not hold all of the answers. The advent of the behavioral variables percentage of attacks while moving (PAM) and attacks while stationary (AWS) may further help quantify the foraging mode continuum. Moreover, it is equally interesting that recent analyses of foraging mode using ancestor reconstruction and independent contracts suggest multiple transitions in search behavior, with different suites of correlated traits in some lineages and striking patterns of convergence in others.

The future of the foraging mode paradigm lies in the willingness of research-ers to depolarize foraging mode. The utility of a polarized, dichotomous view of lizard foraging was evident in generating a research "framework" to attack a problem. That is, by focusing on the extremes of foraging modes, one could test for the existence of differences. However, as more species are studied and better analytical methods are applied to foraging data, we are witnessing a turnover in the strict dichotomous perception of the paradigm. We encourage workers to embrace the variation in foraging modes as exciting, interesting, and deserving of further investigation. It is the now numerous exceptions to the rule (such as tongue projection in scincomorphans) that must drive the utility of the foraging mode paradigm, if we are to see meaningful advances in the field.

In the future, it will be important to keep in mind how foraging modes are defined and that percent time moving (PTM) and moves per minute (MPM) are relative metrics. Even though species may be categorized as SW, there is still substantial variation in the component variables of foraging mode that can make more detailed analyses interesting. An important methodological advance in this regard may be lag sequential analysis (see Butler, 2005). Lag sequential analysis examines whether a behavior of interest (here locomotion) precedes a particular behavior (foraging) more frequently than would be expected at random. Use of this method will be an important step forward in our ability to separate foraging modes and associated behaviors and further explore the natural variation that is inherent in foraging strategies.

Hence, where should we go from here? What are the burning questions identified by the contributors? Without question, future work must emphasize seasonal variation and within-family variation. Definitive ecological and evolutionary correlates may not emerge when conducting broad interfamilial comparisons. Instead, these correlates may be more enlightening when done on a restricted group of species. Hence, one could focus on North American sceloporines or Kalahari lacertid lizards. Even though the taxonomic representation is not as great, studying a focal clade such as these is likely to be a better control for other factors, e.g., seasonality, habitat, etc. Similarly, work needs to be done addressing possible ontogenetic variation in foraging behavior. Granted, the study of small subadult lizards is difficult, yet on theoretical grounds juveniles face very different energetic, social, and predation-risk challenges compared with adults. It is not only possible that ontogenetic shifts in foraging mode exist, but likely. A starting point might be a study of spiny-tailed iguanas, *Ctenosaura pectinata* or *C. similis*. Juveniles of each species are known to eat insects (*C. pectinata*: Durtsche, 2000; *C. similis*: Van Devender 1982) and later switch to herbivory. Thus, do juveniles start out as sit-and-wait insectivores like many iguanians only to later switch to active foraging herbivory? Another obvious example would be *Pedioplanis* or *Meroles* lizards in the Kalahari. Evolutionary transitions in foraging mode are known in these two genera (AF → SW or mixed foraging) and substantial dietary variation exits within some species from season to season (Pianka, 1986). Thus, seasonal variation in foraging mode is possible but has not been examined in detail (see, for example, Whiting, Chapter 13). Furthermore, closely related species that have more recently diverged in foraging modes may be more likely to show signs of significant ontogenetic variation. Should the ontogeny of foraging mode be different among species, heterochronic changes in key traits (energy budget, life history, etc.) are likely to be the mechanism of change. These and many other interesting hypotheses have yet to be addressed.

Beyond new avenues of research, several existing lines of investigation warrant deeper consideration. There is a paucity of information on MPM–PTM and related measures of foraging mode across many squamate taxa. To better define broad evolutionary patterns, more data are required for more species. Methodologically these data are simple and inexpensive to collect. Analytically, the use of lag sequential analysis will improve the description of foraging behavior because this approach directly links the locomotor behaviors to foraging behaviors. Using video to collect foraging data may also prove useful. Video is now lightweight and inexpensive; having a permanent record of the movement behaviors and their provenance is certainly valuable. Video data will also allow for analysis of prey capture methods and handling or processing times on a variety of natural prey types. Recent evidence that only the true wide foragers have lost tongue prehension suggests that feeding studies in many more taxa are warranted to tease out the trade-offs between feeding and chemosensory function in lizard tongues.

Along with a general appeal for more data, we specifically need increased sampling of species *within* clades, among *unstudied* clades and in different habitats (e.g. tropics). Several lineages show a wide variation in foraging modes (e.g. lacertids and scincids) and thus our understanding of the factors that effect evolutionary changes in foraging mode may be best examined in such groups. Furthermore, seemingly well understood clades such as polychrotids are now a superb example of how "presumed" foraging modes may not hold up under scrutiny. At the time of writing, no published accounts of foraging behavior are available for varanid, anguid, or xantusid lizards. Granted, many of these lizards are secretive and hard to study; however, given current technologies in telemetry and remote sensing, studies of even these hard-to-observe lizards are possible.

Shifting phylogenies will clearly affect the immediate and long-term future of interpretations of the evolutionary patterns and significance of foraging mode in lizards. Early studies of foraging mode recognized the strong phylogenetic signal associated with the trait (Pianka, 1966, 1973). Indeed, major preliminary analyses of ecological correlates of foraging mode explicitly included a phylogenetic approach (see, for example, Dunham *et al.*, 1988) or acknowledged the tendency for foraging mode to remain conservative within families (Cooper 1994a,b). It is now overwhelmingly apparent that phylogenetic history has played an important role in the evolution of foraging mode and associated characters. Most contributions in this volume, and for the past 10 or more years, interpreted their data using the phylogeny published by Estes *et al.* (1988). However, new, very different lizard phylogenies (Townsend *et al.*, 2004; Vidal and Hedges, 2005) have appeared recently that may have a dramatic influence on interpretations of foraging evolution described within this volume and elsewhere (both were published late in the preparation of this volume and thus were not seen by many of our authors). At a minimum these new phylogenies change the basal taxa, and thus basal foraging mode states, and lend much stronger support to the independent evolution of many convergent traits in the crown Anguimorpha and Lacertoidea. In addition, if one must consider the Iguania as a crown group rather than the basal and primitive condition (as suggested by new phylogenies), it supports the conclusions of Schwenk and Wagner (2001) and Reilly and McBrayer (Chapter 10) that their radiation demands more respect as a novel, important, and evolutionarily stable strategy among lizards. Like others, we strongly encourage future studies of foraging mode traits to apply all three phylogenies to further explore the utility, and robustness, of phylogenetic hypotheses. Character mapping of foraging mode on the lizard phylogeny also reveals a more complex picture regardless of phylogeny used. There are several transitions in foraging mode from sit-and-wait to active foraging; however, certain families exhibit numerous reversals, e.g. active foraging → sit-and-wait. Future work should be directed toward understanding the factors associated with evolutionary transitions in foraging mode; character mapping is an efficient mechanism to do so.

Finally, future studies should build on the strong database that does exist and add more integrative approaches and thinking. Foraging mode has always represented a composite variable containing several co-adapted traits. The act of searching or watching for prey involves sensory input, integration of that input, and functional and behavioral responses to it. In addition, field conditions, prey biology, community structure, and autecological relevance needs better quantification (Anderson, Chapter 15). As pointed out by Vanhooydonck *et al.* (Chapter 14), quantification of prey availability (field studies) and functionally relevant prey traits (tested in the laboratory, e.g. hardness, escape speed) are necessary to link prey performance to predator performance capacity to patterns of foraging mode. Phenotypes (or species) represent trade-offs among these that have been optimized via natural selection. Thus, studies of foraging mode *require* integrative thinking because the foraging mode concept spans the morphological and physiological parameters of an individual. To use foraging mode as a simple proxy for other ecological traits ignores many of the more interesting trade-offs that are inherent in them.

We encourage future investigators to embrace the integrative nature of foraging mode. Only by doing so can we understand the true pervasiveness and variability that shapes foraging biology in lizards.

References

Butler, M. A. (2005). Foraging mode of the chameleon, *Bradypodion pumilum*: a challenge to the sit-and-wait versus active forager paradigm? *Biol. J. Linn. Soc.* **84**, 797–808.

Cooper, W. E. (1994a). Chemical discrimination by tongue-flicking in lizards: a review with hypotheses on its origin and its ecological and phylogenetic relationships. *J. Chem. Ecol.* **20**, 439–87.

Cooper, W. E. (1994b). Prey chemical discrimination, foraging mode, and phylogeny. In *Lizard Ecology*, ed. L. J. Vitt and E. R. Pianka, pp. 95–116. Princeton, NJ: Princeton University Press.

Cooper, W. E. Jr. (2005a). The foraging mode controversy: both continuous variation and clustering of foraging modes occur. *J. Zool. Lond.* **267**, 179–90.

Cooper, W. E. Jr. (2005b). Duration of movement as a lizard foraging movement variable. *Herpetologica* **61**, 363–72.

Cooper, W. E. Jr., Vitt, L. J., Caldwell, J. P. and Fox, S. F. (2005). Relationships among foraging variables, phylogeny, and foraging modes, with new data for nine North American lizard species. *Herpetologica* **57**, 65–76.

Cooper, W. E. Jr, Vitt, L. J., Caldwell, J. P. and Fox, S. F. (2001). Foraging modes of some American lizards: relationships among measurement variables and discreteness of modes. *Herpetologica* **61**, 250–9.

Cooper, W. E. Jr. and Whiting, M. J. (1999). Foraging modes in lacertid lizards from southern Africa. *Amph.-Rept.* **20**, 299–311.

Dunham, A. E., Miles, D. B. and Reznick, D. N. (1988). Life history patterns in squamate reptiles. In *Biology of the Reptilia*, vol. 16, ed. C. Gans and R. B. Huey, pp. 331–86. New York: A. R. Liss.

Durtsche, R. D. (2000). Ontogenetic plasticity of food habits in the Mexican spiny-tailed iguana, *Ctenosaura pectinata*. *Oecologia* **124**, 185–95.

Eckhardt, R. C. (1979). The adaptive syndromes of two guilds of insectivorous birds in the Colorado Rocky Mountains. *Ecol. Monogr.* **49**, 129–49.

Estes, R., de Queiroz, K. and Gauthier, J. (1988). Phylogenetic relationships within Squamata. In *Phylogenetic Relationships of the Lizard Families: Essays Commemorating Charles L. Camp*, ed. R. Estes and G. Pregill, pp. 119–281. Stanford, CA: Stanford University Press.

Evans, L. T. (1961). Structure as related to behavior in the organization of populations of reptiles. In *Vertebrate Speciation*, ed. W. F. Blair, pp. 148–78. Houston, TX: University of Texas Press.

Huey, R. B. and Pianka, E. R. (1981). Ecological consequences of foraging mode. *Ecology* **62**, 991–9.

Huey, R. B., Bennett, A. F., John-Alder, H. B. and Nagy, K. A. (1984). Locomotor capacity and foraging behaviour of Kalahari lacertid lizards. *Anim. Behav.* **32**, 41–50.

Kennedy, J. P. (1956). Food habits of the rusty lizard *Sceloporus olivaceous* Smith. *Texas J. Sci.* **8**, 328–49.

Magnusson, W. E., de Paiva, L. J., da Rocha, R. M. *et al.* (1985). The correlates of foraging mode in a community of Brazilian lizards. *Herpetologica* **41**, 324–32.

McLaughlin, R. L. (1989). Search modes of birds and lizards: evidence for alternative movement patterns. *Am. Nat.* **133**, 654–70.

Perry, G. (1995). The evolutionary ecology of lizard foraging: a comparative study. Ph.D. dissertation, University of Texas at Austin.

Perry, G. (1999). The evolution of search modes: ecological versus phylogenetic perspectives. *Am. Nat.* **15**, 98–109.

Perry, G., Lampl, I., Lerner, A. *et al.* (1990). Foraging mode in lacertid lizards: variation and correlates. *Amph.-Rept.* **11**, 373–84.

Pietruszka, R. D. (1986). Search tactics of desert lizards: how polarized are they? *Anim. Behav.* **34**, 1742–58.

Pianka, E. R. (1966). Convexity, desert lizards, and spatial heterogeneity. *Ecology* **47**, 1055–9.

Pianka, E. R. (1973). The structure of lizard communities. *Ann. Rev. Ecol.* **4**, 53–74.

Pianka, E. R. (1986). *Ecology and Natural History of Desert Lizards: Analyses of the Ecological Niche and Community Structure.* Princeton, NJ: Princeton University Press.

Rand, A. S. (1967). Ecology and social organization in the iguanid lizard *Anolis lineatopis. Proc. U.S. Natl. Mus.* **122**, 1–79.

Regal, P. J. (1978). Behavioral differences between reptiles and mammals: an analysis of activity and mental capacities. In *Behavior and Neurology of Lizards*, ed. N. Greenberg and P. D. Maclean, pp. 183–202. Washington, D.C.: Department of Health, Education, and Welfare.

Rose, E. R. (2004). Foraging behavior in *Gambelia wislizenii*, the long-nosed leopard lizard, in Harney County, Oregon. Master's Thesis, Western Washington University.

Schoener, T. W. (1969). Models of optimal size for solitary predators. *Am. Nat.* **103**, 277–313.

Schoener, T. W. (1971). Theory of feeding strategies. *Ann. Rev. Ecol. Syst.* **2**, 360–404.

Schwenk, K. (1994). Why snakes have forked tongues. *Science* **263**, 1573–7.

Schwenk, K. and Wagner, G. P. (2001). Function and the evolution of phenotypic stability: Connecting pattern to process. *Am. Zool.* **41**, 552–63.

Townsend, T., Larson, A., Louis, E. and Macey, R. J. (2004). Molecular phylogenetics of Squamata: the position of snakes, Amphisbaenians, and Dibamids and the root of the Squamate tree. *Syst. Biol.* **53**, 735–57.

Van Devender, R. W. (1982). Growth and ecology of spiny-tailed and green iguanas in Costa Rica, with comments on the evolution of herbivory and large body size. In *Iguanas of the World*, ed. G. M. Burghardt and A. S. Rand, pp 162–83. Park Ridge, NJ: Noyes.

Vidal, N. and Hedges, S. B. (2005). The phylogeny of squamate reptiles (lizards, snakes, and amphisbaenians) inferred from nine nuclear protein coding genes. *C. R. Biol.* **328**, 1000–8.

Vitt, L. J. (1983). Tail loss in lizards: the significance of foraging and predator escape modes. *Herpetologica* **39**, 151–62.

Webb, J. K., Brook, B. W. and Shine, R. (2003). Does foraging mode influence life history traits? A comparative study of growth, maturation, and survival of two species of sympatric snakes from south-eastern Australia. *Aust. Ecol.* **28**, 601–10.

Index

Printed in the United States
By Bookmasters